	ϕ
Tensile Yield	0.9
" Fracture	0.75
Compression	0.85
Bending	0.9
Flexural Shear	0.9
Bolted Connection	0.75

Yield $\phi P_n = \phi F_y A_g$
Fracture $\phi P_n = \phi F_u A_e$

For Bolted $A_e = A_n U$
For Welded $A_e = A_g U$

U - values

Plate (welded or fastened)   1

$U = 1 - \dfrac{\overline{x}}{\ell}$

(L-shape)

Longitudinal welds

$U = 1 - \dfrac{\overline{x}}{\ell}$

(L-shape)

Longitudinal welds on plate

$U = \begin{cases} 1.0 & \ell \geq 2w \\ 0.87 & 2w > \ell \geq 1.5w \\ 0.75 & 1.5w > \ell \geq w \end{cases}$

Tension Load transmitted **only** by transverse welds $U = 1$

→ not used in A

see also:
Bolted Connections p.77
Welded Plates p.78
For welded connections: (L-shape)
1) For W, S, or M shapes with width to depth ratio of at least ⅔
(and T shapes cut from them) and connected at the flanges
$U = .9$
2) for all other shapes → $U = .85$

When Fastened on all sides, $U = 1$

staggered holes
for Fracture, $A_n = A_g - \sum d' t_w + \sum \dfrac{s^2}{4g} t_w$

t_w = thickness
d' = diameter of fastener + ⅛"

A_{gv} = thickness·2L

Block Shear
a) if $F_u A_{nt} > .6 F_u A_{nv}$

$\phi R_n = \phi(.6 F_y A_{gv} + F_u A_{nt})$
$\phi R_n \leq \phi(.6 F_u A_{nv} + F_u A_{nt})$ } use smaller

b) if $F_u A_{nt} < .6 F_u A_{nv}$

$\phi R_n = \phi(.6 F_u A_{nv} + F_y A_{gt})$
$\phi R_n = \phi(.6 F_u A_{nv} + F_u A_{nt})$ } use smaller

Structural Steel Design: LRFD Method

THIRD EDITION

Jack C. McCormac
Clemson University, Clemson, South Carolina

and

James K. Nelson, Jr.
*Western Michigan University,
Kalamazoo, Michigan*

Prentice
Hall

Pearson Education, Inc.
Upper Saddle River, New Jersey 07458

Library of Congress Cataloging-in-Publication Data on file

Vice President and Editorial Director, ECS: *Marcia J. Horton*
Acquisitions Editor: *Laura Fischer*
Editorial Assistant: *Erin Katchmar*
Vice President and Director of Production and Manufacturing, ESM: *David W. Riccardi*
Executive Managing Editor: *Vince O'Brien*
Managing Editor: *David A. George*
Production Editor: *Scott Disanno*
Director of Creative Services: *Paul Belfanti*
Creative Director: *Carole Anson*
Art Director: *Jayne Conte*
Cover Designer: *Bruce Kenselaar*
Art Editor: *Greg Dulles*
Manufacturing Manager: *Trudy Pisciotti*
Manufacturing Buyer: *Lisa McDowell*
Marketing Manager: *Holly Stark*

© 2003 by Pearson Education, Inc.
Pearson Education, Inc.
Upper Saddle River, NJ 07458

Printed in the United States of America
10 9 8 7 6 5 4

ISBN 0-13-047959-4

Pearson Education Ltd., *London*
Pearson Education Australia Pty. Ltd., *Sydney*
Pearson Education Singapore, Pte. Ltd.
Pearson Education North Asia Ltd., *Hong Kong*
Pearson Education Canada, Inc., *Toronto*
Pearson Educación de Mexico, S.A. de C.V.
Pearson Education—Japan, *Tokyo*
Pearson Education Malaysia, Pte. Ltd.
Pearson Education, Inc., *Upper Saddle River, New Jersey*

Preface

The authors' major objective in preparing this new edition was to update the text to conform to both the 1999 Load and Resistance Factor (LRFD) Design Specification and the 2002 edition of the LRFD Manual of Steel Construction.

Among the several changes made in the new specification and included in the steel manual are the following:

1. The inclusion of data and equations in both U.S. customary units and metric units.
2. The introduction of two new important ASTM steels, A913 and A992.
3. A few revisions in bolt criteria.
4. Revised design procedure for fatigue loadings.
5. A new section concerning the evaluation of existing structures.

In addition to the revisions in the Specification, several changes and additions have been made in the text concerning the enclosed computer programs. First, the program previously named INSTEP has been updated to the new specifications and has been changed from a DOS format to a Windows format. It is now named INSTEP32. Though the program was written specifically to solve many of the textbook-type problems presented in this book, it has often been used by professional engineers in their design practices.

Despite the practical applications of INSTEP32 and its value in teaching steel design, the authors feel that many professors would like their students to have some experience with at least one of the major commercial steel-design programs on the market today, in hopes that such experience would enable students to cross the bridge more quickly between the classroom and actual design practice. As a result, a student version of SAP 2000 has been included with INSTEP32 on the enclosed disk and presented herein. Use of this software will also enable the student to better understand the relationship between analysis and design.

Chapter 20, a new chapter in the text, presents an introduction to the subject of systems design. In this regard, recent requirements for "capstone" courses in our engineering schools have made the subject of "open-ended" problems a serious component of design studies. Therefore, the topic of systems design along with "open-ended" problems and an introduction to SAP 2000 are included in Chapter 20.

The authors wish to thank the following persons who reviewed this edition: Robert Abendroth, Daniel G. Linzell, Rolla Idriss, W. H. Walker, and Ahmad M. Itani.

They also thank the reviewers and users of the previous editions of this book for their suggestions, corrections, and criticisms. They are always grateful to anyone who takes the time to contact them concerning any part of their book.

Jack C. McCormac
James K. Nelson

Contents

Introduction to Structural Steel Design

1.1 ADVANTAGES OF STEEL AS A STRUCTURAL MATERIAL

A person traveling in the United States might quite understandably decide that steel was the perfect structural material. He or she would see an endless number of steel bridges, buildings, towers, and other structures. After seeing these numerous steel structures, this traveler might be surprised to learn that steel was not economically made in the United States until late in the nineteenth century, and the first wide-flange beams were not rolled until 1908.

The assumption of the perfection of this metal, perhaps the most versatile of structural materials, would appear to be even more reasonable when its great strength, light weight, ease of fabrication, and many other desirable properties are considered. These and other advantages of structural steel are discussed in detail in the paragraphs that follow.

1.1.1 High Strength

The high strength of steel per unit of weight means that the weight of structures will be small. This fact is of great importance for long-span bridges, tall buildings, and structures situated on poor foundations.

1.1.2 Uniformity

The properties of steel do not change appreciably with time, as do those of a rein-forced-concrete structure.

1.1.3 Elasticity

Steel behaves closer to design assumptions than most materials because it follows Hooke's law up to fairly high stresses. The moments of inertia of a steel structure can be accurately calculated, while the values obtained for a reinforced-concrete structure are rather indefinite.

1

Erection of steel joists. (Courtesy of Vulcraft.)

1.1.4   Permanence

Steel frames that are properly maintained will last indefinitely. Research on some of the newer steels indicates that under certain conditions no painting maintenance whatsoever will be required.

1.1.5   Ductility

The property of a material by which it can withstand extensive deformation without failure under high tensile stresses is said to be its *ductility*. When a *mild* or *low-carbon* structural steel member is being tested in tension, a considerable reduction in cross section and a large amount of elongation will occur at the point of failure before the actual fracture occurs. A material that does not have this property is generally unacceptable and is probably hard and brittle, and it might break if subjected to a sudden shock.

In structural members under normal loads, high stress concentrations develop at various points. The ductile nature of the usual structural steels enables them to yield locally at those points, thus preventing premature failures. A further advantage of ductile structures is that when overloaded their large deflections give visible evidence of impending failure (sometimes jokingly referred to as "running time").

1.1.6   Toughness

Structural steels are tough—that is, they have both strength and ductility. A steel member loaded until it has large deformations will still be able to withstand large forces. This is a very important characteristic because it means that steel members can be

subjected to large deformations during fabrication and erection without fracture—thus allowing them to be bent, hammered, sheared, and have holes punched in them without visible damage. The ability of a material to absorb energy in large amounts is called *toughness*.

1.1.7   Additions to Existing Structures

Steel structures are quite well suited to having additions made to them. New bays or even entire new wings can be added to existing steel frame buildings, and steel bridges may often be widened.

1.1.8   Miscellaneous

Several other important advantages of structural steel are as follows: (a) ability to be fastened together by several simple connection devices including welds and bolts, (b) adaptation to prefabrication, (c) speed of erection, (d) ability to be rolled into a wide variety of sizes and shapes as described in Section 1–4, (e) fatigue strength, (f) possible reuse after a structure is disassembled, and (g) scrap value, even though not reusable in its existing form. Steel is the ultimate recyclable material.

1.2   DISADVANTAGES OF STEEL AS A STRUCTURAL MATERIAL

In general, steel has the following disadvantages:

1.2.1   Maintenance Costs

Most steels are susceptible to corrosion when freely exposed to air and water and therefore must be painted periodically. The use of weathering steels, however, in suitable applications tends to eliminate this cost.

1.2.2   Fireproofing Costs

Although structural members are incombustible, their strength is tremendously reduced at temperatures commonly reached in fires when the other materials in a building burn. Many disastrous fires have occurred in empty buildings where the only fuel for the fires was the buildings themselves. Furthermore, steel is an excellent heat conductor—nonfireproofed steel members may transmit enough heat from a burning section or compartment of a building to ignite materials with which they are in contact in adjoining sections of the building. As a result, the steel frame of a building may have to be protected by materials with certain insulating characteristics, and the building may have to include a sprinkler system if it is to meet the building code requirements of the locality in question.

1.2.3   Susceptibility to Buckling

As the length and slenderness of a compression member is increased, its danger of buckling increases. For most structures the use of steel columns is very economical because of their high strength-to-weight ratios. Occasionally, however, some additional steel is needed to stiffen them so they will not buckle. This tends to reduce their economy.

1.2.4   Fatigue

Another undesirable property of steel is that its strength may be reduced if it is subjected to a large number of stress reversals or even to a large number of variations of tensile stress. (Fatigue problems occur only when tension is involved.) The present practice is to reduce the estimated strengths of such members if it is anticipated that they will have more than a prescribed number of cycles of stress variation.

1.2.5   Brittle Fracture

Under certain conditions steel may lose its ductility, and brittle fracture may occur at places of stress concentration. Fatigue type loadings and very low temperatures aggravate the situation.

1.3   EARLY USES OF IRON AND STEEL

Although the first metal used by human beings was probably some type of copper alloy such as bronze (made with copper, tin, and perhaps some other additives), the most important metal developments throughout history have occurred in the manufacture and use of iron and its famous alloy called steel. Today, iron and steel make up nearly 95 percent of all the tonnage of metal produced in the world.[1]

Despite diligent efforts for many decades, archaeologists have been unable to discover when iron was first used. They did find an iron dagger and an iron bracelet in the Great Pyramid in Egypt, which they claim had been there undisturbed for at least 5,000 years. The use of iron has had a great influence on the course of civilization since the earliest times, and may very well continue to do so in the centuries ahead. Since the beginning of the Iron Age in about 1000 B.C., the progress of civilization in peace and war has been heavily dependent on what people have been able to make with iron. On many occasions its use has decidedly affected the outcome of military engagements. For instance, in 490 B.C. in Greece at the Battle of Marathon, the greatly outnumbered Athenians killed 6400 Persians and lost only 192 of their own men. Each of the victors wore 57 pounds of iron armor in the battle. (This was the battle from which the runner Pheidippides ran the approximately 25 miles to Athens and died while shouting news of the victory.) This victory supposedly saved Greek civilization for many years.

According to the classic theory concerning the first production of iron in the world, there was once a great forest fire on Mount Ida in Ancient Troy (now Turkey) near the Aegean Sea. The land surface supposedly had a rich content of iron and the heat of the fire is said to have produced a rather crude form of iron which could be hammered into various shapes. Many historians believe, however, that human beings first learned to use iron that fell to the earth in the form of meterorites. Frequently the iron in meteorites is combined with nickel to produce a harder metal. Perhaps early human beings were able to hammer and chip this material into crude tools and weapons.

Steel is defined as a combination of iron and a small amount of carbon, usually less than 1 percent. It also contains small percentages of some other elements. Although

---

[1]American Iron and Steel Institute, *The Making of Steel* (Washington, D.C., not dated), p. 6.

some steel has been made for at least 2000–3000 years, there was really no economical production method available until the middle of the nineteenth century.

The first steel almost certainly was obtained when the other elements necessary for producing it were accidentally present when iron was heated. As the years went by, steel probably was made by heating iron in contact with charcoal. The surface of the iron absorbed some carbon from the charcoal, which was then hammered into the hot iron. Repeating this process several times resulted in a case-hardened exterior of steel. In this way the famous swords of Toledo and Damascus were produced.

The first large volume process for producing steel was named after Sir Henry Bessemer of England. He received an English patent for his process in 1855, but his efforts to obtain a U.S. patent for the process in 1856 were unsuccessful because it was shown that William Kelly of Eddyville, Kentucky, had made steel by the same process seven years before Bessemer applied for his English patent. Although Kelly was given the patent, the name Bessemer was used for the process.[2]

Kelly and Bessemer learned that a blast of air through molten iron burned out most of the impurities in the metal. Unfortunately, at the same time the blow eliminated some desirable elements such as carbon and manganese. It was later learned that these needed elements could be restored by adding spiegeleisen, which is an alloy of iron, carbon, and manganese. It was further learned that the addition of limestone in the converter resulted in the removal of the phosphorus and most of the sulfur.

Before the Bessemer process was developed, steel was an expensive alloy used primarily for making knives, forks, spoons, and certain types of cutting tools. The Bessemer process reduced production costs by at least 80 percent and allowed for the first time production of large quantities of steel.

The Bessemer converter was commonly used in the United States until after the turn of the century, but since that time it has been replaced with better methods, such as the open-hearth process and the basic oxygen process.

As a result of the Bessemer process, structural carbon steel could be produced in quantity by 1870, and by 1890, steel had become the principal structural metal used in the United States.

Today most of the structural steel shapes and plates produced in the United States are made by melting scrap steel. This scrap steel is obtained from junk cars, and scrapped structural shapes as well as from discarded refrigerators, motors, typewriters, bed springs, and other similar items. The molten steel is poured into molds which have approximately the final shapes of the members. The resulting sections, which are run through a series of rollers to squeeze them into their final shapes, have better surfaces and fewer internal or residual stresses than newly made steel.

The shapes may be further processed by cold rolling, by applying various coatings, and perhaps by the process of *annealing*. This is the process by which the steel is heated to an intermediate temperature range, held at that temperature for several hours, and then allowed to slowly cool to room temperature. It results in steel with less hardness and brittleness, but with greater ductility.

The term *cast iron* refers to materials with very low carbon content materials, while the very high carbon content materials are referred to as *wrought iron*. Steels fall

---

[2]American Iron and Steel Institute, *Steel 76* (Washington, D.C., 1976), pp. 5–11.

in between cast iron and wrought iron and have carbon contents in the range of 0.15 percent to 1.7 percent (as described in Section 1–8 of this chapter).

The first use of metal for a sizable structure occurred in England in Shropshire (about 140 miles northwest of London) in 1779, when cast iron was used for the construction of the 100-ft Coalbrookdale Arch Bridge over the River Severn. It is said that this bridge (which still stands) was a turning point in engineering history because it changed the course of the Industrial Revolution by introducing iron as a structural material. This iron was supposedly four times as strong as stone and thirty times as strong as wood.[3]

A number of other cast-iron bridges were constructed in the following decades, but soon after 1840 the more malleable wrought iron began to replace cast iron. The development of the Bessemer process and subsequent advances such as the open-hearth process permitted the manufacture of steel at competitive prices. This encouraged the beginning of the almost unbelievable developments of the last 120 years with structural steel.

1.4 STEEL SECTIONS

The first structural shapes made in the United States were angle irons rolled in 1819. I-shaped steel sections were first rolled in the United States in 1884, and the first skeleton frame structure (the Home Insurance Company Building in Chicago) was erected that same year. Credit for inventing the "skyscraper" is usually given to engineer William LeBaron Jenny, who planned this building apparently during a bricklayers' strike. Prior to this time, tall buildings in the United States were constructed with load-bearing brick walls that were several feet thick.

For the exterior walls of this 10-story building, Jenny used cast-iron columns encased in brick. The beams for the lower six floors were made from wrought iron, while structural steel beams were used for the upper floors. The first building completely framed with structural steel was the second Rand-McNally building, completed in Chicago in 1890.

An important feature of the 985-ft wrought-iron Eiffel tower constructed in 1889 was the use of mechanically operated passenger elevators. The availability of these machines, along with Jenny's skeleton frame idea, led to the construction of thousands of high-rise buildings throughout the world in the next 100 years.

During these early years the various mills rolled their own individual shapes and published catalogs providing the dimensions, weight, and other properties of these shapes. In 1896, the Association of American Steel Manufacturers (now the American Iron and Steel Institute, AISI) made the first efforts to standardize shapes. Today, nearly all structural shapes are standardized, though their exact dimensions may vary just a little from mill to mill.[4]

Structural steel can be economically rolled into a wide variety of shapes and sizes without appreciably changing its physical properties. Usually the most desirable members

---

[3]M. H. Sawyer, "World's First Iron Bridge," *Civil Engineering* (New York: ASCE, December 1979), pp. 46–49.
[4]W. McGuire, *Steel Structures* (Englewood Cliffs, NJ: Prentice-Hall, 1968), pp. 19–21.

are those with large moments of inertia in proportion to their areas. The **I, T,** and **C** shapes, so commonly used, fall into this class.

Steel sections are usually designated by the shapes of their cross sections. As examples, there are angles, tees, zees, and plates. It is necessary, however, to make a definite distinction between American standard beams (called *S beams*) and wide-flange beams (called *W beams*) as they are both I shaped. The inner surface of the flange of a W section is either parallel to the outer surface or nearly so, with a maximum slope of 1 to 20 on the inner surface, depending on the manufacturer.

Erection of tower for the Holy Name Church, Edensburg, PA. (Courtesy of the Lincoln Electric Company.)

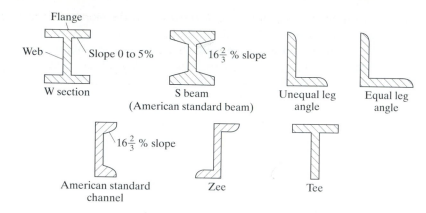

FIGURE 1.1

Rolled-steel shapes.

The S beams, which were the first beam sections rolled in America, have a slope on their inside flange surfaces of 1 to 6. It might be noted that the constant (or nearly constant) thickness of W-beam flanges compared with the tapered S-beam flanges may facilitate connections. Wide-flange beams comprise nearly 50 percent of the tonnage of structural steel shapes rolled today. The W and S sections are shown in Figure 1.1 together with several other familiar steel sections. The uses of these various shapes will be discussed in detail in the chapters to follow.

Constant reference is made throughout this book to the *Manual of Steel Construction Load & Resistance Factor Design* published by the American Institute of Steel Construction (AISC). This manual, which provides detailed information for structural steel shapes, is referred to hereafter as the LRFD Manual or simply the Manual or the Handbook. The 3rd edition of the handbook, is based on "Load and Resistance Factor Design Specification for Structural Steel Buildings" (Dec. 27, 1999). This specification is included in Part 16 of the LRFD Manual.

Structural shapes are abbreviated by a certain system described in the handbook for use in drawings, specifications, and designs. This system is standardized so that all steel mills can use the same identification for purposes of ordering, billing, etc. In addition, so much work is handled today with computers and other automated equipment that it is necessary to have a letter and number system which can be printed out with a standard keyboard (as opposed to the old system where certain symbols were used for angles, channels, etc.). Examples of this abbreviation system are as follows:

1. A W27 × 114 is a W section approximately 27 in deep weighing 114 lb/ft.
2. An S12 × 35 is an S section 12 in deep weighing 35 lb/ft.
3. An HP12 × 74 is a bearing pile section which is approximately 12 in deep weighing 74 lb/ft. Bearing piles are made with the regular W rolls but with thicker webs to provide better resistance to the impact of pile driving. The width and depth of these sections are approximately equal, and the flanges and webs have equal or almost equal thickness.
4. An M8 × 6.5 is a miscellaneous section 8 in deep weighing 6.5 lb/ft. It is one of a group of doubly-symmetrical H-shaped members which cannot by dimensions be

classified as a W, S, or HP section as the slope of their inner flanges is other than $16\frac{2}{3}\%$.

5. A C10 $\times$ 30 is a channel 10 in deep weighing 30 lb/ft.

6. An MC18 $\times$ 58 is a miscellaneous channel 18 in deep weighing 58 lb/ft which cannot be classified as a C shape because of its dimensions.

7. An L6 $\times$ 6 $\times$ $\frac{1}{2}$ is an equal leg angle, each leg being 6 in long and $\frac{1}{2}$ in thick.

8. A WT18 $\times$ 150 is a tee obtained by splitting a W36 $\times$ 300. This type of section is known as a structural tee.

9. Rectangular steel sections are classified as wide *plates* or narrow *bars*. In general, sections wider than 8 in are said to be plates, while the narrower ones are called bars. Detailed information on these sections is provided in Part 1 of the LRFD Manual (Tables 1-19 to 1-22). A plate is usually designated by its thickness times its width times its length as PL$\frac{1}{2}$ $\times$ 10 $\times$ 1 ft 4 in. Actually, the term "plate" is almost universally used today whether a member is fabricated from plate or bar stock.

The student should refer to the LRFD Manual for information concerning other shapes, such as hollow structural sections (HSS). Detailed information on these and other sections will be presented herein as needed.

In Part 1 of the LRFD Manual, the dimensions and properties of W, S, C, and other shapes are tabulated. The dimensions of the members are given in decimal form (for the use of designers) and in fractions to the nearest sixteenth of an inch (for the use of craftsmen and steel detailers or drafters). Also provided for the use of designers are such items as moments of inertia, section moduli, radii of gyration, and other cross-sectional properties discussed later in this text.

There are variations present in any manufacturing process, and the steel industry is certainly no exception. As a result, the cross-sectional dimensions of steel members may vary somewhat from the values specified in the LRFD Manual. Maximum tolerances for the rolling of steel shapes are prescribed by the American Society for Testing and Materials (ASTM) A6 Specification and are presented in Tables 1-54 to 1-61 in the Manual. As a result, calculations can be made on the basis of the properties given in the Manual regardless of the manufacturer.

Some steel sections listed in the Manual are available in the United States from only one or two steel producers and thus, on occasion, may be difficult to obtain promptly. Accordingly, when specifying sections, the designer would be wise to contact a steel fabricator for a list of sections readily available.

Through the years there have been changes in the sizes of steel sections. For instance, there may be insufficient demand to continue rolling a certain shape; an existing shape may be dropped because a similar sized but more efficient shape has been developed, and so forth. Occasionally, designers may need to know the properties of one of the discontinued shapes no longer listed in the latest edition of the Handbook or in other tables normally available to them.

For example, it may be desired to add another floor to an existing building which was constructed with shapes no longer rolled. In 1953 the AISC published a book entitled *Iron and Steel Beams 1873 to 1952*, which provides a complete listing of iron and

Mariners Ballpark Seattle, WA. (Courtesy of Trade ARBED.)

steel beams and their properties rolled in the United States during that period. Since that book was published there have been many additional shape changes. As a result, the wise structural designer will carefully preserve old editions of the Manual so as to have them available when needed. A new up-to-date edition of this book of previously used shapes is being published by the AISC early in 2003.

1.5 METRIC UNITS

Almost all of the examples and homework problems presented in this book make use of U.S. customary units. The author, however, feels that today's designer must be able to perform his or her work in either customary or metric units.

Many people find it rather distracting to read a book in which numbers, units, equations, and so on are given in two sets of units side by side. In an attempt to alleviate this problem, throughout the text we have placed a shaded area around any items pertaining specifically to work in metric units.

If readers are working with customary units, they can completely ignore the shaded areas. On the other hand, the authors hope that the same shaded areas will be helpful to those engineers using metric units for their work.

The problem of working with metric units when performing structural steel design in the United States has almost been eliminated by the AISC. Almost all of their equations are written in a form applicable to both systems. In addition, the metric equivalents of the standard U.S. shapes are provided in Section 17 of the Manual. For instance, a W36 $\times$ 300 section is shown there as a W920 $\times$ 446, where the 920 is mm and the 446 is kg/m.

1.6   COLD-FORMED LIGHT-GAGE STEEL SHAPES

In addition to the hot-rolled steel shapes discussed in the previous section, there are some cold-formed steel shapes. These are made by bending thin sheets of carbon or low-alloy steels into almost any desired cross section, such as the ones shown in Figure 1.2.[5] These shapes—which may be used for light members in roofs, floors, and walls—vary in thickness from about 0.01 in. up to about 0.25 in. The thinner shapes are most often used for some structural panels.

Though cold-working does reduce ductility somewhat, it causes some strength increases. Under certain conditions, design specifications will permit the use of these higher strengths.

Concrete floor slabs are very often cast on formed steel decks which serve as economical forms for the wet concrete and are left in place after the concrete hardens. Several types of decking are available, some of which are shown in Figure 1.3. The sections with the deeper cells have the useful feature that electrical and mechanical conduits can be placed in them. The use of steel decks for floor slabs is discussed in Chapter 16 of this text.

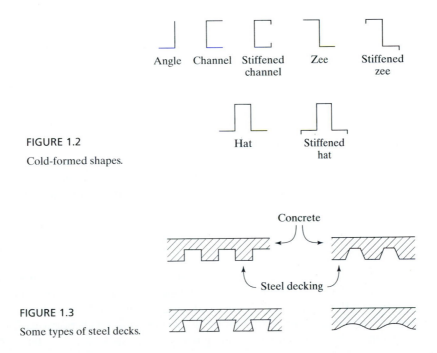

Angle   Channel   Stiffened   Zee   Stiffened
                    channel   zee

Hat   Stiffened
       hat

FIGURE 1.2

Cold-formed shapes.

Concrete

Steel decking

FIGURE 1.3

Some types of steel decks.

[5]Load and Resistance Factor Design, *Specification for the Design of Cold-formed Steel Members* (Washington, D.C.: American Iron and Steel Institute, 1991).

1.7   STRESS–STRAIN RELATIONSHIPS IN STRUCTURAL STEEL

To understand the behavior of steel structures, an engineer must be familiar with the properties of steel. Stress–strain diagrams present valuable information necessary to understand how steel will behave in a given situation. Satisfactory steel design methods cannot be developed unless complete information is available concerning the stress–strain relationships of the material being used.

If a piece of ductile structural steel is subjected to a tensile force, it will begin to elongate. If the tensile force is increased at a constant rate, the amount of elongation will increase linearly within certain limits. In other words, elongation will double when the stress goes from 6000 to 12,000 psi (pounds per square inch). When the tensile stress reaches a value roughly equal to three-fourths of the ultimate strength of the steel, the elongation will begin to increase at a greater rate without a corresponding increase in the stress.

The largest stress for which Hooke's law applies or the highest point on the linear portion of the stress–strain diagram is called the *proportional limit*. The largest stress that a material can withstand without being permanently deformed is called the *elastic limit*. This value is seldom actually measured and for most engineering materials, including structural steel, is synonymous with the proportional limit. For this reason, the term *proportional elastic limit* is sometimes used.

The stress at which there is a significant increase in the elongation or strain without a corresponding increase in stress is said to be the *yield* stress. It is the first point on the stress–strain diagram where a tangent to the curve is horizontal. The yield stress is probably the most important property of steel to the designer, as so many design procedures are based on this value. Beyond the yield stress there is a range in which a considerable increase in strain occurs without increase in stress. The strain that occurs before the yield stress is referred to as the *elastic strain;* the strain that occurs after the yield stress, with no increase in stress, is referred to as the *plastic strain*. Plastic strains are usually from 10 to 15 times the elastic strains.

Yielding of steel without stress increase may be thought to be a severe disadvantage when in actuality it is a very useful characteristic. It has often performed the wonderful service of preventing failure due to omissions or mistakes on the designer's part. Should the stress at one point in a ductile steel structure reach the yield point, that part of the structure will yield locally without stress increase, thus preventing premature failure. This ductility allows the stresses in a steel structure to be redistributed. Another way of describing this phenomenon is to say that very high stresses caused by fabrication, erection, or loading will tend to equalize themselves. It might also be said that a steel structure has a reserve of plastic strain that enables it to resist overloads and sudden shocks. If it did not have this ability, it might suddenly fracture, like glass or other vitreous substances.

Following the plastic strain, there is a range in which additional stress is necessary to produce additional strain. This is called *strain-hardening*. This portion of the diagram is not too important to today's designer because the strains are so large. A familiar stress–strain diagram for mild or low-carbon structural steel is shown in Figure 1.4. Only the initial part of the curve is shown here because of the great deformation which occurs before failure. At failure in the mild steels, the total strains are from 150 to 200 times the elastic strains. The curve will actually continue up to its maximum stress value and then "tail off" before failure. A sharp reduction in the cross section of the member (called "necking") takes place just before the member fractures.

Steel erection for the Chase Manhattan Bank
Building, New York City. (Courtesy of Bethlehem
Steel Corporation.)

The stress–strain curve of Figure 1.4 is typical of the usual ductile structural steel
and is assumed to be the same for members in tension or compression. (The compression
members must be stocky because slender compression members subjected to compression loads tend to buckle laterally, and their properties are greatly affected by the
bending moments so produced.) The shape of the diagram varies with the speed of
loading, the type of steel, and the temperature. One such variation is shown in the figure by the dotted line marked *upper yield*.

This shape stress–strain curve is the result when a mild steel has the load applied
rapidly, while the *lower yield* is the case for slow loading.

You should note that the stress–strain diagram of Figure 1.4 was prepared for a
mild steel at room temperature. During welding operations and during fires, structural
steel members may be subjected to very high temperatures. Stress–strain diagrams prepared for steels with temperatures above 200°F will be more rounded and nonlinear
and will not exibit well defined yield points. Steels (particularly those with high carbon
contents) may actually increase a little in tensile strength as they are heated to a temperature of about 700°F. As temperatures are raised into the 800° to 1000°F range,
strengths are drastically reduced, and at 1200°F little strength is left.

Figure 1.5, which is a copy of Figure 2.8(a) in the LRFD Manual, shows the variation of yield strengths for several grades of steel as their temperatures are raised from
room temperature up to 1800–1900°F. Temperatures of the magnitudes shown can
easily be reached in steel members during fires, in localized area of members when
welding is being performed, in members in foundries over open flame, and so on.

Puerta Europa Madrid, Spain. (Courtesy of Trade ARBED.)

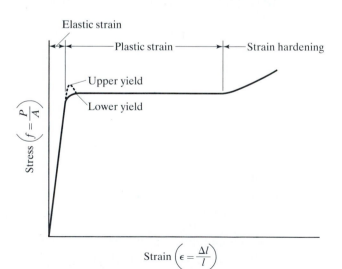

FIGURE 1.4

Typical stress–strain diagram for a mild or low-carbon structural steel at room temperature.

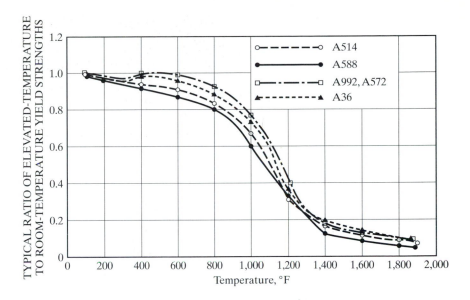

FIGURE 1.5

When steel sections are cooled below 32°F, their strengths will increase a little, but they will have substantial reductions in ductility and toughness.

A very important property of a structure that has not been stressed beyond its yield point is that it will return to its original length when the loads are removed. Should it be stressed beyond this point, it will return only part of the way back to its original position. This knowledge leads to the possibility of testing an existing structure by loading and unloading. If, after the loads are removed, the structure does not resume its original dimensions, it has been stressed beyond its yield point.

Steel is an alloy consisting almost entirely of iron (usually over 98 percent). It also contains small quantities of carbon, silicon, manganese, sulfur, phosphorus, and other elements. Carbon is the material that has the greatest effect on the properties of steel. The hardness and strength of steel increase as the carbon content is increased. A 0.01 percent increase in carbon content will cause steel's yield strength to go up about 0.5 kips per square inch (ksi). Unfortunately, however, more carbon will cause steel to be more brittle and will adversely affect its weldability. If the carbon content is reduced, the steel will be softer and more ductile, but also weaker. The addition of such elements as chromium, silicon, and nickel produces steels with considerably higher strengths. Though frequently quite useful, these steels, however, are appreciably more expensive and often are not as easy to fabricate.

A typical stress–strain diagram for a brittle steel is shown in Figure 1.6. Unfortunately, low ductility or brittleness is a property usually associated with high strengths in steels (although not entirely confined to high-strength steels). As it is desirable to have both high strength and ductility, the designer may have to decide between the two extremes or to compromise. A brittle steel may fail suddenly without warning when overstressed, and during erection could possibly fail due to the shock of erection procedures.

Brittle steels have a considerable range in which stress is proportional to strain, but do not have clearly defined yield stresses. Yet, to apply many of the formulas given

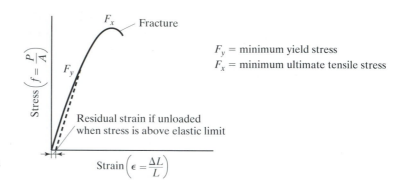

FIGURE 1.6

Typical stress–strain diagram for a brittle steel.

in structural steel design specifications, it is necessary to have definite yield stress values regardless of whether the steels are ductile or brittle.

If a steel member is strained beyond its elastic limit and then unloaded, it will not return to a condition of zero strain. As it is unloaded, its stress–strain diagram will follow a new path (shown by the dotted line in Figure 1.6) parallel to the initial straight line. The result is a permanent or residual strain.

The yield stress for a brittle steel is usually defined as the stress at the point of unloading which corresponds to some arbitrarily defined residual strain (0.002 being the common value). In other words, we increase the strain by a designated amount and draw a line from that point, parallel to the straight-line portion of the stress–strain diagram, until the new line intersects the old. This intersection is the yield stress at that particular strain. If 0.002 is used, the intersection is usually referred to as the yield stress at 0.2 percent offset strain.

1.8   MODERN STRUCTURAL STEELS

The properties of steel can be greatly changed by varying the quantities of carbon present and by adding other elements such as silicon, nickel, manganese, and copper. A steel that has a significant amount of the latter elements is referred to as an *alloy steel*. Although these elements do have a great effect on the properties of steel, the actual quantities of carbon or other alloying elements are quite small. For instance, the carbon content of steel is almost always less than 0.5 percent by weight and is normally from 0.2 to 0.3 percent.

The chemistry of steel is extremely important in its effect on such properties of the steel as weldability, corrosion resistance, resistance to brittle fracture, and so on. The ASTM specifies the exact maximum percentages of carbon, manganese, silicon, etc., which are permissible for a number of structural steels. Although the physical and mechanical properties of steel sections are primarily determined by their chemical composition, they are also influenced to a certain degree by the rolling process and by their stress history and heat treatment.

In the past few decades, a structural carbon steel designated as A36 and having a minimum yield stress $F_y = 36$ ksi was the commonly used structural steel. Today (2002), however, most of the structural steel used in the United States is manufactured

One-half of a 170-ft clear span roof truss for the Athletic and Convention Center, Lehigh University, Bethlehem, PA. (Courtesy of Bethlehem Steel Corporation.)

by melting scrap steel in electric furnaces. With this process, a 50 ksi steel can be produced and sold at almost the same price as 36 ksi steel.

The 50 ksi steels are the predominant ones in use today. In fact, some of the steel mills charge extra for W sections if they are to consist of A36 steel. On the other hand, 50 ksi angles have on occasion been rather difficult to obtain without making special orders to the steel mills. As a result, A36 angles are still frequently used. In addition, 50 ksi plates may cost more than A36 steel.

In recent decades, the engineering and architecture professions have been continually requesting stronger steels—steels with more corrosion resistance, steels with better welding properties, and various other requirements. Research by the steel industry during this period has supplied several groups of new steels which satisfy many of the demands. Today there are quite a few structural steels designated by the ASTM and included in the LRFD Specification.

Structural steels are generally grouped into several major ASTM classifications: the carbon steels A36, A53, A500, A501, and A529; the high-strength low alloy steels

Steel dome. (Courtesy of Trade ARBED.)

A572, A618, A913, and A992, and; the corrosion resistant high-strength low-alloy steels A242, A588, and A847. Considerable information is presented for each of these steels in Part 2 of the Manual. The sections that follow include a few general remarks about these steel classifications.

1.8.1 Carbon Steels

These steels have as their principal strengthening agents carefully controlled quantities of carbon and manganese. Carbon steels have their contents limited to the following maximum percentages: 1.7 percent carbon, 1.65 percent manganese, 0.60 percent silicon, and 0.60 percent copper. These steels are divided into four categories depending on carbon percentages as follows:

1. Low-carbon steel: < 0.15 percent.
2. Mild steel: 0.15 to 0.29 percent. (The structural carbon steels fall into this category.)
3. Medium-carbon steel: 0.30 to 0.59 percent.
4. High-carbon steel: 0.60 to 1.70 percent.

1.8.2 High-Strength Low-Alloy Steels

There are a large number of high-strength low-alloy steels, and they are included under several ASTM numbers. In addition to containing carbon and manganese, these steels obtain their higher strengths and other properties by the addition of one or more alloying agents such as columbium, vanadium, chromium, silicon, copper, and nickel. Included are steels with yield stresses as low as 42 ksi and as high as 70 ksi. These steels generally have much greater atmospheric corrosion resistance than the carbon steels.

The term *low-alloy* is used arbitrarily to describe steels for which the total of all the alloying elements does not exceed 5 percent of the total composition of the steel.

1.8.3   Atmospheric Corrosion-Resistant High-Strength Low-Alloy Structural Steels

When steels are alloyed with small percentages of copper, they become more corrosion-resistant. When exposed to the atmosphere, the surfaces of these steels oxidize and form a very tightly adherent film (sometimes referred to as a "tightly bound patina" or "a crust of rust") which prevents further oxidation and thus eliminates the need for painting. After this process takes place within 18 months to 3 years (depending on the type of exposure—rural, industrial, direct or indirect sunlight, etc.), the steel reaches a deep reddish-brown or black color.

Supposedly the first steel of this type was developed in 1933 by the U.S. Steel Corporation (now the USX Corporation) to provide resistance to the severe corrosive conditions of railroad coal cars.

You can see the many uses that can be made of such a steel, particularly for structures with exposed members that are difficult to paint—bridges, electrical transmission towers, and others. This steel is not considered to be satisfactory for use where it is frequently subject to saltwater sprays or fogs, or continually submerged in water (fresh or salt) or the ground, or where there are severe corrosive industrial fumes. It is also not satisfactory in very dry areas, as in some western parts of the United States. For the patina to form, the steel must be subjected to a wetting and drying cycle. Otherwise it will continue to look like unpainted steel.

Table 1-1 herein, which is Table 2-1 in the LRFD Manual, lists the 12 ASTM steels mentioned earlier in this section together with their specified minimum yield strengths (F_y) and their specified minimum tensile strengths (F_u). In addition, the right hand columns of the table provide information regarding the availability of the shapes in the various grades of steels, as well as the preferred grade to use for each of them. The preferred steel in each case is shown with a black box.

You will note in the table that A36 is the preferred material to use for Ms, Ss, HPs, and others, while A992 is the preferred material for the most common shapes, the Ws. The shaded boxes show the shapes available in grades of steel other than the preferred grades. Before shapes are specified in these grades, the designer should check on their availability from the steel producers. Finally, the blank boxes indicate the grades of steel which are not available for certain shapes. Similar information is provided for plates and bars in Table 2-2 of the Manual. (This table is not shown.)

As stated previously, steels may be made stronger by using special alloys. Another factor affecting steel strengths is thickness. The thinner steel is rolled, the stronger it becomes. Thicker members tend to be more brittle and their slower cooling rates cause steel to have a coarser microstructure. In this regard, you will notice in Table 1-2 (Table 2-4 in the Manual) that W sections are divided into five groups depending on their maximum thicknesses. The Ws with the smallest thicknesses are placed in Group 1, while those in Group 2 have larger maximum thicknesses. Similarly, the maximum thicknesses of the Ws in Groups 3, 4, and 5 are successively larger.

Referring back to Table 1-1, you can see that several of the steels listed are available with different yield and tensile stresses with the same ASTM number. For instance, A572 shapes are available with 42, 50, 55, 60, and 65 ksi yield strengths. Next,

Table 1-1
Applicable ASTM Specifications
for Various Structural Shapes

Steel Type	ASTM Designation		F_y Min. Yield Stress (ksi)	F_u Tensile Stress[a] (ksi)	W	M	S	HP	C	MC	L	HSS Rect.	HSS Round	Steel Pipe
Carbon	A36		36	58–80[b]										
	A53 Gr. B		35	60										
	A500	Gr. B	42	58										
		Gr. B	46	58										
		Gr. C	46	62										
		Gr. C	50	62										
	A501		36	58										
	A529[c]	Gr. 50	50	65–100										
		Gr. 55	55	70–100										
High-Strength Low-Alloy	A572	Gr. 42	42	60										
		Gr. 50	50	65[d]										
		Gr. 55	55	70										
		Gr. 60[e]	60	75										
		Gr. 65[e]	65	80										
	A618[f]	Gr. I & II	50[g]	70[g]										
		Gr. III	50	65										
	A913		50	50[h]	60[h]									
			60	60	75									
			65	65	80									
			70	70	90									
	A992		50–65[i]	65[i]										
Corrosion Resistant High-Strength Low-Alloy	A242		42[j]	63[j]										
			46[k]	67[k]										
			50[l]	70[l]										
	A588		50	70										
	A847[f]		50	70										

■ = Preferred material specification.

▨ = Other applicable material specification, the availability of which should be confirmed prior to specification.

☐ = Material specification does not apply.

[a]Minimum unless a range is shown.
[b]For shapes over 426 lb/ft, only the minimum of 58 ksi applies.
[c]Groups 1 and 2 shapes only. To improve weldability a maximum carbon equivalent can be specified (per ASTM Supplementary Requirement S78). If desired, maximum tensile stress of 90 ksi can be specified (per ASTM Supplementary Requirement S79).
[d]If desired, minimum tensile stress of 70 ksi can be specified (per ASTM Supplementary Requirement S81).
[e]Groups 1, 2, and 3 shapes only.
[f]ASTM A618 can also be specified as corrosion-resistant; see ASTM A618.
[g]Minimum applies for walls nominally ¾-in. thick and under. For wall thicknesses over ¾-in., F_y = 46 ksi and F_u = 67 ksi.
[h]If desired, maximum yield stress of 65 ksi and maximum yield-to-tensile strength ratio of 0.85 can be specified (per ASTM Supplementary Requirement S75).
[i]A maximum yield-to-tensile strength ratio of 0.85 and carbon equivalent formula are included as mandatory in ASTM A992.
[j]Groups 4 and 5 shapes only.
[k]Group 3 shapes only.
[l]Groups 1 and 2 shapes only.

reading the footnotes in Table 1-1 we note that Grades 60 and 65 steels have the footnote letter e by them. This footnote indicates that the only A572 shapes available with these strengths are those thinner ones listed in Groups 1, 2, and 3 in Table 1-2. Similar footnotes are provided in the tables for A992 and other steels.

1.9   USES OF HIGH-STRENGTH STEELS

There are indeed other groups of high-strength steels, such as the ultra-high-strength steels that have yield strengths from 160 to 300 ksi. These steels have not been included in the LRFD Manual because they have not been assigned ASTM numbers.

It is said that today more than 200 steels exist on the market that provide yield stresses in excess of 36 ksi. The steel industry is now experimenting with steels with yield stresses varying from 200 to 300 ksi, and this may be only the beginning. Many people in the steel industry feel that steels with 500 ksi yield strengths will be made available within a few years. The theoretical binding force between iron atoms has been estimated to be in excess of 4000 ksi.[6]

Although the prices of steels increase with increasing yield stresses, the percentage of price increase does not keep up with the percentage of yield-stress increase. The result is that the use of the stronger steels will quite frequently be economical for tension members, beams, and columns. Perhaps the greatest economy can be realized with tension members (particularly those without bolt holes). They may provide a great deal of saving for beams if deflections are not important or if deflections can be controlled by methods described in later chapters. In addition, considerable economy can frequently be achieved with high-strength steels for short- and medium-length stocky columns. Another application which can provide considerable savings is hybrid construction. In this type of construction two or more steels of different strengths are used, the weaker steels being used where stresses are smaller and the stronger steels where stresses are higher.

Among the other factors that might lead to the use of high-strength steels are the following:

1. Superior corrosion resistance.
2. Possible savings in shipping, erection, and foundation costs caused by weight saving.
3. Use of shallower beams permitting smaller floor depths.
4. Possible savings in fireproofing because smaller members can be used.

The first thought of most engineers in choosing a type of steel is the direct cost of the members. Such a comparison can be made quite easily, but determining which strength grade is most economical requires consideration of weights, sizes, deflections, maintenance, and fabrication. To make an accurate general comparison of the steels is probably impossible—rather, it is necessary to consider the specific job.

---

[6]L. S. Beedle, et al., *Structural Steel Design* (New York: Ronald Press, 1964), p. 44.

TABLE 1.2   Tensile Group Classification of Structural Shapes[a]

Shape		Group 1	Group 2	Group 3	Group 4[b]	Group 5[b]
W-Shapes	**W44x**	–	–	230 to 335	–	–
	W40x	–	149 to 249	264 to 331	362 to 593	–
	W36x	–	135 to 210	230 to 300	328 to 798	–
	W33x	–	118 to 169	201 to 291	318 to 387	–
	W30x	–	90 to 211	235, 261	292 to 391	–
	W27x	–	84 to 178	194 to 258	281 to 539	–
	W24x	55, 62	68 to 162	176 to 229	250 to 370	–
	W21x	44 to 57	62 to 147	166 to 201	–	–
	W18x	35 to 71	76 to 143	158, 175	–	–
	W16x	26 to 57	67 to 100	–	–	–
	W14x	22 to 53	61 to 132	145 to 211	233 to 550	605 to 808
	W12x	14 to 58	65 to 106	120 to 190	210 to 336	–
	W10x	12 to 45	49 to 112	–	–	–
	W8x	10 to 48	58, 67	–	–	–
	W6x	8.5 to 25	–	–	–	–
	W5x	16, 19	–	–	–	–
	W4x	13	–	–	–	–
M-Shapes		all	–	–	–	–
S-Shapes		to 35 lb/ft incl.	over 35 lb/ft	–	–	–
HP-Shapes		–	to 102 lb/ft incl.	over 102 lb/ft	–	–
American Standard Channels (C)		to 20.7 lb/ft incl.	over 20.7 lb/ft	–	–	–
Miscellaneous Channels (MC)		to 28.5 lb/ft incl.	over 28.5 lb/ft	–	–	–
Angles (L)		to 1/2-in incl.	over 1/2 in to 3/4 in incl.	over 3/4 in	–	–
Structural Tees (WT, MT, ST)		Structural tees cut from W-, M-, and S-shapes fall into the same group as the structural shapes from which they are cut.				

- Indicates that tensile group number does not apply to that shape or shape range.
[a] This table has been adjusted from the similar table in ASTM A6 to include all shapes listed in ASTM A6 Tables A2-1 through A2-8.
[b] Special requirements may apply, per LRFD Specification Section A3.1c.

1.10   MEASUREMENT OF TOUGHNESS

The fracture toughness of steel is used as a general measure of its impact resistance or its ability to absorb sudden increases in stress at a notch. The more ductile steel is, the greater will be its toughness. On the other hand, the lower its temperature, the higher will be its brittleness.

There are several procedures available for estimating notch toughness, but the Charpy V-notch test is the most commonly used. Although this test (which is described in ASTM Specification A6) is somewhat inaccurate, it does help identify brittle steels. With this test, the energy required to fracture a small bar of rectangular cross section with a specified notch (see Figure 1.7) is measured.

The bar is fractured with a pendulum swung from a certain height. The amount of energy needed to fracture the bar is determined from the height to which the pendulum rises after the blow. The test may be repeated for different temperatures and plotted as a graph, as shown in Figure 1.8. Such a graph clearly shows the relationship among temperature, ductility, and brittleness. The temperature at the point of steepest slope is the *transition temperature*.

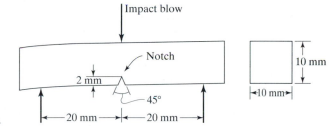

FIGURE 1.7

Specimen for Charpy V-notch test.

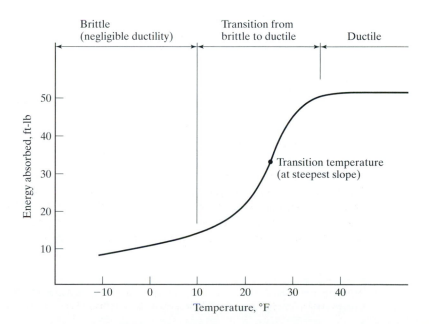

FIGURE 1.8

Results of Charpy V-notch test.

Different structural steels have different specifications for required absorbed energy levels (say 20 ft-lb at 20°F) depending on the temperature, stress, and loading conditions under which they are to be used. The topic of brittleness is continued in the next section.

1.11 JUMBO SECTIONS

In Table 1-2, steel shapes are placed in groups numbered 1 to 5, depending on their flange and web thicknesses. The heavy W sections in groups 4 and 5 (and structural tees cut from them) are often referred to as *jumbo sections*.

Jumbo sections were originally developed for use as compression members and are quite satisfactory for that purpose. However, designers have frequently used them for tension or flexural members. In these uses, flange and web areas have had some serious cracking problems where welding or thermal cutting has been used. These cracks may result in smaller load-carrying capacities and problems related to fatigue.[7]

Thick pieces of steel tend to be more brittle than thin ones. Some of the reasons for this are that the core areas of thicker shapes (shown in Figure 1.9) are subject to less rolling, have higher carbon contents (needed to produce required yield stresses), and have higher tensile stresses from cooling. These topics are discussed in later chapters.

Jumbo sections spliced with welds can be satisfactorily used for axial tension or flexural situations if the procedures listed in Specification A3.1c of the LRFD Specification are carefully followed. Included among the requirements are the following:

1. The steel used must have certain absorbed energy levels as determined by the Charpy V-notch test (20 ft-lb at 70°F). It is absolutely necessary that the tests be made on specimens taken from the core areas (shown in Figure 1.9), where brittle fracture has proved to be a problem.
2. Temperature must be controlled during welding, and the work must follow a certain sequence.
3. Special splice details are required.

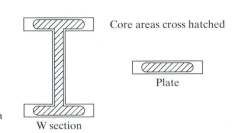

Core areas cross hatched

Plate

W section

FIGURE 1.9

Core areas where brittle failure may be a problem in thick heavy members.

---

[7]R. Bjorhovde, "Solutions for the Use of Jumbo Shapes," *Proceedings 1988 National Steel Construction Conference*, AISC, Chicago, June 8–11, 1988, pp. 2-1 to 2-20.

1.12   LAMELLAR TEARING

The steel specimens used for testing and developing stress–strain curves usually have their longitudinal axes in the direction that the steel was rolled. Should specimens be taken with their longitudinal axes transverse to the rolling direction "through the thickness" of the steel, the results will be lower ductility and toughness. Fortunately, this is a matter of little significance for almost all situations. It can, however, be quite important when thick plates and heavy structural shapes are used in highly restrained welded joints. (It can be a problem for thin members, too, but it is much more likely to give trouble in thick members.)

If a joint is highly restrained, the shrinkage of the welds in the through-the-thickness direction cannot be adequately redistributed, and the result can be a tearing of the steel called *lamellar tearing*. (*Lamellar* means "consisting of thin layers.") The situation is aggravated by the application of external tension. Lamellar tearing may show up as fatigue cracking after a number of cycles of load applications.

The lamellar tearing problem can be eliminated or greatly minimized with appropriate weld details and weld procedures. For example, the welds should be detailed so that shrinkage occurs as much as possible in the direction the steel was rolled. Several steel companies produce steels with enhanced through-the-thickness properties which provide much greater resistance to lamellar tearing. Even if such steels are used for heavy restrained joints, the special joint details mentioned above still are necessary.[8]

1.13   FURNISHING OF STRUCTURAL STEEL

The furnishing of structural steel consists of the rolling of the steel shapes, the fabrication of the shapes for the particular job (including cutting to the proper dimensions and punching the holes necessary for field connections), and their erection. Very rarely will a single company perform all three of these functions, and the average company performs only one or two of them. For instance, many companies fabricate structural steel and erect it, while others may only be steel fabricators or steel erectors. There are approximately 400 to 500 companies in the United States which make up the fabricating industry for structural steel. Most of them do both fabrication and erection.

Steel fabricators normally carry very little steel in stock because of the high interest and storage charges. When they get a job, they may order the shapes to certain lengths directly from the rolling mill, or they may obtain them from service centers. Service centers, which are an increasingly important factor in the supply of structural steel, buy and stock large quantities of structural steel which they buy at the best prices they can find anywhere in the world.

Structural steel is usually designed by an engineer in collaboration with an architectural firm. The designer makes design drawings that show member sizes, controlling dimensions, and any unusual connections. The company that is to fabricate the steel makes the detailed drawings subject to the engineer's approval. These drawings provide

---

[8]"Commentary on Highly Restrained Welded Connections," *Engineering Journal*, AISC, 10, no. 3 (3d quarter, 1973), pp. 61–73.

all the information necessary to fabricate the members correctly. They show the dimensions for each member, the locations of holes, the positions and sizes of connections, and the like. A part of a typical detail drawing for a bolted steel beam is shown in Figure 1.10. There may be a few items included on this drawing that are puzzling to you since you have read only a few pages of this book. However, these items should become clear as you study the chapters to follow.

On actual detail drawings, details probably will be shown for several members. Here, the author has shown only one member just to indicate the information needed so the shop can correctly fabricate the member. The darkened circles and rectangles indicate that the bolts are to be installed in the field, while the nondarkened ones show the connections which are to be made in the shop.

The erection of steel buildings is more a matter of assembly than nearly any other part of construction work. Each of the members is marked in the shop with letters and numbers to distinguish it from the other members to be used. The erection is performed in accordance with a set of erection plans. These plans are not detailed drawings, but are simple line diagrams showing the position of the various members in the building. The drawings show each separate piece or subassembly of pieces together with assigned shipping or erection marks so that the steelworkers can quickly identify and locate members in their correct positions in the structure. (Persons performing

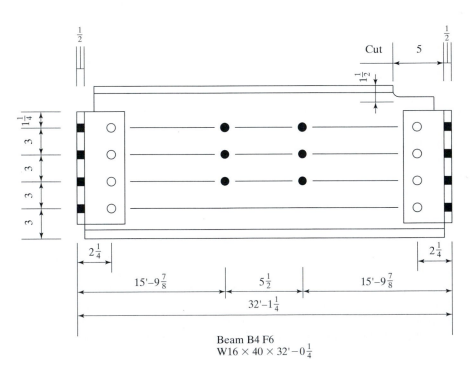

Beam B4 F6
W16 × 40 × 32'−0$\frac{1}{4}$

FIGURE 1.10

Part of a detail drawing.

steel erection often are called *iron-workers*, which is a name held over from the days before structural steel.) Directions (north, south, east, or west) usually are painted on column faces.

Sometimes the erection drawing gives the sizes of the members, but this is not necessary. They may or may not be shown, depending on the practice of the particular fabricator.

Beams, girders, and columns will be indicated on the drawings by the letters B, G, or C respectively, followed by the number of the particular member as B5, G12, and so on. Often, there will be several members of these same designations where members are repeated in the building.

Multistory steel frames often will have several levels of identical or nearly identical framing systems. Thus, one erection plan may be used to serve several floors. For such situations the member designations for the columns, beams, and girders will have the level numbers incorporated in them. For instance, column C15(3–5) is column 15, third to 5th floors, while B4F6, or just B4(6), represents beam B4 for the sixth floor. A portion of a building erection drawing is shown in Figure 1.11.

Next, we describe briefly the erection of the structural steel members for a building. Initially, a group of ironworkers, sometimes called the "raising gang," erects the steel members, installing only a sufficient number of bolts to hold the members in position.

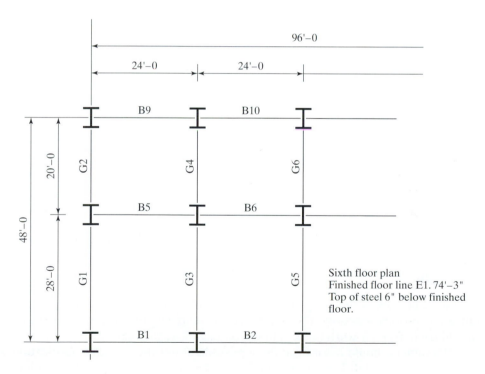

FIGURE 1.11

Part of an erection drawing showing where each member is to be located.

In addition, they place any guy cables where needed for stability and also plumb the steel frame.[9]

Another group of ironworkers, who are sometimes referred to as the "detail gang," install the remaining bolts, carry out any needed field welding, and complete the plumbing of the structure.[9]

After the last two steps are completed, another crew installs the metal decking for the floor and roof slabs. They in turn are followed by the crews who place the necessary concrete reinforcing and the concrete for those slabs.[9]

1.14 THE WORK OF THE STRUCTURAL DESIGNER

The structural designer arranges and proportions structures and their parts so that they will satisfactorily support the loads to which they may feasibly be subjected. It might be said that he or she is involved with the general layout of structures; studies of the possible structural forms that can be used; consideration of loading conditions; analysis of stresses, deflections, and so on; design of parts; and the preparation of design drawings. More precisely, the word *design* pertains to the proportioning of the various parts of a structure after the forces have been calculated, and it is this process which will be emphasized throughout the text, using structural steel as the material.

1.15 RESPONSIBILITIES OF THE STRUCTURAL DESIGNER

The structural designer must learn to arrange and proportion the parts of structures so that they can be practically erected and will have sufficient strength and reasonable economy. These items are discussed briefly below.

1.15.1 Safety

Not only must the frame of a structure safely support the loads to which it is subjected, but it must support them in such a manner that deflections and vibrations are not so great as to frighten the occupants or to cause unsightly cracks.

1.15.2 Cost

The designer needs to keep in mind the factors that can lower cost without sacrifice of strength. These items, which are discussed in more detail throughout the text, include the use of standard-size members, simple connections and details, and members and materials that will not require an unreasonable amount of maintenance through the years.

1.15.3 Practicality

Another objective is the design of structures that can be fabricated and erected without great problems arising. Designers need to understand fabrication methods and should try to fit their work to the fabrication facilities available.

Designers should learn everything possible about the detailing, fabrication, and field erection of steel. The more the designer knows about the problems, tolerances, and clearances in shop and field the more probable it is that reasonable, practical, and

---

[9]A. R. Tamboli, editor, *Steel Design Handbook LRFD Method* (New York: McGraw-Hill, 1997), pp. 12–37.

Steel erection for Transamerica Pyramid, San Francisco, CA. (Courtesy of Kaiser Steel Corporation.)

economical designs will be produced. This knowledge should include information concerning the transportation of the materials to the job site (such as the largest pieces that can be transported practically by rail or truck), labor conditions, and the equipment available for erection. Perhaps the designer should ask, "Could I get this thing together if I were sent out to do it?"

Finally, he or she needs to proportion the parts of the structure so that they will not unduly interfere with the mechanical features of the structure (pipes, ducts, etc.) or the architectural effects.

1.16   ECONOMICAL DESIGN OF STEEL MEMBERS

The design of a steel member involves much more than a calculation of the properties required to support the loads and the selection of the lightest section providing these properties. Although at first glance this procedure would seem to give the most economical designs, many other factors need to be considered.

Today, the labor costs involved in the fabrication and erection of structural steel are thought to run close to 60% of the total costs of steel structures. On the other hand, material costs represent only about 25% of total costs. Thus, we can see that any efforts we make to improve the economy of our work in structural steel should be primarily concentrated in the labor area.

When designers are considering costs, they have a tendency to think only of quantities of materials. As a result, they will sometimes carefully design a structure with the lightest possible members and end up with some very expensive labor situations with only minor material savings. Among the many factors that need to be considered in providing economical steel structures are the following:

1. One of the best ways to achieve economy is to have open communications between designers, fabricators, erectors, and others involved in a particular project. If this is done during the design process, the abilities and experience of each of the parties may be utilized at a time when it is still possible to implement good economical ideas.

2. The designer needs to select steel sections of sizes which are usually rolled. Steel beams and bars and plates of unusual sizes will be difficult to obtain during boom periods and will be expensive during any period. A little study on the designer's part will enable him or her to avoid these expensive shapes. Steel fabricators are constantly supplied with information from the steel companies and the steel warehousers as to the sizes and lengths of sections available. (Most structural shapes can be obtained in lengths from 60 to 75 ft depending on the producer, while it is possible under certain conditions to obtain some shapes up to 120 ft in length.)

3. A blind assumption that the lightest section is the cheapest one may be in considerable error. A building frame designed by the "lightest-section" procedure will consist of a large number of different shapes and sizes of members. Trying to connect these many-sized members and fit them in the building will be quite complicated, and the pound price of the steel will in all probability be rather high. A more reasonable approach would be to smooth out the sizes by selecting many members of the same sizes, although some of them may be slightly overdesigned.

4. The beams usually selected for the floors in buildings will normally be the deeper sections, because these sections for the same weights have the largest moments of inertia and the greatest resisting moments. As building heights increase, however, it may be economical to modify this practice. As an illustration, consider the erection of a 20-story building, for which each floor has a minimum clearance. It is assumed that the depths of the floor beams may be reduced by 6 in without an unreasonable increase in beam weights. The beams will cost more, but the building height will be reduced by 20×6 in $= 120$ in, or 10 ft, with resulting savings in walls, elevator shafts, column heights, plumbing, wiring, and footings.[10]

---

[10]H. Allison, "Low- and Medium-Rise-Steel Buildings" (Chicago: AISC, 1991), pp. 1–5.

5. The costs of erection and fabrication for structural steel beams are approximately the same for light or heavy members. Thus, beams should be spaced as far apart as possible to reduce the number of members which have to be fabricated and erected.

6. Structural steel members should be painted only if so required by the applicable specification. You should realize that steel should not be painted if it is to be in contact with concrete. Furthermore, the various fireproofing materials used for protecting steel members adhere better if the surfaces are unpainted.[11]

7. It is very desirable to keep repeating the same section over and over again. Such a practice will reduce the detailing, fabrication, and erection costs.

8. For larger sections, particularly the built-up ones, the designer needs to have information pertaining to transportation problems. The desired information includes the greatest lengths and depths that can be shipped by truck or rail (see Section 1.18), clearances available under bridges and power lines leading to the project, and allowable loads on bridges. It may be possible to fabricate a steel roof truss in one piece, but is it possible to transport it to the job site and erect it in one piece?

9. Sections should be selected which are reasonably easy to erect and which have no conditions that will make them difficult to maintain. As an example, it is necessary to have access to all exposed surfaces of steel bridge members so that they may be periodically painted (unless one of the special corrosion-resistant steels is being used).

10. Buildings are often filled with an amazing conglomeration of pipes, ducts, conduits, and other items. Every effort should be made to select steel members that will fit in with the requirements made by these items.

11. The members of a steel structure are often exposed to the public, particularly in the case of steel bridges and auditoriums. Appearance may often be the major factor in selecting the type of structure, as where a bridge is desired that will fit in and actually contribute to the appearance of an area. Exposed members may be surprisingly graceful when a simple arrangement and perhaps curved members are used, but other arrangements may create a terrible eyesore. The student has certainly seen illustrations of each case. It is very interesting to know that beautiful structures in steel are usually quite reasonable in cost.

The question is often asked, *How do we achieve economy in structural steel design? The answer is simple: it lies in what the steel fabricator does not have to do.* (In other words, economy can be realized when fabrication is minimized.)

The April 2000 issue of Modern Steel Construction provides a great deal of excellent material on the topic of economy in steel construction.[12]

---

[11]Ibid, pp. 1–5.
[12]*Modern Steel Construction,* April, 2000, vol. 40, no. 4 (Chicago: American Institute of Steel Construction), pp. 6, 25–48, 60.

Steel erection for Transamerica Pyramid, San Francisco, CA. (Courtesy of Kaiser Steel Corporation.)

1.17   FAILURE OF STRUCTURES

Many people who are superstitious do not discuss flat tires or make their wills because they fear they would be tempting fate. These same people would probably not care to discuss the subject of engineering failures. Despite the prevalence of this superstition, an awareness of the items that have most frequently caused failures in the past is invaluable to experienced and inexperienced designers alike. Perhaps a study of past failures is more important than a study of past successes. Benjamin Franklin supposedly made the observation that "a wise man learns more from failures than from success."

The designer of little experience particularly needs to know where the most attention should be given and where outside advice is needed. The vast majority of designers, experienced and inexperienced, select members of sufficient size and strength. The collapse of structures is usually due to insufficient attention to the details of connections, deflections, erection problems, and foundation settlement. Rarely, if ever, do steel structures fail due to faults in the material, but rather due to its improper use.

A frequent fault of designers is that after carefully designing the members of a structure, they carelessly select connections which may or may not be of sufficient size. They may even turn the job of selecting the connections over to drafters, who may not have sufficient understanding of the difficulties that can arise in connection design. Perhaps the most common mistake made in connection design is to neglect some of the forces acting on the connections, such as twisting moments. In a truss for which the members have been designed for axial forces only, the connections may be eccentrically loaded, resulting in moments that cause increasing stresses. These secondary stresses are occasionally so large that they need to be considered in design.

Another source of failure occurs when beams supported on walls have insufficient bearing or anchorage. Imagine a beam of this type supporting a flat roof on a rainy night when the roof drains are not functioning properly. As the water begins to form puddles on the roof, it tends to cause the beam to sag in the middle, causing a pocket to catch more rain, which will cause more beam sag, and so on. As the beam deflects, it pushes out against the walls, possibly causing collapse of walls or slippage of beam ends off the wall. Picture a 60-ft steel beam supported on a wall with only an inch or two of bearing when the temperature drops 50 or 60 degrees overnight. A collapse due to a combination of beam contraction, outward deflection of walls, and vertical deflection caused by precipitation loads is not difficult to visualize; furthermore, actual cases in engineering literature are not difficult to find.

Foundation settlements cause a large number of structural failures, probably more than any other factor. Most foundation settlements do not result in collapse, but they very often cause unsightly cracks and depreciation of the structure. If all parts of the foundation of a structure settle equally, the stresses in the structure theoretically will not change. The designer, usually not able to prevent settlement, has the goal of designing foundations in such a manner that equal settlements occur. Equal settlements may be an impossible goal, and consideration should be given to the stresses that would be produced if settlement variations occurred. The student's background in structural analysis will tell him or her that uneven settlements in statically indeterminate structures may cause extreme stress variations. Where foundation conditions are poor, it is desirable, if feasible, to use statically determinate structures whose stresses are not appreciably changed by support settlements. (The student will learn in subsequent discussions that the ultimate strength of steel structures is usually affected only slightly by uneven support settlements.)

Some structural failures occur because inadequate attention is given to deflections, fatigue of members, bracing against swaying, vibrations, and the possibility of buckling of compression members or the compression flanges of beams. The usual structure when completed is sufficiently braced with floors, walls, connections, and special bracing, but there are times during construction when many of these items are not present. As previously indicated, the worst conditions may well occur during erection, and special temporary bracing may be required.

1.18   HANDLING AND SHIPPING STRUCTURAL STEEL

The following general rules apply to the sizes and weights of structural steel pieces that can be fabricated in the shop, shipped to the job, and erected:

1. The maximum weights and lengths that can be handled in the shop and at a construction site are roughly 90 tons and 120 ft respectively.
2. Pieces as large as 8 ft high, 8 ft wide, and 60 ft long can be shipped on trucks with no difficulty (provided the axle or gross weights do not exceed the permissible values given by public agencies along the designated routes).
3. There are few problems in railroad shipment if pieces are no larger than 10 ft high, 8 ft wide, 60 ft long, and weigh no more than 20 tons.
4. Routes should be carefully studied and carriers consulted for weights and sizes exceeding the values given in 2 and 3 above.

1.19   CALCULATION ACCURACY

A most important point many students with their superb pocket calculators and personal computers have difficulty understanding is that structural design is not an exact science for which answers can confidently be calculated to eight significant figures. Among the reasons for this fact are the methods of analysis are based on partly true assumptions, the strengths of materials used vary appreciably, and maximum loadings can only be approximated. With respect to this last sentence, how many of the users of this book could estimate within 10 percent the maximum load in pounds per square foot that will ever occur on the building floor which they are now occupying? Calculations to more than two or three significant figures are obviously of little value and may actually be harmful in that they mislead the student by giving him or her a fictitious sense of precision.

1.20   COMPUTERS AND STRUCTURAL STEEL DESIGN

The availability of personal computers has drastically changed the way steel structures are analyzed and designed. In nearly every engineering school and design office, computers are used to perform structural analysis problems. Many of the structural analysis programs commercially available also can perform structural design.

Many calculations are involved in structural steel design, and many of these calculations are quite time-consuming. With the use of a computer, the design engineer can greatly reduce the time required to perform these calculations, and likely increase the accuracy of the calculations. In turn, this will then provide the engineer with more time to consider the implications of the design, the resulting performance of the structure, and time to try changes that may improve economy or behavior.

Although computers do increase design productivity, they also tend to reduce the engineer's "feel" for the structure. This can be a particular problem for young engineers with very little design experience. Unless design engineers have this feel for system behavior, the use of computers can result in large costly mistakes. Such situations may arise where anomalies and inconsistences are not immediately apparent to the inexperienced engineer. Theoretically, the computer design of alternative systems for a few projects should substantially improve the engineer's judgment in a short span of time. Without computers, the development of this same judgment would likely require the engineer to work his or her way through numerous projects.

1.21   COMPUTER-AIDED DESIGN IN THIS TEXT

Enclosed in this book is a CD-ROM containing a limited steel design software system called INSTEP32. This software was prepared by the authors to assist students in solving many of the problems presented in this text. The design procedures used in the software are based on the 3rd edition of the LRFD Manual. Most, but not all, of the checks necessary for complete design are performed.

As such, the software is somewhat limited. Nevertheless, through use of INSTEP32, the student can easily perform structural steel design, study the effects of design changes, and develop an appreciation of the significance of the various design parameters.

INSTEP32 is a Windows® based program designed to operate on IBM-compatible personal computers with the Windows9x®, Windows NT 4.0®, or Windows 2000® operating systems. The minimum system requirements are a VGA monitor, a keyboard, a mouse, and 5 Mbytes of free disk space. A printer is required to obtained printed design calculations.

To install the software on your computer, insert the CD-ROM. Then, in Windows Explorer, open the INSTEP32 directory on the CD-ROM and double-click on Setup. The Setup program will guide you through the installation process.

INSTEP32 is started by clicking the INSTEP32 icon on the Start Menu on the desktop. A complete description of the INSTEP32 program, the assumptions and limitations of the procedures used, and the information necessary to perform design with INSTEP32, can be obtained by clicking on Help–Help Topics on the menu bar after starting the software. The different kinds of designs that can be performed are selected from the Design menu item or by clicking on the design icons on the toolbar. Definitions of the various terms and parameters are defined by clicking on the Help button in the different data-entry dialogs. To the greatest extent possible, the notation used is consistent with the LRFD Manual.

Also included on the CD-ROM is the student edition of the SAP2000 commercial structural analysis and design software. SAP2000 will be used in the design of building frames in the last chapters of this textbook. It is installed by double-clicking on Setup in the SAP2000 directory on the CD-ROM.

CHAPTER 2

Specifications, Loads, and Methods of Design

2.1 SPECIFICATIONS AND BUILDING CODES

The design of most structures is controlled by building codes and design specifications. Even if not so controlled, the designer will probably refer to them as a guide. No matter how many structures a person has designed, it is impossible for him or her to have encountered every situation. By referring to specifications, he or she is making use of the best available material on the subject. Engineering specifications that are developed by various organizations present the best opinion of those organizations as to what represents good practice.

Municipal and state governments concerned with the safety of the public have established building codes with which they control the construction of various structures within their jurisdiction. These codes, which are actually laws or ordinances, specify minimum design loads, design stresses, construction types, material quality, and other factors. They vary considerably from city to city, a fact that causes some confusion among architects and engineers.

Several organizations publish recommended practices for regional or national use. Their specifications are not legally enforceable unless they are embodied in the local building code or made a part of a particular contract. Among these organizations are the AISC and AASHTO (American Association of State Highway and Transportation Officials). Nearly all municipal and state building codes have adopted the AISC Specification and nearly all state highway and transportation departments have adopted the AASHTO Specifications.

Readers should note that logical and clearly written codes are quite helpful to design engineers. Furthermore, there are far fewer structural failures in areas that have good building codes that are strictly enforced.

Some people feel that specifications prevent engineers from thinking for themselves—and there may be some basis for the criticism. They say that the ancient engineers who built the great pyramids, the Parthenon, and the great Roman bridges were

South Fork Feather River Bridge in northern California, being erected by use of a 1626-ft-long cableway strung from 210-ft-high masts anchored on each side of the canyon. (Courtesy of Bethlehem Steel Corporation.)

controlled by few specifications, which is certainly true. On the other hand, it should be said that only a few score of these great projects were built over many centuries, and they were apparently built without regard to cost of material, labor, or human life. They were probably built by intuition and by certain rules of thumb developed by observing the minimum size or strength of members which would fail only under given conditions. Their probably numerous failures are not recorded in history; only their successes endured.

Today, however, there are hundreds of projects being constructed at any one time in the United States that rival in importance and magnitude the famous structures of the past. It appears that if all engineers in our country were allowed to design projects such as these without restrictions there would be many disastrous failures. *The important thing to remember about specifications, therefore, is that they are written, not for the purpose of restricting engineers, but for the purpose of protecting the public.*

No matter how many specifications are written, it is impossible for them to cover every possible design situation. As a result, no matter which building code or specification is or is not being used, the ultimate responsibility for the design of a safe structure lies with the structural designer. Obviously the intent of these specifications is that the loading used for design be the one that causes the largest stresses.

Another very important code was published in 2000, the *International Building Code*[1] (IBC), which was developed because of the need for a modern building code

---

[1]International Code Council, Inc., *International Building Code* (Falls Church, VA, 2000).

that emphasizes performance. It is intended to provide a model set of regulations to safeguard the public in all communities.

2.2   LOADS

Perhaps the most important and most difficult task faced by the structural engineer is the accurate estimation of the loads that may be applied to a structure during its life. No loads that may reasonably be expected to occur may be overlooked. After loads are estimated, the next problem is to determine the worst possible combinations of these loads that might occur at one time. For instance, would a highway bridge completely covered with ice and snow be simultaneously subjected to fast-moving lines of heavily loaded trailer trucks in every lane and to a 90-mile lateral wind, or is some lesser combination of these loads more feasible?

Section A.4 of the LRFD Specification states that the nominal loads to be used for structural design shall be the ones stipulated by the applicable code under which the structure is being designed or as dictated by the conditions involved. If there is an absence of a code, the design loads shall be those provided in a publication of the American Society of Civil Engineers entitled *Minimum Design Loads for Buildings and Other Structures.*[2] This publication is commonly referred to as ASCE 7. It was originally published by the American National Standards Institute (ANSI) and referred to as the *ANSI 58.1 Standard.* The ASCE took over its publication in 1988.

In general, loads are classified according to their character and duration of application. As such, they are said to be *dead loads, live loads,* and *environmental loads.* Each of these types of loads are discussed in the next few sections.

2.3   DEAD LOADS

Dead loads are loads of constant magnitude that remain in one position. They consist of the structural frame's own weight and other loads that are permanently attached to the frame. For a steel-frame building, the frame, walls, floors, roof, plumbing, and fixtures are dead loads.

To design a structure, it is necessary for the weights or dead loads of the various parts to be estimated for use in the analysis. The exact sizes and weights of the parts are not known until the structural analysis is made and the members of the structure selected. The weights, as determined from the actual design, must be compared with the estimated weights. If large discrepancies are present, it will be necessary to repeat the analysis and design using better estimated weights.

Reasonable estimates of structure weights may be obtained by referring to similar types of structures or to various formulas and tables available in several locations. The weights of many materials are given in Part 17 of the LRFD Manual. Even more detailed information on dead loads is provided in Tables C1 and C2 of ASCE 7–98. An experienced engineer can estimate very closely the weights of most materials and will spend little time repeating designs because of poor estimates.

---

[2]American Society of Civil Engineers, *Minimum Design Loads for Buildings and Other Structures.* ASCE 7-98 formerly ANSI A58.1 (New York, ASCE, 2000).

TABLE 2.1   Typical Dead Loads for Some Common Building
Materials

Reinforced concrete	150 lb/cu ft
Structural steel	490 lb/cu ft
Movable steel partitions	4 psf
Plaster on concrete	5 psf
Suspended ceilings	2 psf
3-ply ready roofing	1 psf
Hardwood flooring (7/8 in)	4 psf
$2 \times 12 \times 16$ in double wood floors	7 psf
Wood studs with 1/2 in gypsum	8 psf
Clay brick wythes (4 in)	39 psf

The approximate weights of some common building materials for roofs, walls, floors, and so on are presented in Table 2.1.

2.4   LIVE LOADS

Live loads are loads that may change in position and magnitude. They are caused when a structure is occupied, used, and maintained. Live loads that move under their own power, such as trucks, people, and cranes, are said to be *moving loads*. Those loads that may be moved are *movable loads*, such as furniture and warehouse materials. A great deal of information on the magnitudes of these various loads, along with specified minimum values, are presented in ASCE 7-98.

1. *Floor loads*. The minimum gravity live loads to be used for building floors are clearly specified by the applicable building code. Unfortunately, however, the values given in these various codes vary from city to city, and the designer must be sure that his or her designs meet the requirements of the locality in question. A few of the typical values for floor loadings are listed in Table 2.2. These values were adopted from ASCE 7-98. In the absence of a governing code, this is an excellent one to follow.

Quite a few building codes specify concentrated loads that must be considered in design. Section 4-3 of ASCE 7-98 and Section 1607.4 of IBC-2000 are two such examples. The loads specified are considered as alternatives to the uniform loads previously considered herein.

Some typical concentrated loads taken from Table 4-1 of ASCE 7-98 and Table 1607.1 of IBC-2000 are listed in Table 2.3 of this chapter. These loads are to be placed on floors or roofs at the positions where they will cause the most severe conditions. Unless otherwise specified, each of these concentrated loads is spread over an area 2.5×2.5 ft square (6.25 ft^2).

2. *Traffic loads for bridges*. Bridges are subjected to series of concentrated loads of varying magnitude caused by groups of truck or train wheels.

3. *Impact loads*. Impact loads are caused by the vibration of moving or movable loads. It is obvious that a crate dropped on the floor of a warehouse or a truck

TABLE 2.2   Typical Minimum Uniform Live Loads
for Design of Buildings

Type of building	LL (psf)
Apartment houses	
Apartments	40
Public rooms	100
Dining rooms and restaurants	100
Garages (passenger cars only)	50
Gymnasiums, main floors, and balconies	100
Office buildings	
Lobbies	100
Offices	50
Schools	
Classrooms	40
Corridors first floor	100
Corridors above 1st floor	80
Storage warehouses	
Light	125
Heavy	250
Stores (retail)	
First floor	100
Other floors	75

TABLE 2.3   Typical Concentrated Live Loads for Buildings

Hospitals—operating rooms, private rooms, and wards	1000 lb
Manufacturing building (light)	2000 lb
Manufacturing building (heavy)	3000 lb
Office floors	2000 lb
Retail stores (first floors)	1000 lb
Retail stores (upper floors)	1000 lb
School classrooms	1000 lb
School corridors	1000 lb

bouncing on uneven pavement of a bridge cause greater forces than would occur if the loads were applied gently and gradually. Cranes picking up loads and elevators starting and stopping are other examples of impact loads. Impact loads are equal to the difference between the magnitude of the loads actually caused and the magnitude of the loads had they been dead loads.

Section 4.7 of the ASCE 7-98 Specification requires that when structures are supporting live loads that tend to cause impact, it is necessary for those loads to be increased by the percentages given in Table 2.4.

   4. *Longitudinal loads.* Longitudinal loads are another type of load that needs to be considered in designing some structures. Stopping a train on a railroad bridge or a truck on a highway bridge causes longitudinal forces to be applied. It is not difficult to imagine the tremendous longitudinal force developed when the driver of

Cold-storage warehouse, Grand Junction, CO. (Courtesy of the American Institute of Steel Construction, Inc.)

a 40-ton-trailer truck traveling 60 mph suddenly has to apply the brakes while crossing a highway bridge. There are other longitudinal load situations, such as ships bumping a dock during berthing and the movement of traveling cranes that are supported by building frames.

5. *Other live loads.* Among the other types of live loads with which the structural engineer will have to contend are *soil pressures* (such as the exertion of lateral earth pressures on walls or upward pressures on foundations); *hydrostatic static pressures* (as water pressure on dams, inertia forces of large bodies of water during earthquakes, and uplift pressures on tanks and basement structures); *blast loads* (caused by explosions, sonic booms, and military weapons); *thermal forces* (due to changes in temperature resulting in structural deformations and resulting structural forces); and *centrifugal forces* (as those caused on curved bridges by trucks and trains or similar effects on roller coasters, etc.).

TABLE 2.4   Live Load Impact Factors

Elevator machinery	100%
Motor driven machinery	20%
Reciprocating machinery	50%
Hangers for floors or balconies	33%

Hungry Horse Dam and Reservoir, Rocky Mountains, in northwestern Montana. (Courtesy of the Montana Travel Promotion Division.)

2.5 ENVIRONMENTAL LOADS

Environmental loads are caused by the environment in which a particular structure is located. For buildings, environmental loads are caused by rain, snow, wind, temperature change, and earthquakes. Strictly speaking, environmental loads are live loads, but they are the result of the environment in which the structure is located. Even though they do vary with time, they are not all caused by gravity or operating conditions, as is typical with other live loads. A few comments are presented in the paragraphs that follow concerning the different types of environmental loads.

1. *Snow.* In the colder states, snow loads are often quite important. One inch of snow is equivalent to a load of approximately 0.5 psf, but it may be higher at lower elevations where snow is denser. For roof designs, snow loads varying from 10 to 40 psf are commonly used, the magnitude depending primarily on the slope of the roof and, to a lesser degree, on the character of the roof surface. The larger values are used for flat roofs and the smaller ones for sloped roofs. Snow tends to slide off sloped roofs, particularly those with metal or slate surfaces. A load of approximately 10 psf might be used for 45° (degree) slopes and a 40-psf load for flat roofs. Studies of snowfall records in areas with severe winters may indicate the occurrence of snow loads much greater than 40 psf, with values as high as 100 psf in northern Maine.

   Snow is a variable load that may cover an entire roof or only part of it. The snow loads that are applied to a structure are dependent upon many factors, including geographic location, roof pitch, sheltering, and the shape of the roof. Section 7 of ASCE 7–98 provides a great deal of information concerning snow loads, including charts and formulas for estimating their magnitudes. There may be drifts against walls or buildup in valleys or between parapets. Snow may slide off one roof onto a lower one. The wind may blow it off one side of a sloping roof or the snow may crust over and remain in position even during very heavy winds.

Bridges are generally not designed for snow loads, since the weight of the snow is usually not significant in comparison with truck and train loads. In any case, it is doubtful that a full load of snow and maximum traffic would be present at the same time. Bridges and towers are sometimes covered with layers of ice from 1 to 2 in thick. The weight of the ice runs up to about 10 psf. Another factor to be considered is the increased surface area of the ice-coated members as it pertains to wind loads.

2. *Rain.* Though snow loads are a more severe problem than rain loads for the usual roof, the situation may be reversed for flat roofs, particularly those in warmer climates. If water on a flat roof accumulates faster than it runs off, the result is called *ponding* because the increased load causes the roof to deflect into a dish shape that can hold more water, which causes greater deflections, and so on. This process continues until equilibrium is reached or until collapse occurs. Ponding is a serious matter as illustrated by the large number of flat-roof failures that occur during rain storms every year in the United States. It has been claimed that almost 50% of the lawsuits faced by building designers are concerned with roofing systems.[3] Ponding is one of the most common subjects of such litigation.

Ponding will occur on almost any flat roof to a certain degree, even though roof drains are present. Roof drains may very well be used, but they may be inadequate during severe storms, or they may become partially or completely clogged. The best method of preventing ponding is to have an appreciable slope of the roof (1/4 in/ft or more) together with good drainage facilities. In addition to the usual ponding, another problem may occur for very large flat roofs (with perhaps an acre or more of surface area). During heavy rainstorms, strong winds frequently occur. If there is a great deal of water on the roof, a strong wind may very well push a large quantity of it toward one end. The result can be a dangerous water depth as regards the load in psf on that end. For such situations, *scuppers* are sometimes used. These are large holes or tubes in walls or parapets that enable water above a certain depth to quickly drain off the roof.

Section 8 of ASCE 7–98 provides information for estimating the magnitude of rain loads that may accumulate on flat roofs.

3. *Wind Loads.* A survey of engineering literature for the past 150 years reveals many references to structural failures caused by wind. Perhaps the most infamous of these have been bridge failures, such as those of the Tay Bridge in Scotland in 1879 (which caused the deaths of 75 persons) and the Tacoma Narrows Bridge (Tacoma, Washington) in 1940. But there have also been some disastrous building failures due to wind during the same period, such as that of the Union Carbide Building in Toronto in 1958. It is important to realize that a

---

[3]Gary Van Ryzin, 1980, "Roof Design: Avoid Ponding by Sloping to Drain," *Civil Engineering* (New York, ASCE, January), pp. 77–81.

large percentage of building failures due to wind have occurred during their erection.[4]

A great deal of research has been conducted in recent years on the subject of wind loads. Nevertheless, a great deal more work needs to be done as the estimation of these forces can by no means be classified as an exact science. The magnitudes of wind loads vary with geographical locations, heights above ground, types of terrain surrounding the buildings, including other nearby structures, and other factors.

Wind pressures are usually assumed to be uniformly applied to the windward surfaces of buildings and are assumed to be capable of coming from any direction. These assumptions are not very accurate because wind pressures are not uniform over large areas, the pressures near the corners of buildings being probably greater than elsewhere due to wind rushing around the corners, and so on. From a practical standpoint, therefore, all of the possible variations cannot be considered in design, although today's specifications are becoming more and more detailed in their requirements.

When the designer working with large low-rise buildings makes poor wind estimates, the results are probably not too serious, but this is not the case when tall slender buildings (or long flexible bridges) are being designed. For many years, the average designer ignored wind forces for buildings whose heights were not at least twice their least lateral dimensions. For such cases as these, it was felt that the floors and walls provided sufficient lateral stiffnesses to eliminate the need for definite wind bracing systems. A better practice for designers to follow, however, is to consider all the possible loading conditions that a particular structure may have to resist. If one or more of these conditions (as perhaps wind loading) seem of little significance, they may then be neglected. Should buildings have their walls and floors constructed with modern lightweight materials and/or should they be subjected to unusually high wind loads (as in coastal or mountainous areas) they will probably have to be designed for wind loads even if the height/least lateral dimension ratios are less than two.

Building codes do not usually provide for the estimated forces during tornadoes. The average designer considers the forces created directly in the paths of tornadoes to be so violent that it is not economically feasible to design buildings to resist them. This opinion is undergoing some change, however, as it has been found that the wind resistance of structures (even for small buildings, including houses) can be greatly increased at very reasonable costs by using better practices of tying the parts of structures together from the roofs through the walls to the footings, and by making appropriate connections of window frames to walls and perhaps other parts of the structure.[5,6]

Wind forces act as pressures on vertical windward surfaces, pressures or suction on sloping windward surfaces (depending on the slope), and suction on flat surfaces and on leeward vertical and sloping surfaces (due to the creation of negative pressures

[4]Wind Forces on Structures, Task Committee on Wind Forces. Committee on Loads and Stresses, Structural Division, ASCE, Final Report, *Transactions ASCE* 126, Part II (1961): 1124–1125.
[5]P. R. Sparks, "Wind Induced Instability in Low-Rise Buildings," *Proceedings of the 5th U.S. National Conference on Wind Engineering*, Lubbock, TX, November 6–18, 1985.
[6]P. R. Sparks, "The Risk of Progressive Collapse of Single-Story Buildings in Severe Storms," *Proceedings of the ASCE Structures Congress*, Orlando, FL, August 17–20, 1987.

Access bridge, Renton, WA. (Courtesy of the Bethlehem Steel Corporation.)

or vacuums). The student may have noticed this definite suction effect where shingles or other roof coverings have been lifted from the leeward roof surfaces of buildings. Suction or uplift can easily be demonstrated by holding a piece of paper at two of its corners and blowing above it. For some common structures, uplift loads may be as large as 20 to 30 psf or even more.

During the passing of a tornado or hurricane, a sharp reduction in atmospheric pressure occurs. This decrease in pressure is not reflected inside airtight buildings, and the inside pressures, being greater than the external pressures, cause outward forces against the roofs and walls. Nearly everyone has heard stories of the walls of a building "exploding" outward during a storm.

As you can see, the accurate calculation of the most severe wind pressures that need to be considered for the design of buildings and bridges is quite an involved problem. Despite this fact, sufficient information is available today to permit the satisfactory estimation of these pressures in a reasonably efficient manner.

A procedure for estimating the wind pressures applied to buildings is presented in Section 6 of ASCE 7.98. Several factors are involved when we attempt to account for the effects of wind speed, shape, and orientation of the building in question, terrain characteristics around the structure, importance of the building as to human life and welfare, and so on. Though the procedure seems rather complex, it is greatly simplified with the tables presented in the aforementioned specification.

4. *Earthquake Loads.* Many areas of the world fall in "earthquake territory," and in those areas it is necessary to consider seismic forces in design for all types of structures. Through the centuries, there have been catastrophic failures of buildings, bridges, and other structures during earthquakes. It has been estimated that

as many as 50,000 people lost their lives in the 1988 earthquake in Armenia.[7] The 1989 Loma Prieta and 1994 Northridge earthquakes in California caused many billions of dollars of property damage as well as considerable loss of life.

Figure 2.1 shows seismic risk zones in various locations in the United States. These zones were established on the basis of past earthquakes.[8] This map indicates that the most severe risks are in the zones numbered 4 and the least severe risks are in those numbered 0. This somewhat dated map is presented to give the student a feeling for those areas of the country where earthquake risks are greatest. Much more detailed maps are used today to estimate potential seismic forces.

Steel structures can be economically designed and constructed to withstand the forces caused during most earthquakes. On the other hand, the cost of providing seismic resistance to existing structures (called *retrofitting*) can be extremely high. *Recent earthquakes have clearly shown that the average building or bridge that has not been designed for earthquake forces can be destroyed by an earthquake that is not particularly severe.*

During an earthquake there is an acceleration of the ground surface. This acceleration can be broken down into vertical and horizontal components. Usually the vertical component is assumed to be negligible, but the horizontal component can be severe. The structural analysis for the expected effects of an earthquake should include a study of the structure's response to the ground motion caused by the earthquake. However, it is common in design to approximate the effects of ground motion with a set of horizontal static loads acting at each level of the structure. The various formulas used to change the earthquake accelerations into static forces are dependent on the distribution of the mass of the structure, the type of framing, its stiffness, its location, and so on. Such an approximate approach is usually adequate for regular low-rise buildings, but it is not suitable for irregular and high-rise structures. For these latter structures, an overall dynamic analysis is usually necessary.

Some engineers seem to think that the seismic loads to be used in design are merely percentage increases of the wind loads. This surmise is incorrect, however, as seismic loads are different in their action and are not proportional to the exposed area of the building, but to the distribution of the mass of the building above the particular level being considered.

The forces due to the horizontal acceleration increase with the distance of the floor above the ground because of the "whipping effect" of the earthquake. Obviously, towers, water tanks, and penthouses on building roofs occupy precarious positions when an earthquake occurs.

Another factor to be considered in seismic design is the soil condition. Almost all of the structural damage and loss of life in the Loma Prieta earthquake occurred in areas that have soft clay soils. Apparently, these soils amplified the motions of the underlying rock.[9]

---

[7]V. Fairweather, "The Next Earthquake," *Civil Engineering* (New York: ASCE, March 1990), pp. 54–57.
[8]*American Society of Civil Engineers Minimum Design Loads for Buildings and Other Structures*, ASCE 7–88 (New York: ASCE), pp. 33–34.
[9]Fairweather, op. cit.

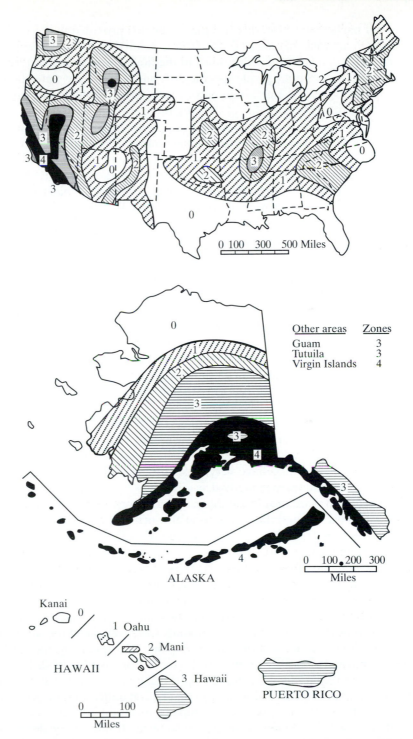

FIGURE 2.1

Map of seismic zones of the United States. The severity of risk ranges from a low
in Zone 0 to a high in Zone 4. (From ASCE Standard 7-88, reprinted with the
permission of the American Society of Civil Engineers.)

Of particular importance are the comments provided in the Commentary to the Seismic Provisions relating to drift. (*Drift* is defined as the movement or displacement of one story of a building with respect to the floor above or below.) Actually, the AISC Specification does not provide specific drift limits. It merely states that limits on story drift are to be used which are in accord with the governing code and shall not be so large as to impair the stability of the structure.

If structures are designed so that computed drifts during an earthquake of specified intensity are limited to certain maximum values, it will be necessary to provide added strength and stiffness to those structures. The result will be structures whose performance is substantially improved during earthquakes. The LRFD Manual does not provide detailed specifications for designing structures subject to seismic loads, but such information is presented in Section 9 of ASCE 7-98.

2.6   LOAD AND RESISTANCE-FACTOR DESIGN

A steel design method called the *load and resistance design method* (LRFD) is presented in this book. For the past decade, it has been the method taught to most of the students in the colleges and universities of the United States. However, a rather large proportion of practicing designers in the United States use an older steel design method called *allowable stress design (ASD)*. As a result, the student should be familiar with both ASD and LRFD. Rather than confusing readers by combining a discussion of both methods at the same time, the LRFD method is presented in the main body of the text, and an introduction to ASD design is presented in Appendix A of the book.

The reader should realize that the AISC is planning to publish a revised specification and handbook in 2005 that will supposedly be a combination of the best features of the two present design methods. A person who has studied one of the two methods (LRFD or ASD) should have very little trouble understanding the other one or the proposed combined design method.

Load and resistance factor design is based on a *limit states* philosophy. The term *limit state* is used to describe a condition at which a structure or some part of that structure ceases to perform its intended function. There are actually two categories of limit states, strength and serviceability.

Strength limit states are based on the safety or load-carrying capacity of structures and include plastic strengths, buckling, fracture, fatigue, overturning, and so on.

Serviceability limit states refer to the performance of structures under normal service loads and are concerned with the uses and/or occupancy of structures including such items as excessive deflections, slipping, vibrations, cracking, and deterioration.

Not only must the structure be capable of supporting the design or ultimate loads, it must also be able to support the service or working loads in such a manner as to meet the requirements of the users or occupants of the structure. For instance, a high-rise building must be designed so that lateral deflections or drift are not excessive during ordinary storms; that is, the occupants must not become uncomfortable (frightened or seasick). As to the strength limit state, the frame will be designed so that it will safely resist the ultimate load occurring during that 50-year storm, although there may be some minor damage to the building and the occupants may suffer some discomfort.

The LRFD Specification[10] concentrates on very specific requirements relating to the strength limit state and allows the designer some freedom of judgment regarding the serviceability area. This doesn't mean that the serviceability limit state is not significant, but by far the most important consideration (as in all structural specifications) is the life and property of the public. As a result, minimum standards of public safety are not left to the judgment of the individual designer.

In LRFD, the working or service loads (Q_i) are multiplied by certain load or safety factors (λ_i) which are almost always larger than 1.0 and the resulting "factored loads" used for designing the structure. The magnitudes of load factors vary depending on the type and combination of the loads and are discussed in detail in the next section of this chapter.

The structure is proportioned to have a design ultimate strength sufficient to support the factored loads. This strength is considered to equal the nominal or theoretical strength of the member (R_n) multiplied by a resistance or capacity reduction factor (ϕ), which is normally less than 1.0, with which the designer attempts to account for the uncertainties in material strengths, dimensions, and workmanship. Furthermore, these factors have been adjusted somewhat to ensure more uniform reliability in design as described in Section 2.10. Values of ϕ factors for different situations are discussed in detail in Section 2.8.

For a particular member, the preceding information can be summarized as follows: (Sum of products of load effects and load factors) $\leq$ (resistance factor) (nominal resistance)

$$\Sigma \lambda_i Q_i \leq \phi R_n.$$

The left-hand side of this equation refers to the load effects on the structure, while the right-hand side refers to the resistance or capacity of the structure.

In Canada, a probability-based limit states design method has been in use for quite a few years for both hot-rolled and cold-formed steel structures. A uniform limit states steel design code is being adopted by nearly every European country. It is called Eurocode (EC) 3.[11] Similar codes have been prepared in Europe for reinforced concrete, timber, and masonry design, while much progress is being made in America toward similar codes.

2.7   LOAD FACTORS

Load factors are used to increase the magnitudes of the calculated loads applied to structures. Their purpose is to increase the loads to account for the uncertainties involved in estimating the magnitudes of dead or live loads. For instance, "How close in percent could you, the reader, estimate the worst wind or snow loads that will ever be applied to the building which you are now occupying?"

---

[10]American Institute of Steel Construction, *Load and Resistance Factor Design* (Chicago: AISC, December, 1999).

[11]A. D. Weller, "An Introduction to EC 3." *The Structural Engineer, Journal of the Institution of Structural Engineers* 71, No. 18 (London, September 21, 1993), pp. 326–331.

The load factor for dead loads is smaller than the one used for live loads because designers can estimate so much more accurately the magnitudes of dead loads than they can the magnitudes of live loads. In this regard, the student will notice that loads which remain in place for long time periods will be less variable in magnitude, while those that are applied for brief periods, such as wind loads, will have larger variations. It is hoped that the LRFD design procedure will make the designer more conscious of load variations than he or she is with the allowable stress design method.

The load factors to be used for a particular design shall be the ones stipulated in the applicable code under which the design is being made. In the absence of such a code, the factors used should be the ones specified in ASCE 7. These factors (given in Section 2.3.2 of that specification) are the ones used in this textbook. It is stated in ASCE 7-98 that structures and their components are to be designed so their strength is at least equal to the values obtained with the load combinations to follow:

1. $1.4(D + F)$
2. $1.2(D + F + T) + 1.6(L + H) + 0.5(L_r \text{ or } S \text{ or } R)$
3. $1.2D + 1.6(L_r \text{ or } S \text{ or } R) + (0.5L \text{ or } 0.8W)$
4. $1.2D + 1.6W + 0.5L + 0.5(L_r \text{ or } S \text{ or } R)$
5. $1.2D + 1.0E + 0.5L + 0.2S$
6. $0.9D + 1.6W + 1.6H$
7. $0.9D + 1.0E + 1.6H$

In these load combinations the following abbreviations are used:

U = the design (ultimate) load
D = dead load
F = fluid load
T = self straining force
L = live load
L_r = roof live load
H = lateral earth pressure load, ground water pressure or pressure of bulk materials
S = snow load
R = rain load
W = wind load
E = earthquake load

The preceding load factors do not vary in relation to the seriousness of failure. The reader may feel that a higher load factor should be used for a hospital than for a cattle barn, but this is not required. It is assumed, however, that the designer will consider the seriousness of failure when he or she specifies the magnitudes of the service loads. It also should be realized that the ASCE 7 load factors are minimum values and the designer is perfectly free to use larger ones if it is deemed prudent.

The following are several additional comments regarding the application of the load combination expressions:

1. It is to be noted that in selecting design loads, adequate allowance must be made for impact conditions before the loads are substituted into the combination expressions.

2. The load combinations (6) and (7) are used to account for the possibilities of uplift. Such a condition is included to cover cases in which tension forces develop owing to overturning moments. It will govern only for tall buildings where high lateral loads are present. In this combination, the dead loads are reduced by 10 percent to take into account situations where they may have been overestimated.

3. It is to be clearly noted that the wind and earthquake forces have signs—that is, they may be compressive or they may be tensile (that is tending to cause uplift). Thus, the signs must be accounted for in substituting into the load combinations. Furthermore, these forces are not plus or minus. For a particular column, the maximum tensile W or E force will, in all probability, be different from its maximum compressive force.

4. In applying combinations (3), (4), and (5), the load factor for L is to be 1.0 for garages, for areas used as places of public assembly, and for areas where the live load is greater than 100 psf. In ASCE 7-98, some other changes in load factors are given for certain circumstances of lateral earth pressures, and wind and earthquake forces.

5. The magnitudes of the loads (D, L, L_r, etc.) should be obtained from the governing building code or from ASCE 7-98. Wherever applicable, the live loads used for design should be the reduced values specified for large floor areas, multistory buildings, and so on.

Examples 2-1 to 2-3 show the calculation of the factored loads using the applicable LRFD load combinations. The largest value obtained in each case is referred to as the *critical* or *governing load combination* and is to be used in design.

Example 2-1

A floor system has W24 $\times$ 55 sections spaced 8 ft on-center supporting a floor dead load of 50 psf and a live load of 80 psf. Determine the governing load in lb/ft which each beam must support.

Solution. Noting that each foot of the beam must support itself (a dead load) plus $8 \times 1 = 8$ ft^2 of the building floor

$$D = 55 + (8)(50) = 455 \text{ lb/ft}$$
$$L = (8)(80) = 640 \text{ lb/ft}$$

Computing factored loads and noting that D and L are the only loads to be supported we find that it is necessary to substitute only into combinations (1) and (2).

1. $U = (1.4)(D + F) = (1.4)(455 + 0) = 637 \text{ lb/ft}$
2. $U = (1.2)(D + F + T) + (1.6)(L + H) + (0.5)(L_r \text{ or } S \text{ or } R)$
    $= (1.2)(455 + 0 + 0) + (1.6)(640 + 0) + (0.5)(0 + 0 + 0)$
    $= 1570 \text{ lb/ft} \leftarrow$
    Governing factored load $= 1570$ lb/ft

Example 2-2

A roof system with W16 × 40 sections spaced 9 ft on-center is to be used to support a dead load of 40 psf, a roof live or snow or rain load of 30 psf, and a wind load of 20 psf. Compute the governing factored load.

Solution.
$$D = 40 + (9)(40) = 400 \text{ lb/ft}$$
$$L = 0$$
$$L_r \text{ or } S \text{ or } R = (9)(30) = 270 \text{ lb/ft}$$
$$W = (9)(20) = 180 \text{ lb/ft}$$

Substituting into the load combination expressions

1. $U = (1.4)(400 + 0) = 560 \text{ lb/ft}$
2. $U = (1.2)(400 + 0 + 0) + (1.6)(0 + 0) + (0.5)(270) = 615 \text{ lb/ft}$
3. $U = (1.2)(400) + (1.6)(0 + 270 + 0) + (0 + 0.8 \times 180) = 1056 \text{ lb/ft} \leftarrow$
4. $U = (1.2)(400) + (1.6)(180) + 0 + (0.5)(270 + 0 + 0) = 903 \text{ lb/ft}$
5. $U = (1.2)(400) + 0 + 0 + (0.2)(270) = 534 \text{ lb/ft}$
6. $U = (0.9)(400) + (1.6)(180) + 0 = 648 \text{ lb/ft}$
7. $U = (0.9)(400) + 0 + 0 = 360 \text{ lb/ft}$

Governing factored load = 1056 lb/ft

Example 2-3

The various axial loads for a building column have been computed according to the applicable building code with the following results: dead load = 200 k, load from roof = 50 k (roof live load), live load from floors (has been reduced as applicable for large floor area and multistory columns) = 250 k, compression wind = 80 k, tensile wind = 65 k, compression earthquake = 60 k, and tensile earthquake = 70 k.
    Determine the critical design load using the ASCE 7 load combinations.

Solution.

1. $U = (1.4)(200 + 0) = 280 \text{ k}$
2. $U = (1.2)(200 + 0 + 0) + (1.6)(250 + 0) + (0.5)(50 + 0 + 0) = 665 \text{ k} \leftarrow$
3. a. $(1.2)(200) + (1.6)(50) + (0.5)(250) = 445 \text{ k}$
   b. $(1.2)(200) + (1.6)(50) + (0.8)(80) = 384 \text{ k}$
   c. $(1.2)(200) + (1.6)(50) + (0.8)(-65) = 268 \text{ k}$
4. a. $(1.2)(200) + (1.6)(80) + (0.5)(250) + (0.5)(50) = 518 \text{ k}$
   b. $(1.2)(200) + (1.6)(-65) + (0.5)(250) + (0.5)(50) = 286 \text{ k}$

5. a. $(1.2)(200) + (1.0)(60) + (0.5)(250) = 425$ k

   b. $(1.2)(200) + (1.0)(-70) + (0.5)(250) = 295$ k

6. a. $U = (0.9)(200) + (1.6)(80) + 0 = 308$ k

   b. $U = (0.9)(200) - (1.6)(65) = 76$ k

7. a. $U = (0.9)(200) + (1.0)(60) + 0 = 240$ k

   b. $U = (0.9)(200) - (1.0)(70) = 110$ k

The critical factored load combination or design strength required for this column is 665 k as determined by load combination (2). It will be noted that the results of combinations (6) and (7) do not indicate an uplift problem.

2.8 RESISTANCE FACTORS

To accurately estimate the ultimate strength of a structure an engineer must take into account the uncertainties in material strengths, dimensions, and workmanship. With a resistance factor, the designer attempts to account for imperfections in analysis theory (remember the imperfect assumptions you made to analyze trusses and other members), variations in material properties, and imperfect dimensions of members.

This is done by multiplying the theoretical ultimate strength (called the *nominal strength* here) of each member by a capacity reduction factor ϕ, which is almost always less than 1.0. These values are 0.85 for columns, 0.75 or 0.90 for tension members, 0.90 for bending or shear in beams, and so on.

Typical resistance factors from the LRFD Specification are shown in Table 2.2. Some terms used in this table (such as groove and fillet welds, slip-resistant bolts, and web crippling) will be defined in later chapters. The magnitudes of the resistance factors given in the LRFD Specification are based on research recommendations from investigators at Washington University in St. Louis.[12,13]

TABLE 2.2   Typical Resistance Factors

Resistance or ϕ factors	Situations
1.00	Bearing on the projected areas of pins, web yielding under concentrated loads, slip-resistant bolt shear values
0.90	Beams in bending and shear, fillet welds with stress parallel to weld axis, groove welds base metal, yielding on gross section of tension members
0.85	Columns, web crippling, edge distance, and bearing capacity at holes
0.80	Shear on effective area of full-penetration groove welds, tension normal to effective area of partial-penetration groove welds
0.75	Bolts in tension, plug, or slot welds, fracture in the net section of tension members
0.65	Bearing on bolts (other than A307)
0.60	Bearing on concrete foundations

[12]M. K. Ravindra and T. V. Galambos, "Load and Resistance Factor Design for Steel," *ASCE Journal of Structural Division,* 104, no. ST9 (September 1978), pp. 1337–1353.
[13]Research Report 45, C.E. Dept., Washington University, St. Louis, MO, May 1976.

2.9 DISCUSSION OF SIZES OF LOAD AND RESISTANCE FACTORS

Students may feel that it is foolish and uneconomical to design structures with such large load factors and such small resistance factors. As the years go by, however, they will learn that these factors are subject to so many uncertainties that they may spend sleepless nights wondering if those they have used are sufficient (and they may join other designers in calling them "factors of ignorance"). Some of the uncertainties affecting these factors are as follows:

1. Material strengths may initially vary appreciably from their assumed values, and they will vary more with time due to creep, corrosion, and fatigue.

2. The methods of analysis are often subject to appreciable errors.

3. The so-called vagaries of nature or acts of God (hurricanes, earthquakes, etc.) cause conditions difficult to predict.

4. The stresses produced during fabrication and erection are often severe. Laborers in shop and field seem to treat steel shapes with reckless abandon. They drop them. They ram them. They force the members into position to line up the bolt holes. In fact, the stresses during fabrication and erection may exceed those that occur after the structure is completed. The floors for the rooms of apartment houses and office buildings are probably designed for service live loads varying from 40 to 80 psf (pounds per square foot). During the erection of such buildings, the contractor may have 10 ft of bricks or concrete blocks or other construction materials or equipment piled up on some of the floors (without the knowledge of the structural engineer), causing loads of several hundred pounds per square foot. This discussion is not intended to criticize the practice (not that it is a good one), but rather to make the student aware of the things that happen during construction. (It is probable that the majority of steel structures are overloaded somewhere during construction, but hardly any of them fail. On many of these occasions, the ductility of the steel has surely saved the day.)

5. There are technological changes that affect the magnitude of live loads. The constantly increasing traffic loads applied to bridges through the years is one illustration. The wind also seems to blow harder as the years go by, or at least building codes keep raising the minimum design wind pressures as more is learned about the subject.

6. Although the dead loads of a structure can usually be estimated quite closely, the estimate of the live loads is more inaccurate. This is particularly true in estimating the worst possible combination of live loads occurring at any one time.

7. Other uncertainties are the presence of residual stresses and stress concentrations, variations in dimensions of member cross sections, etc.

2.10 RELIABILITY AND THE LRFD SPECIFICATION

Probably fewer than 1 in 10 undergraduate engineering students take a course in statistics. Thus, these statistically untrained students are often a little dismayed when they run into a statistically based subject such as this one. In the pages which follow, these students will find that such worrying is unnecessary, as the LRFD theory and calculations are usually rather simple to follow. Furthermore, they may actually gain a little statistical knowledge along the way and thereby better understand the theory behind the development of LRFD.

The word *reliability,* as used in this textbook, refers to the estimated percentage of times that the strength of a structure will equal or exceed the maximum loading applied to that structure during its estimated life (say 50 years).

In this section, we discuss the following:

1. how LRFD investigators developed a procedure for estimating the reliability of given designs;
2. how they set what to them were desirable reliability percentages for different situations; and
3. how they were able to adjust resistance or ϕ factors so steel designers are able to obtain the reliability percentages established in Item 2 of this list.

Before proceeding with this discussion, a few comments are presented concerning the word *failure* as it is used in the discussion of reliability. Let us assume that a designer states that his or her designs are 99.7 percent reliable (and this is the approximate value obtained with most LRFD designs). This means that if this person were to design 1000 different structures, three of them would probably be overloaded at some time during their estimated 50-year lives and thus would fail. The reader probably and quite reasonably thinks this to be an unacceptably high rate of failure.

A 99.7 percent reliability doesn't mean that 3 of 1000 structures are going to fall flat on the ground. Rather, it means that those structures at some time will be loaded into the plastic range and perhaps the strain-hardening range. As a result, deformations may be quite large during the overloading, and some slight damage may occur. It is not anticipated that any of these structures will completely collapse. (The reader who is unfamiliar with statistics might say he or she wants 100 percent reliability in design, but this is an impossible goal statistically, as we will see in the paragraphs that follow.)

For this discussion, it is assumed that we make a study of the reliability of a large number of steel structures designed at various times and with different past editions of the AISC Specification. To do this, we will compute the resistance or strength, R, of each structure as well as the maximum loading, Q, expected during the life of the structure. The structure will be deemed to be safe if $R \geq Q$.

The actual values of R and Q are random variables, and it is therefore impossible to say with 100 percent certainty that R is always equal to or larger than Q for a particular structure. No matter how carefully a structure is designed and constructed, there will always be some chance that Q will be greater than R or that the strength limit state will be exceeded. The goal of the preparers of the LRFD Specification was to keep this chance to a very small and consistent percentage.

Thus, the magnitudes of both resistances and loads are uncertain. If we were to plot a curve of $l_n(\frac{R}{Q})$ values for a large number of structures, the result would be a typical bell-shaped probability curve with mean values R_m and Q_m and a standard deviation. If, at any location, $R < Q$, we will have exceeded the strength limit state.

Such a curve is plotted in Figure 2.2. It is to be remembered that the logarithm of 1.0 is 0 and thus, if $\ln R/Q < 0$ the strength limit state has been exceeded. Such a situation is represented by the shaded area on the diagram. Another way of expressing this

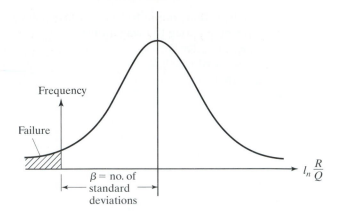

FIGURE 2.2

Definition of reliability index β.

fact is to say that the larger the number of standard deviations from the mean values to the shaded area, the greater is the reliability. In the figure, the number of standard deviations is represented by β and is called the *reliability index*.

Even though the probable values of R and Q are not known very well, a formula has been developed with which values of β can be reasonably calculated.[14,15,16] The formula follows:

$$\beta = \frac{\ln (R_m/Q_m)}{\sqrt{V_R^2 + V_Q^2}}$$

In this expression, R_m and Q_m are, respectively, the mean resistance and load effects, while V_R and V_Q are the corresponding coefficients of variation.

As a result of the preceding work, it is now possible to design a particular element in accordance with a certain edition of the AISC Specification and with the appropriate statistical information, compute the value of β for the design. This process is called *calibration*.

The results of our study of the designs of these steel structures will show that the percentage of structures for which the design strengths equal or exceed the worst anticipated loading will vary as we examine designs made in accordance with the requirements of different editions of the AISC Specification. Furthermore, our calculations will show that this reliability will vary for the designs of different types of members (such as columns and beams) made with the same edition of the AISC Specification.

Based upon the calculations of reliability described here, the investigators decided to use consistent β values in this new specification:

---

[14] C. A. Cornell, "A Probability-Based Structural Code," *Journal, American Concrete Institute*, vol. 66, no. 12 (December 1969).

[15] B. Ellingwood, T. V. Galambos, J. G. MacGregor, and C. A. Cornell, "Development of a Probability-Based Load Criterion for American National Standard A58," National Bureau of Standards Special Publication 577 (June 1980).

[16] M. K. Ravindra and T. V. Galambos, "Load and Resistance Factor Design for Steel," *Journal of the Structural Division*, vol. 104, no. ST9, (September 1978), pp. 1337–1353.

1. $\beta = 3.00$ for members subject to gravity loading.
2. $\beta = 4.50$ for connections. (This value reflects the usual practice that connections should be stronger than the members they connect.)
3. $\beta = 2.5$ for members subject to gravity plus wind loadings. (This value reflects an old practice that safety factors do not have to be as large for cases where lateral loads, which are of shorter duration, are involved.)
4. $\beta = 1.75$ for members subject to gravity loads plus earthquake loads.

Then the values of ϕ for the various parts of the specification were adjusted so the β values shown here would be obtained in design. In effect, this causes most LRFD designs to be almost identical with those obtained by the allowable stress method when the live to dead load ratio is three.

2.11 ADVANTAGES OF LRFD

The average person looking at this material might ask, "Will LRFD save money compared with allowable stress design (ASD)?" The answer is that it probably will, particularly if the live loads are small compared with the dead loads.

It should be noted, however, that the AISC has introduced LRFD not for the specific purpose of obtaining immediate economic advantages, but because it helps provide a more uniform reliability for all steel structures whatever the loads, and it is written in a form that facilitates the introduction of the advances in knowledge that will occur through the years in structural steel design.

Despite this stated objective of the AISC, the author would like to continue for a little while with the topic of possible steel weight savings with LRFD. In allowable stress design, the same safety factor was used for dead loads and live loads, whereas in LRFD a much smaller safety or load factor is used for dead loads (because they generally can be determined so much more accurately than can live loads). As a result of this fact, a weight of steel comparison for ASD and LRFD designs will necessarily depend on the ratio of the live loads to dead loads.

For the usual building, the live load to dead load ratio varies from approximately 0.25 to as high as 4.0 or even a little higher for some very light structures. Low-rise steel buildings and preengineered buildings generally will fall in the upper range of these ratios. In allowable stress design we use the same safety factors for dead and live loads regardless of their ratio to each other. Thus, with ASD, heavier members will result and the factor of safety will increase as the live load to dead load ratio decreases.

It can be shown that for the lower range of L to D values, that is, less than about three, there can be steel weight savings with LRFD, perhaps as much as one-sixth for tension members and columns and perhaps as much as one-tenth for beams. On the other hand, if we have very high L and D ratios there is a slight increase in steel weight with LRFD compared with ASD.[17]

---

[17]J. A. Edinger, "Introduction to the Proposed AISC Load and Resistance Factor Design Specification," *Engineering Journal*, AISC, Vol. 21, no. 1 (first quarter, 1984), pp. 62–65.

2.12 COMPUTER EXAMPLE

Now that we understand the principles of load factors and load combinations, we will use the computer and the INSTEP32 design software to perform the necessary calculations for load combinations. To perform the calculations for load combinations, start INSTEP32 and then select Design—Load Combinations from the menu bar. After you have done this, the data entry dialog shown in the following screen will appear:

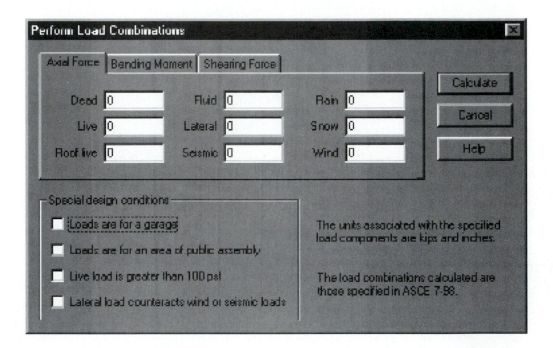

Through this dialog, you can enter data for the fundamental axial and flexural load effects. There is a separate tab for each of these load effects. You can also indicate whether any of the special conditions of ASCE 7-98 apply, such as loads for a parking garage, or where the live load is greater than 100 psf.

After entering all of the loads components, click on the Calculate button to perform the actual load combinations. The resulting load combinations are displayed on the screen. These calculations can be printed in the form of archival design calculations by clicking on File–Print on the menu bar in the main window or by clicking on the printer icon on the toolbar.

Example 2-4

Using the computer program INSTEP32, compute the governing factored loads for each of the following: $D = 200$ k, $L_r = 50$ k, $L = 250$ k, $W = 80$ k, and $E = 60$ k. (These were the values previously considered in Example 2-3.)

Solution.

Specified Load Components

	Axial	Bending	Shear
Dead Load D	200.000000		0.000000
Live Load L	250.000000		0.000000
Roof Live Load L	50.000000	0.000000	0.000000
Fluid Load F	0.000000	0.000000	0.000000
Lateral Load H	0.000000	0.000000	0.000000
Earthquake Load E	60.000000	0.000000	0.000000
Rain Load R	0.000000	0.000000	0.000000
Snow Load S	0.000000	0.000000	0.000000
Wind Load W	80.000000	0.000000	0.000000

Calculated Combined Loading Using ASCE 7-98

Group 1 Load Combination	Axial	Bending	Shear
1.4(D + F)	280.000000	0.000000	0.000000

Group 2 Load Combinations	Axial	Bending	Shear
1.2(D + F) + 1.6(L + H) + 0.5Lr	665.000000	0.000000	0.000000
1.2(D + F) + 1.6(L + H) + 0.5S	640.000000	0.000000	0.000000
1.2(D + F) + 1.6(L + H) + 0.5R	640.000000	0.000000	0.000000

Group 3 Load Combinations	Axial	Bending	Shear
1.2D + 1.6Lr + 0.5L	445.000000	0.000000	0.000000
1.2D + 1.6S + 0.5L	365.000000	0.000000	0.000000
1.2D + 1.6R + 0.5L	365.000000	0.000000	0.000000
1.2D + 1.6Lr + 0.8W	384.000000	0.000000	0.000000
1.2D + 1.6S + 0.8W	304.000000	0.000000	0.000000
1.2D + 1.6R + 0.8W	304.000000	0.000000	0.000000

Group 4 Load Combinations	Axial	Bending	Shear
1.2D + 1.6W + 0.5L + 0.5Lr	518.000000	0.000000	0.000000
1.2D + 1.6W + 0.5L + 0.5S	493.000000	0.000000	0.000000
1.2D + 1.6W + 0.5L + 0.5R	493.000000	0.000000	0.000000

Group 5 Load Combination	Axial	Bending	Shear
1.2D + 1.0E + 0.5L + 0.2S	425.000000	0.000000	0.000000

Group 6 Load Combination	Axial	Bending	Shear
0.9D + 1.6W + 1.6H	308.000000	0.000000	0.000000

Group 7 Load Combination	Axial	Bending	Shear
0.9D + 1.0E + 1.6H	240.000000	0.000000	0.000000

PROBLEMS

2-1. Structural steel beams are to be placed 12 ft on-center under a reinforced concrete floor slab. If they are to support a dead working load $D = 100$ psf of floor area and a live working load $L = 60$ psf of floor area, determine the factored uniform load per foot that each beam must support. (*Ans.* 2592 lb/ft)

2-2. For the design of a roof, the estimated working or service loads are: dead load $D = 40$ psf, snow $S = 25$ psf, and wind $W = 30$ psf. Calculate the factored loads in psf to be used for design.

2-3. A column is to support the following service or working loads: $D = 100$ k axial compression dead load, $L = 40$ k floor axial compression live load, and $W = 50$ k axial compression or 40 k tension wind load. Compute the required design strength of the member. (*Ans.* 220 k)

2-4. Determine the required design strength for a column for which dead load $D = 120$ k, floor live load $L = 80$ k, roof live load $L_r = 40$ k, and wind $W = 60$ k compression or 70 k tension.

2-5. The estimated service or working axial loads and bending moments for a beam-column are as follows: $D = 110$ k, $L = 30$ k, $M_D = 20$ ft-k, and $M_L = 12$ ft-k. Compute the axial load and moment values that must be used in design. (*Ans.* 180 k and 43.2 ft-k)

2-6. The following axial service or working loads have been estimated for the design of a particular column: dead load $D = 80$ k, floor live load $L = 60$ k, roof live load $L_r = 25$ k, and wind load $W = 30$ k compression or 35 k tension. Determine the required design strength for the member.

2-7. The working or service loads have been determined for a particular building column as required by the applicable building code with the following results: $D = 300$ k, $L = 200$ k, L_r or S or R from roof $= 40$ k, $W = 150$ k compression or 130 k tension, and $E = 60$ k compression or 75 k tension. Determine the critical factored design load for the column. (*Ans.* 720 k)

2-8 to 2-10. Using the enclosed computer program, INSTEP32, solve the following problems:

2-8. Prob. 2-2.

2-9. Prob. 2-4. (*Ans.* + 300 k or −4 k)

2-10. Prob. 2-7.

C H A P T E R   3

Analysis of Tension Members

3.1 INTRODUCTION

Tension members are found in bridge and roof trusses, towers, bracing systems, and in situations where they are used as tie rods. The selection of a section to be used as a tension member is one of the simplest problems encountered in design. As there is no danger of buckling, the designer needs only to compute the factored force to be carried by the member and divide that force by a design stress to determine the effective cross-sectional area required. Then it is necessary to select a steel section that provides the required area. Though these introductory calculations for tension members are quite simple, they do serve the important tasks of getting students started with design ideas and getting their "feet wet" regarding the massive LRFD Manual.

One of the simplest forms of tension members is the circular rod, but there is some difficulty in connecting it to many structures. The rod has been used frequently in the past, but has only occasional uses today in bracing systems, light trusses, and in timber construction. One important reason rods are not popular with designers is that they have been used improperly so often in the past that they have a bad name; however, if designed and installed correctly, they are satisfactory for many situations.

The average size rod has very little stiffness and may quite easily sag under its own weight, injuring the appearance of the structure. The threaded rods formerly used in bridges often worked loose and rattled. Another disadvantage of rods is the difficulty of fabricating them with the exact lengths required and the consequent difficulties of installation.

When rods are used in wind bracing, it is a good practice to produce initial tension in them, as this will tighten up the structure and reduce rattling and swaying. To obtain initial tension, the members may be detailed shorter than their required lengths, a method that gives the steel fabricator very little trouble. A common rule of thumb used is to detail the rods about 1/16 in short for each 20 ft of length. (Approximate stress $f = \epsilon E = [\frac{1}{16}/(12)(20)](29 \times 10^6) = 7550$ psi.) Another very satisfactory method involves tightening the rods with some sort of sleeve nut or turnbuckle. Part 15 of the LRFD Manual provides detailed information for these devices.

The preceding discussion on rods should illustrate why rolled shapes such as angles have supplanted rods for most applications. In the early days of steel structures, tension members consisted of rods, bars, and perhaps cables. Today, although the use of cables is increasing for suspended-roof structures, tension members usually consist of single angles, double angles, tees, channels, W sections, or sections built up from plates or rolled shapes. These members look better than the old ones, are stiffer, and are easier to connect. Another type of tension section often used is the welded tension plate or flat bar, which is very satisfactory for use in transmission towers, signs, foot bridges, and similar structures.

The tension members of steel roof trusses may consist of single angles as small as 2 1/2 × 2 × 1/4 for minor members. A more satisfactory member is made from two angles placed back to back with sufficient space between them to permit the insertion of plates (called gusset plates) for connection purposes. Where steel sections are used back-to-back in this manner, they should be connected every 4 or 5 ft to prevent rattling, particularly in bridge trusses. Single angles and double angles are probably the most common types of tension members in use. Structural tees make very satisfactory chord members for welded trusses because web members can conveniently be connected to them.

For bridges and large roof trusses, tension members may consist of channels, W or S shapes, or even sections built up from some combination of angles, channels, and plates. Single channels are frequently used, as they have little eccentricity and are conveniently connected. Although, for the same weight, W sections are stiffer than S sections, they may have a connection disadvantage in their varying depths. For instance, the W12 × 79, W12 × 72, and W12 × 65 all have slightly different depths (12.4 in, 12.3 in, and 12.1 in, respectively), while the S sections of a certain nominal size all have the same depths. For instance, the S12 × 50, the S12 × 40.8, and the S12 × 35 all have 12.00 in depths.

Although single structural shapes are a little more economical than built-up sections, the latter are occasionally used when the designer is unable to obtain sufficient area or rigidity from single shapes. Where built-up sections are used it is important to remember that field connections will have to be made and paint applied; therefore, sufficient space must be available to accomplish these things.

Members consisting of more than one section need to be tied together. Tie plates (also called tie bars) located at various intervals or perforated cover plates serve to hold the various pieces in their correct positions. These plates serve to correct any unequal distribution of loads between the various parts. They also keep the slenderness ratios of the individual parts within limitations and they may permit easier handling of the built-up members. Long individual members such as angles may be inconvenient to handle due to flexibility, but when four angles are laced together into one member, as shown in Fig. 3.1, the member has considerable stiffness. None of the intermittent tie plates may be considered to increase the effective areas of the sections. As they do not

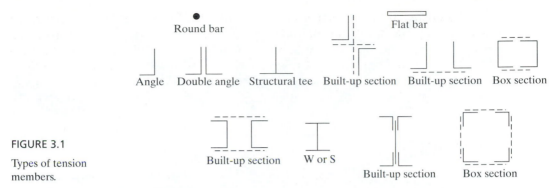

FIGURE 3.1

Types of tension members.

theoretically carry portions of the force in the main sections, their sizes are usually governed by specifications and perhaps by some judgment on the designer's part. Perforated cover plates (see Fig. 6.9) are an exception to this rule, as part of their areas can be considered as being effective in resisting axial load.

A building framework under construction. (Courtesy of Bethlehem Steel Corporation.)

The skeleton of the roof of a Ford building under construction. (Courtesy of Bethlehem Steel Corporation.)

A few of the various types of tension members in general use are illustrated in Fig. 3.1. In this figure, the dotted lines represent the intermittent tie plates or bars used to connect the shapes.

Steel cables are made with special steel alloy wire ropes that are cold-drawn to the desired diameter. The resulting wires with strengths of about 200,000 to 250,000 psi can be economically used for suspension bridges, cable supported roofs, ski lifts, and other similar applications.

Normally, to select a cable tension member the designer uses a manufacturer's catalog. The yield stress of the steel and the cable size required for the design force are determined from the catalog . It is also possible to select clevises or other devices to use for connectors at the cable ends.

3.2   DESIGN STRENGTH OF TENSION MEMBERS

A ductile steel member without holes subject to a tensile load can resist without fracture a load larger than its gross cross-sectional area times its yield stress because of strain hardening. However, a tension member loaded until strain hardening is reached will lengthen a great deal before fracture, a fact that will, in all probability, take away the member's usefulness and may even cause failure of the structural system of which the member is a part.

If, on the other hand, we have a tension member with bolt holes, it can possibly fail by fracture at the net section through the holes. This failure load may very well be smaller than the load required to yield the gross section away from the holes. It is to be realized that the portion of the member where we have a reduced cross-sectional area due to the presence of holes normally is very short compared with the total length of the member. Though the strain-hardening situation is quickly reached at the net section portion of the member, yielding there may not really be a limit state of significance because the overall change in length of the member due to yielding in this small part of the member length may be negligible.

As a result of the preceding information, the LRFD Specification (D1) states that the design strength of a tension member, $\phi_t P_n$, is to be the smaller of the values obtained by substituting into the following two expressions:

For the limit state of yielding in the gross section (which is intended to prevent excessive elongation of the member),

$$P_n = F_y A_g$$
$$P_u \leq \phi_t F_y A_g \text{ with } \phi_t = 0.90.$$

(LRFD Equation D1-1)

For fracture in the net section where bolt or rivet holes are present,

$$P_n = F_u A_e$$
$$P_u \leq \phi_t F_u A_e \text{ with } \phi_t = 0.75.$$

(LRFD Equation D1-2)

In the preceding expression, F_u is the specified minimum tensile stress and A_e is the effective net area that can be assumed to resist tension at the section through the holes. This area may be somewhat smaller than the actual net area, A_n, because of stress concentrations and other factors that are discussed in Section 3.5 of this chapter. Values of F_y and F_u are provided in Table 1.1 of Chapter 1 of this text for the ASTM structural steels on the market today.

For tension members consisting of steel shapes, there actually is a third limit state, block shear, a topic presented in Section 3.7 of this chapter.

The design strengths presented here are not applicable to threaded steel rods or to members with pin holes (as in eyebars). These situations are discussed in Sections 4.3 and 4.4.

It is not likely that stress fluctuations will be a problem in the average building frame because the changes in load in such structures usually occur only occasionally and produce relatively minor stress variations. Full design wind or earthquake loads occur so infrequently that they are not considered in fatigue design. Should there, however, be frequent variations or even reversals in stress, the matter of fatigue must be considered. This subject is presented in Section 4.5.

3.3   NET AREAS

The presence of a hole obviously increases the unit stress in a tension member, even if the hole is occupied by a bolt. (When fully tightened high-strength bolts are used, there may be some disagreement with this statement under certain conditions.) There is less area of steel to which the load can be distributed and there will be some concentration of stress along the edges of the hole.

Tension is assumed to be uniformly distributed over the net section of a tension member, although photoelastic studies show there is a decided increase in stress intensity around the edges of holes, sometimes equaling several times the stresses if the holes were not present. For ductile materials, however, a uniform stress distribution assumption is reasonable when the material is loaded beyond its yield stress. Should the fibers around the holes be stressed to their yield stress, they will yield without further stress increase, with the result that there is a redistribution or balancing of stresses. At ultimate load, it is reasonable to assume a uniform stress distribution. The importance of ductility on the strength of bolted tension members has been clearly demonstrated in tests. Tension members (with bolt holes) made from ductile steels have proved to be as much as one-fifth to one-sixth stronger than similar members made from brittle steels with the same strengths. We have already shown in Chapter 1 that it is possible for steel to lose its ductility and become subject to brittle fracture. Such a condition can be created by fatigue-type loads and by very low temperatures.

This initial discussion is applicable only for tension members subjected to relatively static loading. Should tension members be designed for structures subjected to fatigue-type loadings, considerable effort should be made to minimize the items causing stress concentrations such as points of sudden change of cross section and sharp corners. In addition, as described in Section 4.5, the members may have to be enlarged.

The term "net cross-sectional area," or simply "net area," refers to the gross cross-sectional area of a member minus any holes, notches, or other indentations. In considering the area of such items as these, it is important to realize that it is usually necessary to subtract an area a little larger than the actual hole. For instance, in fabricating structural steel that is to be connected with bolts, the long-used practice was to punch holes with a diameter 1/16 in larger than that of the bolts. When this practice was followed, the punching of a hole was assumed to damage or even destroy 1/16 in more of the surrounding metal. As a result, the diameter of the hole subtracted was 1/8 in larger than the diameter of the bolt. The area of the hole was rectangular and equalled the diameter of the bolt times the thickness of the metal.

Today, drills made from very much improved steels enable fabricators to drill very large numbers of holes without resharpening. As a result, a large proportion of bolt holes are now prepared with numerically controlled drills. Though it seems reasonable to add only 1/16 in to the bolt diameters for such holes, to be consistent, we add 1/8 in for all standard bolt holes in this text. (Should the holes be slotted as described in Chapter 12, the usual practice is to add 1/16 in to the actual width of the holes.)

For steel much thicker than bolt diameters, it is difficult to punch out the holes to the full sizes required without excessive deformation of the surrounding material. These holes may be subpunched (with diameters 3/16 in undersized) and then reamed out to full size after the pieces are assembled. Very little material is damaged by this quite expensive process, as the holes are even and smooth and it is considered unnecessary to subtract the 1/16 in for damage to the sides.

It may be necessary to have an even greater latitude in meeting dimensional tolerances during erection and for high-strength bolts larger than 5/8 in in diameter. For such a situation, holes larger than the standard ones may be used without reducing the performance of the connections. These oversized holes can be short-slotted or long-slotted as described in Section 12.9.

Example 3-1 illustrates the calculations necessary for determining the net area of a plate type of tension member.

Example 3-1

Determine the net area of the 3/8 × 8-in plate shown in Fig. 3.2. The plate is connected at its end with two lines of 3/4-in bolts.

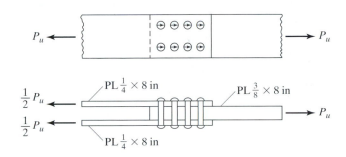

FIGURE 3.2

Solution

$$\text{Net area} = A_n = \left(\frac{3}{8}\right)(8) - (2)\left(\frac{3}{4} + \frac{1}{8}\right)\left(\frac{3}{8}\right) = 2.34 \text{ in}^2 \, (1510 \text{ mm}^2)$$

The connections of tension members should be arranged so that no eccentricity is present. (An exception to this rule is permitted by the LRFD Specification for certain bolted and welded connections as described in Chapter 14.) If this arrangement is possible, the stress is assumed to be spread uniformly across the net section of a member. Should the connections have eccentricities, moments will be produced that will cause additional stresses in the vicinity of the connection. Unfortunately, it is often quite difficult to arrange connections without eccentricity. Although specifications cover some situations, the designer may have to give consideration to eccentricities in some cases by making special estimates.

The lines of action of truss members meeting at a joint are assumed to coincide. Should they not coincide, ecentricity is present and secondary stresses are the result. The centers of gravity (c.g.s) of truss members are assumed to coincide with the lines of action of their respective forces. No problem is present in a symmetrical member as its center of gravity is at its center line, but for unsymmetrical members the problem is a little more difficult. For these members the center line is not the center of gravity, but the usual practice is to arrange the members at a joint so their gage lines coincide. If a member has more than one gage line, the one closest to the actual center of gravity of the member is used in detailing. Figure 3.3 shows a truss joint in which the c.g.s coincide.

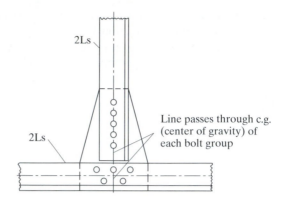

FIGURE 3.3

Lining up c.g.s of members.

3.4   EFFECT OF STAGGERED HOLES

Should there be more than one row of bolt holes in a member, it is often desirable to stagger them in order to provide as large a net area as possible at any one section to resist the load. In the preceding paragraphs, tensile members have been assumed to fail transversely as along line *AB* in either Fig. 3.4(a) or 3.4(b). Figure 3.4(c) shows a member in which a failure other than a transverse one is possible. The holes are staggered, and failure along section *ABCD* is possible unless the holes are a large distance apart.

To determine the critical net area in Fig. 3.4(c), it might seem logical to compute the area of a section transverse to the member (as *ABE*) less the area of one hole and

Trans-World Dome, St. Louis, MS. (Courtesy of Trade ARBED.)

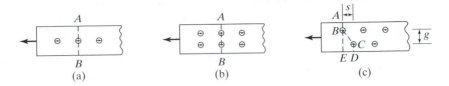

FIGURE 3.4

Possible failure sections in plates.

then the area along section $ABCD$ less two holes. The smallest value obtained along these sections would be the critical value. This method is at fault, however. Along the diagonal line from B to C, there is a combination of direct stress and shear and a somewhat smaller area should be used. The strength of the member along section $ABCD$ is obviously somewhere between the strength obtained by using a net area computed by subtracting one hole from the transverse cross-sectional area and the value obtained by subtracting two holes from section $ABCD$.

Tests on joints show that little is gained by using complicated theoretical formulas to consider the staggered-hole situation, and the problem is usually handled with an empirical equation. The LRFD Specification (B2) and other specifications use a very simple method for computing the net width of a tension member along a zigzag section.[1] The method is to take the gross width of the member regardless of the line along which failure might occur, subtract the diameter of the holes along the zigzag section being considered, and add for each inclined line the quantity given by the expression $s^2/4g$. (Since this simple expression was introduced in 1922, many investigators have proposed other, often quite complicated rules. However, none of them seems to provide significantly better results.)

In this expression, s is the longitudinal spacing (or pitch) of any two holes and g is the transverse spacing (or gage) of the same holes. The values of s and g are shown in Fig. 3.4(c). There may be several paths, any one of which may be critical at a particular joint. Each possibility should be considered, and the one giving the least value should be used. The smallest net width obtained is multiplied by the plate thickness to give the net area, A_n. Example 3-2 illustrates the method of computing the critical net area of a section which has three lines of bolts. (For angles, the gage for holes in opposite legs is considered to be the sum of the gages from the back of the angle minus the thickness of the angle.)

Holes for bolts and rivets are normally drilled or punched in steel angles at certain standard locations. These locations or gages are dependent on the angle-leg widths and on the number of lines of holes. Table 3.1, which is taken from Fig. 10.6 of the LRFD Manual, shows these gages. It is unwise for the designer to require different gages from those given in the table unless unusual situations are present, because of the appreciably higher fabrication costs that will result.

---

[1]V. H. Cochrane, "Rules for Riveted Hole Deductions in Tension Members," *Engineering News-Record* (New York, November 16, 1922), pp. 847–848.

TABLE 3.1   Usual Gages for Angles, in Inches

	Leg	8	7	6	5	4	$3\frac{1}{2}$	3	$2\frac{1}{2}$	2	$1\frac{3}{4}$	$1\frac{1}{2}$	$1\frac{3}{8}$	$1\frac{1}{4}$	1
	g	$4\frac{1}{2}$	4	$3\frac{1}{2}$	3	$2\frac{1}{2}$	2	$1\frac{3}{4}$	$1\frac{3}{8}$	$1\frac{1}{8}$	1	$\frac{7}{8}$	$\frac{7}{8}$	$\frac{3}{4}$	$\frac{5}{8}$
	g_1	3	$2\frac{1}{2}$	$2\frac{1}{4}$	2										
	g_2	3	3	$2\frac{1}{2}$	$1\frac{3}{4}$										

Example 3-2

Determine the critical net area of the 1/2-in-thick plate shown in Fig. 3.5 using the LRFD Specification (Section B2). The holes are punched for 3/4-in bolts.

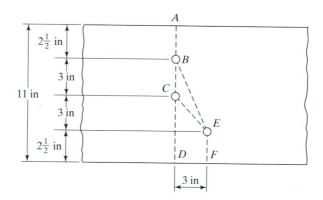

FIGURE 3.5

Solution. The critical section could possibly be *ABCD, ABCEF*, or *ABEF*. Hole diameters to be subtracted are 3/4 + 1/8 = 7/8 in. The net widths for each case are as follows:

$$ABCD = 11 - (2)\left(\frac{7}{8}\right) = 9.25 \text{ in}$$

$$ABCEF = 11 - (3)\left(\frac{7}{8}\right) + \frac{(3)^2}{(4)(3)} = 9.125 \text{ in (controls)}$$

$$ABEF = 11 - (2)\left(\frac{7}{8}\right) + \frac{(3)^2}{(4)(6)} = 9.625 \text{ in}$$

The reader should note that it is a waste of time to check route *ABEF* for this plate. Two holes need to be subtracted for routes *ABCD* and *ABEF*. As *ABCD* is a shorter route it obviously controls over *ABEF.*

$$A_n = (9.125)\left(\frac{1}{2}\right) = 4.56 \text{ in}^2 \qquad\qquad Ans.$$

The problem of determining the minimum pitch of staggered bolts such that no more than a certain number of holes need be subtracted to determine the net section is handled in Example 3-3.

Example 3-3

For the two lines of bolt holes shown in Fig. 3.6, determine the pitch that will give a net area $DEFG$ equal to the one along ABC. The problem may also be stated as follows: determine the pitch that will give a net area equal to the gross area less one bolt hole. The holes are punched for 3/4 in bolts.

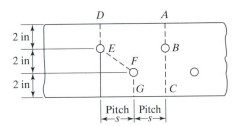

FIGURE 3.6

Solution

$$ABC = 6 - (1)\left(\frac{7}{8}\right) = 5.125 \text{ in}$$

$$DEFG = 6 - (2)\left(\frac{7}{8}\right) + \frac{s^2}{(4)(2)} = 4.25 + \frac{s^2}{8}$$

$$ABC = DEFG$$

$$5.125 = 4.25 + \frac{s^2}{8}$$

$$s = 2.65 \text{ in} \qquad\qquad\qquad\qquad Ans.$$

The $s^2/4g$ rule is merely an approximation or simplification of the complex stress variations that occur in members with staggered arrangements of bolts. Steel specifications can only provide minimum standards and designers will have to logically apply such information to complicated situations which the specifications could not cover in their attempts at brevity and simplicity. The next few paragraphs present a discussion and numerical examples of the $s^2/4g$ rule applied to situations not specifically addressed in the LRFD Specification.

The LRFD Specification does not include a method to be used for determining the net widths of sections other than plates and angles. For channels, W sections, S sections, and others the web and flange thicknesses are not the same. As a result, it is necessary to work with net areas rather than net widths. If the holes are placed in straight lines across such a member the net area can simply be obtained by subtracting the cross-sectional areas of the holes from the gross area of the member. If the holes are staggered it is necessary to multiply the $s^2/4g$ values by the applicable thickness to change it to an area. Such a procedure is illustrated for a W section in Example 3-4 where bolts pass through the web only.

Example 3-4

Determine the net area of the W12 × 16($A_g = 4.71$ in^2) shown in Fig. 3.7 assuming the holes are for 1-in bolts.

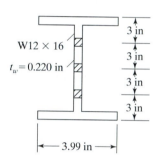
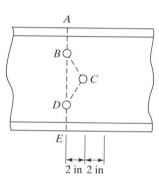

FIGURE 3.7

Solution. Net areas:

$$ABDE = 4.71 - (2)\left(1\frac{1}{8}\right)(0.220) = 4.21 \text{ in}^2$$

$$ABCDE = 4.71 - (3)\left(1\frac{1}{8}\right)(0.220) + (2)\frac{(2)^2}{(4)(3)}(0.220) = 4.11 \text{ in}^2 \leftarrow$$

If the zigzag line goes from a web hole to a flange hole, the thickness changes at the junction of the flange and web. In Example 3-5, the authors have computed the net area of a channel that has bolt holes staggered in its flanges and web. The channel is assumed to be flattened out into a single plate as shown in parts (b) and (c) of Fig. 3.8. The net area along route $ABCDEF$ is determined by taking the area of the channel minus the area of the holes along the route in the flanges and web plus the $s^2/4g$ values for each zigzag line times the appropriate thickness. For line CD, $s^2/4g$ has been multiplied by the thickness of the web. *For lines BC and DE (which run from holes in the web to holes in the flange) an approximate procedure has been used in which the $s^2/4g$ values have been multiplied by the average of the web and flange thicknesses.*

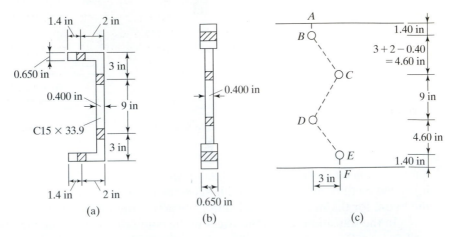

FIGURE 3.8

Example 3-5

Determine the net area along route $ABCDEF$ for the $C15 \times 33.9(A_g = 9.95 \text{ in}^2)$ shown in Fig. 3.8. Holes are for 3/4-in bolts.

Solution

Approximate net A along

$$ABCDEF = 9.95 - (2)\left(\frac{7}{8}\right)(0.650)$$

$$- (2)\left(\frac{7}{8}\right)(0.400)$$

$$+ \frac{(3)}{(4)(9)}(0.4000)$$

$$+ (2)\frac{(3)^2}{(4)(4.60)}\left(\frac{0.650 + 0.400}{2}\right)$$

$$= 8.726 \text{ in}^2 \leftarrow$$

3.5   EFFECTIVE NET AREAS

When a member other than a flat plate or bar is loaded in axial tension until failure occurs across its net section, its actual tensile failure stress will probably be less than the coupon tensile strength of the steel *unless all of the various elements that make up the section are connected so stress is transferred uniformly across the section.*

If the forces are not transferred uniformly across a member cross section, there will be a transition region of uneven stress running from the connection out into the member for some distance. This is the situation shown in Fig. 3.9(a), where a single

Staggered Fasteners:
Problem 2) Determine the tensile design strength of the L6 x 3½ x ⁵/₁₆. The bolts are ¾"
in diameter. If A36 steel is used, is the member adequate for a service dead load of 31
kips and a service live load of 31 kips?

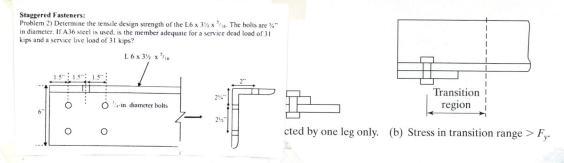

cted by one leg only.   (b) Stress in transition range $> F_y$.

angle tension member is connected by one leg only. At the connection more of the load
is carried by the connected leg, and it takes the transition distance shown in part (b) of
the figure for the stress to spread uniformly across the whole angle.

In the transition region the stress in the connected part of the member may very
well exceed F_y and go into the strain-hardening range. Unless the load is reduced the
member may fracture prematurely. The further we move out from the connection, the
more uniform the stress becomes. In the transition region, the shear transfer has
"lagged" and the phenomenon is referred to as *shear lag*.

In such a situation, the flow of tensile stress between the full member cross section and the smaller connected cross section is not 100 percent effective. As a result,
the LRFD Specification (B3) states that the effective net area, A_e, of such a member is
to be determined by multiplying an area A (which is the net area or the gross area or
the directly connected area as described in the next few pages) by a reduction factor U.
The use of a factor such as U accounts for the nonuniform stress distribution in a simple manner. An explanation of the way in which U factors are determined follows.

$$A_e = AU$$ (LRFD Equation B3-1)

The angle shown in Fig. 3.10(a) is connected at its ends to only one leg. You can
easily see that its area effective in resisting tension can be appreciably increased by
shortening the width of the unconnected leg and lengthening the width of the connected leg as shown in Fig. 3.10(b).

Investigators have found that a convenient measure of the effectiveness of a
member such as an angle connected by one leg is the distance $\bar{x}$ measured from the
plane of the connection to the centroid of the area of the whole section.[2,3] The smaller
the value of $\bar{x}$, the larger is the effective area of the member, and thus the larger is the
member's design strength. The specification in effect reduces the length L of a connection with sl ... le of U then equals L'/L or
$1 - \bar{x}/L$. Sev ... inder of this section is devoted to th ... bolted and welded tension
members.

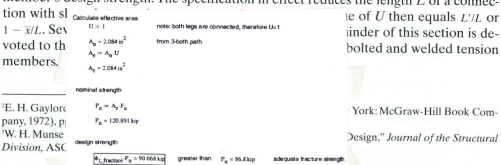

Calculate effective area

$U := 1$ note: both legs are connected, therefore U=1

$A_n = 2.084 \text{ in}^2$ from 3-both path

$A_e := A_n \cdot U$

$A_e = 2.084 \text{ in}^2$

nominal strength

$P_n := A_e \cdot F_u$

$P_n = 120.891 \text{ kip}$

design strength

$\phi_{t\_fracture} \cdot P_n = 90.668 \text{ kip}$ greater than $P_u = 86.8 \text{ kip}$ adequate fracture strength

[2]E. H. Gaylor ... York: McGraw-Hill Book Company, 1972), p ...
[3]W. H. Munse ... Design," *Journal of the Structural*
Division, ASC ...

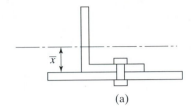 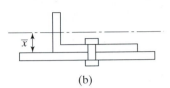

FIGURE 3.10

Reducing shear lag by reducing length of unconnected leg and and thus $\bar{x}$.

(a)

(b)

3.5.1   Bolted Members

Should a tension load be transmitted by bolts, A equals the net area A_n of the member, and U is computed as follows:

$$U = 1 - \frac{\bar{x}}{L} \leq 0.9.$$

The length L used in this expression is equal to the distance between the first and last bolts in the line. When there are two or more lines of bolts, it is the length of the line with the maximum number of bolts. Should the bolts be staggered, it is the out-to-out dimension between the extreme bolts in a line. You will note that the longer the connection (L) becomes the larger U will become as will the effective area of the member. (On the other hand, we will learn in the connection chapters of this text that the effectiveness of connectors is somewhat reduced if very long connections are used.) Insufficient data are available for the case in which only one bolt is used in each line. It is thought that a conservative approach for this case is to let $A_e = A_n$ of the connected element.

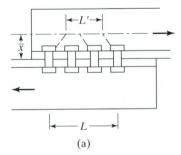

(a)

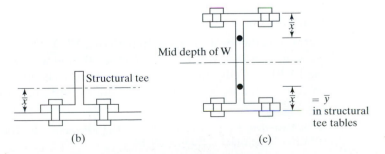

FIGURE 3.11

Values of $\bar{x}$ for different shapes.

(b)

(c)

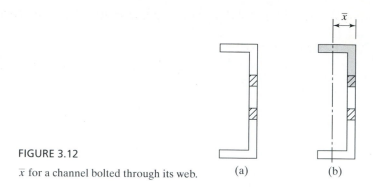

FIGURE 3.12

$\bar{x}$ for a channel bolted through its web.   (a)   (b)

In order to calculate U for a W section connected by its flanges only, we will assume that the section is split into two structural tees. Then the value of $\bar{x}$ used will be the distance from the outside edge of the flange to the c.g. of the structural tee as shown in part (c) of Fig. 3.11. Parts (b) and (c) of Fig. C-B3.1 of the LRFD Commentary illustrate the recommended procedures for calculating $\bar{x}$ values for channel and I-shaped sections where the loads are transferred by means of bolts passing through the member webs only.

The LRFD Specification permits the designer to use larger values of U than obtained from the equation if such values can be justified by tests or other rational criteria.

Section B3 of the LRFD Commentary provides suggested $\bar{x}$ values for use in the equation for U for several situations not addressed in the Specification. Included are values for W and C sections bolted only through their webs. Also considered are single angles with two lines of staggered bolts in one of their legs. The basic idea for computing $\bar{x}$ for these cases is presented in the following paragraph.[4]

The channel of Fig. 3.12(a) is connected with two lines of bolts through its web. The "angle" part of this channel above the center of the top bolt is shown darkened in part (b) of the figure. This part of the channel is unconnected. For shear lag purpose, we can determine the horizontal distance from the outside face of the web to the channel centroid. This distance, which is given in the Manual shape tables, will be the $\bar{x}$ used in the equation. It is felt that with this idea in mind the reader will be able to understand the values shown in the Commentary for other sections.

Example 3-6 illustrates the calculations necessary for determining the effective net area of a bolted W section connected only to its flanges. In addition, the design strength of the member is computed.

Example 3-6

Determine the tensile design strength of a W10 × 45 with two lines of 3/4-in-diameter bolts in each flange using A572 Grade 50 steel with $F_y = 50$ ksi and $F_u = 65$ ksi and the LRFD Specification. There are assumed to be at least three bolts in each line 4 in on center, and the bolts are not staggered with respect to each other.

---

[4]W. S. Easterling and L. G. Giroux, "Shear Lag Effects in Steel Tension Members," *Engineering Journal*, AISC, no. 3 (3rd Quarter, 1993), pp. 77–89.

Solution. Using a W10 × 45 ($A_g = 13.3$ in², $d = 10.10$ in, $b_f = 8.020$ in, $t_f = 0.620$ in)

(a) Gross-section yield
$$\phi_t P_n = \phi F_y A_g = (0.90)(50)(13.3) = 598.5 \text{ k}$$

(b) Net-section fracture
$$A_n = 13.3 - (4)\left(\frac{7}{8}\right)(0.620) = 11.13 \text{ in}^2 = A$$

Referring to the tables for half of a W10 × 45 (or to a WT5 × 22.5), we find $\bar{x} = 0.907$ in. Then,

$$U = 1 - \frac{\bar{x}}{L} = 1 - \frac{0.907}{8} = 0.89 < 0.9 \qquad \text{(OK)}$$

$$A_e = UA = (0.89)(11.13) = 9.91 \text{ in}^2$$

$$\phi_t P_n = \phi_t F_u A_e = (0.75)(65)(9.91) = 483.1 \text{ k}$$

Design strength $\phi_t P_n = 483.1$ k ←

It can be shown that if $\dfrac{A_e}{A_g} \geq \dfrac{0.9 F_y}{0.75 F_u}$, tension yielding will control over tension fracture. The right hand side of this expression equals 0.923 for the commonly used steels with $F_y = 50$ ksi and $F_u = 65$ ksi. Tables 3.1 to 3.7 at the end of Part 3 of the Manual provide the tension yield strengths and tensile fracture strengths of various sections including Ws, angles, structural tees, and others for cases where $A_e = 0.75 A_g$ and using the preferred steel grade for each as recommended in Table 1.1 of this text (Table 2.1 in the Manual).

The 1986 LRFD Specification presented a set of standard U values that could be used for bolted members instead of substituting into the $1 - \frac{x}{L}$ expression. These values, which are listed in Table 3.2 and in Section B.3 of the Commentary to the LRFD Specification, are still acceptable. (For Example 3-6, note that U from the table is 0.90 as $b_f/d = 8.020/10.10 > 2/3$.) The tabulated values are particularly useful for situations in which initial design size selections are being made and we have insufficient information to calculate U values. We will encounter this situation in the first two examples of Chapter 4.

TABLE 3.2 Permissible U Values for Bolted Connections

a. *W, M*, or *S* shapes with flange widths not less than two-thirds the depth, and structural tees cut from these shapes, provided the connection is to the flanges and has no fewer than three fasteners per line in the direction of stress, $U = 0.90$.

b. *W, M*, or *S* shapes not meeting the conditions of subparagraph **a**, structural tees cut from these shapes, and all other shapes including built-up cross sections, provided the connection has no fewer than three fasteners per line in the direction of stress, $U = 0.85$.

c. All members having only two fasteners per line in the direction of stress, $U = 0.75$.

Should fastener holes be located away from member ends, shear lag effects will not apply, and U values will equal 1.0.

3.5.2 Welded Members

When tension loads are transferred by welds, the following rules from LRFD Specification B.3 are to be used to determine values for A and U. (A_e as for bolted connections $= AU$.)

1. Should the load be transmitted only by longitudinal welds to other than a plate member, or by longitudinal welds in combination with transverse welds, A is to equal the gross area of the member A_g.
2. Should a tension load be transmitted only by transverse welds, A is to equal the area of the directly connected elements and U is to equal 1.0.
3. Tests have shown that when flat plates or bars connected by longitudinal fillet welds (a term to be described in Chapter 14) are used as tension members, they may fail prematurely by shear lag at the corners if the welds are too far apart. Therefore, the LRFD Specification states that when such situations are encountered the length of the welds may not be less than the width of the plates or bars. The letter A represents the area of the plate and UA is the effective net area. For such situations, the following values of U are to be used (LRFD Specification B3.2(d)).

When $l \geq 2w$ $\qquad\qquad\qquad\qquad\qquad\qquad\qquad\qquad\qquad\qquad$ $U = 1.0$
When $2w > l \geq 1.5w$ $\qquad\qquad\qquad\qquad\qquad\qquad\qquad\qquad\quad$ $U = 0.87$
When $1.5w > l \geq w$ $\qquad\qquad\qquad\qquad\qquad\qquad\qquad\qquad\quad$ $U = 0.75$

where $l =$ weld length, in
$\qquad\quad w =$ plate width (distance between welds), in

For combinations of longitudinal and tranverse welds, l is to be used equal to the length of the longitudinal weld because the transverse weld has little or no effect on the shear lag (that is, it does little to get the load into the unattached parts of the member).

Examples 3-7 and 3-8 illustrate the calculations of the effective areas and design strengths of two welded members.

Example 3-7

The 1×6 in plate shown in Fig. 3.13 is connected to a 1×10 in plate with longitudinal fillet welds to transfer a tensile load. Determine the design strength $\phi_t P_n$ of the member if $F_y = 50$ ksi and $F_u = 65$ ksi.

Solution. Considering the smaller PL

(a) Gross-section yield

$$\phi_t P_n = \phi_t F_y A_g = (0.90)(50)(1 \times 6) = 270 \text{ k}$$

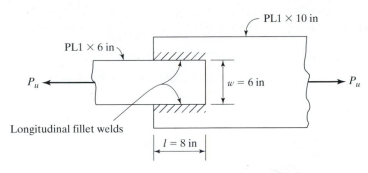

FIGURE 3.13

(b) Net-section fracture

$$A = A_g = 1 \times 6 = 6 \text{ in}^2$$
$$1.5w = 9 \text{ in} > l = 8 \text{ in} > w = 6 \text{ in}$$
$$\therefore U = 0.75$$
$$A_e = AU = (6.0)(0.75) = 4.50 \text{ in}^2$$
$$\phi_t P_n = \phi_t F_u A_e = (0.75)(65)(4.50) = 219.4 \text{ k} \leftarrow$$
$$\text{Design strength } \phi_t P_n = 219.4 \text{ k}$$

Example 3-8

Compute the design strength $\phi_t P_n$ for the angle shown in Fig. 3.14. It is welded on the ends and sides of the 8 in leg only and $F_y = 50$ ksi and $F_u = 70$ ksi.

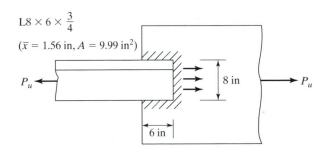

FIGURE 3.14

Angle welded to 8-in leg only.

Solution. As only one leg of the angle is connected, a reduced effective area needs to be computed.

(a) $\phi_t P_n = \phi_t F_y A_g = (0.9)(50)(9.99) = 449.6 \text{ k}$

(b) $\qquad U = 1 - \dfrac{\overline{x}}{L} \le 0.90 = \left(1 - \dfrac{1.56}{6.00}\right) = 0.74 < 0.90$

$A_e = AU = (9.99)(0.74) = 7.39 \text{ in}^2$

$\phi_t P_n = \phi_t F_u A_e = (0.75)(70)(7.39) = 388 \text{ k} \leftarrow$

Design strength $\phi_t P_n = 388 \text{ k}$

3.6 CONNECTING ELEMENTS FOR TENSION MEMBERS

When splice or gusset plates are used as statically loaded tensile connecting elements, their strength shall be determined as follows:

For yielding of welded or bolted connection elements

$$\phi = 0.90$$
$$R_n = A_g F_y. \qquad \text{(LRFD Equation J5-1)}$$

For fracture of bolted connection elements

$$\phi = 0.75$$
$$R_n = A_n F_u \text{ with } A_n \le 0.85 A_g. \qquad \text{(LRFD Equation J5-2)}$$

The net area A_n to be used in the second of these expressions may not exceed 85 percent of A_g. Tests have shown for decades that bolted tension connection elements rarely have an efficiency greater than 85 percent, even if the holes represent a very small percentage of the gross area of the elements. In Example 3-9 the strength of a pair of tensile connecting plates is computed.

Example 3-9

The tension member ($F_y = 50$ ksi and $F_u = 65$ ksi) of Example 3-6 is assumed to be connected at its ends with two 3/8 × 12-in plates, as shown in Fig. 3.15. If two lines of 3/4-in bolts are used in each plate, determine the design tensile force which the plates can transfer.

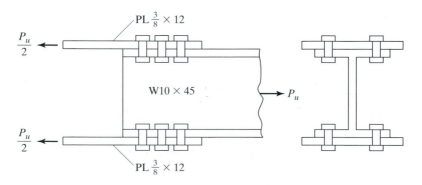

FIGURE 3.15

Solution

$$\phi_t F_y A_g = (0.9)(50)\left(2 \times \frac{3}{8} \times 12\right) = 405 \text{ k}$$

$$A_n \text{ of 2 plates} = \left[\left(\frac{3}{8} \times 12\right) - \left(\frac{7}{8} \times 2 \times \frac{3}{8}\right)\right]2 = 7.69 \text{ in}^2$$

$$0.85A_g = (0.85)\left(2 \times \frac{3}{8} \times 12\right) = 7.65 \text{ in}^2 < 7.69 \text{ in}^2 \quad \therefore A_n = 7.65 \text{ in}^2$$

$$\phi_t P_n = \phi_t F_u A_n = (0.75)(65)(7.65) = 372.9 \text{ k} \leftarrow$$

$$\phi_t P_n = 372.9 \text{ k}$$

3.7   BLOCK SHEAR

The design strength of a tension member is not always controlled by $\phi_t F_y A_g$ or $\phi_t F_u A_g$ or by the strength of the bolts or welds with which the member is connected. It may instead be controlled by its *block shear* strength as described in this section.

The failure of a member may occur along a path involving tension on one plane and shear on a perpendicular plane as shown in Fig. 3.16, where several possible block shear failures are shown. For these situations it is possible for a "block" of steel to tear out.

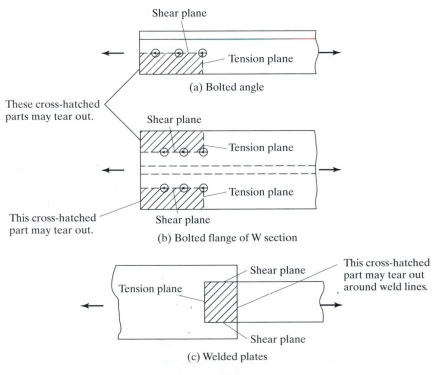

(a) Bolted angle

(b) Bolted flange of W section

(c) Welded plates

FIGURE 3.16

Block shear.

When a tensile load applied to a particular connection is increased, the fracture strength of the weaker plane will be approached. That plane will not fall then because it is restrained by the stronger plane. The load can be increased until the fracture strength of the stronger plane is reached. During this time, the weaker plane is yielding. The total strength of the connection equals the fracture strength of the stronger plane plus the yield strength of the weaker plane.[5] Thus, it is not realistic to add the fracture strength of one plane to the fracture strength of the other plane to determine the block shear resistance of a particular member. *You can see that block shear is a tearing or rupture situation and not a yielding situation.*

The member shown in Fig. 3.17(a) has a large shear area and a small tensile area, thus the primary resistance to a block shear failure is shearing and not tensile. The LRFD Specification states that it is logical to assume that when shear fracture occurs on this large shear-resisting area, the small tensile area has yielded.

Part (b) of Fig. 3.17 shows a free body of the block that tends to tear out of the angle of part (a). You can see in this sketch that the block shear is caused by the bolts bearing on the back of the bolt holes.

In part (c) of Fig. 3.17 a member is shown which, so far as block shear goes, has a large tensile area and a small shear area. The LRFD feels that for this case the primary

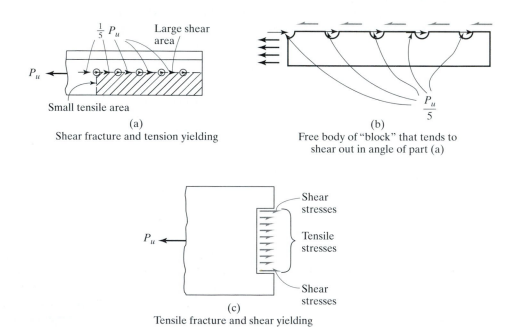

(a)
Shear fracture and tension yielding

(b)
Free body of "block" that tends to shear out in angle of part (a)

(c)
Tensile fracture and shear yielding

FIGURE 3.17

Block shear.

[5]L. B. Burgett, "Fast Check for Block Shear," *Engineering Journal*, AISC, *29*, no. 4 (4th Quarter, 1992), pp. 125–127.

resisting force against a block shear failure will be tensile and not shearing. Thus, a block shear failure cannot occur until the tensile area fractures. At that time it seems logical to assume that the shear area has yielded.

Based on the preceding discussion, the LRFD Specification (J4.3) states that the block shear design strength of a particular member is to be determined by (1) computing the tensile fracture strength on the net section in one direction and adding to that value the shear yield strength on the gross area on the perpendicular segment and (2) computing the shear fracture strength on the gross area subject to tension and adding it to the tensile yield strength on the net area subject to shear on the perpendicular segment. The expression to use is the one with the larger rupture term.

Test results show that this procedure gives good results. Furthermore, it is consistent with the calculations previously used for tension members where gross areas are used for one limit state of yielding ($\phi_t F_y A_g$) and net area for the fracture limit state ($\phi_t F_y A_e$). The LRFD Specification (J4.3) states that the block shear rupture design strength is to be determined as follows:

1. If $F_u A_{nt} \geq 0.6\, F_u A_{nv}$, we will have shear yielding and tension fracture and thus should use the equation to follow:

   $$\phi R_n = \phi[0.6\, F_y A_{gv} + F_u A_{nt}] \leq \phi[0.6\, F_u A_{nv} + F_u A_{nt}].\quad \text{(LRFD Equation J4-3a)}$$

2. If $F_u A_{nt} < 0.6\, F_u A_{nv}$, we will have tension yielding and shear fracture and thus should use the following equation:

   $$\phi R_n = \phi[0.6\, F_u A_{nv} + F_y A_{gt}] \leq \phi[0.6\, F_u A_{nv} + F_u A_{nt}] \quad \text{(LRFD Equation J4-3b)}$$
   in which $\phi = 0.75$

   $$A_{gv} = \text{gross area subjected to shear, in}^2(\text{mm}^2)$$
   $$A_{gt} = \text{gross area subjected to tension, in}^2(\text{mm}^2)$$
   $$A_{nv} = \text{net area subjected to shear, in}^2(\text{mm}^2)$$
   $$A_{nt} = \text{net area subjected to tension, in}^2(\text{mm}^2).$$

Examples 3-10 to 3-12 illustrate the determination of the block shear strengths for three members. This topic of block shear is continued in the connection chapters of this text, where we will find that it is absolutely necessary to check beam connections where the top flange of the beams are coped or cut back as illustrated in Figs. 10.3(c) and 15.6(b).

Example 3-10

The A572 Grade 50 ($F_u = 65$ ksi) tension member shown in Fig. 3.18 is connected with three 3/4-in bolts. Determine the block shearing strength of the member and its tensile strength.

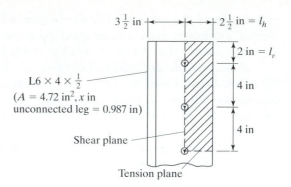

FIGURE 3.18

Solution

$$A_{gv} = (10)\left(\frac{1}{2}\right) = 5.0 \text{ in}^2$$

$$A_{gt} = (2.5)\left(\frac{1}{2}\right) = 1.25 \text{ in}^2$$

$$A_{nv} = \left(10 - 2.5 \times \frac{7}{8}\right)\left(\frac{1}{2}\right) = 3.91 \text{ in}^2$$

$$A_{nt} = \left(2.50 - \frac{1}{2} \times \frac{7}{8}\right)\left(\frac{1}{2}\right) = 1.03 \text{ in}^2$$

$$F_u A_{nt} = (65)(1.03) = 66.9 \text{ k} < 0.6 F_u A_{nv} = (0.6)(65)(3.91) = 152.5 \text{ k}$$

$$\therefore \text{ Use LRFD Equation J4-3b}$$

$$\phi R_n = (0.75)[(0.6)(65)(3.91) + (50)(1.25)] = 161 \text{ k}$$

$$< 0.75[(0.6)(65)(3.91) + (65)(1.03)] = 164.6 \text{ k}$$

Tensile strength of angle

(a) $\phi_t P_n = \phi_t F_y A_g = (0.9)(50)(4.72) = 212.4 \text{ k}$

(b) $A_n = 4.72 - (1)\left(\frac{7}{8}\right)\left(\frac{1}{2}\right) = 4.28 \text{ in}^2 = A$

$$U = 1 - \frac{0.987}{8} = 0.88$$

$$A_e = UA = (0.88)(4.28) = 3.77 \text{ in}^2$$

$$\phi_t P_n = \phi_t F_u A_c = (0.75)(65)(3.77) = 183.8 \text{ k}$$

$\phi_t P_n$ for member smaller of $\phi R_n = 161 \text{ k}$ or of $P_u = 183.8 \text{ k}$

$$\phi_t P_n = 161 \text{ k} \leftarrow$$

Tables are available in Part 9 of the LRFD Manual with which the block shear strengths of coped W beams can be determined. In Table 9.3, values of $\phi F_u A_{nt}$ are tabulated per inch of material thickness, and then in Table 9.4 values of $\phi(0.6F_y A_{gv})$ per inch of material thickness are given.

Example 3-11

Determine the block shear design strength of the A36 welded member shown in Fig. 3.19. ($F_y = 36$ ksi, $F_u = 58$ ksi)

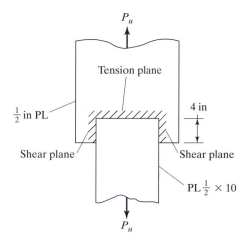

FIGURE 3.19

Solution

$$A_{gv} = \left(\frac{1}{2}\right)(8) = 4.0 \text{ in}^2$$

$$A_{gt} = \left(\frac{1}{2}\right)(10) = 5.0 \text{ in}^2$$

$$A_{nv} = \left(\frac{1}{2}\right)(8) = 4.0 \text{ in}^2$$

$$A_{nt} = \left(\frac{1}{2}\right)(10) = 5.0 \text{ in}^2$$

$$F_u A_{nt} = (58)(5.0) = 290 \text{ k} > 0.6F_u A_{nv} = (0.6)(58)(4.0) = 139 \text{ k}$$

$$\therefore \text{ Must use LRFD Equation J4-3a}$$

$$\phi R_n = 0.75[(0.6)(36)(4.0) + (58)(5.0)]$$

$$= 282 \text{ k} < (0.75)[(0.6)(58)(4.0) + (58)(5)] = 322 \text{ k}$$

Tensile strength of plate

$$\phi_t P_n = \phi F_y A_g = (0.9)(36)\left(\frac{1}{2} \times 10\right) = 162 \text{ k} \leftarrow$$

Design strength of plate $= 162$ k

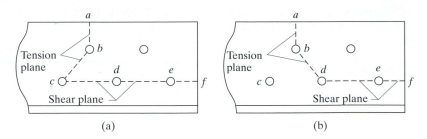

FIGURE 3.20

Sometimes cases are encountered where it is not altogether clear what sections should be considered for block shear calculations. For such situations, designers will have to use their own judgment. One such case is shown for a structural tee in Fig. 3.20. In part (a) of the figure, it is first assumed that web tear-out will occur along the lines *abcdef*. An alternate tear-out possibility for the same member along lines *abdef* is shown in part (b) of the figure. For this connection it is assumed that the load to be resisted is distributed equally among the five bolts. Thus, when web tear out is considered for case (b), we will assume only $4/5P_u$ is carried by the section in question, because one of the bolts is outside of the tear-out area.

Notice that the total block shear strength of the member will equal the block shear strength along path *abdef* plus the strength of bolt C, as it also must fail. To compute the width of the tension planes *abc* and *abd* for these cases, it seems logical to make use of the $s^2/4g$ expression for the tensile area previously presented in Section 3.4.

Example 3-12

Determine the tensile design strength of the Grade 50 ($F_u = 65$ ksi) section used in Fig. 3.21 if 7/8-in bolts are used. Include block shear calculations for the flanges.

Solution

(a) $\phi_t P_n = \phi_t F_y A_g = (0.90)(50)(8.79) = 395.5$ k

(b) $A_n = 8.79 - (4)\left(\dfrac{7}{8} + \dfrac{1}{8}\right)(0.440) = 7.03$ in^2

$$U = 0.85 \text{ from Table 3.2 since } \frac{b_f}{d} < \frac{2}{3}$$

$$\phi_t P_n = \phi_t F_u A_e = \phi_t F_u U A_n = (0.75)(65)(0.85)(7.03) = 291.3 \text{ k}$$

(c) Checking block shear considering both flanges

$$A_{gv} = (4)(10)(0.440) = 17.60 \text{ in}^2$$

$$A_{gt} = (4)(1.51)(0.440) = 2.66 \text{ in}^2$$

$$A_{nv} = (4)\left[10 - (2.5)\left(\frac{7}{8} + \frac{1}{8}\right)\right](0.440) = 13.20 \text{ in}^2$$

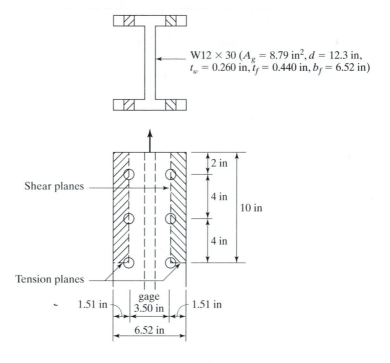

W12 × 30 (A_g = 8.79 in², d = 12.3 in,
t_w = 0.260 in, t_f = 0.440 in, b_f = 6.52 in)

FIGURE 3.21

$$A_{nt} = (4)\left[1.51 - \left(\frac{1}{2}\right)\left(\frac{7}{8} + \frac{1}{8}\right)\right](0.440) = 1.78 \text{ in}^2$$

$F_u A_{nt} = (65)(1.78) = 116 \text{ k}$

$< 0.6 F_u A_{nv} = (0.6)(65)(13.20) = 515 \text{ k}$

$\therefore$ Use LRFD Equation J4-36

$\phi R_n = 0.75[(0.6)(65)(13.2) + (50)(2.66)] = 486 \text{ k}$

$\leq 0.75[(0.6)(65)(13.2) + (65)(1.78)] = 473 \text{ k}$

$\phi_t P_n$ = smallest of 395.5 k, 291.3 k or 473 k

$\phi_t P_n = 291.3 \text{ k}$

3.8   COMPUTER EXAMPLE

In Example 3-13, the INSTEP32 program is used to determine the design tensile strength of a *W* section. Notice that the units used in inputting data are inches and kips. The following steps are taken:

1. Click on the tension member sign ⤢ .
2. Go to "Design Parameter" and input the net area, length, and *U* values given.
3. Go to "Steel and Shape," and input the grade of steel and the *W* shape used. Here it will be necessary to click the shape that appears on the screen and then scroll up or down the list of shapes shown until the one in question is found and clicked.
4. Finally click "OK" and the solution that follows will be provided.

Example 3-13

Using the program INSTEP32, determine the design tensile strength of a 12-ft long W12 × 136 consisting of A572 Grade 50 steel if the net area is assumed to be 35.52 in^2 and $U = 0.9$.

Solution

Input:

$P_u = 0$ kips

Net area $= 35.52$ sq. in

Length $= 144$ in

$U = 0.9$

Output:

Axial Tension Design Summary

Section: W12 × 136

Check yield criteria

$P_n = 50[39.9] = 1995$ kips

$0.9P_n = 0.9[1995] = 1795.5$ kips

Check fracture criteria

$P_n = 0.9[35.52][65] = 2077.92$ kips

$0.75P_n = 0.75[2077.92] = 1558.44$ kips

Check slenderness criteria

$L/R = 144/3.16 = 45.5696 \leftarrow$

PROBLEMS (Use standard-size bolt holes for all problems.)

3-1 to 3-11.   *Compute the net area of each of the given members.*

3-1.   (Ans. 13.23 in^2)

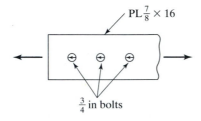

FIGURE P3-1

3-2.

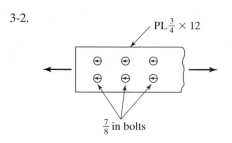

$PL\frac{3}{4} \times 12$

$\frac{7}{8}$ in bolts

FIGURE P3-2

3-3. (Ans. 10.80 in^2)

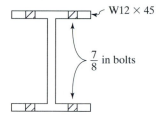

W12 × 45

$\frac{7}{8}$ in bolts

FIGURE P3-3

3-4.

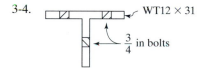

WT12 × 31

$\frac{3}{4}$ in bolts

FIGURE P3-4

3-5. An L8 × 4 × 5/8 with one line of 7/8-in bolts in each leg. (Ans. 5.91 in^2)

3-6. A pair of Ls 7 × 4 × 3/4 with two rows of 3/4-in bolts in the long legs and one row in the short legs.

3-7. A W18 × 50 with two holes in each flange and two in the web, all for 3/4-in bolts. (Ans. 12.08 in^2)

3-8. The built-up section shown in the illustration for which 1-in bolts are used.

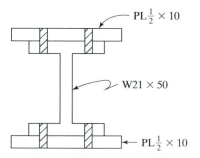

$PL\frac{1}{2} \times 10$

W21 × 50

$PL\frac{1}{2} \times 10$

3-9.   The 3/4 × 9 plate shown in the illustration. The holes are for 3/4-in bolts. (Ans. 5.68 in²)

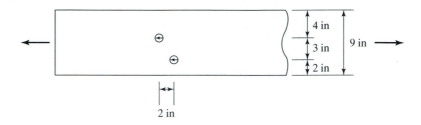

3-10.   The 3/4 × 12 plate shown in the illustration. The holes are for 3/4-in bolts.

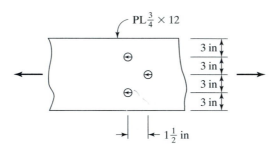

3-11.   The 1 × 12 plate shown in the illustration. The holes are for 7/8-in bolts. (Ans. 9.65 in²)

3-12.   The 8 × 6 × 3/4 angle shown has one line of 3/4-in bolts in each leg. The bolts are 3 in on center in each line and are staggered 1.5 in with respect to each other.

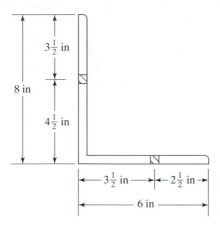

FIGURE P3-12

3-13.   For the plate shown, find the minimum pitch s for which only two bolts need to be subtracted at any one section in calculating the net area. Bolts are 7/8 in (Ans. 2.24 in)

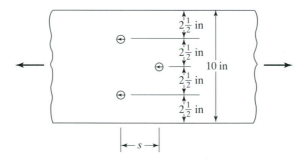

FIGURE P3-13

3-14.   Find the minimum pitch s in the plate of Prob. 3-13 so that only two and one-half bolts need be subtracted at any one section.

3-15.   An L8 × 8 × 5/8 is used as a tension member with one gage line of 1-in bolts in each leg at the usual gage location. (See Table 3.1.) What is the minimum amount of stagger necessary so that only one bolt need be subtracted from the gross area of the angle? Compute the net area of this member if the holes are staggered at 2 in (Ans. $S = 6.14$ in, $A_n = 8.27$ in^2)

3-16. An L7 × 4 × 3/4 is shown. Two rows of 7/8-in bolts are used in the long leg and one in the short leg. Determine the minimum stagger (or pitch s in the figure) necessary so that only two holes need be subtracted in determining the net area.

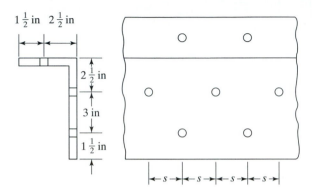

FIGURE P3–16

3-17. An 8 × 6 × 3/4 angle has one line of holes in each leg for 3/4-in bolts. Determine the minimum pitch so that only one and one-half holes need be deducted to obtain the net area. (Use the usual gage for angles as given in Table 3.1 of this chapter.) (*Ans.* 3.57 in)

3-18. Determine the effective net cross-sectional area of the C15 × 50 shown. Holes are for 1-in bolts.

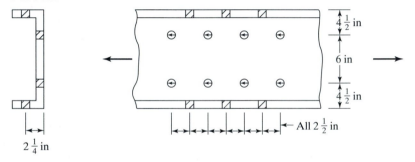

FIGURE P3-18

3-19. Compute the effective net area of the built-up section shown if the holes are punched for 7/8-in bolts. Assume at least three bolts in each line. Assume $U = 0.85$ (*Ans.* 43.32 in²)

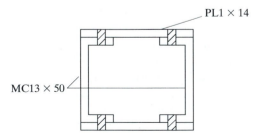

FIGURE P3-19

3-20 to 3-22. *Determine the effective net areas of the sections shown using the U values given in Table 3.2 of this chapter. Assume three bolts in a line.*

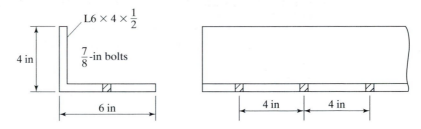

FIGURE P3-20

3-21.   Determine the effective net area of the MC12 × 31 shown. Assume the holes are for 1-in bolts. (Ans. 6.87 in^2)

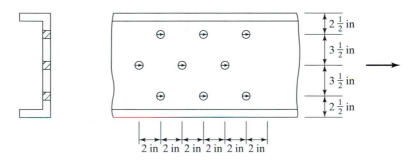

FIGURE P3-21

3-22.   A C15 × 40. is connected through its web with three gage lines of 3/4-in bolts. The gage lines are 3 in on centers, and the bolts are spaced 4 in on centers along the gage line. If the center row of bolts is staggered with respect to the outer row, determine the effective net cross-sectional area of the channel. Assume there are three bolts in each line.

3-23.   (Ans. 8.02 in^2)

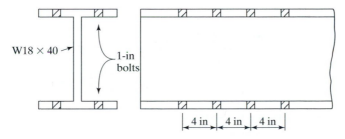

FIGURE P3-23

Block Shear:
Problem 3) Problem 3-32 (page 95, McCormac and Nelson, 3rd Ed.)

3-24 to 3-33. *Use the formula for U.*

3-24. Repeat Prob. 3-20.

3-25. Repeat Prob. 3-21. (Ans 6.99 in²)

3-26. Repeat Prob. 3-22.

3-27. Using A992 steel determine the tensile design strength $\phi_t P_n$ of a W12 × 45 with two lines of 3/4-in bolts in each flange (three bolts in each line 3 in on center). Neglect block shear strength. (Ans. 439.2 k)

3-28. Determine the tensile design strength $\phi_t P_n$ of a W18 × 119 (A992 steel) if it has two lines of 1-in bolts in each flange (at least four bolts in each line 3 in on center). Neglect block shear.

3-29. A single-angle tension member (7 × 4 × 3/4) has two gage lines in its long leg and one in the short leg for 3/4-in bolts arranged as shown. Determine the tensile strength $\phi_t P_n$ of this member if A36 steel is used and if block shear is neglected. (Ans. 249.5 k)

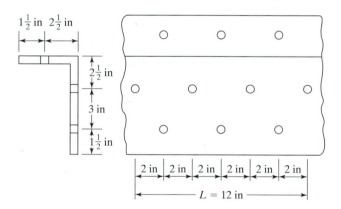

FIGURE P3-29

3-30. Compute the design strength $\phi_t P_n$ of the bolted connection shown neglecting block shear. The angle is made of A36 steel and the bolts are 3/4 in.

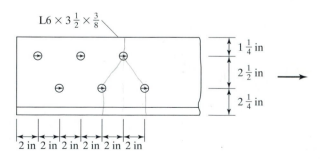

FIGURE P3-30

3-31.   Determine the tensile design strength $\phi_t P_n$ of the pair of 6 × 6 × 3/4 angles shown, made from A36 steel. Standard gages are to be used as determined from Table 3.1 in this chapter for the 3/4-in bolts. Neglect block shear. (Ans. 548.2 k)

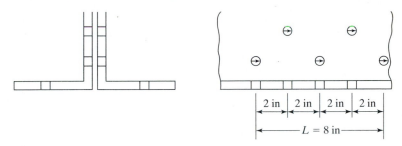

2 in   2 in   2 in   2 in

$L = 8$ in

FIGURE P3-31

3-32.   The 7 × 4 × 3/8 angle shown is connected with three 1-in bolts. If the angle consists of A36 steel, determine its block shear strength. Compare the results with the tensile design strength of the member.

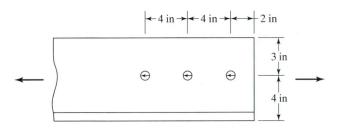

4 in   4 in   2 in

3 in

4 in

FIGURE P3-32

3-33.   A W16 × 31 is connected at its ends with the plates shown. Determine the block shear strength of the member if A992 steel is used and if it is connected with six 7/8-in bolts in each flange as shown. Compare the result with the tensile design strength of the member. Do not check plate strength. (Ans. 238.4 k)

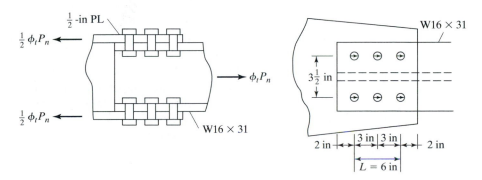

$\frac{1}{2}$ -in PL

$\frac{1}{2}\,\phi_t P_n$

$\phi_t P_n$

$\frac{1}{2}\,\phi_t P_n$

W16 × 31

W16 × 31

$3\frac{1}{2}$ in

2 in   3 in   3 in   2 in

$L = 6$ in

FIGURE P3-33

3-34.   Repeat Prob. 3-27 if A242 Grade 50 steel is used and block shear is included.

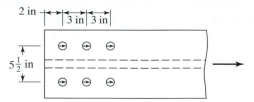

FIGURE P3-34

3-35.   Compute the tensile design strength of the 6 × 6 × 1/2 angle shown if it consists of a steel with F_y = 50 ksi and F_u = 65 ksi. Consider block shear as well as the tensile strength of the angle. (Ans. 203.3 k)

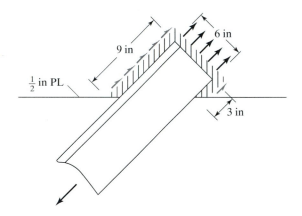

FIGURE P3-35

3-36.   A W14 × 61 has two lines of 3/4-in bolts (three in a line 4 in on center) in each flange. If it consists of A992 steel, determine the maximum dead and live service tensile loads (P_D and P_L) it can support if 30% of the service load is dead and 70% is live. Use the U value given in Table 3-2 of this chapter.

3-37.   Repeat Prob. 3-36 if the member is a C12 × 30 (F_y = 50 ksi, F_u = 65 ksi) with three lines of parallel 3/4-in bolts (four in a line 3 in on center) in the web. Use Table 3-2 for U. (Ans. 62.8 k, 146.4 k)

3-38.   An A992 WT15 × 74 has transverse welds to its flange only at its ends. Determine its design tensile strength $\phi_t P_n$ using the LRFD Spec. B.3 expression for U.

3-39.   The two MC18 × 42.75 shown have transverse welds along their webs only Compute the design tensile strength $\phi_t P_n$ of the member if A36 steel is used. (Ans. 704.7 k)

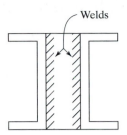

FIGURE P3-39

3-40 to 3-44.   Using the enclosed computer program where $A_e = 0.75A_g$ solve the following problems.

3-40.   Determine the design tensile strength of a 30-ft long W12 × 58 with two lines of four 7/8-in bolts 4 in on center in each flange. Use A992 steel.

3-41.   Same question and data as for Prob. 3-40 except the member is W14 × 90. (Ans. 968.9 k)

3-42.   Repeat Problem 3.41 if A36 steel is used.

3-43.   Determine the design tensile strength of a 24-ft long MC12 × 40 with one line of three 3/4-in bolts in each flange 4 in on center. Use A36 steel. (Ans. 382.3 k)

3-44.   Repeat Prob. 3-43 using A572 Grade 60 steel.

C H A P T E R   4

Design of Tension Members

4.1 SELECTION OF SECTIONS

The determination of the design strengths of various tension members was presented in Chapter 3. In this chapter, the selection of members to support given tension loads is described. Although the designer has considerable freedom in the selection, the resulting members should have the following properties: (a) compactness, (b) dimensions that fit into the structure with reasonable relation to the dimensions of the other members of the structure, and (c) connections to as many parts of the sections as possible to minimize shear lag.

The choice of member type is often affected by the type of connections used for the structure. Some steel sections are not very convenient to bolt together with the required gusset or connection plates, while the same sections may be welded together with little difficulty. Tension members consisting of angles, channels, and W or S sections will probably be used when the connections are made with bolts, while plates, channels, and structural tees might be used for welded structures.

Various types of sections are selected for tension members in the examples to follow, and in each case where bolts are used some allowance is made for holes. Should the connections be made *entirely* by welding, no holes have to be added to the net areas to give the required gross area. *The student should realize, however, that very often welded members may have holes punched in them for temporary bolting during field erection before the permanent field welds are made. These holes need to be considered in design.* It is also to be remembered that in LRFD Equation D1-2 ($P_n = F_u A_e$) the value of A_e may be less than A_g even though there are no holes, depending on the arrangement of welds and on whether all of the parts of the members are connected.

The slenderness ratio of a member is the ratio of its unsupported length to its least radius of gyration. Steel specifications give preferable maximum values of slenderness ratios for both tension and compression members. The purpose of such limitations for tension members is to ensure the use of sections with sufficient stiffness to prevent undesirable lateral deflections or vibrations. Although tension members are not subject to buckling under normal loads, stress reversal may occur during shipping and erection and perhaps due to wind or earthquake. The specifications recommend that slenderness ratios be kept below certain maximum values in order that some minimum compressive strengths be provided in the members. For tension members other than rods, LRFD Specification B7 recommends maximum slenderness ratios of 300. Members whose design is controlled by tension loads, but which may be subjected to some compression due to other loading conditions, are not required to meet the preferable maximum slenderness ratio requirement for compression members, which is 200. (For slenderness ratios greater than 200, design compressive stresses will be very small—in fact, less than 5.33 ksi. This will be discussed further later in the text.)

It should be noted that out-of-straightness does not affect the strength of tension members very much because the tension loads tend to straighten the members. (The same statement cannot be made for compression members.) For this reason, the LRFD specification is a little more liberal in its consideration of tension members, including those subject to some compressive forces due to transient loads such as wind or earthquake.

The *recommended* maximum slenderness ratio of 300 is not applicable to tension rods. Maximum L/r values for rods are left to the designer's judgment. If a maximum value of 300 was specified for them, they would seldom be used because of their extremely small radii of gyration.

The 1996 AASHTO Specifications provide mandatory maximum slenderness ratios of 200 for main tension members and 240 for secondary members. (A *main member* is defined by the AASHTO as one in which stresses result from dead and/or live loads, while *secondary members* are those used to brace structures or to reduce the unbraced length of other members—main or secondary.) *No such distinction is made in the LRFD Specification between main and secondary members.*

Example 4-1 illustrates the design of a bolted tension member with a W section, while Example 4-2 illustrates the selection of a bolted single-angle tension member. In both cases, the LRFD Specification is used. The design strength $\phi_t P_n$ is the lesser of (a) $\phi_t F_y A_g$, (b) $\phi_t F_u A_e$, or (c) the block shear strength ϕR_n.

a. To satisfy the first of these expressions, the minimum gross area must be at least equal to the following:

$$\text{min } A_g = \frac{P_u}{\phi_t F_y}. \tag{4.1}$$

b. To satisfy the second expression, the minimum value of A_e must be at least

$$\text{min } A_e = \frac{P_u}{\phi_t F_u}.$$

And since $A_e = U A_n$ for a bolted member, the minimum value of A_n is

$$\text{min } A_n = \frac{\text{min } A_e}{U} = \frac{P_u}{\phi_t F_u U}.$$

Transfer truss, 150 Federal Street, Boston, MA. (Courtesy Owen Steel Company, Inc.)

Then the minimum A_g for the second expression must at least equal the minimum value of A_n plus the estimated hole areas.

$$\min A_g = \frac{P_u}{\phi_t F_u U} + \text{estimated hole areas} \qquad (4.2)$$

 c. The third expression can be evaluated once a trial shape has been selected, and the other parameters related to the block shear strength are known.

The designer can substitute into Equations (1) and (2), taking the larger value of A_g so obtained for an initial size estimate. It is, however, well to notice that the maximum preferable slenderness ratio L/r is 300. From this value it is easy to compute the smallest preferable value of r with respect to each principal axis of the cross section for a particular design, that is, the value of r for which the slenderness ratio will be exactly 300. It is undesirable to consider a section whose least r is less than this value because its slenderness ratio would exceed the preferable maximum value of 300

$$\min r = \frac{L}{300} \qquad (4.3)$$

For the two examples to follow, it is necessary to substitute only into the first two load factor expressions 4-1 and 4-2 because only dead and live loads are involved. As there are no other loads, these expressions are reduced to the following abbreviated expressions:

$$P_u = 1.4D$$
$$P_u = 1.2D + 1.6L$$

As the first of these expressions will not control unless the dead load is more than eight times as large as the live load, the first expression is omitted for the remaining problems in this text (unless $D > 8L$).

In Example 4-1, a W section is selected for a given set of tensile loads. For this first application of the tension design formulas, the author has narrowed the problem down to one series of W shapes so the reader can concentrate on the application of the formulas and not become lost in considering W8s, W10s, W14s, and so on. Exactly the same procedure can be used for trying these other series, as is used here for the W12.

Example 4-1

Select a 30-ft-long W12 section of A992 steel to support a tensile service dead load $P_D = 130$k and a tensile service live load $P_L = 110$k. As shown in Figure 4.1, the member is to have two lines of bolts in each flange for 7/8-in bolts (at least three in a line 4 in on center).

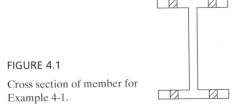

FIGURE 4.1

Cross section of member for Example 4-1.

Solution. Considering the two ordinary load conditions

$$P_u = 1.4P_D = (1.4)(130) = 182 \text{ k}$$

$$P_u = 1.2P_D + 1.6P_L = (1.2)(130) + (1.6)(110) = 332 \text{ k} \leftarrow$$

Computing the minimum A_g required

1. $\min A_g = \dfrac{P_u}{\phi_t F_y} = \dfrac{332}{(0.90)(50)} = 7.38 \text{ in}^2$

2. $\min A_g = \dfrac{P_u}{\phi_t F_u U} + \text{estimated hole areas}$

Bridge over Allegheny River at Kittaning, PA. (Courtesy American Bridge Company)

Assume $U = 0.90$ from Table 3.2 and assume flange thickness is about 0.380 in after looking at W12 sections in the LRFD Manual which have areas of 7.38 in² or more.

$$\text{min } A_g = \frac{332}{(0.75)(65)(0.90)} + (4)(1.00)(0.520) = 9.65 \text{ in}^2 \leftarrow$$

3. Preferable min $r = \dfrac{L}{300} = \dfrac{(12)(30)}{300} = 1.2$ in

Try W12 × 35 ($A_g = 10.3$ in²,

$d = 12.50$ in, $b_f = 6.56$ in, $t_f = 0.520$, $r_y = 1.54$ in)

Checking

1. $\phi_t P_n = \phi_t F_y A_g = (0.90)(50)(10.3) = 463.5$ k > 332 k   (OK)

2. $\overline{x}$ for half of W12 × 35 or that is a WT6 × 17.5 = 1.30 in

   $L = (2)(4) = 8$ in

   $U = \left(1 - \dfrac{\overline{x}}{L}\right) \leq 0.90 = \left(1 - \dfrac{1.30}{8}\right) = 0.84 < 0.90$

   $A_n = 10.3 - (4)(1.00)(0.520) = 8.22$ in²

   $\phi_t P_n = \phi_t F_u A_e = (0.75)(65)(0.84 \times 8.22) = 336.6$ k > 332 k   (OK)

3. $\dfrac{L_y}{r_y} = \dfrac{(12)(30)}{1.54} = 234 < 300$ (OK)

   Use W12 × 35

In Example 4-2, a broader example is presented in that the lightest satisfactory angle in the LRFD Manual is selected for a given set of tensile loads.

Example 4-2

Design a 9-ft single-angle tension member to support a dead tensile working load of 30 k and a live tensile working load of 40 k. The member is to be connected to one leg only with 7/8-in bolts (at least three in a line 3 in on center). Assume that only one bolt is to be located at any one cross section. Use A36 steel.

Discussion. There are many different angles listed in the LRFD Manual which will support the load for the conditions described. As a result, it may seem rather difficult to arrive at the absolutely lightest satisfactory section. It is possible, however, to set up a table with which the various possible sections may be methodically considered for various angle thicknesses. In the table presented with this solution the author has listed various possible thicknesses of angles which will provide sufficient areas. He has then found the lightest angles for each thickness which provides the estimated area needed. Finally, a glance at the completed table reveals the lightest angle or the one with the smallest cross-sectional area. A similar process could be followed for designing tension members from other steel sections.

Solution. This problem is one which can be conveniently handled with the use of a table, and such a procedure is followed here for a few angle sizes. The limiting values are calculated as follows:

$$P_u = (1.2)(30) + (1.6)(40) = 100 \text{ k}$$

1. $\min A_g$ required $= \dfrac{P_u}{\phi_t F_y} = \dfrac{100}{(0.90)(36)} = 3.09 \text{ in}^2$

2. From Table 3.2 assume $U = 0.85$

   $\min A_n$ required $= \dfrac{P}{\phi_t F_u U} = \dfrac{100}{(0.75)(58)(0.85)} = 2.70 \text{ in}^2$

3. $\min r = \dfrac{L}{300} = \dfrac{(12)(9)}{300} = 0.36 \text{ in}$

Angle t (in)	Area of one 1-in bolt hole (in²)	Gross area required = larger of $P_u/\phi_t F_y$ or $P_u/\phi_t F_u U$ + est. hole area (in²)	Lightest angles available, their areas (in²) and least radii of gyration (in)
$\frac{5}{16}$	0.312	3.09	$6 \times 6 \times \frac{5}{16} (A = 3.67,\ r_z = 1.19)$
$\frac{3}{8}$	0.375	3.09	$6 \times 3\frac{1}{2} \times \frac{3}{8} (A = 3.41,\ r_z = 0.764)$
			$5 \times 3 \times \frac{7}{16} (A = 3.31,\ r_z = 0.644)$
$\frac{7}{16}$	0.438	3.14	$4 \times 4 \times \frac{7}{16} (A = 3.31,\ r_z = 0.777)$
			$4 \times 3 \times \frac{1}{2} (A = 3.25,\ r_z = 0.633) \leftarrow$
$\frac{1}{2}$	0.500	3.20	$3\frac{1}{2} \times 3\frac{1}{2} \times \frac{1}{2} (A = 3.27,\ r_z = 0.679) \leftarrow$
$\frac{5}{8}$	0.625	3.33	$4 \times 3 \times \frac{5}{8} (A = 3.99,\ r_z = 0.631)$

Use L4 $\times$ 3 $\times \frac{1}{2}$

In Part 3 of the Manual, example designs similar to those of Examples 4-1 and 4-2 are presented for a number of different shapes, including Ws, single Ls, WTs, HSS sections, and so on. To calculate the U values for the HSS sections, it is necessary to obtain some of their cross-sectional properties, available in HSS specifications, but not included in this text or in the Manual.

4.2   BUILT-UP TENSION MEMBERS

Sections D2 and J3.5 of the LRFD Specification provide a definite set of rules describing how the different parts of built-up tension members are to be connected together.

1. When a tension member is built up from elements in continuous contact with each other, such as a plate and a shape, or two plates, the longitudinal spacing of connectors between those elements must not exceed 24 times the thickness of the thinner plate, or 12 in if the member is to be painted or if it is not to be painted and not to be subjected to corrosive conditions.

2. Should the member consist of unpainted weathering steel elements in continuous contact and be subject to atmospheric corrosion, the maximum permissible connector spacings are 14 times the thickness of the thinner plate, or 7 in.

3. Should a tension member be built up from two or more shapes separated by intermittent fillers, the shapes must be connected to each other at intervals such that the
slenderness ratio of the individual shapes between the fasteners does not exceed 300.

4. The distance from the center of any bolts to the nearest edge of the connected
part under consideration may not be larger than 12 times the thickness of the
connected part, or 6 in.

Example 4-3 illustrates the review of a tension member that is built up from two
channels that are separated from each other. Included in the problem is the design of
tie plates or tie bars to hold the channels together as shown in Figure 4.2(b). These
plates which are used to connect the parts of built-up members on their open sides result in more uniform stress distribution among the various parts. Section D-2 of the
LRFD Specification provides empirical rules for their design. (Perforated cover plates
may also be used.) The rules are based on many decades of experience with built-up
tension members.

In the "Dimensions and Properties" section of Part 1 of the Manual, the usual positions for placing bolts in the flanges of Ws, Cs, WTs, etc., are listed under the heading
"Workable Gage." For the channels used in this example, the gage g is given as $1\frac{3}{4}$ in
and is shown in Figure 4.2.

In Figure 4.2 the distance between the lines of bolts connecting the tie plates in
the channels can be seen to equal 8.50 in. The LRFD Specification (D2) states that the
length of tie plates (lengths are always measured parallel to the long direction of the
members in this text) may not be less than two-thirds the distance between the lines of
connectors. Furthermore, their thickness may not be less than one-fiftieth of this distance.

Members with rectangular cross sections are classified either as plates or bars. In
general, sections $\leq$ 8-in wide are referred to as *bars*, while those with greater widths
are called *plates*. (Detailed information concerning these elements is given in Tables
1.19 and 1.20 of Part 1 of the Manual.) The preferred practice for both plates and bars
is that they should be selected to the nearest 1/4 in as to width and the nearest 1/8 in as
to thickness. Sheets and strips are distinguished from plates and bars by their dimensional characteristics, as shown in the Manual's Table 1.22.

The minimum permissible width of tie plates (not mentioned in the specification)
is the width between the lines of connectors plus the necessary edge distance on each
side to keep the bolts from splitting the plate. For this example, this minimum edge distance is taken as 11/2 in from Table J3.4 of the LRFD Specification. (Detailed information concerning edge distances for bolts is provided in Chapter 12.) The plate
dimensions are rounded off to agree with the plate sizes available from the steel mills
as given in the Bars and Plates section of Part 1 of the LRFD Manual. It is much cheaper to select standard thicknesses and widths rather than picking odd ones that will require cutting or other operations.

The LRFD Specification (D2) provides a maximum spacing between tie plates by
stating that the L/r of each individual component of a built-up member running along
by itself between tie plates should preferably not exceed 300. If the designer substitutes
into this expression ($L/r = 300$), the least r of an individual component of the built-up
member, the value of L may be computed. This will be the maximum spacing of the tie
plates preferred by the LRFD Specification for this member.

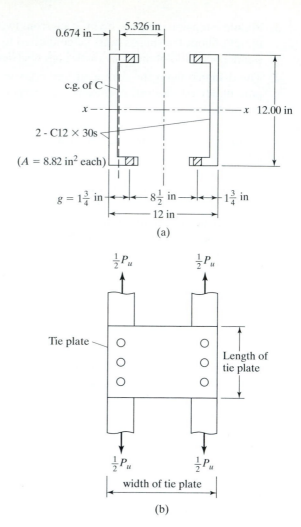

FIGURE 4.2

Built-up section for Example 4-3.

Example 4-3

Two C12 × 30s (see Figure 4.2) have been selected to support a dead tensile working load of 120 k and a 240-k live tensile working load. The member is 30 ft long, consists of A36 steel, and has one line of at least three 7/8-in bolts in each channel flange 3 in on center. Using the LRFD Specification, determine whether the member is satisfactory and design the necessary tie plates. Assume centers of bolt holes are 1.75 in from the backs of the channels.

Solution.   Using C12 × 30s $(A_g = 8.81$ in^2 each, $t_f = 0.501$ in, $I_x = 162$ in^4 each, $I_y = 5.12$ in^4 each, y axis 0.674 in from back of C, $r_y = 0.762$ in)

Load to be resisted

$$P_u = (1.2)(120) + (1.6)(240) = 528 \text{ k}$$

Design strengths

$$\phi_t P_n = \phi_t F_y A_g = (0.90)(36)(2 \times 8.81) = 570.9 \text{ k} > 528 \text{ k} \qquad \text{(OK)}$$

$$A_n = [8.81 - (2.0)(\tfrac{7}{8} + \tfrac{1}{8})(0.501)]2 = 15.62 \text{ in}^2$$

$$U = 0.85 \text{ from Table 3-2}$$

$$\phi_t P_n = \phi_t F_u A_n U = (0.75)(58)(15.62)(0.85) = 577.5 \text{ k} > 528 \text{ k} \qquad \text{(OK)}$$

Slenderness ratio

$$I_x = (2)(162) = 324 \text{ in}^4$$

$$I_y = (2)(5.12) + (2)(8.81)(5.33)^2 = 511 \text{ in}^4$$

$$r_x = \sqrt{\frac{324}{17.62}} = 4.29 \text{ in} < r_y = \sqrt{\frac{511}{17.64}} = 5.38 \text{ in}$$

$$\frac{L_x}{r_x} = \frac{(12 \times 30)}{4.29} = 83.9 < 300$$

Design of tie plates (LRFD Specification D2)

Distance between lines of bolts $= 12.00 - (2)(1\tfrac{3}{4}) = 8.50$ in

Minimum length of tie plates $= (\tfrac{2}{3})(8.50) = 5.67$ in (*say 6 in*)

Minimum thickness of tie plates $= (\tfrac{1}{50})(8.50) = 0.17$ in (*say $\tfrac{3}{16}$ in*)

Minimum width of tie plates $= 8.50 + (2)(1\tfrac{1}{2}) = 11.5$ in (*say 12 in*)

Maximum preferable spacing of tie plates least r of one $C = 0.762$ in

Maximum preferable $\dfrac{L}{r} = 300$

$$\frac{(12)(L)}{0.762} = 300$$

$$L = 19.05 \text{ ft}$$

Use $\tfrac{3}{16} \times 6 \times 1$ ft 0 in tie plate 15 ft 0 in o.c. (on center)

4.3   RODS AND BARS

When rods and bars are used as tension members, they may be simply welded at their ends, or they may be threaded and held in place with nuts. The LRFD nominal tensile design stress for threaded rods is given in their Table J3.2 and equals $\phi 0.75 F_u$ and is to be applied to the gross area of the rod A_D computed with the major thread diameter, that is, the diameter to the outer extremity of the thread. The area required for a particular tensile load can then be calculated from the following expression:

$$A_D \geq \frac{P_u}{\phi 0.75 F_u} \text{ with } \phi = 0.75.$$

In Table 7.4 of the Manual entitled "Threading Dimensions for High-Strength and Non-High-Strength Bolts" properties of standard threaded rods are presented. Example 4-4 illustrates the selection of a rod using this table. *It will be noted that the LRFD Specification (Section J1.7) states that the factored load P_u used for connection design may not be less than 10 k except for lacing, sag rods, or girts.*

Example 4.4

Using A36 steel and the LRFD Specification, select a standard threaded rod of A36 steel to support a tensile working dead load of 10 k and a tensile working live load of 20 k.

Solution

$$P_u = (1.2)(10) + (1.6)(20) = 44 \text{ k} \leftarrow$$

$$A_D \geq \frac{P_u}{\phi 0.75 F_u} = \frac{44}{(0.75)(0.75)(58)} = 1.35 \text{ in}^2$$

Use $1\frac{3}{8}$-in-diameter rod with six threads per in. $(A_D = 1.49 \text{ in}^2)$

As shown in Figure 4.3, upset rods sometimes are used, where the rod ends are made larger than the regular rod and the threads are placed in the upset ends. Threads obviously reduce the cross-sectional area of a rod. If a rod is upset and the threads are placed in that part of the rod, the result will be a larger cross-sectional area at the root of the thread than we would have if the threads were placed in the regular part of the rod.

Table J3.2 in the LRFD Specification states that the nominal tensile strength of the threaded portion of the upset ends is equal to $0.75 F_u A_D$, where A_D is the cross-sectional area of the rod at its major thread diameter. This value must be larger than the nominal body area of the rod (before upsetting) times F_y, so that the net section fracture strength exceeds the gross section yield strength.

FIGURE 4.3

A round upset rod.

Upsetting permits the designer to use the entire area of the regular part of the bar for strength calculations. Nevertheless, the use of upset rods probably is not economical and should be avoided unless a large order is being made.

One situation in which tension rods are sometimes used is in steel-frame industrial buildings with purlins running between their roof trusses to support the roof surface. These types of buildings will also frequently have girts running between the columns along the vertical walls. (Girts are horizontal beams used on the sides of buildings, usually industrial, to resist lateral bending due to wind. They also are often used to support corrugated or other types of siding.) Sag rods may be required to provide support for the purlins parallel to the roof surface and vertical support for the girts along the walls. For roofs with steeper slopes than one vertically to four horizontally, sag rods are often considered necessary to provide internal support for the purlins, particularly where the purlins consist of steel channels. Steel channels are commonly used as purlins, but they have very little resistance to lateral bending. Although the resisting moment needed parallel to the roof surface is small, an extremely large channel is required to provide such a moment. The use of sag rods for providing lateral support to purlins made from channels usually is economical because of the bending weakness of channels about their y axes. For light roofs (as where trusses support corrugated steel roofs) sag rods will probably be needed at the one-third points if the trusses are more than 20 ft on centers. Sag rods at the midpoints are usually sufficient if the trusses are less than 20 ft on centers. For heavier roofs such as those made of slate, cement tile, or clay tile, sag rods will probably be needed at closer intervals. The one-third points will probably be necessary if the trusses are spaced at greater intervals than 14 ft, and the midpoints will be satisfactory if truss spacings are less than 14 ft. Some designers assume that the load components parallel to the roof surface can be taken by the roof, particularly if it consists of corrugated steel sheets, and that tie rods are unnecessary. This assumption, however, is open to some doubt and definitely should not be followed if the roof is very steep.

New Albany Bridge crossing the Ohio River between Louisville, KY, and New Albany, Ind. (Courtesy of the Lincoln Electric Company.)

Designers have to use their own judgment in limiting the slenderness values for rods, as they will usually be several times the limiting values mentioned for other types of tension members. A common practice of many designers is to use rod diameters no less than 1/500th of their lengths to obtain some rigidity, even though design calculations may permit smaller sizes.

It is usually desirable to limit the minimum size of sag rods to 5/8 in because smaller rods than these are often damaged during construction. The threads on smaller rods are quite easily damaged by overtightening, which seems to be a frequent habit of construction workers. Sag rods are designed for the purlins of a roof truss in Example 4-5. The rods are assumed to support the simple beam reactions for the components of the gravity loads (roofing, purlins, snow and ice) parallel to the roof surface. Wind forces are assumed to act perpendicular to the roof surfaces and theoretically will not affect the sag rod forces. The maximum force in a sag rod will occur in the top sag rod because it must support the sum of the forces in the lower sag rods. It is theoretically possible to use smaller rods for the lower sag rods, but this reduction in size will probably be impractical.

Example 4-5

Design the sag rods for the purlins of the truss shown in Figure 4.4. Purlins are to be supported at their one-third points between the trusses, which are spaced 21 ft on centers. Use A36 steel and assume a minimum size rod of 5/8 in is permitted. A clay tile roof weighing 16 psf (0.77 kN/m^2) of roof surface is used and supports a snow load of 20 psf (0.96 kN/m^2) of horizontal projection of roof surface. Details of the purlins and the sag rods and their connections are shown in Figures 4.4 and 4.5. In these figures, the dotted lines represent a common practice of using ties and struts in the end panels in the plane of the roof to give greater resistance to loads located on one side of the roof (a loading situation which might occur when snow is blown off one side of the roof during a severe windstorm).

Solution. Gravity loads in psf of roof surface are as follows.

Average weight in psf of the seven purlins on each side of the roof

$$= \frac{(7)(11.5)}{37.9} = 2.1 \text{ psf}$$

$$\text{Snow} = 20\left(\frac{3}{\sqrt{10}}\right) = 19.0 \text{ psf}$$

Tile roofing = 16.0 psf

$$w_u = (1.2)(2.1 + 16.0) + (0.5)(19.0) = 31.22 \text{ psf}$$
$$w_u = (1.2)(2.1 + 16.0) + (1.6)(19.0) = 52.12 \text{ psf} \leftarrow$$

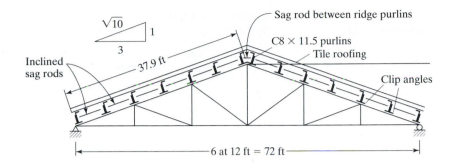

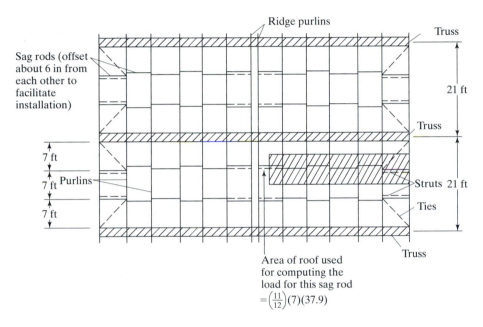

FIGURE 4.4

Plan of two bays of roof.

$$\text{Component of loads parallel to roof surface} = \left(\frac{1}{\sqrt{10}}\right)(52.12) = 16.48 \text{ psf}$$

You can see in Figures 4.4 and 4.5 that half of the load component parallel to the roof surface between the top two purlins on each side of the truss is carried directly to the horizontal sag rod between the purlins. In this example there are seven purlins (with six spaces between them) on each side of the truss. Thus, 1/12th of the total inclined load goes directly to the horizontal sag rod.

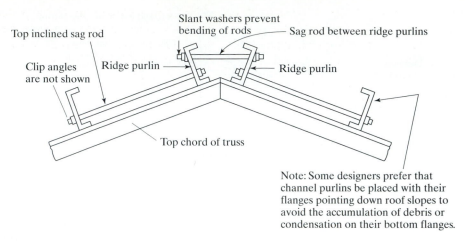

FIGURE 4.5

Details of sag rod connections.

Load on top inclined sag rod $= \left(\dfrac{11}{12}\right)(37.9)(7)(16.48) = 4008$ lb

$$A_D \geq \frac{P_u}{\phi 0.75 F_u} = \frac{4.008}{(0.75)(0.75)(58)} = 0.123 \text{ in}^2$$

Use 5/8-in rod (minimum practical size),

11 threads per in $(A_D = 0.307 \text{ in}^2)$

Force in tie rod between ridge purlins

$$T = (37.9)(7)(16.48)\frac{\sqrt{10}}{3} = 4608 \text{ lb}$$

or it equals $(4.008)\left(\dfrac{\sqrt{10}}{3}\right) + \left(\dfrac{1}{12}\right)(37.9)(7)(16.48)\left(\dfrac{\sqrt{10}}{3}\right) = 4608$ lb

$$A_D \geq \frac{4.608}{(0.75)(0.75)(58)} = 0.141 \text{ in}^2$$

Use 5/8-in rod

4.4   PIN-CONNECTED MEMBERS

Until the early years of the twentieth century, nearly all bridges in the United States were pin-connected, but today pin-connected bridges are used infrequently because of the advantages of bolted and welded connections. One trouble with the old pin-connected trusses was the wearing of the pins in the holes, which caused looseness of the joints.

An eyebar is a special type of pin-connected member whose ends where the pin holes are located are enlarged, as shown in Figure 4.6. Though just about obsolete today, eyebars at one time were very commonly used for the tension members of bridge trusses.

Pin-connected eyebars are used occasionally today as tension members for long-span bridges and as hangars for some types of bridges and other structures where they are normally subjected to very large dead loads. As a result, the eyebars are usually prevented from rattling and wearing under live loads.

Eyebars are generally made not by forging them, but by thermally cutting them from plates. As stated in the LRFD Commentary (D3), extensive testing has shown that thermally cut members result in more balanced designs. The loads of eyebars are specially shaped so as to provide optimum stress flow around the holes. These proportions are based on long experience and testing with forged eyebars, and the results are rather conservative for today's thermally cut members.

The LRFD Specification (D3) provides detailed requirements for pin-connected members as to strength and proportions of the pins and plates. The design strength of such a member is the lowest value obtained from the following equations. Reference is made to Figure 4.7.

1. *Tensile strength on the effective net area. See Figure 4.7(a).*

$$\phi = \phi_t = 0.75$$
$$P_n = 2tb_{\text{eff}}F_u \qquad \text{(LRFD Equation D3-1)}$$

In which t = plate thickness and $b_{\text{eff}} = 2t + 0.63$, but may not exceed the distance from the hole edge to the edge of the part measured perpendicular to the line of force.

2. *Shear design strength on the effective area. See Figure 4.7(b).*

$$\phi = \phi_{sf} = 0.75$$
$$P_n = 0.6A_{sf}F_u \qquad \text{(LRFD Equation D3-2)}$$

In which $A_{sf} = 2t(a + d/2)$ where a is the shortest distance from the edge of the pin hole to the member edge measured parallel to the force.

3. *Strength of surfaces in bearing. See Figure 4.7(c).*

$$\phi = 0.75$$
$$R_n = 1.8F_yA_{pb} \qquad \text{(LRFD Equation J8-1)}$$

FIGURE 4.6

End of an eyebar.

In which A_{pb} = projected bearing area = dt. Notice that LRFD Equation J8-1 applies to milled surfaces, pins in reamed, drilled, or bored holes, and ends of fitted bearing stiffeners. (LRFD Specification J8 also provides other equations for determining the bearing strength for expansion rollers and rockers.)

4. *Tensile strength of the gross section. See Figure 4.7(d).*

$$\phi = 0.90$$
$$P_n = F_y A_g \qquad \text{(LRFD Equation D1-1)}$$

The LRFD Specification (D3) says that thicknesses $< 1/2$ in for both eyebars and pin-connected plates are only permissible when external nuts are provided to tighten the pin plates and filler plates into snug contact. The bearing design strength of such plates are provided in LRFD Specification J8.

In addition to the other requirements mentioned, LRFD Specification D3 specifies certain proportions between the pins and the eyebars. These values are based on long experience in the steel industry, and experimental work by B. G. Johnston.[1] It has been found that when eyebars and pin-connected members are made from steels with yield stresses greater than 70 kips per square inch (ksi) there may be a possibility of

$$P_n = (2t)(2t + 0.63)(F_u)$$

(a) Tensile strength on net section

$$P_n = (0.6)(2t)\left(a + \frac{d}{2}\right)(F_u)$$

(b) Shear design strength on effective area

$$P_n = 1.8 \, F_y dt$$

(c) Bearing strength of surface (this is bearing on projected rectangular area behind bolt)

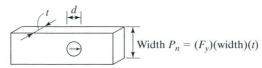

Width $P_n = (F_y)(\text{width})(t)$

(d) Tensile strength of gross section

FIGURE 4.7

Strength of pin-connected tension members.

[1]B. G. Johnston, "Pin-Connected Plate Links," *Transactions ASCE* 104 (1939).

"dishing" (a complicated inelastic stability failure where the head of the eyebar tends to curl laterally into a dish-like shape). For this reason, the LRFD Specification requires stockier member proportions for such situations (hole diameter not to exceed five times the plate thickness and the width of the eyebar reduced accordingly).

4.5   DESIGN FOR FATIGUE LOADS

It is not likely that fatigue stresses will be a problem in the average building frame because the changes in load in such structures usually occur only occasionally and produce relatively minor stress variations. Should there, however, be frequent variations or even reversals of stress, the matter of fatigue must be considered. Fatigue can be a problem in buildings when crane runway girders or heavy vibrating or moving machinery or equipment are supported.

If steel members are subjected to loads that are applied and then removed or changed significantly many thousands of times, cracks may occur, and they may spread so much as to cause failure. The steel must be subjected to either stress reversals or to variations in tension stress because fatigue problems occur only when tension is present. (Sometimes, however, fatigue cracks may occur in members that are subjected only to calculated compression stresses if parts of those members have high residual tension stresses.) The results of fatigue loading is that steel members may fail at stresses well below the stresses at which they would fail if they were subject to static loads. The fatigue strength of a particular member is dependent on the number of cycles of stress change, the range of stress change, and the size of flaws.

Through the years, many thousands of fatigue tests have been conducted, and the results have been discussed in innumerable papers on the subject. Despite all of this work, however, a clear understanding of fatigue behavior is not yet available to the steel design profession. As a result, steel design for fatigue situations is based almost entirely on test results.

In Appendix K.3 of the LRFD Specification, a simple design method is presented for considering fatigue stresses. For this discussion, the term *stress range* is defined as the magnitude of the change in stress in a member due to the application and removal of unfactored live loads. Should there be stress reversal, the stress range equals the numerical sum of the maximum repeated tensile and compressive stresses.

The fatigue life of members increases as the stress range is decreased. Furthermore, at very low stress ranges, the fatigue life is very large. In fact, there is a range at which life appears to be infinite. This range is called the *threshold fatigue stress range*.

If it is anticipated that there will be fewer than 20,000 cyles of loading, no consideration needs to be given to fatigue. (Note that three cycles per day for 25 years equals 27,375 cycles.) If the number of cycles is greater than 20,000, an allowable or *design stress range* is calculated as specified in Appendix K.3 of the Specifications. Should a member be selected and found to have a design stress range less than the actual stress range, it will be necessary to select a larger member.

The following two additional notes pertain to the AISC fatigue design procedure:

1. The design stress range determined in accordance with the AISC requirements is only applicable for the following situations:

a. Structures for which the steel has adequate corrosion protection for the conditions expected in that locality.

b. Structures for which temperatures do not exceed 300°F.

2. The provisions of the Specification apply to stresses which are calculated with unfactored loads and the maximum permitted stress due to these loads is $0.66 F_y$.

Formulas are given in Appendix K.3 of the Specification for computing the design stress range. For most cases,

$$F_{SR} = \left(\frac{C_f}{N} \right)^{0.333} \geq F_{TH}. \qquad \text{(LRFD Equation A-K3.1)}$$

In metric units,

$$F_{SR} = \left(\frac{C_f \times 327}{N} \right)^{0.333} \geq F_{TH} \qquad \text{(LRFD Equation A-K3.1M)}$$

where

F_{SR} = design stress range, ksi (MPa),

C_f = constant from Table A-K3.1 in LRFD Appendix,

N = number of stress fluctuations in design life,

and

F_{TH} = threshold fatigue stress range from Table A-K3.1 in LRFD Appendix, ksi (MPa).

Example 4-6 presents the design of a tension member subjected to fluctuating loads using Appendix K of the LRFD Specification. Stress fluctuations and reversals are an everyday problem in the design of bridge structures. The AASHTO Specifications provide allowable stress ranges determined in a manner very similar to those of the LRFD Specification.

Example 4-6

A tension member is to consist of a W12 section (F_y = 50 ksi) with fillet-welded end connections. The service dead load is 40 k, while it is estimated that the service live load will vary from a compression of 20 k to a tension of 90 k fifty times per day for an estimated design life of 25 years. Select the section.

Solution

$$P_u = (1.2)(40) + (1.6)(90) = 192 \text{ k}$$

Estimated section size for gross section yield

$$A_g \geq \frac{P_u}{\phi_t F_y} = \frac{192}{(0.9)(50)} = 4.26 \text{ in}^2$$

Try $W12 \times 16(A_g = 4.71 \text{ in}^2)$

$$N = (50)(365)(25) = 456{,}250$$

Referring to Table A-K3.1 in Appendix K of LRFD Specification, the member falls into Section 1 of the table and into stress category A.

$$C_f = 250 \times 10^8 \text{ from table}$$

$$F_{TH} = 24 \text{ ksi from table}$$

$$F_{SR} = \left(\frac{C_f}{N}\right)^{0.333} = \left(\frac{250 \times 10^8}{456,250}\right)^{0.333} = 37.84 \text{ ksi}$$

$$\text{Max service load tension} = \frac{40 + 90}{4.71} = 27.60 \text{ ksi}$$

$$\text{Min service load tension} = \frac{40 - 20}{4.71} = 4.25 \text{ ksi}$$

$$\text{Actual stress range} = 27.60 - 4.25 = 23.35 \text{ ksi}$$

$$< F_{SR} = 37.84 \text{ ksi} \hspace{3cm} \text{(OK)}$$

$$\text{Use W12} \times 16$$

4.6 COMPUTER EXAMPLE

The program INSTEP32 may be used to select the lightest section for a given tensile load, or it may be used to select the lightest shape of a given type (as W, C, MC, etc.) or it may be used to select the lightest section in a particular series as W12 or W14. The user must realize, however, that the program is set to pick sections only for the case in which $A_e = 0.75A_g$.

In the example to follow (4–7), we desire to select the lightest W12 for a given situation. The usual data is input to the computer including the steel grade, load to be supported, length, and so on. In addition, under "Design Parameters" it is necessary to click on "Find lightest section" and "In the same series." To select the lightest section in a particular series, it is necessary to scroll up or down the list of shapes until we reach the W12. Then we can click "OK," and the lightest suitable section in the desired series will be selected.

Example 4-7

Using the computer diskette INSTEP32, select a 12-ft W12 section of A572 Grade 50 steel to support a tensile load = 1440 k. Assume that the effective net area of the member equals 0.75 percent of its gross area and $u = 1.0$.

Solution

Input:

Factored design loads

$P_u = 1440$ kips

Component details

Net area $= 29.925$ sq. in

Length $= 144$ in

$U = 1$

Output:

Axial Tension Design Summary
Lightest Section W12 × 136 in the W12 family
Check yield criteria
Pn $= 50(39.9) = 1995$ kips
0.9Pn $= 0.9(1995) = 1795.5$ kips
Check fracture criteria
Pn $= 1(29.925)(65) = 1945.12$ kips
0.75Pn $= 0.75(1945.12) = 1458.84$ kips
Check slenderness criteria
L/R $= 144/3.16 = 45.5696 \leftarrow$

PROBLEMS

4-1 to 4-8. *Select sections for the conditions described using $F_y = 50$ ksi and $F_u = 65$ ksi unless otherwise noted, and neglecting block shear. $U = 1 - \frac{\bar{x}}{L} \leq 0.90$ except for Prob. 4-8.*

4-1. Select the lightest W14 section available to support working tensile loads of $P_D = 220$ k and $P_L = 250$ k. The member is to be 30 ft long and is assumed to have two lines of holes for 1-in bolts in each flange. There will be at least three bolts in each line 4 in on center. (*Ans.* W14 × 68)

4-2. Repeat Prob. 4-1 selecting a W12 section.

4-3. Select the lightest W12 available to support a factored tensile load P_u of 380 k. Assume there are two lines of 3/4-in bolts in each flange (at least three bolts in each line 4 in on center). The member is to be 28 ft long. (*Ans.* W12 × 40)

4-4. Select the lightest S section that will safely support the service tensile loads $P_D = 70$ k and $P_L = 110$ k. The member is to be 18 ft long and is assumed to have one line of holes for 1-in bolts in each flange. Assume there are at least three holes in each line 4 in on center. Use A36 steel.

4-5. Repeat Prob. 4-4 if an MC is to be used. (*Ans.* MC 10 × 28.5)

4-6. Select the lightest available S section to support the service tensile loads $P_D = 80$ k and $P_L = 100$ k. The member is to be 20 ft long, and is assumed to have two lines of 7/8-in bolts in each flange (4 in a line 3 in on center). $F_y = 50$ ksi. $F_u = 65$ ksi

4-7. Repeat Prob. 4-6 if the member is to consist of A36 steel. (*Ans.* S 12 × 31.8)

4-8. A welded tension member is to support a design load $P_u = 650$ k and is to consist of two channels placed 12 in out to out with the flanges turned in. Select the lightest standard channels available. Assume $U = 0.87$. The member is to be 30 ft long. Use A36 steel.

4-9 to 4-16. *Select the lightest section for each of the situations described. Assume bolts are 4 in. on center. Do not consider block shear. Determine U from Table 3.2 in text except for Prob. 4-11.*

4-17. A36 steel is to be used in selecting a single angle member to resist service tensile loads of $P_D = 70$ k and $P_L = 80$ k. The member is to be 20 ft long and is assumed to be connected with one line of four 7/8-in bolts 3.5-in o.c. in the longer leg if an unequal leg angle is used. Neglect block shear.

$$U = 0.85. \ (Ans. \ \text{L8} \times 6 \times \tfrac{1}{2})$$

TABLE 4.1

Prob. no.	Section	P_D (kips)	P_L (kips)	Length (ft)	Steel	End connection	
4–9	W12	120	170	22	A992	Two lines of $\frac{3}{4}$-in bolts (3 in a line) each flange	(*Ans.* W12 × 40)
4–10	W14	180	220	24	A992	Two lines of $\frac{7}{8}$-in bolts (3 in a line) each flange	
4–11	W10	100	90	18	A992	Longitudinal welds to flanges only with $U = 0.87$	(*Ans.* W10 × 22)
4–12	W12	400	150	28	A36	Two lines of $\frac{3}{4}$-in bolts (2 in a line) each flange	
4–13	W10	90	110	20	A36	Transverse welds to flanges only	(*Ans.* W10 × 33)
4–14	S	60	100	18	A36	One line of 1-in bolts (3 in a line) each flange	
4–15	WT7	80	90	16	A992	Transverse welds to flange only	(*Ans.* WT7 × 26.5)
4–16	WT6	100	140	15	A992	Longitudinal welds to flange only 10 in long and not connected to a plate	

4-18.   Select a pair of C8s for the condition shown. Use A572 Grade 50 steel and assume transverse welds at the ends along the webs only. $L = 24$ ft, $P_u = 300$ k. Do not consider block shear.

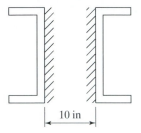

10 in

FIGURE P4-18

4-19.   Repeat Prob. 4-17 using a pair of angles. Assume angle legs are touching, and assume one hole for a 7/8-in bolt is to be taken out of each angle. Also, assume $U = 0.85$. (*Ans.* 2Ls $6 \times 3\frac{1}{2} \times \frac{3}{8}$)

4-20.   Design member L_2L_3 of the truss shown in the accompanying illustration. It is to consist of a pair of angles with a 3/8-in gusset plate between the angles at each joint. Use A36 steel and assume two lines of three 3/4-in bolts in each vertical angle leg 4 in on center. Consider only the angles shown in the double-angle tables of the Manual. For each load $P_D = 20$ k and $P_r = 12$ k (roof load). Do not consider block shear. Obtain U from Table 3.2 in text.

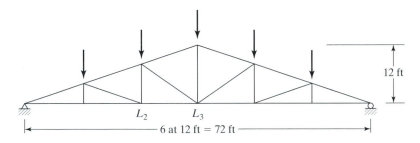

L_2 L_3

6 at 12 ft = 72 ft

12 ft

FIGURE P4–20

4-21.   Select a single-angle tension member to resist the service loads $P_D = 80$ k and $P_L = 70$ k. The member is to be 18 ft long and is to be connected in the longer leg with one line of four 7/8-in bolts 4 in on center. Assume $F_r = 42$ ksi and $F_u = 60$ ksi. Neglect block shear. $U = 1 - \frac{\bar{x}}{L}$. (*Ans.* L8 × 6 × $\frac{7}{16}$)

4-22.   Repeat Prob. 4-8, assuming that two lines of 7/8-in bolts will be used in web with at least three bolts in each line 4 in on center. Also design tie plates. U is to be determined from LRFD Specification B3.

4-23.   A tension member is to consist of four equal leg angles arranged as shown in the accompanying illustration to support the service loads $P_D = 180$ k and $P_L = 320$ k. The member is assumed to be 30 ft long and is to have one line of three 7/8-in bolts in each leg. Design the member with 50 ksi steel, $F_u = 65$ ksi, including the necessary tie plates. Neglect block shear. ($Ans.$ 4 Ls 5 × 5 × $\frac{1}{2}$)

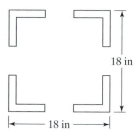

FIGURE P4-23

4-24. Select a standard threaded round rod as a hanger to resist the service loads $P_D = 10$ k and $P_L = 12$ k using A36 steel.

4-25. Select a standard threaded rod to resist service loads $P_D = 12$ k and $P_L = 15$ k using A36 steel. ($Ans.$ Use $1\frac{1}{4}$-in. rod 7 threads per in.)

4-26. The horizontal thrust at the base of the three-hinged arch shown in the accompanying illustration is to be resisted by a tie rod of A36 steel. What size threaded round tie rod should be used if the arch supports the working loads shown in the figure?

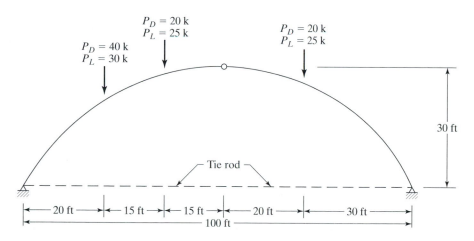

FIGURE P4-26

4-27 to 4-29. Repeat the problems indicated using INSTEP32. Remember the program is set to use $A_e = 0.75 A_g$. Assume U = 0.9.

4-27. Prob. 4-1. ($Ans.$ W14 × 68)

4-28. Prob. 4-2.

4-29. Prob. 4-4. ($Ans.$ S12 × 31.8)

Introduction to Axially Loaded Compression Members

5.1   GENERAL

There are several types of compression members, the column being the best known. Among the other types are the top chords of trusses and various bracing members. In addition, many members have compression in some of their parts. These include the compression flanges of rolled beams and built-up beam sections, and members that are subjected simultaneously to bending and compressive loads. Columns are usually thought of as being straight vertical members whose lengths are considerably greater than their thicknesses. Short vertical members subjected to compressive loads are often called struts, or simply compression members; however, the terms *column* and *compression member* will be used interchangeably in the pages that follow.

There are three general modes by which axially loaded columns can fail. These are flexural buckling, local buckling, and torsional buckling. These modes of buckling are briefly defined as follows:

1. *Flexural buckling* (also called Euler buckling) is the primary type of buckling discussed in this chapter. Members are subject to flexure or bending when they become unstable.
2. *Local buckling* occurs when some part or parts of the cross section of a column are so thin that they buckle locally in compression before the other modes of buckling can occur. The susceptibility of a column to local buckling is measured by the width-thickness ratios of the parts of its cross section. This topic is addressed in Section 5.7 of this chapter.
3. *Torsional buckling* may occur in columns that have certain cross-sectional configurations. These columns fail by twisting (torsion) or by a combination of torsional and flexural buckling. This topic is initially addressed in Section 6.8 of this text.

The longer a column becomes for the same cross section, the greater becomes its tendency to buckle and the smaller becomes the load it will support. The tendency of a member to buckle is usually measured by its *slenderness ratio*, which has previously been defined as the ratio of the length of the member to its least radius of gyration. This tendency to buckle is also affected by such factors as the types of end connections, eccentricity of load application, imperfection of column material, initial crookedness of columns, residual stresses from manufacture, etc.

The loads supported by a building column are applied by the column section above and by the connections of other members directly to the column. The ideal situation is for the loads to be applied uniformly across the column, with the center of gravity of the loads coinciding with the center of gravity of the column. Furthermore, it is desirable for the column to have no flaws, to consist of a homogeneous material, and to be perfectly straight, but these situations are obviously impossible to achieve.

Loads that are exactly centered over a column are referred to as *axial* or *concentric loads*. The dead loads may or may not be concentrically placed over an interior building column, and the live loads may never be centered. For an outside column, the load situation is probably even more eccentric, as the center of gravity of the loads will usually fall well on the inner side of the column. In other words, it is doubtful that a perfect axially loaded column will ever be encountered in practice.

The other desirable situations are also impossible to achieve because of the following: imperfections of cross-sectional dimensions, residual stresses, holes punched for bolts, erection stresses, and transverse loads. It is difficult to take into account all of these imperfections in a formula.

Slight imperfections in tension members and beams can be safely disregarded as they are of little consequence. On the other hand, slight defects in columns may be of major significance. A column that is slightly bent at the time it is put in place may have significant bending moments equal to the column load times the initial lateral deflection. Mill straightness tolerances, as taken from ASTM A6, are presented in Table 1-54 of the LRFD Manual.

Obviously, a column is a more critical member in a structure than is a beam or tension member because minor imperfections in materials and dimensions mean a great deal. This fact can be illustrated by a bridge truss that has some of its members damaged by a truck. The bending of tension members probably will not be serious as the tensile loads will tend to straighten those members; but the bending of any compression members is a serious matter, as compressive loads will tend to magnify the bending in those members.

The preceding discussion should clearly show that column imperfections cause them to bend, and the designer must consider stresses due to those moments as well as those due to axial loads. Chapters 5 to 7 are limited to a discussion of axially loaded columns, while members subjected to a combination of axial loads and bending loads are discussed in Chapter 11.

The spacing of columns in plan establishes what is called a *bay*. For instance, if the columns are 20 ft on center in one direction and 25 ft in the other direction, the bay size is 20 ft × 25 ft. Larger bay sizes increase the user's flexibility in space planning. As to

Two International Place, Boston, MA. (Courtesy of Owen Steel Company, Inc.)

economy, a detailed study by John Ruddy[1] indicates that when shallow spread footings are used, bays with length-to-width ratios of about 1.25 to 1.75, and areas of about 1000 sq ft are the most cost efficient. When deep foundations are used, his study shows that larger bay areas are more economical.

5.2   RESIDUAL STRESSES

Research at Lehigh University has shown that residual stresses and their distribution are very important factors affecting the strength of axially loaded steel columns. These stresses are of particular importance for columns with slenderness ratios varying from approximately 40 to 120, a range that includes a very large percentage of practical columns. A major cause of residual stress is the uneven cooling of shapes after hot-rolling. For instance, in a W shape the outer tips of the flanges and the middle of the web cool quickly, while the areas at the intersection of the flange and web cool more slowly.

The quicker cooling parts of the sections when solidified resist further shortening, while those parts that are still hot tend to shorten further as they cool. The net result is that the areas that cooled more quickly have residual compressive stresses, while the slower cooling areas have residual tensile stresses. The magnitude of these stresses varies from about 10 to 15 ksi (69 to 103 MPa) although some values greater than 20 ksi (138 MPa) have been found.

When rolled-steel column sections with their residual stresses are tested, their proportional limits are reached at P/A values of only a little more than half of their yield stresses and the stress-strain relationship is nonlinear from there up to the yield stress. Because of the early localized yielding occurring at some points of the column cross sections, buckling strengths are appreciably reduced. Reductions are greatest for columns with slenderness ratios varying from approximately 70 to 90 and may possibly be as high as 25 percent.[2]

As a column load is increased, some parts of the column will quickly reach the yield stress and go into the plastic range because of residual compression stresses. The stiffness of the column will be reduced and become a function of the part of the cross section that is still elastic. A column with residual stresses will behave as though it has a reduced cross section. This reduced section or elastic portion of the column will change as the applied stresses change. The buckling calculations for a particular column with residual stresses can be handled by using an effective moment of inertia I_e of the elastic portion of the cross section or by using the tangent modulus. For the usual sections used as columns the two methods give fairly close results.

Welding can produce severe residual stresses in columns that actually can approach the yield point in the vicinity of the weld. Another important fact is that columns may actually be appreciably bent by the welding process, decidedly affecting their load-carrying ability. Figure 5.1 shows the effect of residual stresses (due to cooling and fabrication) on the stress-strain diagram for a hot-rolled W shape.

[1]J. L. Ruddy, "Economics of Low-Rise Steel-Framed Structures," *Engineering Journal,* AISC, *20,* no. 3 (3rd quarter, 1983), pp. 107–118.

[2]L. S. Beedle and L. Tall, "Basic Column Strength," *Proc. ASCE 86* (July 1960), pp. 139–173.

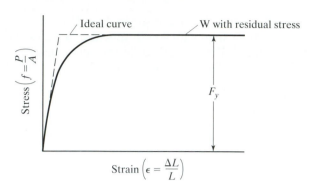

FIGURE 5.1

Effect of residual stresses on
column stress-strain diagram.

The welding together of built-up shapes frequently causes even higher residual stresses than those caused by the uneven cooling of hot-rolled H-shaped sections.

Residual stresses may also be caused during fabrication when *cambering* is performed by cold bending and due to cooling after welding. Cambering is the bending of a member in one direction so that it won't look so bad when the service loads bend it in the other direction. For instance, we may bend a beam initially upward so that it will be approximately straight when its normal gravity loads are applied.

5.3   SECTIONS USED FOR COLUMNS

Theoretically, an endless number of shapes can be selected to safely resist a compressive load in a given structure. From a practical viewpoint, however, the number of possible solutions is severely limited by such considerations as sections available, connection problems, and type of structure in which the section is to be used. The paragraphs that follow are intended to give a brief résumé of the sections which have proved to be satisfactory for certain conditions. These sections are shown in Fig. 5.2 and the letters in parentheses in the paragraphs to follow refer to the parts of this figure.

The sections used for compression members usually are similar to those used for tension members with certain exceptions. The exceptions are caused by the fact that the strengths of compression members vary in some inverse relation to the slenderness ratios and stiff members are required. Individual rods, bars, and plates usually are too slender to make satisfactory compression members, unless they are very short and lightly loaded.

Single-angle members (a) are satisfactory for bracing and compression members in light trusses. Equal-leg angles may be more economical than unequal-leg angles because their least r values are greater for the same area of steel. The top chord members of bolted roof trusses might consist of a pair of angles back to back (b). There will often be a space between them for the insertion of a gusset or connection plate at the joints necessary for connections to other members. An examination of this section will show that it is probably desirable to use unequal-leg angles with the long legs back-to-back to give a better balance between the r values about the x and y axes.

If roof trusses are welded, gusset plates may be unnecessary and structural tees (c) might be used for the top chord compression members because the web members

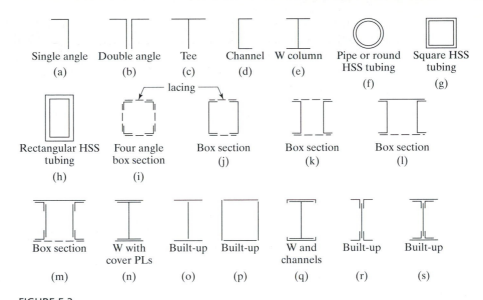

FIGURE 5.2

Types of compression members.

can be welded directly to the stems of the tees. Single channels (d) are not satisfactory for the average compression member because of their almost negligible r values about their web axes. They can be used if some method of providing extra lateral support in the weak direction is available. The W shapes (e) are the most common shapes used for building columns and for the compression members of highway bridges. Their r values, although far from being equal about the two axes, are much more nearly balanced than for channels.

Several famous bridges constructed during the nineteenth century (such as the Firth of Forth Bridge in Scotland and Ead's Bridge in St. Louis) made extensive use of tube-shaped members. Their use, however, declined due to connection problems and manufacturing costs, but with the development of economical welded tubing their use is again increasing (although the tube shapes of today are very small compared with the giant ones used in those early steel bridges).

Hollow structural sections (square, rectangular, or round) and steel pipe are very valuable sections for buildings, bridges, and other structures. These clean, neat-looking sections are easily fabricated and erected. For small and medium loads, the round sections (f) are quite satisfactory. They are often used as columns in long series of windows, as short columns in warehouses, as columns for the roofs of covered walkways, in the basements and garages of residences, etc. Round columns have the advantage of being equally rigid in all directions and are usually very economical unless moments are too large for the sizes available. The LRFD Manual furnishes the sizes of these sections and classifies them as being round HSS sections, or as standard, extra strong, or double extra strong steel pipe.

Square and rectangular tubing (g) and (h) have not been used to a great extent for columns until recently. In fact, for many years, only a few steel mills manufactured steel tubing for structural uses. Perhaps the major reason tubing was not used to a great extent was the difficulty of making connections with rivets or bolts. This problem has been fairly well eliminated, however, by the advent of modern welding. The use of tubing for structural purposes by architects and engineers in the years to come will probably be greatly increased for several reasons. These include

1. The most efficient compression member is one that has a constant radius of gyration about its centroid, a property available in round HSS tubing and pipe sections. Square tubing is the next most efficient compression member.

2. Four-sided and round sections are much easier to paint than are the six-sided open W, S, and M sections. Furthermore, the rounded corners make it easier to apply paint or other coatings uniformly around the sections.

3. They have less surface area to paint or fireproof.

4. They have excellent torsional resistance.

5. The surfaces of tubing are quite attractive.

6. When exposed, the wind resistance of round sections is only about two-thirds of that of flat surfaces of the same width.

7. If cleanliness is important, hollow structural tubing doesn't have the problem of dirt collecting between the flanges of open structural shapes.

A slight disadvantage that comes into play in certain cases is that the ends of tube and pipe sections that are subject to corrosive atmospheres may have to be sealed to protect their inaccessible inside surfaces from corrosion. Although making very attractive exposed members for beams, these sections are at a definite weight disadvantage compared with W sections, which have so much larger resisting moments for the same weights.

For many column situations, the weight of square or rectangular tube sections—usually referred to as hollow structural sections (HSS)—can be less than one half the weights required for open-profile sections (W, M, S, channel, and angle sections). It is

Bents fabricated from tubing sections in New Jersey. (Courtesy of Bethlehem Steel Corporation.)

true that tube and pipe sections may cost perhaps 25 percent more per pound than open sections, but this still allows the possibility of up to 20 percent savings in many cases.[3]

Hollow structural sections are available with yield strengths up to 50 ksi and can be obtained with enhanced atmospheric corrosion resistance. Today they make up about 3 percent of the structural steel fabricated for buildings and bridges in the United States. In Japan and Europe, the values are respectively 15 percent and 25 percent and still growing. Thus, it is probable that their use will continue to rise in the United States in the years to come.[4] Detailed information, including various tables on hollow structural sections, can be obtained from the American Institute of Hollow Steel Sections, 929 McLaughlin Run Road, Suite 8, Pittsburgh, PA 15017.

Where compression members are designed for very large structures, it may be necessary to use built-up sections. Built-up sections are needed where the members are long and support very heavy loads and/or when there are connection advantages. Generally speaking, a single shape such as a W section is more economical than a built-up section having the same cross-sectional area. With heavy column loads, high-strength steels can frequently be used with very economical results if their increased strength permits the use of W sections rather than built-up members.

When built-up sections are used, they must be connected on their open sides with some type of lacing (also called lattice bars) to hold the parts together in their proper positions and to assist them in acting together as a unit. The ends of these members are connected with *tie* plates (also called *batten* plates or *stay* plates). Several types of lacing for built-up compression members are shown in Fig. 6.9.

The dashed lines in Fig. 5.2 represent lacing or discontinuous parts, and the solid lines represent parts that are continuous for the full length of the members. Four angles are sometimes arranged as shown in (i) to produce large r values. This type of member may often be seen in towers and in crane booms. A pair of channels (j) is sometimes used as a building column or as a web member in a large truss. It will be noted that there is a certain spacing for each pair of channels at which their r values about the x and y axes are equal. Sometimes the channels may be turned out, as shown in (k).

A section well suited for the top chords of bridge trusses is a pair of channels with a cover plate on top (1) and with lacing on the bottom. The gusset or connection plates at joints are conveniently connected to the insides of the channels and may also be used as splices. When the largest channels available will not produce a top chord member of sufficient strength, a built-up section of the type shown in (m) may be used.

When the rolled shapes do not have sufficient strength to resist the column loads in a building or the loads in a very large bridge truss, their areas may be increased by adding plates to the flanges (n). In recent years, it has been found that for welded construction a built-up column of the type shown in part (o) is a more satisfactory shape than a W with welded cover plates (n). It seems that in bending (as where a beam frames into the flange of a column) it is difficult to efficiently transfer tensile force through the cover plate to the column without pulling the plate away from the column. For very heavy column loads a welded box section of the type shown in (p) has proved

---

[3]"High Design, Low Cost," *Modern Steel Construction,* (Chicago: AISC, March-April 1990), pp. 32–34.
[4]Ibid., p. 34.

to be quite satisfactory. Some other built-up sections are shown in parts (q) through (s). The built-up sections shown in parts (n) through (q) have an advantage over those shown in parts (i) through (m) in that they do not require the expense of the lattice work necessary in the former. Lateral shearing forces are negligible for the single column shapes and for the nonlatticed built-up sections, *but they are definitely not negligible for the built-up latticed columns.*

Today *composite columns* are being increasingly used. These columns usually consist of steel pipe and structural tubing filled with concrete or of W shapes encased in concrete usually square or rectangular in cross section. (These columns are discussed in Chapter 17.)

5.4   DEVELOPMENT OF COLUMN FORMULAS

The use of columns dates to before the dawn of history, but it was not until 1729 that a paper was published on the subject by Pieter van Musschenbroek, a Dutch mathematician.[5] He presented an empirical column formula for estimating the strength of rectangular columns. A few years later, in 1757, Leonhard Euler, a Swiss mathematician, wrote a paper of great value concerning the buckling of columns. He was probably the first person to realize the significance of buckling. The Euler formula, the most famous of all column expressions, is derived in Appendix B of this text. This formula, which is discussed in the next section, marked the real beginning of theoretical and experimental investigation of columns.

Engineering literature is filled with formulas developed for ideal column conditions, but these conditions are not encountered in actual practice. Consequently, practical column design is based primarily on formulas that have been developed to fit with reasonable accuracy test-result curves. The reasoning behind this procedure is simply that the independent derivation of column expressions does not produce formulas that give results comparing closely with test-result curves for all slenderness ratios.

The testing of columns with various slenderness ratios results in a scattered range of values such as those shown by the broad band of dots in Fig. 5.3. The dots will not fall on a smooth curve even if all of the testing is done in the same laboratory because of the difficulty of exactly centering the loads, lack of perfect uniformity of the materials, varying dimensions of the sections, residual stresses, end restraint variations, etc. The usual practice is to attempt to develop formulas which give results represented by an approximate average of the test results. The student should also realize that laboratory conditions are not field conditions and column tests probably give the limiting values of column strengths.

The magnitudes of the yield stresses of the sections tested are quite important for short columns as their failure stresses are close to those yield stresses. For columns with intermediate slenderness ratios, the yield stresses are of lesser importance in their effect on failure stresses, and they are of no significance for long slender columns. For intermediate range columns, residual stresses have more effect on the results while the failure stresses for long slender columns are very sensitive to end support conditions.

---

[5]L. S. Beedle et al., *Structural Steel Design* (New York: Ronald Press, 1964), p. 269.

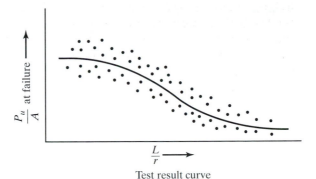

FIGURE 5.3

Test result curve.

Test result curve

In addition to residual stresses and nonlinearity of material, another dominant factor in its effect on column strength is member out-of-straightness.

5.5 THE EULER FORMULA

The stress at which a column buckles obviously decreases as the column becomes longer. After it reaches a certain length, that stress will have fallen to the proportional limit of the steel. For that length and greater lengths, the buckling stress will be elastic.

For a column to buckle elastically it will have to be long and slender. Its buckling load P can be computed with the Euler formula that follows:

$$P = \frac{\pi^2 EI}{L^2}$$

This formula usually is written in a slightly different form which involves the column's slenderness ratio. Since $r = \sqrt{I/A}$, we can say that $I = Ar^2$. Substituting this value into the Euler formula and dividing both sides by the cross-sectional area, the Euler buckling stress is obtained. This value is represented by F_e in the LRFD Manual:

$$\frac{P}{A} = \frac{\pi^2 E}{(L/r)^2} = F_e$$

Example 5-1 illustrates the application of the Euler formula to a steel column. If the value obtained for a particular column exceeds the steel's proportional limit, the elastic Euler formula is not applicable.

Example 5-1

a. A W10 × 22 is used as a 15-ft-long pin-connected column. Using the Euler expression, determine the column's critical or buckling load. Assume the steel has a proportional limit of 36 ksi.

b. Repeat part (a) if the length is changed to 8 ft.

Solution

     a.   Using a W10 × 22 (A = 6.49 in², r_x = 4.27 in, r_y = 1.33 in)

         Minimum $r = r_y$ = 1.33 in

$$\frac{L}{r} = \frac{(12)(15)}{1.33} = 135.34$$

$$\text{Critical or buckling stress } F_e = \frac{(\pi^2)(29 \times 10^3)}{(135.34)^2}$$

$$= 15.63 \text{ ksi} < \text{ the proportional limit of 36 ksi}$$

OK column is in elastic range

Critical or buckling load = (15.63)(6.49) = 101.4k

     b.   Using an 8-ft W10 × 22

$$\frac{L}{r} = \frac{(12)(8)}{1.33} = 72.18$$

$$\text{Critical or buckling stress } F_e = \frac{(\pi^2)(29 \times 10^3)}{(72.18)^2} = 54.94 \text{ ksi} > 36 \text{ ksi}$$

∴   column is in inelastic range and Euler equation is not applicable

---

     The student should carefully note that the buckling load determined from the Euler equation is independent of the strength of the steel used.

     The equation is only useful if the end support conditions are carefully considered. The results obtained by application of the formula to specific examples compare very well with test results for centrally loaded, long, slender columns with rounded ends. Designers, however, do not encounter perfect columns of this type. The columns with which they work do not have rounded ends and are not free to rotate because their ends are bolted or welded to other members. These practical columns have different amounts of restraint against rotation, varying from slight restraint to almost fixed conditions. For the actual cases encountered in practice where the ends are not free to rotate, different length values can be used in the formula and more realistic buckling stresses will be obtained.

     To successfully use the Euler equation for practical columns, the value of L should be the distance between points of inflection in the buckled shape. This distance is referred to as the *effective length* of the column. For a pinned-end column (whose ends can rotate but cannot translate) the points of inflection or zero moment are located at the ends a distance L apart. For columns with different end conditions, the effective lengths may be entirely different. Effective lengths are discussed extensively in the next section.

5.6   END RESTRAINT AND EFFECTIVE LENGTHS OF COLUMNS

End restraint and its effect on the load-carrying capacity of columns is a very important subject indeed. Columns with appreciable rotational and translational end restraint can support considerably more load than those with little rotational end restraint as at hinged ends.

450 Lexington Ave., New York City. (Courtesy of Owen Steel Company, Inc.)

The effective length of a column was defined in the last section as the distance between points of zero moment in the column, that is, the distance between its inflection points. In steel specifications, the effective length of a column is referred to as *KL,* where *K* is the *effective length factor. K* is the number that must be multiplied by the length of the column to obtain its effective length. Its magnitude depends on the rotational restraint supplied at the ends of the column and upon the resistance to lateral movement provided.

The concept of effective lengths is simply a mathematical method of taking a column, whatever its end and bracing conditions, and replacing it with an equivalent pinned-end braced column. A complex buckling analysis could be made for a frame to determine the critical stress in a particular column. The *K* factor is determined by finding the pinned-end column with an equivalent length that provides the same critical stress. The *K* factor procedure is a method of making simple solutions for complicated frame buckling problems.

Columns with different end conditions have entirely different effective lengths. For this initial discussion, it is assumed that no sidesway or joint translation is possible between the member ends. Sidesway or joint translation means that one or both ends of a column can move laterally with respect to each other. Should a column be connected with frictionless hinges as shown in Fig. 5.4(a), its effective length would be

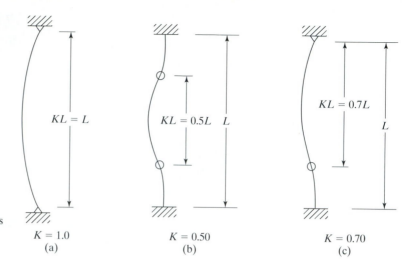

FIGURE 5.4

Effective lengths for columns in braced frames (sidesway prevented).

equal to the actual length of the column and K would equal 1.0. If there were such a thing as a perfectly fixed-ended column, its points of inflection (or points of zero moment) would occur at its one-fourth points and its effective length would equal $L/2$ as shown in Fig. 5.4(b). As a result, its K value would equal 0.50.

Obviously, the smaller the effective length of a particular column, the smaller its danger of lateral buckling and the greater its load-carrying capacity. In Fig. 5.4(c), a column is shown with one end fixed and one end pinned. The K for this column is theoretically equal to 0.70.

Neither perfect pin connections nor any perfect fixed ends exist, and the usual column falls in between the two extremes. This discussion would seem to indicate that column effective lengths always vary from an absolute minimum of $L/2$ to an absolute maximum of L, but there are many exceptions to this rule. An example is given in Fig. 5.5(a), where a simple bent is shown. The base of each of the columns is pinned and the other end is free to rotate and move laterally (called *sidesway*). Examination of this figure will show that the effective length will exceed the actual length of the column as the elastic curve will theoretically take the shape of the curve of a pinned-end column of twice its length and K will theoretically equal 2.0. Notice in part (b) of the figure how much smaller the lateral deflection of column AB would be if it were pinned both top and bottom so as to prevent sidesway.

Structural steel columns serve as parts of frames, and these frames are sometimes *braced* and sometimes *unbraced*. A braced frame is one for which sidesway or joint translation is prevented by means of bracing, shear walls, or lateral support from adjoining structures. An unbraced frame does not have any of these types of bracing supplied and must depend on the stiffness of its own members and the rotational rigidity of the joints between the frame members to prevent lateral buckling. For braced frames K values can never be greater than 1.0, but for unbraced frames the K values will always be greater than 1.0 because of sidesway.

Table C-C2.1 of the "Commentary on the LRFD Specification" gives recommended effective length factors when ideal conditions are approximated. This table is

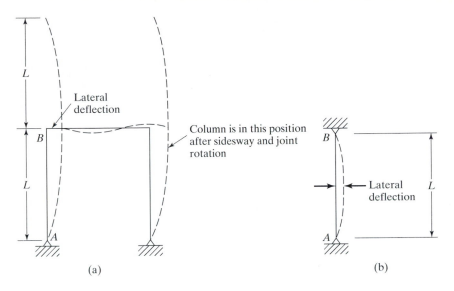

Lateral deflection

Column is in this position after sidesway and joint rotation

Lateral deflection

(a)

(b)

FIGURE 5.5

reproduced here as Table 5.1 with the permission of the AISC. Two sets of K values are provided in the table, one being the theoretical values and the other being the recommended design values based on the fact that perfectly pinned and fixed conditions are not possible. If the ends of the column of Fig. 5.4(b) were not quite fixed, the column would be a little freer to bend laterally and its points of inflection would be farther apart. The recommended design K given is 0.65, while the theoretical value is 0.5. As no column ends are perfectly fixed or perfectly hinged, the designer may wish to interpolate between the values given in the table, the interpolation to be based on his or her judgment of the actual restraint conditions.

The values in Table 5.1 are very useful for preliminary designs. When using this table we will almost always apply the design values and not the theoretical values. In fact, the theoretical values should be used only for those very rare situations where fixed ends are really almost perfectly fixed and/or when simple supports are almost perfectly frictionless (*this means almost never*).

You will note in the table that for cases (a), (b), (c), and (e) the design values are greater than the theoretical values, but that is not the situation for cases (d) and (f), where the values are the same. The reason for this in each of these two latter cases is that if the pinned conditions are not perfectly frictionless the K values will become smaller, not larger. Thus, by making the design values the same as the theoretical ones, we are staying on the safe side.

The K values in Table 5.1 are probably very satisfactory to use for designing isolated columns, but for the columns in continuous frames, they are probably only satisfactory for making preliminary or approximate designs. Such columns are restrained at their ends by their connections to various beams, and the beams themselves are connected to other columns and beams at their other ends and thus also restrained. These connections can appreciably affect the K values. As a result, for most situations the values in Table 5.1 are not sufficient for final designs.

TABLE 5.1   Column Effective Lengths

	(a)	(b)	(c)	(d)	(e)	(f)
Buckled shape of column is shown by dashed line						
Theoretical K value	0.5	0.7	1.0	1.0	2.0	2.0
Recommended design value when ideal conditions are approximated	0.65	0.80	1.2	1.0	2.10	2.0

End condition code

⌐ Rotation fixed and translation fixed

Ⴗ Rotation free and translation fixed

⌐ Rotation fixed and translation free

ᵒ Rotation free and translation free

Source: *Load and Resistance Factor Design Specification for Structural Steel Buildings, December 27, 1999* (Chicago: AISC).

For continuous frames, it is necessary to use a more accurate method for computing K values. Usually this is done by using the alignment charts which will be presented in the first section of Chapter 7. There we will find charts for determining K values for columns in frames braced against sidesway and for frames not braced against sidesway. These charts should almost always be used for final column designs.

5.7   STIFFENED AND UNSTIFFENED ELEMENTS

Up to this point in the text, the author has only considered the overall stability of members, and yet it is entirely possible for the thin flanges or webs of a column or beam to buckle locally in compression well before the calculated buckling strength of the whole member is reached. When thin plates are used to carry compressive stresses, they are particularly susceptible to buckling about their weak axes due to the small moments of inertia in those directions.

The LRFD Specification (Section B5) provides limiting values for the width-thickness ratios of the individual parts of compression members and for the parts of beams in their compression regions. The student should be well aware of the lack of stiffness of thin pieces of cardboard or metal or plastic with free edges. If, however, one

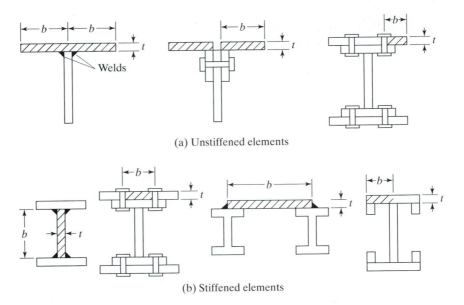

(a) Unstiffened elements

(b) Stiffened elements

FIGURE 5.6

of these elements is folded or restrained, its stiffness is appreciably increased. For this reason, two categories are listed in the LRFD Manual: *stiffened elements* and *unstiffened elements*.

An unstiffened element is a projecting piece with one free edge parallel to the direction of the compression force, while a stiffened element is supported along the two edges in that direction. These two types of elements are illustrated in Fig. 5.6. In each case the width, *b*, and the thickness, *t*, of the elements in question are shown.

Depending on the ranges of different width-thickness ratios for compression elements, and depending on whether the elements are stiffened or unstiffened, the elements will buckle at different stress situations.

For establishing width-thickness ratio limits for the elements of compression members, the LRFD Specification divides members into three classifications as follows: compact sections, noncompact sections, and slender compression elements. These classifications, which decidedly affect the design compression stresses to be used for columns, are discussed in the paragraphs to follow.

5.7.1   Compact Sections

A compact section is one that has a sufficiently stocky profile so that it is capable of developing a fully plastic stress distribution before buckling. The term *plastic* means stressed throughout to the yield stress and is discussed at length in Chapter 8. For a section to be compact, its compression elements must have width-thickness ratios equal to or less than the limiting values given in Table 5.2, which is Table B5.1 of the LRFD specification. (In some areas of the United States where earthquake dangers seem greatest, larger inelastic rotation capacities may be required than those furnished using these LRFD limiting width-thickness ratios.)

5.7.2   Noncompact Sections

A noncompact section is one for which the yield stress can be reached in some but not all of its compression elements before buckling occurs. It is not capable of reaching a fully plastic stress distribution. In Table 5.2 the noncompact sections are those which have width-thickness ratios greater than λ_p but not greater than λ_r.

5.7.3   Slender Compression Elements

A slender element with a cross section that does not satisfy the width-thickness requirements of Table 5.2 can still be used as a column, but the procedure for doing so is very complex. Furthermore, the reduction in design stress is severe. As a result, it is usually more economical to thicken members to take them out of the slender range.

Almost all of the W, M, and S shapes listed in the LRFD Manual are compact for 50 ksi yield stress steels. A few of them are noncompact (and are so indicated in the column and beam tables of the Manual). The values in the tables reflect the reduced design stresses available for noncompact and slender sections.

Should the width-thickness limits for noncompact sections be exceeded, Appendix B5.3 of the LRFD Specification must be consulted. The formulas presented there are rather complex and tedious to apply. Slender compression elements are discussed in detail and a numerical example is presented in Appendix C of this book.

TABLE 5.2 Limiting Width-Thickness Ratios for Compression Elements			
Description of Element	Width Thickness Ratio	Limiting Width-Thickness Ratios	
		λ_p (compact)	λ_r (noncompact)
Unstiffened Elements			
Flanges of I-shaped rolled beams and channels in flexure	b/t	$0.38\sqrt{E/F_y}$ [c]	$0.83\sqrt{E/F_L}$ [e]
Flanges of I-shaped hybrid or welded beams in flexure	b/t	$0.38\sqrt{E/F_{yf}}$	$0.95\sqrt{E/(F_L/k_c)}$ [e], [f]
Flanges projecting from built-up compression members	b/t	NA	$0.64\sqrt{E/(F_y/k_c)}$ [f]
Flanges of I-shaped sections in pure compression, plates projecting from compression elements; outstanding legs of pairs of angles in continuous contact; flanges of channels in pure compression	b/t	NA	$0.56\sqrt{E/F_y}$
Legs of single angle struts; legs of double angle struts with separators; unstiffened elements, i.e., supported along one edge	b/t	NA	$0.45\sqrt{E/F_y}$
Stems of tees	d/t	NA	$0.75\sqrt{E/F_y}$

TABLE 5.2   (cont.) Limiting Width-Thickness Ratios for Compression Elements

	Description of Element	Width Thickness Ratio	Limiting Width-Thickness Ratios	
			λ_p (compact)	λ_r (noncompact)
Stiffened Elements	Flanges of rectangular box and hollow structural sections of uniform thickness subject to bending or compression; flange cover plates and diaphragm plates between lines of fasteners or welds	b/t		
	for uniform compression		$1.12\sqrt{E/F_y}$	$1.40\sqrt{E/F_y}$
	for plastic analysis		$0.939\sqrt{E/F_y}$	-
	Unsupported width of cover plates perforated with a succession of access holes [b]	b/t	NA	$1.86\sqrt{E/F_y}$
	Webs in flexural compression [a]	h/t_w	$3.76\sqrt{E/F_y}$ [c], [g]	$5.70\sqrt{E/F_y}$ [h]
	Webs in combined flexural and axial compression	h/t_w	for $P_u/\phi_b P_y \le 0.125$ [c], [g] $$3.76\sqrt{\frac{E}{F_y}}\left(1 - \frac{2.75 P_u}{\phi_b P_y}\right)$$ for $P_u/\phi_b P_y > 0.125$ [c],[g] $$1.12\sqrt{\frac{E}{F_y}}\left(2.33 - \frac{P_u}{\phi_b P_y}\right)$$ $$\ge 1.49\sqrt{\frac{E}{F_y}}$$	[h] $$5.70\sqrt{\frac{E}{F_y}}\left(1 - 0.74\frac{P_u}{\phi_b P_y}\right)$$
	All other uniformly compressed stiffened elements, i.e., supported along two edges	b/t h/t_w	NA	$1.49\sqrt{E/F_y}$
	Circular hollow sections In axial compression In flexure	D/t	[d] NA $0.07 E/F_y$	$0.11 E/F_y$ $0.31 E/F_y$

[a] For hybrid beams, use the yield strength of the flange F_{yf} instead of F_y.

[b] Assumes net area of plate at widest hole.

[c] Assumes an inelastic rotation capacity of 3 radians. For structures in zones of high seismicity, a greater rotation capacity may be required.

[d] For plastic design use $0.045 E/F_y$.

[e] F_L = smaller of $(F_{yf} - F_r)$ or F_{yw}, ksi (MPa)
    F_r = compressive residual stress in flange
    = 10 ksi (69 MPa) for rolled shapes
    = 16.5 ksi (114 MPa) for welded shapes

[f] $k_c = \dfrac{4}{\sqrt{h/t_w}}$ and $0.35 \le k_c \le 0.763$

[g] For members with unequal flanges, use h_p instead of h when comparing to λ_p.

[h] For members with unequal flanges, see Appendix B5.1.

5.8 LONG, SHORT, AND INTERMEDIATE COLUMNS

A column subject to an axial compression load will shorten in the direction of the load. If the load is increased until the column buckles, the shortening will stop and the column will suddenly bend or deform laterally and may at the same time twist in a direction perpendicular to its longitudinal axis.

The strength of a column and the manner in which it fails are greatly dependent on its effective length. A very short stocky steel column may be loaded until the steel yields and perhaps on into the strain hardening range. As a result it can support about the same load in compression that it can in tension.

As the effective length of a column increases, its buckling stress will decrease. If the effective length exceeds a certain value, the buckling stress will be less than the proportional limit of the steel. Columns in this range are said to fail *elastically*.

As previously shown in Section 5.5, very long steel columns will fail at loads that are proportional to the bending rigidity of the column (EI) and independent of the strength of the steel. For instance, a long column constructed with a 36 ksi yield stress steel will fail at just about the same load as one constructed with a 100 ksi yield stress steel.

Columns are sometimes classed as being long, short, or intermediate. A brief discussion of each of these classifications is presented in the paragraphs to follow.

5.8.1 Long Columns

The Euler formula predicts very well the strength of long columns where the axial buckling stress remains below the proportional limit. Such columns will buckle *elastically*.

5.8.2 Short Columns

For very short columns the failure stress will equal the yield stress and no buckling will occur. (For a column to fall into this class it would have to be so short as to have no practical application. Thus, no further reference is made to them here.)

5.8.3 Intermediate Columns

For intermediate columns some of the fibers will reach the yield stress and some will not. The members will fail by both yielding and buckling and their behavior is said to be *inelastic*. Most columns fall into this range. (For the Euler formula to be applicable for such columns it would have to be modified according to the reduced modulus concept or the tangent modulus concept to account for the presence of residual stresses.)

In Section 5.9, formulas are presented with which the LRFD estimates the strength of columns in these different ranges.

5.9 COLUMN FORMULAS

The LRFD Specification provides one equation (it's the Euler equation) for long columns with elastic buckling and an empirical parabolic equation for short and intermediate columns. With these equations a critical or buckling stress, F_{cr}, is determined for a compression member. Once this stress is computed for a particular compression member it is multiplied by the cross-sectional area of the member to obtain the member's nominal strength. The design strength of the member can then be determined as follows:

$$P_n = A_g F_{cr}$$

$$\phi_c P_n \le \phi_c A_g F_{cr} \quad \text{with } \phi = 0.85$$

(LRFD Equation E2-1)

One LRFD equation for F_{cr} is for inelastic flexural buckling and the other is for elastic flexural buckling. In both equations λ_c in easily remembered form is $\sqrt{F_y/F_e}$, where F_e is the Euler stress $\pi^2 E/(KL/r)^2$. Substituting this value for F_e, we get the form of λ_c given in the LRFD Specification.

$$\lambda_c = \frac{KL}{r\pi}\sqrt{\frac{F_y}{E}}$$

(LRFD Equation E2-4)

Both equations for F_{cr} include the estimated effects of residual stresses and initial out-of-straightness of the members. The inelastic formula that follows is an empirical or test result formula.

$$F_{cr} = (0.658^{\lambda_c^2})\, F_y \text{ for } \lambda_c \le 1.5$$

(LRFD Equation E2-2)

The other equation is for elastic or Euler buckling and is the familiar Euler equation multiplied by 0.877 to estimate the effect of out-of-straightness.

$$F_{cr} = \left(\frac{0.877}{\lambda_c^2}\right) F_y \text{ for } \lambda_c > 1.5$$

(LRFD Equation E2-3)

These equations are represented graphically in Fig. 5.7.

After looking at these equations, the reader might think that their use would be very tedious and time-consuming with a pocket calculator. Such calculations, however, rarely have to be made because the LRFD Manual provides computed $\phi_c F_{cr}$ values for

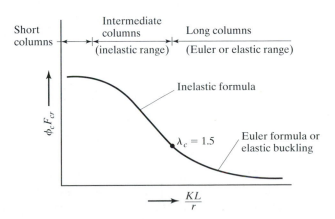

FIGURE 5.7

steels with $F_y = 36$ ksi and 50 ksi for KL/r values from 1 to 200 and has shown the results in Tables 3.36 and 3.50 of the LRFD Specification located in Part 16 of the Manual. Furthermore, there is another table (Table 4 in the LRFD Specification) from which the user may obtain values for steels with any F_y values.

5.10   MAXIMUM SLENDERNESS RATIOS

Section B7 of the LRFD Specification states that compression members *preferably* should be designed with KL/r ratios not exceeding 200. The reader might note from LRFD Tables 3.36 and 3.50 that design stresses $\phi_c F_{cr}$ for KL/r values of 200 are both 5.33 ksi. Should slenderness ratios larger than these be used, the $\phi_c F_{cr}$ values will be very small and will require the user to substitute into the column formulas provided in Section 5.9

5.11   EXAMPLE PROBLEMS

In this section, three simple numerical column problems are presented. In each case, the design strength of a column is calculated. In Example 5-2(a) we determine the strength of a W section. The value of K is calculated as described in Section 5.6, the effective slenderness ratio is computed, and the design stress of the member $\phi_c F_{cr}$ is selected from the appropriate LRFD table and multiplied by the cross-sectional area of the column.

It will be noted that the LRFD Manual in its Part 4 has further simplified the calculations required by computing the column design strength $\phi_c F_{cr} A_g$ for each of the shapes normally used as columns for the commonly used effective lengths or KL values. These were determined with respect to the least radius of gyration for each section, and the F_y values used were the preferred ones given in Table 1.1 of Chapter 1. (It was Table 2.1 in Part 2 of the Manual.)

Example 5-2

 a. Using the column design stress values shown in Table 3.50, Part 16 of the LRFD Manual, determine the design strength ($\phi_c P_n$) of the $F_y = 50$ ksi axially loaded column shown in Fig. 5.8.
 b. Repeat the problem using the column tables of Part 2 of the Manual.
 c. Check local buckling for the section selected using the appropriate values from Table 5.2.

Solution

 a. Using   a   W12 × 72 ($A = 21.1$ in², $r_x = 5.31$ in, $r_y = 3.04$ in, $d = 12.3$ in, $b_f = 12.00$ in, $t_f = 0.670$ in, $k = 1.27$ in, $t_w = 0.430$ in)
 $K = 0.80$ from Table 5.1 in this Chapter.
 Obviously $(KL/r)_y > (KL/r)_x$ and thus controls

$$\left(\frac{KL}{r}\right)_y = \frac{(0.80)(12 \times 15)}{3.04} = 47.37$$

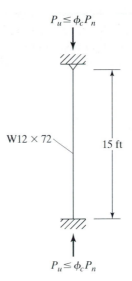

$$P_u \leq \phi_c P_n$$

W12 × 72

15 ft

$$P_u \leq \phi_c P_n$$

FIGURE 5.8

$\phi_c F_{cr} = 36.09$ ksi from Table 3-50, Part 16, LRFD Manual

$$\phi_c P_n = \phi_c F_{cr} A_g = (36.09)(21.1) = 761.5 \text{ k}$$

b. Entering column tables Part 2 of Manual with $K_y L_y$ in feet.

$$K_y L_y = (0.80)(15) = 12 \text{ ft}$$
$$P_u \leq \phi_c P_n = 761 \text{ k}$$

c. Checking W12 × 72 for compactness
For flange

$$\frac{h}{2t_f} = \frac{12.0}{(2)(0.670)} = 8.96 < 0.56\sqrt{\frac{E}{F_y}} = 0.56\sqrt{\frac{29 \times 10^3}{50}} = 13.49 \quad \text{(OK)}$$

For web, noting that $h = d - 2k = 12.3 - (2)(1.27) = 9.76$ in

$$\frac{h}{t_w} = \frac{9.76}{0.430} = 22.70 < 1.49\sqrt{\frac{E}{F_y}} = 1.49\sqrt{\frac{29 \times 10^3}{50}} = 35.88 \quad \text{(OK)}$$

Though width/thickness ratios should be checked for all of the column and beam sections used in design, we have frequently left out these necessary calculations in this text to conserve space. To make this check in most situations, it is only necessary to refer to the Manual column tables where noncompact standard sections are clearly indicated for common shapes. The reader will note that when the enclosed computer program, INSTEP32 is used it shows the width/thickness ratios for each section.

In Example 5-3, the author illustrates the computations necessary to determine the design strength of a built-up column section. Several special requirements for built-up column sections are described in Chapter 6.

Example 5-3

Determine the design strength $\phi_c P_n$ of the axially loaded column shown in Fig. 5.9 if $KL = 19$ ft and 50 ksi steel is used.

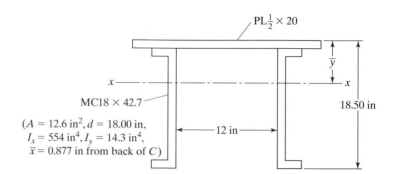

FIGURE 5.9

MC18 × 42.7

$(A = 12.6 \text{ in}^2, d = 18.00 \text{ in},$
$I_x = 554 \text{ in}^4, I_y = 14.3 \text{ in}^4,$
$\bar{x} = 0.877 \text{ in from back of } C)$

PL$\frac{1}{2}$ × 20

12 in

18.50 in

Solution

$$A = (20)\left(\tfrac{1}{2}\right) + (2)(12.6) = 35.2 \text{ in}^2$$

$$\bar{y} \text{ from top} = \frac{(10)(0.25) + (2)(12.6)(9.50)}{35.2} = 6.87 \text{ in}$$

$$I_x = (2)(554) + (2)(12.6)(9.50 - 6.87)^2 + \left(\frac{1}{12}\right)(20)\left(\frac{1}{2}\right)^3 + (10)(6.87 - 0.25)^2$$

$$= 1721 \text{ in}^4$$

$$I_y = (2)(14.3) + (2)(12.6)(6.87)^2 + \left(\tfrac{1}{12}\right)\left(\tfrac{1}{2}\right)(20)^3 = 1554 \text{ in}^4$$

$$r_x = \sqrt{\frac{1721}{35.2}} = 6.99 \text{ in}$$

$$r_y = \sqrt{\frac{1554}{35.2}} = 6.64 \text{ in}$$

$$\left(\frac{KL}{r}\right)_x = \frac{(12)(19)}{6.99} = 32.62$$

$$\left(\frac{KL}{r}\right)_y = \frac{(12)(19)}{6.64} = 34.34 \leftarrow$$

$$\phi_c F_{cr} = 39.03$$

$$\phi_c P_n = (39.03)(35.2) = 1374 \text{ k}$$

To determine the design compression stress to be used for a particular column, it is theoretically necessary to compute both $(KL/r)_x$ and $(KL/r)_y$. The reader will notice, however, that for most of the steel sections used for columns, r_y will be much less than r_x. As a result, only $(KL/r)_y$ is calculated for most columns and used in the applicable column formulas.

For some columns, particularly the long ones, bracing is supplied perpendicular to the weak axis, thus reducing the slenderness or the length free to buckle in that direction. This may be accomplished by framing braces or beams into the sides of a column. For instance, horizontal members called *girts* running parallel to the exterior walls of a building frame may be framed into the sides of columns. The result is stronger columns and ones for which the designer needs to calculate both $(KL/r)_x$ and $(KL/r)_y$. The larger ratio obtained for a particular column indicates the weaker direction and will be used for calculating the design stress $\phi_c F_{cr}$ for that member.

Bracing members must be capable of providing the necessary lateral forces without buckling themselves. The forces to be taken are quite small and are often conservatively estimated to equal 0.02 times the column design loads. These members can be selected as are other compression members. A bracing member must be connected to other members that can transfer the horizontal force by shear to the next restrained level. If this is not done, little lateral support will be provided for the original column in question.

If the lateral bracing were to consist of a single bar or rod ( ) it would not prevent twisting and torsional buckling of the column (see Chapter 6). As torsional buckling is a difficult problem to handle, we should provide lateral bracing that prevents lateral movement and twist.[6]

Steel columns may also be built into substantial masonry walls in such a manner that they are substantially supported in the weaker direction. The designer, however, should be quite careful in assuming complete lateral support parallel to the wall, because a poorly built wall will not provide 100 percent lateral support.

Example 5-4 illustrates the calculations necessary to determine the design strength of a column with two unbraced lengths.

Example 5-4

    a. Using Table 3.50 of Part 16 of the LRFD Manual, determine the design strength $\phi_c P_n$ of the 50 ksi axially loaded W14 × 90 shown in Fig. 5.10. Because of its considerable length this column is braced perpendicular to its weak or y axis at the points shown in the figure. These connections are assumed to permit rotation of the member in a plane parallel to the plane of the flanges. At the same time, however, they are assumed to prevent translation or sidesway and twisting of the cross section about a longitudinal axis passing through the shear center of the cross section. (The shear center is the point in the cross section of a member through which the resultant of the transverse loads must pass so no torsion will occur. See Chapter 10.)

    b. Repeat part (a) using the column tables of Part 4 of the Manual.

---

[6]J. A. Yura, "Elements for Teaching Load and Resistance Factor Design" (New York: AISC, August 1987), p. 20.

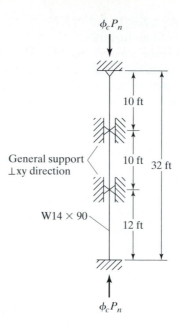

FIGURE 5.10

Solution

a. Using W14 $\times$ 90 (A = 26.5 in^2, r_x = 6.14 in, r_y = 3.70 in)
Determining effective lengths

$$K_x L_x = (0.80)(32) = 25.6 \text{ ft}$$
$$K_y L_y = (1.0)(10) = 10 \text{ ft} \leftarrow \text{governs for } K_y L_y$$
$$K_y L_y = (0.80)(12) = 9.6 \text{ ft}$$

Computing slenderness ratios

$$\left(\frac{KL}{r}\right)_x = \frac{(12)(25.6)}{6.14} = 50.03 \leftarrow$$
$$\left(\frac{KL}{r}\right)_y = \frac{(12)(10)}{3.70} = 32.43$$
$$\phi_c F_{cr} = 35.39 \text{ ksi}$$
$$\phi_c P_n = (35.39)(26.5) = 938 \text{ k}$$

b. Noting from part (a) solution that there are two different KL values

$$K_x L_x = 25.6 \text{ ft}$$
$$K_y L_y = 10 \text{ ft}$$

We would like to know which of these two values is going to control. This can easily be done by determining a value of K_xL_x which is equivalent to K_yL_y. The slenderness ratio in the x direction is equated to an equivalent value in the y direction as follows:

$$\frac{K_xL_x}{r_x} = \text{Equivalent } \frac{K_yL_y}{r_y}$$

$$\text{Equivalent } K_yL_y = r_y\frac{K_xL_x}{r_x} = \frac{K_xL_x}{\dfrac{r_x}{r_y}}$$

Thus, the controlling K_yL_y for use in the tables is the larger of the real $K_yL_y = 10$ ft, or the equivalent K_yL_y.

$$\frac{r_x}{r_y} \text{ for W14} \times 90 \text{ from bottom of column tables} = 1.66$$

$$\text{Equivalent } K_yL_y = \frac{25.6}{1.66} = 15.42 \text{ ft} > K_yL_y \text{ of } 10 \text{ ft}$$

From column tables with $K_yL_y = 15.42$ ft we find by interpolation that $\phi_cP_n = 938$ k.

5.12   COMPUTER EXAMPLE

The axial design compression strength of a W section is determined in Example 5-5 with the computer program INSTEP32. (This program will not handle shapes with slender elements. See Appendix C.)

Example 5-5

Using the enclosed diskette determine the design compression strength of a W14 × 132 which consists of A992 steel. $K_x = K_y = 1.0$ and $L_x = L_y = 20$ ft.

Solution. The user clicks the symbol for axially loaded columns $\boxed{\downarrow}$, after which the design load P is specified equal to zero and the known data are inputted (F_y, K_s, L_s). After clicking "OK" or Calculate, the design capacity of the column will be furnished.

Input:

Factored design loads   Unbraced length parameters
$P_u = 0$ kips   Strong axis
                                            Lx = 240 in
                                            Kx = 1
                                         Weak axis
                                            Ly = 240 in
                                            Ky = 1

Output:

Axial Compression Design Summary
Section: W14 × 132

b/t = 7.15 [Limit 13.4866]

h/t = 17.7 [Limit 35.884]

$[KL/r]_x = 1[240]/6.28 = 38.217$

$[KL/r]_y = 1[240]/3.76 = 63.83$

$L_c = [63.8298/Pi]sqrt[50/29000] = 0.843645$

$F_{cr} = 50[0.658^[0.843645*0.843645]] = 37.1189$

$P_n = 38.8[37.1189] = 1440.21$ kips

$0.85P_n = 0.85[1440.21] = 1224.18$ kips

PROBLEMS

5-1 to 5-4.   *Determine the critical buckling load for each of the columns using the Euler equation.*
$E = 29 × 10^6$ *psi. Proportional limit* = 36,000 *psi. Assume simple ends and maximum permissible L/r = 200.*

5-1.   A solid square bar 1.0 in × 1.0 in

    a. L = 4 ft 0 in (*Ans.* 10.38 k)
    b. L = 5 ft 6 in (*Ans.* Euler equation not applicable)

5-2.   The pipe section shown.

    a. L = 30 ft 0 in
    b. L = 20 ft 0 in
    c. L = 15 ft 0 in

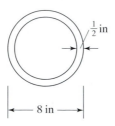

FIGURE P5-2

5-3.   A W12 × 87. L = 30 ft 0 in (*Ans.* 533 k)

5-4.   The two C12 × 30s shown for $L = 40$ft 0 in

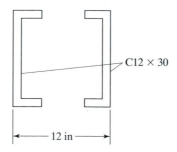

C12 × 30

12 in

FIGURE P5-4

5-5 to 5-8.   Determine the design strength of each of the compression members shown using the LRFD Specification and a steel with $F_y = 50$ ksi except for Prob. 5-8, where it is 46 ksi.

5-5.

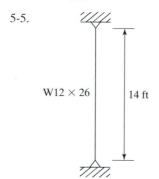

W12 × 26   14 ft

FIGURE P5-5 (*Ans.* 131.7 k)

5-6.

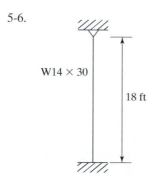

W14 × 30

18 ft

FIGURE P5-6

5-7.

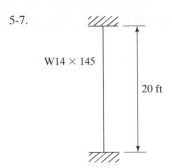

W14 × 145

20 ft

FIGURE P5-7 (*Ans.* 1621 k)

5-8.   HSS
$12 \times 10 \times \frac{1}{2}$

$F_y = 46$ ksi

16 ft

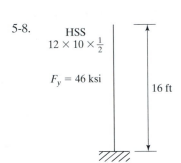

FIGURE P5-8

5-9.   Using 50 ksi steel and the LRFD Specification determine the design strength $\phi_c P_n$ for each of the compression members.

a. A W14 × 53 with KL = 15 ft (*Ans.* 349 k)
b. A W12 × 106 with KL = 20 ft (*Ans.* 858 k)
c. An S12 × 50 with KL = 14 ft (*Ans.* 117.1 k)

5-10.   Using the LRFD Specification determine the design strength $\phi_c P_n$ for each of the following columns. F_y = 50 ksi except for parts (d) and (e).

a. A W14 × 53 with fixed ends, L = 15 ft 6 in
b. A W12 × 45 with pinned ends, L = 14 ft 0 in
c. A W10 × 33 with one end fixed and the other pinned, L = 16 ft 6 in
d. A W14 × 109 with fixed ends, L = 24 ft 0 in, F_y = 65 ksi
e. A pipe 12 × strong with pinned ends, L = 18 ft 0 in, F_y = 35 ksi

5-11.   A W12 × 79 with a 3/4 × 16-in cover plate bolted to each flange is to be used for a column with KL = 20 ft. Find its design strength $\phi_c P_n$ if F_y = 50 ksi (*Ans.* P_u = 1528 k).

5-12.   Using F_y = 50 ksi, determine the design strength $\phi_c P_n$ of the axially loaded compression members shown.

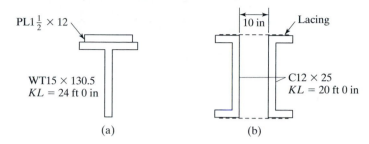

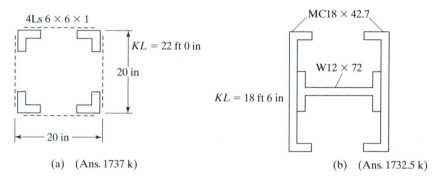

(a)   (b)

FIGURE P5-12

5-13.   Using F_y = 50 ksi determine the design strength $\phi_c P_n$ of the axially loaded compression members shown.

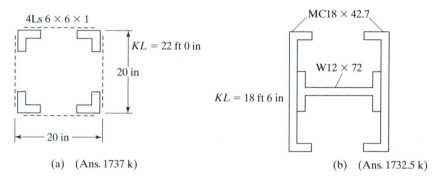

(a)   (Ans. 1737 k)   (b)   (Ans. 1732.5 k)

FIGURE P5-13

5-14.   Using a steel with F_y = 50 ksi, calculate the design strength $\phi_c P_n$ of the concentrically loaded columns shown.

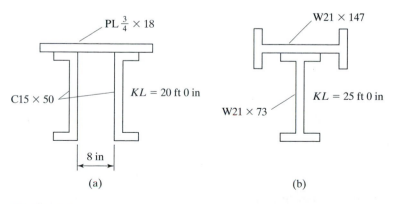

(a)   (b)

FIGURE P5-14

5-15.   Using $F_y = 50$ ksi, determine $\phi_c P_n$ for the axially loaded columns shown.

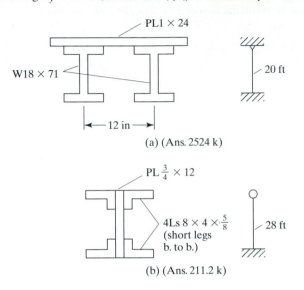

(a) (Ans. 2524 k)

(b) (Ans. 211.2 k)

FIGURE P5-15

5-16.   Determine the design strength $\phi_c P_n$ for each of the axially loaded column shown. $F_y = 50$ ksi.

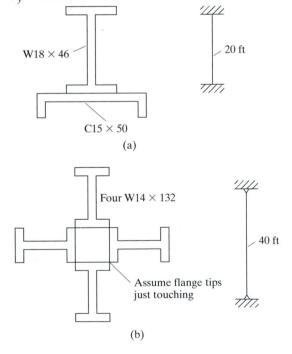

FIGURE P5-16

5-17.   A 24-ft axially loaded W 12 × 87 column is laterally supported perpendicular to its *y* axis at middepth. Determine $\phi_c P_n$ for this column if $F_y = 50$ ksi and if effective length factors = 1.0 are assumed in each case. (*Ans.* 881.6 k)

5-18.   Determine the maximum service live load that the column shown can support if the load is 1/3 dead load and 2/3 live load. $K_x L_x = 24$ ft and $K_y L_y = 12$ ft. $F_y = 50$ ksi.

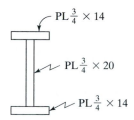

PL$\frac{3}{4}$ × 14

PL$\frac{3}{4}$ × 20

PL$\frac{3}{4}$ × 14

FIGURE P5-18

5-19.   Compute the maximum total service live load that can be applied to the A572, Grade 50 section shown if $K_x L_x = 15$ ft and $K_y L_y = 10$ ft. Assume the service load is 30 percent dead and 70 percent live. (*Ans.* $P_L = 287.2$ k)

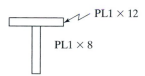

PL1 × 12

PL1 × 8

FIGURE P5-19

5–20 to 5–22.   *Using the computer diskette enclosed with this book, repeat the following problems.*

5-20.   Prob. 5-9

5-21.   Parts (a) to (d) of Prob. 5–10 (*Ans.* 496 k, 324 k, 253 k and 1392 k)

5-22.   Prob. 5-17

Design of Axially Loaded Compression Members

6.1 INTRODUCTION

In this chapter, the designs of several axially loaded columns are presented. Included are the selections of single shapes, W sections with cover plates, and built-up sections constructed with channels. Also included are the designs of sections whose unbraced lengths in the x and y directions are different, as well as the sizing of lacing and tie plates for built-up sections with open sides. Another topic considered is the flexural-torsional buckling of sections.

The design of columns with formulas involves a trial-and-error process. The design stress $\phi_c F_{cr}$ is not known until a column size is selected and vice versa. Once a trial section is assumed, the r values for that section can be obtained and substituted into the appropriate column equation to determine its design stress. Examples 6-1, 6-3, and 6-4 illustrate this procedure.

The designer may assume a design stress, divide that stress into the factored column load to give an estimated column area, select a column section with approximately that area, determine its design stress, multiply that stress by the cross-sectional area of the section to obtain the member's design strength to see if the section selected is over- or underdesigned, and if appreciably so, try another size. The student may feel that he or she does not have sufficient background or knowledge to make reasonable initial design stress assumptions. If this student will read the information contained in the next few paragraphs, however, he or she will immediately be able to make excellent estimates.

The effective slenderness ratio (KL/r) for the average column of 10- to 15-ft length will generally fall between about 40 and 60. If, for a particular column a KL/r somewhere in this approximate range is assumed and substituted into the appropriate column equation (in the LRFD this usually means looking into the tables where the design stresses have already been calculated for KL/r values from 0 to 200), the result will usually be a very satisfactory design stress estimate.

Georgia Railroad Bank and Trust Company Building, Atlanta, GA. (Courtesy of Bethlehem Steel Corporation.)

In Example 6-1, a column with $KL = 10$ ft is selected using the LRFD formulas. An effective slenderness ratio of 50 is assumed, the design stress for that value is determined from Table 3.50 of Part 16 of the Manual, and the resulting stress is divided into the factored column load to obtain an estimated column area. After a trial section is selected with approximately that area, its actual slenderness ratio and design strength are determined. The first estimated size in Example 6-1, though quite close, is a little too small, but the next larger section in that series of shapes is satisfactory.

To estimate the effective slenderness ratio for a particular column, the designer may estimate a value a little higher than 40 to 60 if the column is appreciably longer than the 10- to 15-ft range and vice versa. A very heavy factored column load—say in the 750- or 1000-k range or higher—will require a rather large column for which the radii of gyration will be larger and the designer may estimate a little smaller value of KL/r. For lightly loaded bracing members, he or she may estimate high slenderness ratios of 100 or more.

Example 6-1

Using $F_y = 50$ ksi, select the lightest W14 available for the service column loads $P_D = 130$ k and $P_L = 210$ k. $KL = 10$ ft.

Solution

$$P_u = (1.2)(130) + (1.6)(210) = 492 \text{ k}$$

Assume $\dfrac{KL}{r} = 50$

$\phi_c F_{cr}$ from Table 3.50 (Part 16 of Manual) $= 35.4$ ksi

$$A \text{ required} = \frac{492}{35.4} = 13.90 \text{ in}^2$$

Try W14 × 48($A = 14.1$ in^2, $r_x = 5.85$ in, $r_y = 1.91$ in)

Obviously $(KL/r)_y > (KL/r)_x$ and $\therefore$ controls

$$(KL/r)_y = \frac{(12)(10)}{1.91} = 62.83$$

$\phi_c F_{cr}$ from Manual Table 3.50 $= 31.85$ ksi

$$\phi_c P_n = (31.85)(14.1) = 449 \text{ k} \ < \ 492 \text{ k} \tag{NG}$$

$\therefore$ try next larger W14.

Try W14 × 53($A = 15.6$ in^2, $r_x = 5.89$ in, $r_y = 1.92$ in)

$$(KL/r)_y = \frac{(12)(10)}{1.92} = 62.5$$

$\phi_c F_{cr} = 31.95$ ksi

$$\phi_c P_n = (31.95)(15.6) = 498 \text{ k} \ > \ 492 \text{ k} \tag{OK}$$

Checking width/thickness ratios for W14 × 53($b_f = 8.060$ in,

$t_f = 0.660$ in, $k = 1.25$ in, $d = 13.9$ in, $t_w = 0.370$ in)

$$\frac{b_f}{2t_f} = \frac{8.060}{(2)(0.660)} = 6.11 < 0.56\sqrt{\frac{E}{F_y}} = 0.56\sqrt{\frac{29 \times 10^3}{50}} = 13.49 \tag{OK}$$

$$\frac{h}{t_w} = \frac{13.9 - (2)(1.25)}{0.370} = 30.81 < 1.49\sqrt{\frac{E}{F_y}} = 1.49\sqrt{\frac{29 \times 10^3}{50}} = 35.88 \text{ (OK)}$$

Use W14 × 53, $F_y = 50$ ksi.

6.2 LRFD DESIGN TABLES

For Example 6-2, Part 4 of the LRFD Manual is used to select various column sections from tables without the necessity of using a trial-and-error process. These tables provide axial design strengths $(\phi_c P_n)$ for various practical effective lengths of the steel sections commonly used as columns. The values are given with respect to the least radii of gyration for Ws and WTs with 50 ksi steel. Other grade steels are commonly used for other types of sections as shown in the Manual and listed here. These include 35 ksi for steel pipe, 36 ksi for Ls, 42 ksi for round HSS sections, and 46 ksi for square and rectangular HSS sections.

For most columns consisting of single steel shapes, the effective slenderness ratio with respect to the y axis $(KL/r)_y$ is larger than the effective slenderness ratio with respect to the x axis $(KL/r)_x$. As a result, the controlling or smaller design stress is for the y axis. Because of this, the LRFD tables provide design strengths of columns with respect to their y axes. We will learn in the pages to follow how to handle situations in which $(KL/r)_x$ is larger than $(KL/r)_y$.

The resulting tables are very simple to use. The designer takes the KL value for the minor principal axis in feet, enters the table in question from the left-hand side, and moves horizontally across the table. Under each section is listed the design strength $\phi_c P_n$ for that KL and the steel yield stress. As an illustration it is assumed that we have a factored design load $P_u = 1200$ k, $K_y L_y = 12$ ft, and we want to select the lightest available W14 section using 50 ksi steel. We enter the tables with $KL = 12$ ft in the left column and read from left to right for 50 ksi steel the numbers 9440, 8530, 7760, 7030 k, and so on until several pages later where the consecutive values 1220 and 1110 k are found. The 1110 k value is not sufficient, and we go back to the 1220 k which falls under the W14 × 109. A similar procedure can be followed for the other available shapes.

Example 6-2 illustrates the selection of various possible sections to be used for a particular column. Among the sections chosen are the steel pipe sections shown in Table 4.8 of the Manual. It is possible to support a given load with a standard pipe (labeled S in the table); with an extra strong pipe (XS), which has a smaller diameter but thicker walls and thus is heavier and more expensive; or with a double extra strong pipe (XSS), which has an even smaller diameter and even thicker walls and greater weight.

Example 6-2

Using the LRFD column tables with their given yield strengths

    a. Select the lightest W section available for the loads, steel, and KL of Example 6-1. $F_y = 50$ ksi.

b. Select the lightest satisfactory standard (S), extra strong (XS), and double extra strong (XXS) pipe columns for the situation described in part (a) of this problem. $F_y = 35$ ksi.

c. Select the lightest satisfactory rectangular and square HSS sections for the situation in part (a). $F_y = 46$ ksi.

d. Select the lightest round HSS section for part (a). $F_y = 42$ ksi.

Solution

a. Enter tables with $K_y L_y = 10$ ft, $P_u = 492$ k, and $F_y = 50$ ksi
   Lightest suitable section in each W series

$$W14 \times 53(\phi_c P_n = 498 \text{ k})$$
$$W12 \times 53(\phi_c P_n = 559 \text{ k})$$
$$W10 \times 49(\phi_c P_n = 520 \text{ k}) \leftarrow$$
   Use W10 × 49

b. Pipe Columns

   S none available
   XS12 × 0.500(65.5 lb/ft) = 549 k
   XXS8 × 0.875(72.5 lb/ft) = 575 k

c. Rectangular and square HSS sections

$$\text{HSS14} \times 14 \times \tfrac{5}{16}(57.3 \text{ lb/ft}) = 530 \text{ k}$$
$$\text{HSS12} \times 10 \times \tfrac{3}{8}(52.9 \text{ lb/ft}) = 537 \text{ k}$$

d. Round HSS section

   HSS16 × 0.312(52.3 lb/ft) = 500 k

An axially loaded column is laterally restrained in its weak direction in Fig. 6.1. Example 6-3 illustrates the design of such a column with its different unsupported lengths in the x and y directions. The student can easily solve this problem by trial and error. A trial section can be selected as described in Section 6.1 of this chapter, the slenderness values $(KL/r)_x$ and $(KL/r)_y$ computed, $\phi_c F_{cr}$ determined for the larger value and multiplied by A_g to obtain $\phi_c P_n$. Then, if necessary, another size can be tried, and so on.

For this discussion, it is assumed that K is the same in both directions. Then, if we are to have equal strengths about the x and y axes, the following relation must hold:

$$\frac{L_x}{r_x} = \frac{L_y}{r_y}$$

For L_y to be equivalent to L_x we would have

$$L_x = L_y \frac{r_x}{r_y}$$

If $L_y(r_x/r_y)$ is less than L_x, then L_x controls; if greater than L_x, then L_y controls.

Eversharp, Inc., Building at Milford, CN. (Courtesy of Bethlehem Steel Corporation.)

Based on the preceding information, the LRFD Manual provides a method with which a section can be selected from its tables with little trial and error when the unbraced lengths are different. The designer enters the appropriate table with K_yL_y, selects a shape, takes the r_x/r_y value given in the table for that shape, and multiplies it by L_y. If the result is larger than K_xL_x, then K_yL_y controls and the shape initially selected is the correct one. If the result of the multiplication is less than K_xL_x, then K_xL_x controls and the designer will reenter the tables with a larger K_yL_y equal to $K_xL_x/(r_x/r_y)$ and select the final section.

Example 6-3 illustrates the two procedures described here for selecting a W section which has different effective lengths in the x and y directions.

Example 6-3

Select the lightest satisfactory W12 for the following conditions: $F_y = 50$ ksi, $P_u = 900$ k, $K_xL_x = 26$ ft, and $K_yL_y = 13$ ft.

① check for compactness $\lambda < \lambda_p$

$F_y = 50$ ksi $E = 29000$ ksi $b_f = 9.96$ $k = 1.24$ $d = 26.7$ in $t_f = .640$ in $t_w = .460$ in

Flange: $\dfrac{b_f}{2t_f} = 7.781 < .38\sqrt{\dfrac{E}{F_y}} = 9.152$ true, so ok

Web: $\dfrac{h}{t_w} = \dfrac{d-2k}{t_w} = 52.652 < 3.76\sqrt{\dfrac{E}{F_y}} = 90.553$ true, so ok

② check deflection limit due to service dead + live

$W_{DL} = 2^{k}/ft + .084 \, k/ft = 2.084 \, k/ft$

$P_{LL} = 40$ kip

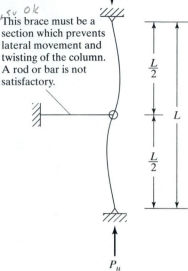

This brace must be a section which prevents lateral movement and twisting of the column. A rod or bar is not satisfactory.

FIGURE 6.1

A column laterally restrained at middepth in its weaker direction.

  a. By trial and error
  b. Using LRFD tables

Solution

  a. Using trial and error

$$\text{Assume } \frac{KL}{r} = 50$$

$$\phi_c F_{cr} = 35.40 \text{ ksi}$$

$$A \text{ required} = \frac{900}{35.40} = 25.42 \text{ in}^2$$

Try W12 × 87($A = 25.6$ in², $r_x = 5.38$ in, $r_y = 3.07$ in)

$$\left(\frac{KL}{r}\right)_x = \frac{(12)(26)}{5.38} = 57.99 \leftarrow$$

$$\left(\frac{KL}{r}\right)_y = \frac{(12)(13)}{3.07} = 50.81$$

$$\phi_c F_{cr} = 33.2 \text{ ksi}$$

$$\phi_c P_n = (33.2)(25.6) = 850 \text{ k} < 900 \text{ k} \tag{NG}$$

A subsequent check of the next larger W section (W12 × 96) shows it will work.
  Use W12 × 96

b. Using LRFD tables
Enter tables with $K_yL_y = 13$ ft, $F_y = 50$ ksi, and $P_u = 900$ k

Try W12 $\times$ 87$\left(\dfrac{r_x}{r_y} = 1.75\right)$ with $\phi_c P_n$ based on K_yL_y

Equivalent $K_yL_y = \dfrac{K_xL_x}{\dfrac{r_x}{r_y}} = (13)(1.75) = 22.75$ ft $< K_xL_x$

Therefore, K_xL_x controls.

Reenter tables with new $K_yL_y = \dfrac{K_xL_x}{r_x/r_y} = \dfrac{26}{1.75} = 14.86$

Use W12 $\times$ 96, $F_y = 50$ Ksi

6.3   COLUMN SPLICES

For multistory buildings, column splices are desirably placed 4 feet above finished floors to permit the attachment of safety cables to the columns as may be required at floor edges or openings. This offset also enables us to keep the splices from interfering with the beam and column connections.

Typical column splices are shown in Fig. 6.2. The column ends are usually milled so they can be placed firmly in contact with each other for purposes of load transfer. When the contact surfaces are milled, a large part of the axial compression (if not all) can be transferred through the contacting areas. It is obvious that splice plates are necessary even though full contact is made between the columns and only axial loads are involved. For instance, the two column sections need to be held together during erection and afterward. What is necessary to hold them together is primarily based on the experience and judgment of the designer. Splice plates are even more necessary when consideration is given to the shears and moments existing in practical columns subjected to off-center loads, lateral forces, moments, and so on.

There is obviously a great deal of difference between tension splices and compression splices. In tension splices, all load has to be transferred through the splice, whereas in splices for compression members, a large part of the load can be transferred directly in bearing between the columns. The splice material is then needed to transfer only the remaining part of the load.

The amount of load to be carried by the splice plates is difficult to estimate. Should the column ends not be milled, the plates should be designed to carry 100 percent of the load. When the surfaces are milled and axial loads only are involved, the amount of load to be carried by the plates might be estimated to be from 25 to 50 percent of the total load. If bending is involved, perhaps 50 to 75 percent of the total load may have to be carried by the splice material.

The bridge specifications spell out very carefully splice requirements for compression members, but the LRFD Specification does not. A few general requirements are given in Section J1.4 of the LRFD Specification. Such splices and a detailed discussion is presented in Section 14 of the Manual.

Design Strength:

Problem 1) The compression member shown in Figure 1 has lateral support at 13 ft from the bottom in the weak direction only. Find the design strength assuming F_y = 50 ksi.

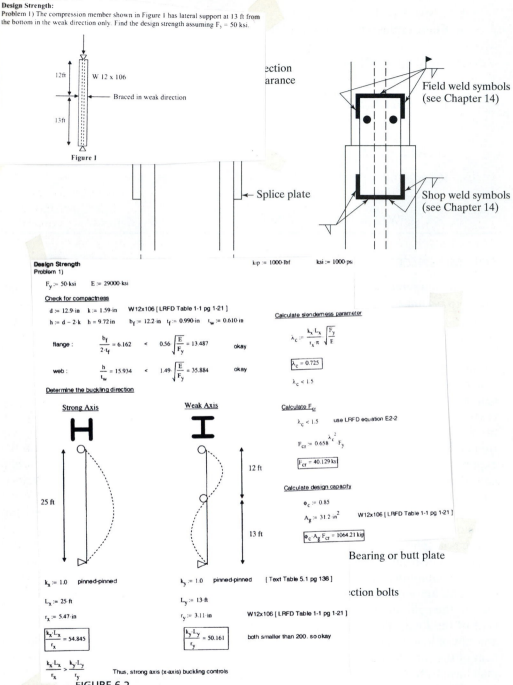

FIGURE 6.2

Column splices. (a) Columns from same W series with total depths close to each other ($d_{lower} < 2$ in greater than d_{upper}). (b) Columns from different W series.

Figure 6.2(a) shows a splice that may be used for columns with substantially the same nominal depths. The student will notice in the LRFD Manual that the W shapes of a given nominal size are divided into groups that are rolled with the same set of rolls. Because of the fixed dimensions of each set of rolls, the clear distances between the flanges are constant for each shape in that group, although their total depths may vary greatly. For instance, the inside distance for each of the 29 shapes (running from the W14 × 61 to the W14 × 808) is approximately 12.60 in, although their total depths vary from 13.9 in to 22.8 in.

It is most economical to use the simple splices of Figure 6.2(a). This can easily be accomplished by using one series of shapes for as many stories of a building as possible. For instance, we may select a particular W14 column section for the top story or top two stories of a building and then keep selecting heavier and heavier W14s for that column as we come down in the building. We may also switch to higher strength steel columns as we move down in the building, thus enabling us to stay with the same W series for even more floors. It will be necessary to use filler plates between the splice plates and the upper column if the upper column has a total depth significantly less than that of the lower column.

Figure 6.2(b) shows a type of splice that can be used for columns of equal or different nominal depths. For this type of splice, the butt plate is shop-welded to the lower column, and the clip angles used for field erection are shop-welded to the upper column. In the field, the erection bolts shown are installed, and the upper column is field-welded to the butt plate. The horizontal welds on this plate resist shears and moments in the columns.

Welded column splice for Colorado State Building, Denver, CO. (Courtesy of Lincoln Electric Company.)

Sometimes splices are used on all four sides of columns. The web splices are bolted in place in the field and field-welded to the column webs. The flange splices are shop-welded to the lower column and field-welded to the top column. The web plates may be referred to as *shear plates* and the flange plates as *moment plates*.

For multistory buildings, the columns may be fabricated for one or more stories. Theoretically, column sizes can be changed at each floor level so that the lightest total column weight is used. The splices needed at each floor will be quite expensive, however, and as a result it is usually more economical to use the same column sizes for at least two stories, even though the total steel weight will be higher. Seldom are the same sizes used for as many as three stories because three-story columns are so difficult to erect. The two-story heights work out very well most of the time.

6.4   BUILT-UP COLUMNS

As previously described in Section 5.3, compression members may be constructed with two or more shapes built-up into a single member. They may consist of parts in contact with each other, such as cover-plated sections ⊤ , or they may consist of parts in near contact with each other, such as pairs of angles ⅂⌐ that may be separated by a small distance from each other equal to the thickness of the end connections or gusset plates between them. They may consist of parts that are spread well apart, such as pairs of channels ⌷ or four angles ⌐ ⌐, and so on.

Two-angle sections probably are the most common type of built-up member. (For example, they frequently are used as the members of light trusses.) When a pair of angles are used as a compression member, they need to be fastened together so they will act as a unit. Welds may be used at intervals (with a spacer bar between the parts if the angles are separated) or they may be connected with "stitch bolts." When the connections are bolted, washers or "ring fills" are placed between the parts to keep them at the proper spacing if the angles are to be separated.

For long columns, it may be convenient to use built-up sections where the parts of the columns are spread out or widely separated from each other. Before heavy W sections were made available, such sections were very commonly used in both buildings and bridges. Today these types of built-up columns are commonly used for crane booms and for the compression members of various kinds of towers. The widely spaced parts of these types of built-up members must be carefully laced or tied together.

Sections 6.5 and 6.6 of this chapter concern compression members that are built up from parts in direct contact (or nearly so) with each other. Section 6.7 addresses built-up compression members whose parts are spread widely apart.

6.5   BUILT-UP COLUMNS WITH COMPONENTS IN CONTACT WITH EACH OTHER

Should a column consist of two equal size plates, as shown in Fig. 6.3, and should those plates not be connected together, each plate will act as a separate column, and each will resist approximately half of the total column load. In other words, the total moment of

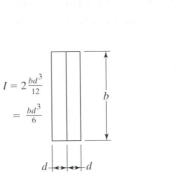

$$I = 2\frac{bd^3}{12}$$

$$= \frac{bd^3}{6}$$

(a) Column cross section

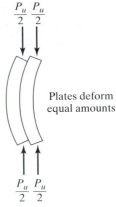

Plates deform equal amounts

(b) Deformed shape of column

FIGURE 6.3

Column consisting of two plates not connected to each other.

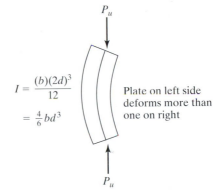

$$I = \frac{(b)(2d)^3}{12}$$

$$= \frac{4}{6}bd^3$$

Plate on left side deforms more than one on right

FIGURE 6.4

Column consisting of two plates fully connected to each other.

inertia of the column will equal two times the moment of inertia of one plate. The two "columns" will act the same and have equal deformations, as shown in part (b) of the figure.

Should the two plates be connected together sufficiently to prevent slippage on each other, as shown in Fig. 6.4, they will act as a unit. Their moment of inertia may be computed for the whole built-up section as shown in the figure and will be four times as large as it was for the column of Fig. 6.3, where slipping between the plates was possible. The reader should also notice that the plates of the column of Fig. 6.4 will deform different amounts as the column bends laterally.

Should the plates be connected in a few places, it would appear that the strength of the resulting column would be somewhere in between the two cases just described.

Reference to Fig. 6.3(b) shows that the greatest displacement between the two plates tends to occur at the ends and the least displacement tends to occur at middepth. As a result, connections placed at column ends which will prevent slipping between the parts have the greatest strengthening effect, while those placed at middepth have the least effect.

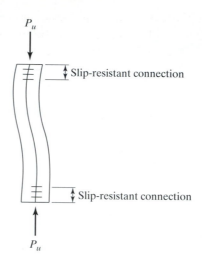

FIGURE 6.5

Column consisting of two plates
connected at its ends only.

Should the plates be fastened together at their ends with slip-resistant connectors, those ends will deform together and the column will take the shape shown in Fig. 6.5. As the plates are held together at the ends, the column will bend in an S shape as shown in the figure.

If the column were to bend in the S shape shown, its K factor would theoretically equal 0.5 and its KL/r value would be the same as the one for the continuously connected column of Fig. 6.4.[1]

$$\frac{KL}{r} \text{ for the column of Fig. 6.4} = \frac{(1)(L)}{\sqrt{\frac{4}{6}bd^3/2bd}} = 1.732L$$

$$\frac{KL}{r} \text{ for the end-fastened column of Fig. 6.5} = \frac{(0.5)(L)}{\sqrt{\frac{1}{6}bd^3/2bd}} = 1.732L$$

Thus, the design stresses are equal for the two cases and the columns would carry the same loads. This is true for the particular case described here, but is not applicable for the common case where the parts of Fig. 6.5 begin to separate.

6.6 CONNECTION REQUIREMENTS FOR BUILT-UP COLUMNS WHOSE COMPONENTS ARE IN CONTACT WITH EACH OTHER

Several requirements concerning built-up columns are presented in LRFD Specification E4. When such columns consist of different components that are in contact with each other and that are bearing on base plates or milled surfaces, they must be connected at their ends with bolts or welds. If welds are used, the weld lengths must at least equal the maximum width of the member. If bolts are used, they may not be spaced

---

[1] J. A. Yura, "Elements for Teaching Load and Resistance Factor Design" (Chicago: AISC, July 1987), pp. 17–19.

longitudinally more than four diameters on center, and the connection must extend for a distance at least equal to $1\frac{1}{2}$ times the maximum width of the member.

The LRFD Specification also requires the use of welded or bolted connections between the end ones described in the last paragraph. These must be sufficient to provide for the transfer of calculated stresses. If it is desired to have a close fit over the entire faying surfaces between the components, it may be necessary to place the connectors even closer than is required for shear transfer.

When the component of a built-up column consists of an outside plate, the LRFD Specification provides specific maximum spacings for fastening. If intermittent welds are used along the edges of the components or if bolts are provided along all gage lines at each section, their maximum spacing may not be greater than the thickness of the thinner outside plate times $0.75\sqrt{E/F_y}$, nor 12 in. Should the fasteners be staggered, the maximum spacing along each gage line shall not be greater than the thickness of the thinner outside plate times $1.12\sqrt{E/F_y}$, nor 18 in (LRFD Specification Section E4.2).

In Chapter 12, high-strength bolts are referred to as being *snug-tight* or *slip-critical*. Snug-tight bolts are those that are tightened until all the plies of a connection are in firm contact with each other. This usually means the tightness obtained by the full manual effort of a worker with a spud wrench, or the tightness obtained after a few impacts with a pneumatic wrench.

Slip-critical bolts are tightened much more than are snug-tight bolts. They are tightened until their bodies or shanks have very high tensile stresses (approaching the lower bound of their yield stress). Such bolts clamp the connected parts of a connection so tightly together between the bolt and nut heads that loads are resisted by friction and slippage is nil. (We will see in Chapter 12 that where slippage is potentially a problem slip-critical bolts should be used. For example, they should be used if the working or service loads cause a large number of stress changes resulting in a possible fatigue situation in the bolts.)

In the following discussion, the letter a represents the distance between connectors and r_i is the least radius of gyration of an individual component of the column. If compression members consisting of two or more shapes are used, they must be connected together at intervals such that the effective slenderness ratio Ka/r_i of each of the component shapes between the connectors is not larger than 3/4 times the governing or controlling slenderness ratio of the whole built-up member. The end connections must be made with welds or slip-critical bolts with clean mill scale or blasted cleaned faying surfaces with Class A coatings. (A Class A coating is one that has a mean slip coefficient not less than 0.33. See Section J3.8a of the LRFD Specifications.) Class A surfaces are typical in structural steel work. Snug-tight bolts may be used for interior bolts.

The design strength of compression members built up from two or more shapes in contact with each other will be determined with the usual applicable LRFD Equations (E2-1, E2-2, and E2-3) with one exception. Should the column tend to buckle in such a manner that relative deformations in the different parts cause shear forces in the connectors between the parts, it will be necessary to modify the KL/r value for that axis of buckling. This modification is required by Section E4 of the LRFD Specification.

Reference is made here to the cover-plated column of Fig. 6.6. If this section tends to buckle about its y axis, the connectors between the W shape and the plates are

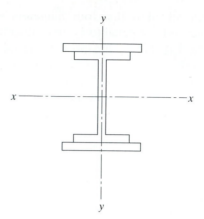

FIGURE 6.6

not subjected to any calculated load. If, on the other hand, it tends to buckle about its x axis, the connectors are subjected to shearing forces. The flanges of the W section and the cover plates will have different stresses and thus different deformations. The result will be shear in the connection between these parts, and $(KL/r)_x$ will have to be modified by LRFD Equation E4-1 or E4-2 as described below. Equation E4-1 is based upon test results that supposedly account for shear deformations in the connectors. Equation 4-2 is based upon theory and was checked by means of tests.

a. For intermediate connectors that are snug-tight bolted:

$$\left(\frac{KL}{r}\right)_m = \sqrt{\left(\frac{KL}{r}\right)_0^2 + \left(\frac{a}{r_i}\right)^2} \quad \text{(LRFD Equation E4-1)}$$

Note that it is important to remember that the design strength of a built-up column will be reduced if the spacing of connectors is such that one of the components of the column can buckle before the whole column buckles.

b. For intermediate connectors that are welded or have fully tensioned bolts as required for slip-critical joints:

$$\left(\frac{KL}{r}\right)_m = \sqrt{\left(\frac{KL}{r}\right)_0^2 + 0.82 \frac{\alpha^2}{(1 + \alpha^2)}\left(\frac{a}{r_{ib}}\right)^2} \quad \text{(LRFD Equation E4-2)}$$

In these two equations

$\left(\dfrac{KL}{r}\right)_o =$ column slenderness of the whole built-up member acting as a unit

$\left(\dfrac{KL}{r}\right)_m =$ modified slenderness of built-up member

$a =$ distance between connectors, in

$r_i =$ minimum radius of gyration of individual component, in

$r_{ib} =$ radius of gyration of individual component relative to its centroidal axis parallel to the member axis of buckling, in

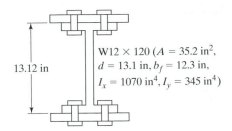

13.12 in

W12 × 120 ($A = 35.2$ in^2, $d = 13.1$ in, $b_f = 12.3$ in, $I_x = 1070$ in^4, $I_y = 345$ in^4)

FIGURE 6.7

W section used as column with cover plates.

$h =$ distance between centroids of individual components perpendicular to the member axis of buckling, in.

$\alpha =$ separation ratio $= \dfrac{h}{2r_{ib}}$

For the case in which the column tends to buckle about an axis such as to cause shear in the connection between the column parts, it will be necessary to compute a modified slenderness ratio $(KL/r)_m$ for that axis and to check to see whether that value will cause a change in the design strength of the member. If it does, it may be necessary to revise sizes and repeat the steps just described.

LRFD Equation E4-1 is used to compute the modified slenderness ratio $(KL/r)_m$ about the major axis to see if it is greater than the slenderness ratio about the minor axis. If it is, that value should be used for determining the design strength of the member.

Section E4 of the LRFD Commentary states that based on judgment and experience, the longitudinal spacing of the connectors for built-up compression members must be such that the slenderness ratios of the individual parts of the members do not exceed three-fourths of the slenderness ratio of the entire member.

Example 6-4 illustrates the design of a column consisting of a W section with cover plates bolted to its flanges as shown in Fig. 6.7. Even though snugtight bolts are used for this column, you should realize that LRFD Specification E4 states that the end bolts must be slip-critical or the ends must be welded. This is required so the parts of the built-up section will not slip with respect to each other and thus will act as a unit in resisting loads. (As a practical note, the usual steel company required to tighten the end bolts to a slip-critical condition will probably just go ahead and tighten them all to that condition.)

As this type of built-up section is not shown in the column tables of the LRFD Manual, it is necessary to use a trial-and-error design procedure. An effective slenderness ratio is assumed. Then, $\phi_c F_{cr}$ for that slenderness ratio is determined and divided into the column design load to estimate the column area required. The area of the W section is subtracted from the estimated total area to obtain the estimated cover plate area. Cover plate sizes are then selected to provide the required estimated area.

Example 6-4

It is desired to design a column for $P_u = 2375$ k using $F_y = 50$ ksi and $KL = 14$ ft. A W12 × 120 (for which $\phi_c P_n = 1220$ k from the Part 4 tables of the Manual) is on hand. Design cover plates to be snug-tight bolted at 6 in spacings to the W section shown in Fig. 6.7 to enable the column to support the required load.

Solution

$$\text{Assume } \frac{KL}{r} = 50$$

$$\phi_c F_{cr} = 35.40 \text{ ksi}$$

$$A \text{ required} = \frac{2375}{35.40} = 67.09 \text{ in}^2$$

$-A$ of W12 × 120 = −35.30

Estimated A of 2 plates = 31.79 in² or 15.90 in² each.

Try 1 PL1 × 16 each flange

$$A = 35.30 + (2)(1)(16) = 67.30 \text{ in}^2$$

$$I_x = 1070 + (2)(16)\left(\frac{13.1 + 1.00}{2}\right)^2 = 2660 \text{ in}^4$$

$$r_x = \sqrt{\frac{2660}{67.30}} = 6.29 \text{ in}$$

$$\left(\frac{KL}{r}\right)_x = \frac{(12)(14)}{6.29} = 26.71$$

$$I_y = 345 + (2)\left(\frac{1}{12}\right)(1)(16)^3 = 1027.7 \text{ in}^4$$

$$r_y = \sqrt{\frac{1027.7}{67.30}} = 3.91 \text{ in}$$

$$\left(\frac{KL}{r}\right)_y = \frac{(12)(14)}{3.91} = 42.97 \leftarrow$$

Computing the modified slenderness ratio yields

$$r_i = \sqrt{\frac{I}{A}} = \sqrt{\frac{\left(\frac{1}{12}\right)(16)(1)^3}{(1)(16)}} = 0.289 \text{ in}$$

$$\frac{a}{r_i} = \frac{6}{0.289} = 20.76$$

$$\left(\frac{KL}{r}\right)_x = \sqrt{\left(\frac{KL}{r}\right)_0^2 + \left(\frac{a}{r_i}\right)^2} = \sqrt{(26.71)^2 + (20.76)^2}$$

$$= 33.83 < 42.97 \ \therefore \text{ does not control}$$

Checking the slenderness ratio of the plates, we have

$$\frac{k_a}{r_i} = 20.76 < \frac{3}{4}\left(\frac{KL}{r}\right)_y = \left(\frac{3}{4}\right)(42.97) = 32.23 \qquad \text{(OK)}$$

$$\phi_c P_n = (37.1)(67.30) = 2497 \text{ k} > P_u = 2375 \text{ k} \qquad \text{(OK)}$$

A subsequent check of $\dfrac{b}{t}$ ratios is satisfactory.

Use W12 × 120 with 1 cover plate 1 × 16 each flange, $F_y = 50$ ksi) (Note: Many other plate sizes could have been selected.)

6.7   BUILT-UP COLUMNS WITH COMPONENTS NOT IN CONTACT WITH EACH OTHER

Example 6-5 presents the design of a member built up from two channels that are not in contact with each other. The parts of such members need to be connected or laced together across their open sides. The design of lacing is discussed immediately after this example and is illustrated in Example 6-6.

Example 6-5

Select a pair of 12-in standard channels for the column and load shown in Fig. 6.8 using $F_y = 50$ ksi. For connection purposes, the back-to-back distance of the channels is to be 12 in.

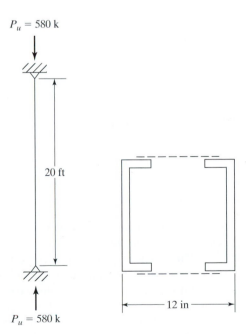

$P_u = 580$ k

20 ft

$P_u = 580$ k

12 in

FIGURE 6.8

Column built up from two channels.

Solution

Assume $\dfrac{KL}{r} = 50$

$$\phi_c F_{cr} = 35.40 \text{ ksi}$$

$$A \text{ required} = \frac{580}{35.40} = 16.38 \text{ in}^2$$

Try 2C12 × 30s (For each channel $A = 8.81$ in^2, $I_x = 162$ in^4, $I_y = 5.12$ in^4, $\bar{x} = 0.674$ in)

$$I_x = (2)(162) = 324 \text{ in}^4$$

$$I_y = (2)(5.12) + (2)(8.81)(5.326)^2 = 510 \text{ in}^4$$

$$r_x = \sqrt{\frac{324}{(2)(8.81)}} = 4.29 \text{ in}$$

$$KL = (1.0)(20) = 20 \text{ ft}$$

$$\frac{KL}{r} = \frac{(12)(20)}{4.29} = 55.94$$

$$\phi_c F_{cr} = 33.82 \text{ ksi}$$

$$\phi_c P_n = (33.82)(17.64) = 597 \text{ k} > 580 \text{ k} \qquad \text{(OK)}$$

Checking width/thickness ratios ($d = 12.00$ in, $b_f = 3.170$ in, $t_f = 0.501$ in, $t_w = 0.510$ in, $k = 1\frac{1}{8}$ in)

$$\frac{b_f}{t_f} = \frac{3.170}{0.501} = 6.33 < 0.56\sqrt{\frac{E}{F_y}} = \sqrt{\frac{29 \times 10^3}{50}} = 13.49 \qquad \text{(OK)}$$

$$\frac{h}{t_w} = \frac{12.00 - (2)(1.125)}{0.510} = 19.12 < 1.49\sqrt{\frac{E}{F_y}} = 1.49\sqrt{\frac{29 \times 10^3}{50}} = 35.88$$

Use 2C12 × 30s, $F_y = 50$ ksi. \qquad (OK)

The open sides of compression members that are built up from plates or shapes may be connected together with continuous cover plates with perforated holes for access purposes, or they may be connected together with lacing and tie plates.

The purpose of the perforated cover plates and the lacing or lattice work is to hold the various parts parallel and the correct distance apart and to equalize the stress distribution between the various parts. The student will understand the necessity for lacing if he or she considers a built-up member consisting of several sections (such as the four-angle member of Fig. 5.2) which supports a heavy compressive load. Each of the parts will tend to individually buckle laterally unless they are tied together to act as

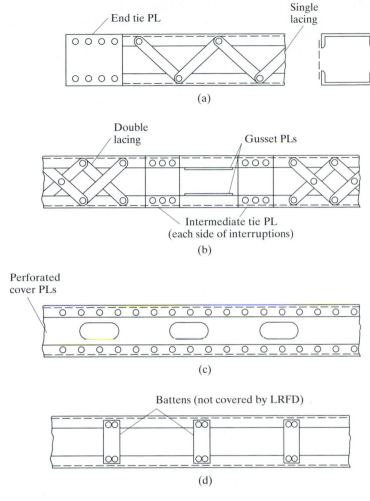

FIGURE 6.9

Lacing and perforated cover plates.

a unit in supporting the load. In addition to lacing, it is necessary to have tie plates (also called stay plates or batten plates) as near the ends of the member as possible and at intermediate points if the lacing is interrupted. Parts (a) and (b) of Fig. 6.9 show arrangements of tie plates and lacing. Other possibilities are shown in parts (c) and (d) of the same figure.

The failure of several structures in the past has been attributed to inadequate lacing of built-up compression members. Perhaps the best-known example was the failure of the Quebec Bridge in 1907. Following its collapse, the general opinion was that the lattice work of the compression chords was too weak and resulted in failure.

If continuous cover plates perforated with access holes are used to tie the members together, the LRFD Specification E4 states that (a) they must comply with the limiting width-thickness ratios specified for compression elements in Section B5.1 of the LRFD Specification; (b) the ratio of the access hole length (in the direction of stress) to the hole width may not exceed 2; (c) the clear distance between the holes in the direction of stress may not be less than the transverse distance between the nearest lines of connecting fasteners or welds; and (d) the periphery of the holes at all points must have a radius no less than $1\frac{1}{2}$ in. Stress concentrations and secondary bending stresses are usually neglected, but lateral shearing forces must be checked as they are for other types of lattice work. (The unsupported width of such plates at access holes is assumed to contribute to the design strength $\phi_c P_n$ of the member if the conditions as to sizes, width-thickness ratios, etc., described in LRFD Specification E4 are met.) Perforated cover plates are attractive to many designers because of several advantages they possess.

1. They are easily fabricated with modern gas cutting methods.
2. Some specifications permit the inclusion of their net areas in the effective section of the main members, provided the holes are made in accordance with their empirical requirements, which have been developed on the basis of extensive research.
3. Painting of the members is probably simplified, compared with ordinary lacing bars.

Dimensions of tie plates and lacing are usually controlled by specifications. Section E4 of the LRFD Specification states that tie plates shall have a thickness at least equal to one-fiftieth of the distance between the connection lines of welds or other fasteners.

Lacing may consist of flat bars, angles, channels, or other rolled sections. These pieces must be so spaced that the individual parts being connected will not have L/r values between connections which exceed the governing value for the entire built-up member. (The governing value is KL/r for the whole built-up section.) Lacing is assumed to be subjected to a shearing force normal to the member equal to not less than 2 percent of the compression design strength $\phi_c P_n$ of the member. The LRFD column formulas are used to design the lacing in the usual manner. Slenderness ratios are limited to 140 for single lacing and 200 for double lacing. Double lacing or single lacing made with angles should preferably be used if the distance between connection lines is greater than 15 in.

Example 6-6 illustrates the design of lacing and end tie plates for the built-up column of Example 6-5. Bridge specifications are somewhat different in their lacing requirements from the LRFD, but the design procedures are much the same.

Example 6-6

Using the LRFD Specification and 36 ksi steel design bolted single lacing for the column of Example 6-5. Reference is made to Fig. 6.10. Assume 3/4-in bolts.

Solution. Distance between lines of bolts is 8.5 in < 15 in; therefore, single lacing is OK.

Assume lacing bars are inclined at 60° with axis of member. Length of channels between lacing connections is 8.5/cos 30° = 9.8 in, and L/r of 1 channel between connections is 9.8/0.763 = 12.9 < 55.94, which is L/r of main member previously determined in Example 6-5.

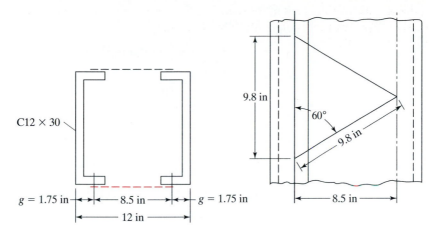

FIGURE 6.10

Two-channel column section with lacing.

Force on lacing bar:

V_u = 0.02 times design compressive strength of member (from Example 6-5)
V_u = (0.02)(580) = 11.6 k

$\frac{1}{2}V_u$ = 5.8 k = shearing force on each plane of lacing

Force in bar (with reference to bar dimensions in Fig. 6.10):

$$\left(\frac{9.8}{8.5}\right)(5.8) = 6.69 \text{ k}$$

Properties of flat bar:

$$I = \tfrac{1}{12}bt^3$$
$$A = bt$$

$$r = \sqrt{\frac{\frac{1}{12}bt^3}{bt}} = 0.289t$$

Design of bar:

$$\text{Assume } \frac{L}{r} = \text{maximum value of 140}$$

$$\frac{9.8}{0.289t} = 140$$

$$t = 0.242 \text{ in} \left(\text{try } \frac{1}{4}\text{-in flat bar} \right)$$

$$\frac{L}{r} = \frac{9.8}{(0.289)(0.250)} = 136$$

$$\phi_c F_{cr} = 11.5 \text{ ksi}$$

$$\text{Area required} = \frac{6.69}{11.5} = 0.582 \text{ in}^2 (2.33 \times \tfrac{1}{4} \text{ needed})$$

$$\text{Minimum edge distance if } \tfrac{3}{4}\text{-in bolt used} = 1\tfrac{1}{4} \text{ in}$$

$$\therefore \text{ Minimum length of bar} = 9.8 + (2)(1\tfrac{1}{4}) = 12.3 \text{ in say } 14 \text{ in}$$

Use $\tfrac{1}{4} \times 2\tfrac{1}{2} \times 1$ ft 2 in bars $F_y = 36$ ksi

Design of end tie plates:

$$\text{Minimum length} = 8.5 \text{ in}$$

$$\text{Minimum } t = (\tfrac{1}{50})(8.5) = 0.17 \text{ in}$$

$$\text{Minimum width} = 8.5 + (2)(1\tfrac{1}{4}) = 11 \text{ in}$$

$$\text{Use } \tfrac{3}{16} \times 8\tfrac{1}{2} \times 0 \text{ ft 12 in end tie plates.}$$

6.8   INTRODUCTORY REMARKS CONCERNING FLEXURAL-TORSIONAL BUCKLING OF COMPRESSION MEMBERS

Axially loaded compression members can theoretically fail in four different fashions: by local buckling of elements that form the cross section, by flexural buckling, by torsional buckling, or by flexural-torsional buckling.

Flexural buckling (also called *Euler buckling* when elastic behavior occurs) is the situation considered up to this point in our column discussions where we have computed slenderness ratios for the principal column axes and determined $\phi_c F_{cr}$ for the highest ratios so obtained. Doubly symmetrical column members (such as W sections) are subject only to local buckling, flexural buckling, and torsional buckling.

Because torsional buckling can be very complex, it is very desirable to prevent its occurrence. This may be done by careful arrangements of the members and by providing bracing to prevent lateral movement and twisting. If sufficient end supports and intermediate lateral bracing are provided, flexural buckling will always control over torsional buckling. The column design strengths given in the LRFD column tables for W, M, S, tube, and pipe sections are based on flexural buckling.

Open sections such as Ws, Ms, and channels have little torsional strength, but box beams have a great deal. Thus, if a torsional situation is encountered, it may be well to use box sections or to make box sections out of W sections by adding welded side plates (| |). Another way in which torsional problems can be reduced is to shorten the lengths of members that are subject to torsion.

For a singly symmetrical section such as a tee or double angle, Euler buckling may occur about the x or y axis. For equal-leg single angles, Euler buckling may occur about the z axis. For all these sections, flexural-torsional buckling is definitely a possibility and may control. (It will always control for unequal-leg single-angle columns.) The values given in the LRFD column load tables for double-angle and structural tee sections were computed for buckling about the weaker of the x or y axis and for flexural-torsional buckling.

The average designer does not consider the torsional buckling of symmetrical shapes or the flexural-torsional buckling of unsymmetrical shapes. The feeling is that these conditions don't control the critical column loads, or at least don't affect them very much. This assumption can be far from the truth. Should we have unsymmetrical columns or even symmetrical columns made up of thin plates, however, we will find that torsional buckling or flexural-torsional buckling may significantly reduce column capacities.

A flexural-torsional example is presented in Appendix D of this book.

6.9   SINGLE-ANGLE COMPRESSION MEMBERS

You will note that we have not discussed the design of single-angle compression members up to this point. The AISC has long been concerned about the problems involved in loading such members fairly concentrically. It can be done rather well if the ends of the angles are milled and if the loads are applied through bearing plates. In practice, however, single-angle columns are often used in such a manner that rather large eccentricities of load applications are present. The sad result is that it is somewhat easy to greatly underdesign such members.

In Part 4 of the LRFD Manual, a special specification is provided for the design of single-angle compression members. Though this specification includes information for tensile, shear, compressive, flexural, and combined loadings, the present discussion is concerned only with the compression case. Some rather complicated formulas are presented in the specification for computing the axial design strengths of single angles.

These formulas were developed to account for three different limit states that might occur in single-angle columns: flexural buckling, local buckling of thin angle legs, and flexural torsional buckling. Using these expressions, the design strengths of concentrically loaded single angles are computed for 36 ksi steel and are shown in Table 4.4 of Part 4 of the Manual. In addition, a numerical example (4.3) is presented.

6.10   COMPUTER EXAMPLE

In Example 6-7, INSTEP32 is used to design an axially loaded column. The user may select the lightest section available in the Manual or may specify a specific series such as W14, W12, etc. Once again, to select a section in a specific series, it is necessary to scroll up or down the list of sections until the desired series is obtained.

Design of Compression Members:
Problem 3) problem 6-1 (page 178, McCormac and Nelson, 3rd Ed.). Follow class notes
or example 6-1 of text. Remember to check local buckling. Compare your final design
strength using AISC Table 4-2.

~~Select the lightest~~ W12 section to support an axial compression design load $P_u = 720$ k.
Use 50 ksi steel and assume $K_x = K_y = 1.0$ and $L_x = 20$ ft and $L_y = 10$ ft.

Solution

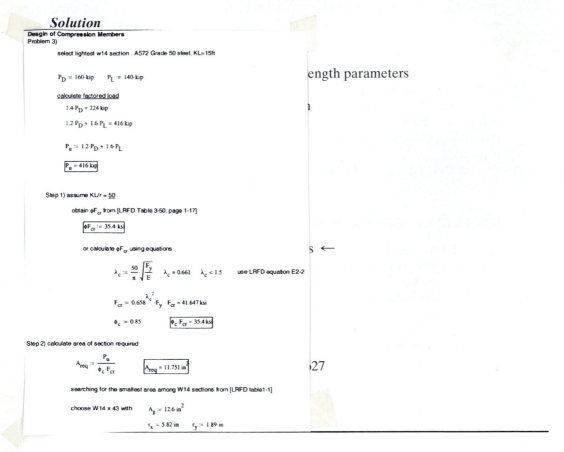

Desgin of Compression Members
Problem 3)

    select lightest w14 section . A572 Grade 50 steel, KL=15ft

    $P_D := 160 \cdot kip$ $P_L := 140 \cdot kip$

    calculate factored load

      $1.4 \cdot P_D = 224\, kip$

      $1.2 \cdot P_D + 1.6 \cdot P_L = 416\, kip$

      $P_u := 1.2 \cdot P_D + 1.6 \cdot P_L$

      $\boxed{P_u = 416\, kip}$

Step 1) assume KL/r = <u>50</u>

    obtain ϕF_{cr} from [LRFD Table 3-50, page 1-17]

    $\boxed{\phi F_{cr} := 35.4 \cdot ksi}$

    or calculate ϕF_{cr} using equations

      $\lambda_c := \dfrac{50}{\pi} \cdot \sqrt{\dfrac{F_y}{E}}$ $\lambda_c = 0.661$ $\lambda_c < 1.5$ use LRFD equation E2-2

      $F_{cr} = 0.658^{\lambda_c^{\,2}} \cdot F_y$ $F_{cr} = 41.647\, ksi$

      $\phi_c := 0.85$ $\boxed{\phi_c \cdot F_{cr} = 35.4\, ksi}$

Step 2) calculate area of section required

    $A_{req} := \dfrac{P_u}{\phi_c \cdot F_{cr}}$ $\boxed{A_{req} = 11.751\, in^2}$

    searching for the smallest area among W14 sections from [LRFD table1-1]

    choose W 14 x 43 with $A_g := 12.6 \cdot in^2$

      $r_x := 5.82 \cdot in$ $r_y := 1.89\, in$

ength parameters

s ←

327

PROBLEMS

All of the columns for these problems are assumed to be in frames that are braced against
sidesway.

6-1 to 6-3. *Use a trial-and-error procedure in which a KL/r value is estimated, a value of $\phi_c F_{cr}$
is determined from the appropriate LRFD table (3.36 or 3.50 in Part 16 of the Manual),
the estimated column area is determined, a trial section is selected, its P_u is computed,
and then another size is tried if necessary, and so on.*

6-1. Select the lightest available W14 section to support the axial compression loads
$P_D = 160$ k and $P_L = 140$ k if $KL = 15$ ft and A572 Grade 50 steel is used.
(*Ans.* W14 × 61)

6-2. Select the lightest available W12 section to support the axial loads $P_D = 220$ k and
$P_L = 280$ k if $KL = 14$ ft and $F_y = 50$ ksi.

6-3. Repeat Prob. 6-2 if $F_y = 36$ ksi. (*Ans.* W12 × 96)

6-4 to 6-20.   *Take advantage of all available column tables in the Manual, particularly those in Part 4.*

6-4.   Repeat Prob. 6-1.

6-5.   Repeat Prob. 6-2. (*Ans.* W12 × 72)

6-6.   Repeat Prob. 6-2 if P_u = 920 k.

6-7.   Several building columns are to be designed using A992 steel and the LRFD Specification. Select the lightest available W sections for these columns that are described as follows:

   a. P_u = 500 k, L = 14 ft, pinned end supports. (*Ans.* W12 × 58)

   b. P_u = 360 k, L = 12 ft, fixed end supports. (*Ans.* W10 × 39)

   c. P_u = 700 k, L = 16 ft 6 in, fixed at bottom, pinned at top. (*Ans.* W12 × 72)

   d. P_u = 1400 k, L = 15 ft, pinned end supports. (*Ans.* W14 × 145)

6-8.   Select the lightest W section available in 50 ksi steel to serve as a pinned end column to support the following axial loads: P_D = 300 k and P_W due to wind = 350 k. Assume KL is 12 ft.

6-9.   A W section is to be selected to support an axial compressive load P_u = 1600 k. The member, which is to be 24 ft long and is to be pinned top and bottom, has lateral support (pinned) supplied in the weak direction at middepth. Select the lightest W12 or 14 section using A572 Grade 50 steel. (*Ans.* W14 × 159)

6-10.   Repeat Prob. 6-9 if P_u = 1000 k.

6-11.   Repeat Prob. 6-9 if lateral support is provided in the weak direction at the one-third points and the total column length is changed to 30 ft. Use 50 ksi steel. (*Ans.* W14 × 176)

6-12.   A 24 ft column is laterally supported in the weak direction at its middepth. Select the lightest W section that can adequately support the axial gravity loads P_D = 300 k and P_L = 180 k using F_y = 50 ksi. Assume all ks are 1.0.

6-13.   A 16 ft column is to be built into a wall in such a manner that it will be continuously braced in its weak direction but not in its strong direction. If the member is to consist of 50 ksi steel and is assumed to have pinned ends, select the lightest satisfactory W14 section available using the LRFD Specification. P_u = 800 k. (*Ans.* W14 × 74)

6-14.   Repeat Prob. 6-13 if P_u = 1200 k.

6-15.   A W14 section of 50 ksi steel is to be selected to support the axial compressive loads P_D = 360 k and P_L = 300 k. The member which is to be 24 ft long is to be fixed top and bottom and is to have lateral support at its one-third points perpendicular to the y axis (pinned). (*Ans.* W14 × 82)

6-16.   Using the steels for which column tables are provided in part 4 of the manual, select the lightest available rolled sections (W, HP, square, rectangular, or round tubing) that are adequate for the following situations:

   a. P_u = 240 k, L = 16 ft, pinned ends.

   b. P_u = 400 k, L = 14 ft, fixed ends.

   c. P_u = 800 k, L = 18 ft, one end pinned and the other fixed.

6-17.   Assuming axial loads only, select W sections for an interior column of the laterally braced frame shown in the accompanying illustration. Use F_y = 50 ksi and the

LRFD Specification. Each column section can be used for one or two stories before it is spliced, whichever seems advisable. Miscellaneous data: concrete weighs 150 lb/ft³. LL on roof = 30 psf. Roofing = 6 psf. LL on interior floors = 80 psf. Partition load on interior floors = 15 psf. All joints are assumed to be pinned so as to simplify calculations. Frames 30 ft on center. (*Ans.* W14 × 53 top two stories and W14 × 90 bottom two stories)

4-in reinforced concrete slab

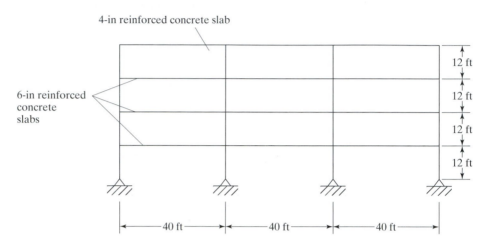

6-in reinforced concrete slabs

FIGURE P6-17

6-18. It is desired to design a column for P_u = 2875 k using A992 steel with KL = 12 ft. A W14 × 145 is on hand with a good supply of 3/4-in thick plates. Design cover plates to be welded to the flanges of the W section to enable the column to support the required load.

6-19. Determine the design compressive strength of the section shown if snug-tight bolts 4 ft on center are used to connect the A36 angles. $K_xL_x = K_yL_y$ = 24 ft. (*Ans.* 112.1 k)

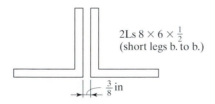

2Ls 8 × 6 × $\frac{1}{2}$
(short legs b. to b.)

$\frac{3}{8}$ in

FIGURE P6-19

6-20. Repeat Prob. 6-19 if the angles are welded together with their long legs back to back at intervals of 6 ft.

6-21. Four 4 × 4 × .5 angles are used to form the member shown in the accompanying illustration. The member is 30 ft long, has pinned ends, and consists of 50 ks. steel.

Determine the design compressive strength of the member. Design single lacing and end tie plates, assuming connections are made to the angles with 3/4-in bolts.

$P_u = 547.9$ k   9/32 × 13 × 1 ft

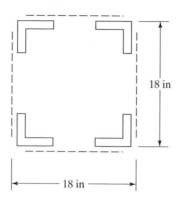

18 in

18 in

FIGURE P6-21

(*Ans.*   ; end tie plates   4 in; single lacing at 60° 3/8 × 2.5 × 1 ft 5.5 in)

6-22. Select a pair of standard channels to support an axial compressive load of $P_u = 925$ k. The member is to be 24 ft long with both ends pinned and is to be arranged as shown in the accompanying illustration. Use $F_y = 50$ ksi, and design

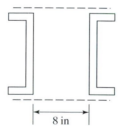

8 in

FIGURE P6-22

single lacing and end tie plates assuming 3/4-in bolts are to be used for connections. Assume bolts located 2.25-in from back of channels.

6-23 to 6-27. *Using the enclosed computer diskette, repeat the following problems*:

6-23. Prob. 6-1. (*Ans.* W14 × 61)

6-24. Prob. 6-3.

6-25. Prob. 6-7. (*Ans.* W12 × 58, W10 × 39, W12 × 72, W14 × 145)

6-26. Prob. 6-9.

6-27. Prob. 6-11. (*Ans.* W14 × 176)

Design of Axially Loaded Compression Members, Continued

7.1 FURTHER DISCUSSION OF EFFECTIVE LENGTHS

The subject of effective lengths was introduced in Chapter 5, and some suggested K factors were presented in Table 5.1. These factors were developed for columns with certain idealized conditions of end restraint which may be very different from practical design conditions. The table values are usually quite satisfactory for preliminary designs and for situations in which sidesway is prevented by bracing. Should the columns be part of a continuous frame subject to sidesway, however, it is often advantageous to make a more detailed analysis as described in this section. To a lesser extent, this is also desirable for columns in frames braced against sidesway.

Perhaps a few explanatory remarks should be made at this point, defining sidesway as it pertains to effective lengths. For this discussion, sidesway refers to a type of buckling. In statically indeterminate structures, sidesway occurs where the frames deflect laterally due to the presence of lateral loads or unsymmetrical vertical loads or where the frames themselves are unsymmetrical. Sidesway also occurs in columns whose ends can move transversely when they are loaded until buckling occurs.

Should frames with diagonal bracing or rigid shear walls be used, the columns will be prevented from sidesway and provided with some rotational restraint at their ends. For these situations, pictured in Fig. 7.1, the K factors will fall somewhere between cases (a) and (d) of Table 5.1.

The LRFD Specification (C2) states that $K = 1.0$ should be used for columns in frames with sidesway inhibited, unless an analysis shows that a smaller value can be used. A specification like $K = 1.0$ is often quite conservative, and an analysis made as described herein may result in some savings.

The true effective length of a column is a property of the whole structure of which the column is a part. In many existing buildings, it is probable that the masonry

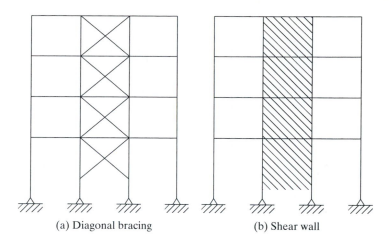

FIGURE 7.1

Sidesway inhibited   (a) Diagonal bracing   (b) Shear wall

walls provide sufficient lateral support to prevent sidesway. When light curtain walls are used, however, as they often are in modern buildings, there is probably little resistance to sidesway. Sidesway is also present in tall buildings in appreciable amounts unless a definite diagonal bracing system or shear walls are used. For these cases, it seems logical to assume that resistance to sidesway is primarily provided by the lateral stiffness of the frame alone.

Theoretical mathematical analyses may be used to determine effective lengths, but such procedures are usually too lengthy and perhaps too difficult for the average designer. The usual procedure is to use either Table 5.1, interpolating between the idealized values as the designer feels appropriate, or the alignment charts that are described in this section.

The most common method for obtaining effective lengths is to use the charts shown in Fig. 7.2. They were developed by O. G. Julian and L. S. Lawrence and frequently are referred to as the Jackson and Moreland charts after the firm where they worked.[1,2] They were developed from a slope-deflection analysis of the frames that included the effect of column loads. One chart was developed for columns braced against sidesway and one for columns subject to sidesway. Their use enables the designer to obtain good K values without struggling through lengthy trial-and-error procedures with the buckling equations.

To use the alignment charts, it is necessary to have preliminary sizes for the girders and columns framing into the column in question before the K factor can be determined for that column. In other words, before the chart can be used, we have to either assume some member sizes or carry out a preliminary design.

---

[1]O. G. Julian and L. S. Lawrence, "Notes on J and L Monograms for Determination of Effective Lengths" (1959). Unpublished.
[2]Structural Stability Research Council, *Guide to Stability Design Criteria for Metal Structures*, 4th ed. T. V. Galambos, ed. (New York: Wiley, 1988).

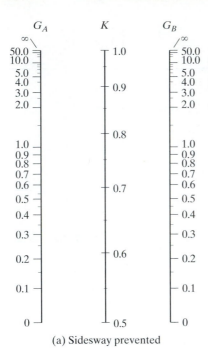

(a) Sidesway prevented

The subscript A and B refer to the joints at the two ends of the comumn section being considered. G is defined as

$$G = \frac{\Sigma \dfrac{I_c}{L_c}}{\Sigma \dfrac{I_g}{L_g}}$$

in which Σ indicates a summation of all members rigidly connected to that joint and lying in the plane in which buckling of the column is being considered, I_c is the moment of inertia and L_c the unsupported length of a column section, and I_g is the moment of inertia and L_g the unsupported length of a girder or other restraining member. I_c and I_g are taken about axes perpendicular to the plane of buckling being considered.

"For column ends supported by, but not rigidly connected to, a footing or foundation, G is theoretically infinity, but, unless actually designed as a true friction free pin, may be taken as '10' for practical designs. If the column end is rigidly attached to a properly designed footing, G may be taken as 1.0. Smaller values may be used if justified by analysis."

From American Institute of Steel Construction, *Manual of Steel Construction Load & Resistance Factor Design*, Volume I, 2d. ed. (Chicago: AISC, 1994), p. 6-186.

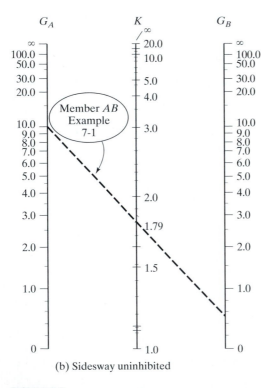

(b) Sidesway uninhibited

FIGURE 7.2

Jackson and Moreland alignment charts for effective lengths of columns in continous frames.

Shearson Lehman/American Express Information Services Center,
New York City. (Courtesy of Owen Steel Company, Inc.)

When we say sidesway is *inhibited*, we mean there is something present other than just columns and girders to prevent sidesway or the horizontal translation of the joints. That means we have a definite system of lateral bracing, or we have shear walls. If we say that sidesway is *uninhibited*, we are saying that resistance to horizontal translation is supplied only by the bending strength and stiffness of the girders and beams of the frame in question with its continuous joints.

The resistance to rotation furnished by the beams and girders meeting at one end of a column is dependent on the rotational stiffnesses of those members. The moment needed to produce a unit rotation at one end of a member if the other end of the member is fixed is referred to as its *rotational stiffness*. From our structural analysis studies, this works out to be equal to $4EI/L$ for a homogeneous member of constant cross section. Based on the preceding, we can say that the rotational restraint at the end of a particular column is proportional to the ratio of the sum of the column stiffnesses to the girder stiffnesses meeting at that joint[3]

$$G = \frac{\sum \dfrac{4EI}{L} \text{ for columns}}{\sum \dfrac{4EI}{L} \text{ for girders}} = \frac{\sum \dfrac{I_c}{L_c}}{\sum \dfrac{I_g}{L_g}}$$

In applying the charts, the G factors at the column bases are quite variable. It is recommended that the following two rules be followed to obtain their values:

1. For pinned columns, G is theoretically infinite, such as when a column is connected to a footing with a frictionless hinge. Recognizing that such a connection is not frictionless, it is recommended that G be made equal to 10 where such nonrigid supports are used.
2. For rigid connections of columns to footings, G theoretically approaches zero, but from a practical standpoint, a value of 1.0 is recommended because no connections are perfectly rigid.

The determination of K factors for the columns of a steel frame using the alignment charts is illustrated in Examples 7-1 and 7-2. The following steps are taken:

1. Select the appropriate chart (sidesway prevented or sidesway uninhibited).
2. Compute G at each end of the column and label the values G_A and G_B as desired.
3. Draw a straight line on the chart between the G_A and G_B values, and read K where the line hits the center K scale.

When G factors are being computed for a rigid frame structure (rigid in both directions), the torsional resistance of the perpendicular girders is generally neglected in the calculations. With reference to Fig. 7.3, it is assumed that we are calculating G for the joint

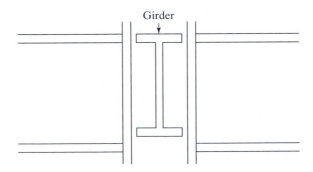

Girder

FIGURE 7.3

[3]W. T. Segui, *LRFD Steel Design* (Boston: PWS Publishing Company, 1994), pp. 96–97.

shown for buckling in the plane of the paper. For such a case, the torsional resistance of the girder shown, which is perpendicular to the plane being considered, is probably neglected.

If the girders at a joint are very stiff (that is, they have very large I/L values) the value of $G = \Sigma \, (I_c/L_c)/\Sigma(I_g/L_g)$ will approach zero and the K factors will be small. If G is very small, the column moments cannot rotate the joint very much; thus, the joint is close to a fixed end situation. Usually, however, G is appreciably larger than zero, resulting in significantly larger values of K.

The effective lengths of each of the columns of a frame are estimated with the alignment charts in Example 7-1. (When sidesway is possible, it will be found that the effective lengths are always greater than the actual lengths, as is illustrated in this example. When frames are braced in such a manner that sidesway is not possible, K will be less than 1.0.) An initial design has provided preliminary sizes for each of the members in the frame. After the effective lengths are determined, each column can be redesigned. Should the sizes change appreciably, new effective lengths can be determined, the column designs repeated, and so on. Several tables are used in the solution of this example. These should be self-explanatory after the clear directions given on the alignment chart are examined.

7.2 FRAMES MEETING ALIGNMENT CHART ASSUMPTIONS

The Jackson and Moreland charts were developed on the basis of a certain set of assumptions, a complete list of which is given in Section C.2 of the Commentary of the LRFD Specification. Among these assumptions are the following:

1. The members are elastic, have constant cross sections and are connected with rigid joints.
2. All columns buckle simultaneously.
3. For braced frames, the rotations at opposite ends of each beam are equal in magnitude and each beam bends in single curvature.
4. For unbraced frames, the rotations at opposite ends of each beam are equal in magnitude, but each beam bends in double curvature.

The frame of Figure 7.4 is assumed to meet all of the assumptions on which the alignment charts were developed. Using the charts, the column effective length factors are determined as shown in Example 7-1.

Example 7-1

Determine the effective length factor for each of the columns of the frame shown in Fig. 7.4 if the frame is not braced against sidesway. Use the alignment charts of Fig. 7.2(b).

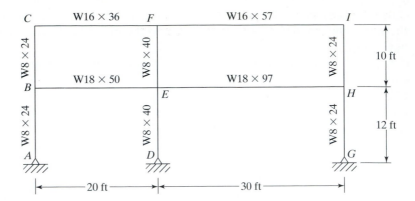

FIGURE 7.4

Solution. Stiffness factors:

Member	Shape	I	L	I/L
AB	W8 × 24	82.7	144	0.574
BC	W8 × 24	82.7	120	0.689
DE	W8 × 40	146	144	1.014
EF	W8 × 40	146	120	1.217
GH	W8 × 24	82.7	144	0.574
HI	W8 × 24	82.7	120	0.689
BE	W18 × 50	800	240	3.333
CF	W16 × 36	448	240	1.867
EH	W18 × 97	1750	360	4.861
FI	W16 × 57	758	360	2.106

G factors for each joint:

Joint	$\Sigma(I_c/L_c)/\Sigma(I_g/L_g)$	G
A	See Fig. 7.2(b)	10.0
B	$\dfrac{0.574 + 0.689}{3.333}$	0.379
C	$\dfrac{0.689}{1.867}$	0.369
D	See Fig. 7.2(b)	10.0
E	$\dfrac{1.014 + 1.217}{(3.333 + 4.861)}$	0.272
F	$\dfrac{1.217}{(1.867 + 2.106)}$	0.306
G	See Fig. 7.2(b)	10.0
H	$\dfrac{0.574 + 0.689}{4.861}$	0.260
I	$\dfrac{0.689}{2.106}$	0.327

Column K factors from chart (Fig. 7-2(b)):

Column	G_A	G_B	$K*$
AB	10.0	0.379	1.78
BC	0.379	0.369	1.12
DE	10.0	0.272	1.77
EF	0.272	0.306	1.11
GH	10.0	0.260	1.76
HI	0.260	0.328	1.11

* It is a little difficult to read the charts to the three places shown by the author. He has used a larger copy of Fig. 7.2 for his work. For all practical design purposes, the K values can be read to two places which can easily be accomplished with this figure.

For most buildings, the values of K_x and K_y should be examined separately. The reason for such individual study lies in the different possible framing conditions in the two directions. Many multistory frames consist of rigid frames in one direction and conventionally connected frames with sway bracing in the other. In addition, the points of lateral support may often be entirely different in the two planes.

There is available a set of rather simple equations for computing effective length factors. On some occasions the designer may find these expressions very convenient to use compared with the alignment charts just described. Perhaps the most useful situation is for computer programs. You can see that it would be rather inconvenient to stop occasionally in the middle of a computer design to read K factors from the charts and input them to the computer. The equations, however, can easily be included in the programs, eliminating the necessity of using alignment charts.[4]

The alignment chart of Fig. 7.2(b) for frames with sidesway uninhibited always indicates that $K \geq 1.0$. In fact, calculated K factors of 2.0 to 3.0 are common, and even larger values are occasionally obtained. To many designers, such large factors seem completely unreasonable. If the designer obtains seemingly high K factors, he or she should carefully review the numbers used to enter the chart (that is, the G values) as well as the basic assumptions used in preparing the charts. These assumptions are discussed in detail in Sections 7.3 and 7.4.

7.3 FRAMES NOT MEETING ALIGNMENT CHART ASSUMPTIONS AS TO JOINT ROTATIONS

In this section, a few comments are presented regarding frames whose joint rotations (and thus their beam stiffnesses) are not in agreement with the assumptions used for developing the charts. It should be noted that this topic is not presented in the LRFD Manual. *As a result, the authors do not use the corrections presented in this section for*

---

[4]P. Dumonteil, "Simple Equations for Effective Length Factors," *Engineering Journal*, AISC, *29*, no. 3 (3rd quarter, 1992), pp. 111–115.

TABLE 7.1 Multipliers for Rigidly Attached Members

Condition at Far End of Girder	Sidesway Prevented, Multiply by:	Sidesway Uninhibited, Multiply by:
Pinned	1.5	0.5
Fixed against rotation	2.0	0.67

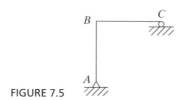

FIGURE 7.5

any other problems in the book, including the solutions for the home problems at the end of this chapter. One exception is Example 7-4.

It can be shown by structural analysis that the rotation at point B in the frame of Figure 7.5 is twice as large as the rotation at B assumed in the construction of the nomographs. Therefore, beam BC in the figure is only one half as stiff as the value assumed for the development of the alignment charts.

The Jackson and Moreland charts can be accurately used for situations in which the rotations are different from those assumed by making adjustments to the computed beam stiffnesses before the chart values are read. Relative stiffnesses for situations other than the one shown in Figure 7.5 also can be determined by structural analysis. Table 7.1, which follows, presents correction factors which are to be multiplied by calculated beam stiffnesses for situations where the beam end conditions are different from those assumed for the development of the charts.

Example 7-2 shows how the correction factors can be applied to a building frame where the rotations at the ends of some of the beams vary from the assumed conditions of the charts.

Example 7-2

Determine K factors for each of the columns of the frame shown in Figure 7.6. W sections have been tentatively selected for each of the members of the frame and their $\frac{I}{l}$ values determined and shown on the figure.

Solution. First the G factors are computed for each joint in the frame. In doing this, the $\frac{I}{l}$ values for members FI and GJ are multiplied by the appropriate factors from Table 7.1.

1. For member FI, $\frac{I}{l}$ value is multiplied by 2.0 because its far end is fixed and as there is no sidesway on that level.

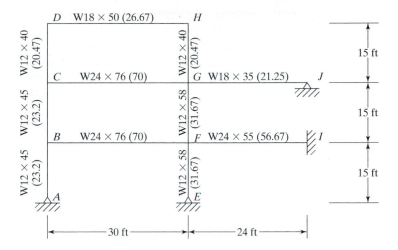

FIGURE 7.6

Steel shapes including their $\dfrac{I}{l}$ values.

2.   For member, GJ, $\frac{I}{l}$ is multiplied by 1.5

$$G_A = 10 \text{ as described in Section 7-1}$$

$$G_B = \frac{23.2 + 23.2}{70} = 0.663$$

$$G_C = \frac{23.2 + 20.47}{70} = 0.624$$

$$G_D = \frac{20.47}{26.67} = 0.767$$

$$G_E = 1.0 \text{ as described in Section 7.1}$$

$$G_F = \frac{31.67 + 31.67}{70 + (2.0)(56.25)} = 0.347$$

$$G_G = \frac{31.67 + 20.47}{70 + (1.5)(21.25)} = 0.512$$

$$G_H = \frac{20.47}{26.67} = 0.768$$

Finally, the k factors are selected from the appropriate alignment chart of Figure 7.2.

Column	G Factors	Chart used	K Factors
AB	10 and 0.663	7-2 (a) no sidesway	0.83
BC	0.663 and 0.624	7-2 (a) no sidesway	0.72
CD	0.624 and 0.767	7-2 (b) has sidesway	1.21
EF	1.0 and 0.347	7-2 (a) no sidesway	0.70
FG	0.347 and 0.512	7-2 (a) no sidesway	0.67
GH	0.512 and 0.768	7-2 (b) has sidesway	1.22

7.4 STIFFNESS-REDUCTION FACTORS

As previously mentioned, the alignment charts were developed on the basis of a set of idealized conditions that are seldom if ever completely met in a real structure. Included among those conditions are the following: column behavior is purely elastic, all columns buckle simultaneously, all members have constant cross sections, all joints are rigid, and so on.

If the actual conditions are different from these assumptions, unrealistically high K factors may be obtained from the charts, and overconservative designs may result. A large percentage of columns will fall in the inelastic range, but the alignment charts were prepared assuming elastic failure. This situation, previously discussed in Chapter 5, is illustrated in Fig. 7.7. For such cases, the chart K values are too conservative and should be corrected as described in this section.

In the elastic range, the stiffness of a column is proportional to EI, where $E = 29,000$ ksi; in the inelastic range, its stiffness is more accurately proportional to $E_T I$, where E_T is a reduced or tangent modulus.

The buckling strength of columns in framed structures was shown in the alignment charts to be related to

$$G = \frac{\text{column stiffness}}{\text{girder stiffness}} = \frac{\Sigma (EI/L) \text{ columns}}{\Sigma (EI/L) \text{girders}}$$

If the columns behave elastically, the modulus of elasticity will be canceled from the preceding expression for G. If the column behavior is inelastic, however, the column stiffness factor will be smaller and will equal $E_T I/L$. As a result, the G factor used to enter the alignment chart will be smaller, and the K factor selected from the chart will be smaller.

Though the alignment charts were developed for elastic column action, they may be used for an inelastic column situation if the G value is multiplied by a correction factor called the *stiffness reduction factor* (SRF). This reduction factor equals the tangent modulus over the elastic modulus (E_T/E) and is approximately equal to

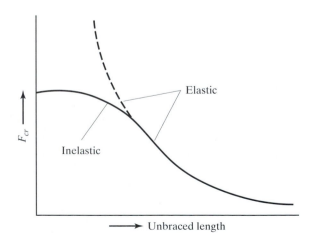

FIGURE 7.7

$F_{cr\,inelastic}/F_{cr\,elastic} \approx (P_u/A)/F_{cr\,elastic}$. Values of this correction factor are shown for various P_u/A values in Table 7.2 which is Table 4.1 in the LRFD Manual. A direct design method for considering inelastic buckling is presented in the Manual. It involves the following steps:

1. Calculate P_u and select a trial column size.
2. Calculate P_u/A and pick the SRF from Table 7.2. (If P_u/A is less than the values given in the table the column is in the elastic range and no reduction needs to be made.)
3. The value of $G_{elastic}$ is computed and multiplied by the SRF and K is picked from the chart.
4. The effective slenderness ratio KL/r is computed, and $\phi_c F_{cr}$ obtained from the Manual is multiplied by the column area to obtain P_u. If this value is appreciably different from the value computed in Step 1, another trial column size is attempted and the four steps are repeated.

Example 7-3 illustrates these steps for the design of a column in a frame subject to sidesway. *In this example, note that the author has only considered in-plane behavior and only bending about the x axis.* As a result of inelastic behavior, the effective length factor is appreciably reduced.

Example 7-3

a. Select a W section for column AB of the unbraced frame shown in Fig. 7.8 assuming that we have elastic behavior and that all of the other assumptions on which the alignment charts were developed are met. $P_u = 1660$ k. $F_y = 50$ ksi. Assume the columns above and below AB are the same size as AB.
b. Repeat part (a) if inelastic column behavior is considered.

Solution

a. Assuming column in elastic range and selecting a trial section based on an estimated $K_y L_y = 12$ ft

$$\text{Try W12} \times 170 \ (A = 50.0 \text{ in}^2, \ I_x = 1650 \text{ in}^4, \ r_x = 5.74 \text{ in})$$

$$G_A = G_B = \frac{\Sigma(I_c/L_c)}{\Sigma(I_g/L_g)} = \frac{(2)(1650/12)}{(2)(800/30)} = 5.16$$

$K = 2.30$ from Fig. 7.1 alignment chart

$$\left(\frac{KL}{r}\right)_x = \frac{(2.30)(12 \times 12)}{5.74} = 57.70$$

$\phi_c F_{cr} = 33.31$ ksi

$\phi_c P_n = (33.31)(50.0) = 1665 \text{ k} > P_u = 1660 \text{ k}$ (OK)

Use W12 × 170 $F_y = 50$ ksi

TABLE 7.2   Stiffness Reduction Factor τ

P_u/A_g	F_y, ksi				
	35	36	42	46	50
43	–	–	–	–	–
42	–	–	–	–	0.0319
41	–	–	–	–	0.0944
40	–	–	–	–	0.155
39	–	–	–	0.00696	0.215
38	–	–	–	0.0756	0.273
37	–	–	–	0.142	0.329
36	–	–	–	0.207	0.383
35	–	–	0.0529	0.270	0.436
34	–	–	0.127	0.331	0.486
33	–	–	0.198	0.390	0.535
32	–	–	0.267	0.447	0.582
31	–	–	0.334	0.501	0.627
30	–	0.0529	0.398	0.554	0.670
29	0.0678	0.139	0.460	0.604	0.711
28	0.155	0.221	0.519	0.651	0.749
27	0.240	0.301	0.576	0.697	0.785
26	0.321	0.377	0.629	0.739	0.819
25	0.398	0.450	0.680	0.779	0.850
24	0.472	0.519	0.727	0.816	0.879
23	0.542	0.585	0.772	0.850	0.905
22	0.608	0.646	0.813	0.882	0.929
21	0.670	0.704	0.850	0.910	0.949
20	0.727	0.757	0.884	0.934	0.966
19	0.780	0.806	0.914	0.955	0.981
18	0.828	0.850	0.941	0.973	0.991
17	0.871	0.890	0.963	0.987	0.999
16	0.909	0.924	0.980	0.996	1.00
15	0.941	0.952	0.993	1.00	
14	0.966	0.975	1.00		
13	0.986	0.991			
12	0.998	1.00			
11	1.00				
10					
9	↓	↓	↓	↓	↓
8					

– indicates stiffness reduction factor is not applicable because P_u exceeds $\phi_c F_y A_g$ (column design strength for $Kl/r = 0$).

From American Institute of Steel Construction, *Manual of Steel Construction Load & Resistance Factor Design*, 3d. ed. (Chicago, AISC, 2002), p. 4-20. Reprinted with the permission of AISC.

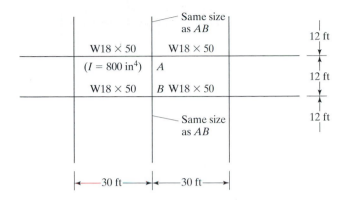

FIGURE 7.8

b.   Inelastic solution

Try the next lightest section W12 × 152
$(A = 44.7 \text{ in}^2, I_x = 1430 \text{ in}^4, r_x = 5.66 \text{ in})$

$$\frac{P_u}{A} = \frac{1660}{44.7} = 37.14 \text{ ksi}$$

SRF = 0.320 from Table 3.1 in Manual

∴ Column is in inelastic range

$$G_A = G_B = \frac{\Sigma(I_c/L_c)}{\Sigma(I_g/L_g)} (\text{SRF})$$

$$= \frac{(2)(1430/12)}{(2)(800/30)}(0.320) = 1.43$$

$K = 1.48$ from Fig. 7.2(b) alignment chart

$$\frac{KL}{r} = \frac{(1.48)(12 \times 12)}{5.66} = 37.65$$

$\phi_c F_{cr} = 38.31 \text{ ksi}$
$\phi_c P_n = (38.31)(44.7) = 1712 \text{ k} > P_u = 1660 \text{ k}$ (OK)
Use W12 × 152 $F_y = 50 \text{ ksi}$

7.5   COLUMNS LEANING ON EACH OTHER FOR IN-PLANE DESIGN

When we have an unbraced frame with beams rigidly attached to columns, it is safe to design each column individually using the sidesway uninhibited alignment chart to obtain the K factors (which will probably be appreciably larger than 1.0).

A column cannot buckle by sidesway unless all of the columns on that story buckle by sidesway. One of the assumptions on which the alignment chart of Fig. 7.2(b) was prepared is that all of the columns on the story in question would buckle at the same time. If this assumption is correct, the columns cannot support or brace each other, because if one gets ready to buckle, they all supposedly are ready to buckle.

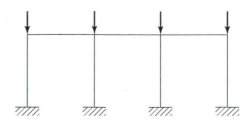

FIGURE 7.9

In some situations, however, certain columns in a frame have some excess buckling strength. If, for instance, the buckling loads of the exterior columns of the unbraced frame of Fig. 7.9 have not been reached when the buckling loads of the interior columns are reached, the frame will not buckle. The interior columns in effect will lean against the exterior columns, that is, the exterior columns will brace the interior ones. For this situation, shear resistance is provided in the exterior columns which resist the sidesway tendency.[5]

A pin-ended column which does not help provide lateral stability to a structure is referred to as a "leaning" column. Such a column depends on the other parts of the structure to provide lateral stability. Section C2.2 of the LRFD Specification states that the effects of gravity loaded "leaning" columns shall be included in the design of moment frame columns.

There are many practical situations in which some columns have excessive buckling strength. This might happen when the designs of different columns on a particular story are controlled by different loading conditions. For such cases, failure of the frame will occur only when the gravity loads are increased sufficiently to offset the extra strength of the lightly loaded columns. As a result, the critical loads for the interior columns of Fig. 7.9 are increased and, in effect, their effective lengths are decreased. In other words, if the exterior columns are bracing the interior ones against sidesway, the K factors for those interior columns are approaching 1.0. Yura[6] says that the effective length of some of the columns in a frame subject to sidesway can be reduced to 1.0 in this type of situation even though there is no apparent bracing system present.

The net effect of the information presented here is that the total gravity load that an unbraced frame can support equals the sum of the strength of the individual columns. In other words, the total gravity load that will cause sidesway buckling in a frame can be split up among the columns in any proportion so long as the maximum load applied to any one column does not exceed the maximum load that column could support if it were braced against sidesway with $K = 1.0$.

For this discussion, the unbraced frame of Fig. 7.10(a) is considered. It is assumed that each column has a $K = 2.0$ and will buckle under the loads shown.

When sidesway occurs, the frame will lean to one side as shown in part (b) of the figure and $P\Delta$ moments equal to 200Δ and 700Δ will be developed.

[5]J. A. Yura, "The Effective Length of Columns in Unbraced Frames," *Engineering Journal*, AISC, 8, no. 2 (second quarter, 1971), pp. 37–42.
[6]Ibid., pp. 39–40.

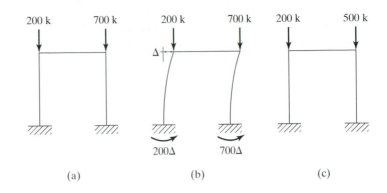

FIGURE 7.10   (a)   (b)   (c)

Suppose that we load the frame with 200 k on the left-hand column and 500 k on the right-hand column (or 200 k less than we had before). We know that for this situation, which is shown in part (c) of the figure, the frame will not buckle by sidesway until we reach a moment of 700Δ at the right-hand column base. This means that our right-hand column can take an additional moment of 200Δ. Thus, as Yura says, the right-hand column has a reserve of strength that can be used to brace the left-hand column and prevent its sidesway buckling.

Obviously, the left-hand column is now braced against sidesway, and sidesway buckling will not occur until the moment at its base reaches 200Δ. Therefore, it can be designed with a K factor less than 2.0 and can support an additional load of 200 k, giving it a total load of 400 k—*but this load must not be greater than would be its capacity if it were braced against sidesway with K = 1.0.* It should be mentioned that the total load the frame can carry still is 900 k, as in part (a) of the figure.

The advantage of the frame behavior described here is illustrated in Figure 7.11, where it is assumed that the interior columns of a frame are braced against sidesway by the exterior columns. As a result the interior columns are assumed to each have K factors equal to 1.0. They are designed for the factored loads shown (660 k each). Then the K factors are determined for the exterior columns with the sidesway uninhibited chart of Fig. 7.11, and they are each designed for column loads equal to 440 + 660 = 1100 k.

Example 7-4

For the frame of Fig. 7.11, which consists of A36 steel, beams are rigidly connected to the exterior columns, while all other connections are simple. The columns are braced top and bottom against sidesway out of the plane of the frame so that K = 1.0 in that direction. Sidesway is possible in the plane of the frame. Design the interior columns assuming K = 1.0 and the exterior columns with K as determined from the alignment chart and P_u = 1100 k. (With this approach to column buckling, the interior columns could carry no load at all since they appear to be unstable under sidesway conditions.)

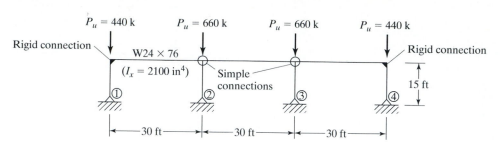

FIGURE 7.11

Solution. *Design of interior columns*

Assume $K = 1.0$, $KL = (1.0)(15) = 15$ ft, $P_u = 660$ k
Use W14 × 90 A36

Design of exterior columns

Out of plane $K_y = 1.0$, $P_u = 440$ k

In plane $P_u = 440 + 660 = 1100$ k, K_x to be determined from alignment chart. Estimating a column size a little larger than would be required for $P_u = 1100$ k. Try W14 × 159 ($A = 46.7$ in², $I_x = 1900$ in⁴, $r_x = 6.38$ in)

$$G_{top} = \frac{1900/15}{2100/30 \times 0.5} = 3.62$$

(noting that girder stiffness is multiplied by 0.5 since sidesway is permitted and far end of girder is hinged.)

$$G_{bottom} = 10$$

$K_x = 2.25$ from Fig. 7.2(b)

$$\frac{K_x L_x}{r_x} = \frac{(2.25)(12 \times 15)}{6.38} = 63.48 \qquad \text{(OK)}$$

$\phi_c F_{cr} = 24.75$ ksi

$\phi_c P_n = (24.75)(46.7) = 1156$ k $> P_u = 1100$ k

Use W14 × 159 A36

Space is not taken in this example to apply column stiffness reduction factors (as described in Section 7.4) to the columns of this frame. Section C2.2 of the LRFD Specification states that *stiffness reduction adjustment due to column inelasticity is permitted when the "leaning" column theory has been applied.*

It is rather frightening to think of additions to existing buildings and the leaner column theory. If we have a building (represented by the solid lines in Fig. 7.12) and we decide to add onto it (indicated by the dashed lines in the same figure), we may think that we can use the old frame to brace the new one and that we can keep expanding

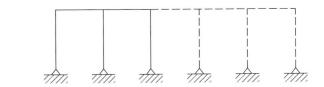

FIGURE 7.12

laterally with no effect on the existing building. Sadly, we may be in for quite a surprise. The leaning of the new columns may cause one of the old ones to fail.

7.6 BASE PLATES FOR CONCENTRICALLY LOADED COLUMNS

The design compressive stress in a concrete or other type of masonry footing is much smaller than it is in a steel column. When a steel column is supported by a footing, it is necessary for the column load to be spread over a sufficient area to keep the footing from being overstressed. Loads from steel columns are transferred through a steel base plate to a fairly large area of the footing below. (It will be noted that a footing performs a related function in that it spreads the load over an even larger area so that the underlying soil will not be overstressed.)

The base plates for steel columns can be welded directly to the columns or they can be fastened by means of some type of bolted or welded lug angles. These connection methods are illustrated in Fig. 7.13. A base plate welded directly to the column is shown in part (a) of the figure. For small columns, these plates are probably shop-welded to the columns, but for larger columns it may be necessary to ship the plates separately and set them to the correct elevations. For this second case, the columns are connected to the footing with anchor bolts that pass through the lug angles which have been shop-welded to the columns. This type of arrangement is shown in part (b) of the figure. Some designers like to use lug angles on both flanges and web. (The reader should be aware of new OSHA regulations for the safe erection of structural steel,

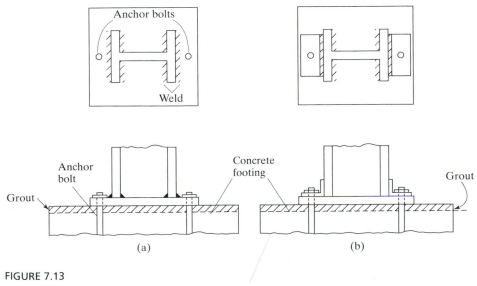

FIGURE 7.13

Column base plates.

200

Effective Area:
Problem 1) Compute the effective area for each of the following cases in two ways:
(1) use the AISC specification equation for $U = 1 - \frac{x}{l} \leq 0.9$
(2) use the average value of U as given in the AISC Commentary. (see Table 3.2 page 77, McCormac and Nelson, 3rd Ed).

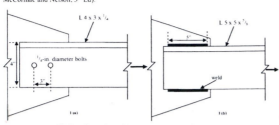

ior bolts for each column. These bolts

ilding is the proper positioning of col-
heir correct elevations, serious stress
the steel frame. One of the following
the erection of a column to its proper
base plates. An article by Ricker[7] de-

For small-to-medium base plates (up to 20 to 22 in), approximately .25-in-thick leveling plates with the same dimensions as the base plates (or a little larger) are shipped to the job and carefully grouted in place to the proper elevations. Then the columns with their attached base plates are set on the leveling plates.

As these leveling plates are very light and can be handled manually, they are set by the foundation contractor. This is also true for the lighter base plates. On the other hand, large base plates which have to be lifted with a derrick or crane are usually set by the steel erector.

For larger base plates up to about 36 in. some types of leveling nuts are used to ility during erection, these nuts must

n, the columns with the attached base ficult to ship them together. For such

Effective Area
Problem 1)
(a) (1) Using AISC specific equation:

$kip := 1000 \cdot lbf$ $ksi := 1000 \cdot psi$

Section properties for L 4 x 3 x $\frac{1}{4}$ [AISC Table 1-7 page 1-36]

$A_g = 1.69 \cdot in^2$ gross area

$x_{bar} = 0.725 \cdot in$ centroid (from longer leg)

$t = \frac{1}{4} \cdot in$ web thickness

Diameter of the holes (for standard holes)

$d_{hole} := \frac{1}{4} \cdot in + \frac{1}{8} \cdot in$

$d_{hole} = 0.375 in$

Calculate net area

$A_n = A_g - d_{hole} \cdot t$

$A_n := 1.69 \cdot in^2 - 0.375 \cdot in \cdot \left(\frac{1}{4} \cdot in\right)$

$A_n = 1.596 in^2$

Calculate U

$U = 1 - \frac{x_{bar}}{L}$

$U := 1 - \frac{0.725 \cdot in}{3 \cdot in}$

$U = 0.758$ smaller than 0.9

effective area

$A_e := A_n \cdot U$

$\boxed{A_e = 1.21 in^2}$

(a) (2) Using AISC commentary

$U := 0.75$ [Table 3.2 page 77, McCormac and Nelson] case c, two fasteners per line

$A_e := A_n \cdot U$

$\boxed{A_e = 1.20 in^2}$

(b) (1) Using AISC specific equation:

Section properties for L 5 x 5 x $\frac{5}{8}$ [AISC Table 1-7 page 1-34]

$A_g = 5.90 \cdot in^2$ gross area

$x_{bar} = 1.47 \cdot in$ centroid

Calculate U
note: See page 78, McCormac and Nelson, this is case 1 of section 3.5.2 of text. Use the same equation as the bolted connections.

$U = 1 - \frac{x_{bar}}{L}$

$U := 1 - \frac{1.47 \cdot in}{5 \cdot in}$

$U = 0.706$ smaller than 0.9

Effective area

$A_e = A_g \cdot U$

$A_e := 5.90 \cdot in^2 \cdot (0.706)$

$\boxed{A_e = 4.17 in^2}$

(a) (2) Using AISC commentary

The average values for welded connections:
Case 1) For W-, M-, or S-shapes with width-to-depth ratio of at least 2/3 (and tee-shapes cut from them) and connected at the flanges,
$U = 0.9$
Case 2) For all other shapes
$U = 0.85$

$U := 0.85$ use case 2, for all other shapes.

$A_e = A_g \cdot U$

$A_e := 5.90 \cdot in^2 \cdot (0.85)$

$\boxed{A_e = 5.01 in^2}$

in must be milled if they are beyond the fl
Part 1 of the LRFD Manual entitled "Permi

---

[7] D. T. Ricker, "Some Practical Aspects of Column Bases," *Engineering Journal*, AISC, *26*, no. 3 (3d quarter, 1989), pp. 81–89.

Column base plate with anchor bolts at each corner. Column welded to base plate. Grout placed below base plate to transfer compression load evenly to reinforced concrete foundation. (Courtesy American Institute of Steel Construction.)

If the bottom surfaces of the plates are to be in contact with cement grout to ensure full bearing contact on the foundation, they do not have to be milled. Furthermore, the top surfaces of plates thicker than 4 in. do not have to be milled if full-penetration welds (to be described in Chapter 14) are used. Notice that when finishing is required as described here, the plates will have to be ordered a little thicker than required for their final dimensions to allow for the cuts.

Initially, columns will be considered that support average-size loads. Should the loads be very small, so that base plates are very small, the design procedure will have to be revised as described later in this section.

The LRFD Specification does not specify a particular method for designing column base plates. The method presented here is based upon example problems presented in Part 14 of the LRFD Manual.

To analyze the base plate shown in Fig. 7.14, the column is assumed to apply a total load equal to P_u to the base plate. Then the load P_u is assumed to be transmitted uniformly through the plate to the footing below with a pressure equal to P_u/A where A is the area of the base plate. The footing will push back with a pressure of P_u/A and will tend to curl up the cantilevered parts of the base plate outside of the column, as shown in the figure. This pressure will also tend to push up the base plate between the flanges of the column.

With reference to Fig. 7.14, the LRFD Manual suggests that maximum moments in a base plate occur at distances $0.80\, b_f$ and $0.95\, d$ apart. The bending moment can be calculated at each of these sections and the larger value used to determine the plate thickness needed. This method of analysis is only a rough approximation of the true conditions because the actual plate stresses are caused by a combination of bending in two directions.

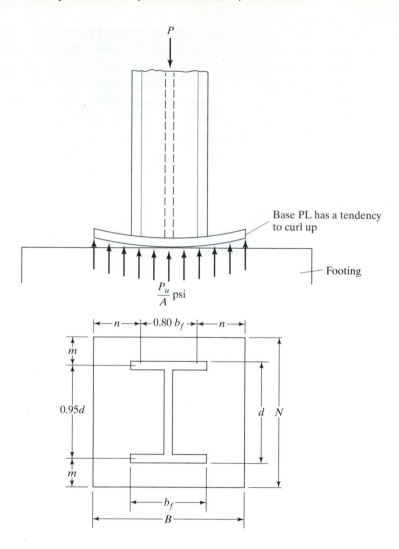

FIGURE 7.14

7.6.1   Plate Area

The design strength of the concrete in bearing beneath the base plate must at least equal the load to be carried. When the base plate covers the entire area of the concrete support, this strength equals ϕ_c (0.60 for bearing on concrete) times the nominal strength of the concrete 0.85 f_c' times A_1 (where f_c' is the 28-day compression strength of the concrete in ksi and A_1 is the area of the plate).

$$P_u \leq \phi_c P_p = \phi_c(0.85\, f_c' A_1) \qquad \text{(from LRFD Equation J9-1)}$$

$$A_1 \leq \frac{P_u}{\phi_c\, 0.85\, f_c'}$$

Should the full area of the concrete support not be covered by the plate, the concrete underneath the plate, surrounded by concrete outside, will be somewhat stronger. For this situation, the LRFD Specification (J9b) permits the design strength above $(\phi 0.85\, f_c'A_1)$ to be increased by multiplying it by $\sqrt{A_2/A_1}$. In the resulting expression, A_2 is the maximum area of the portion of the supporting concrete, which is geometrically similar to and concentric with the loaded area. The value of $\sqrt{A_2/A_1}$ is limited to a maximum value of 2 as shown in the expression below.

$$P_u \le \phi_c P_p = \phi_c(0.85\, f_c'A_1)\sqrt{\frac{A_2}{A_1}} \quad \text{with} \quad \sqrt{\frac{A_2}{A_1}} \le 2 \qquad \text{(from LRFD Equation J9-2)}$$

Then

$$A_1 \ge \frac{P_u}{\phi_c(0.85\, f_c')\sqrt{\dfrac{A_2}{A_1}}}, \quad \text{where} \quad \sqrt{\frac{A_2}{A_1}} \text{ may not be } > 2$$

Furthermore, A_1 may not be less than the depth of the column times its flange width.

$$A_1 = b_f d$$

After the controlling value of A_1 is determined as described above, the plate dimensions B and N (shown in Fig. 7.14, are selected to the nearest 1 or 2 in. so that the values of m and n shown in the figure are roughly equal. Such a procedure will make the cantilever moments in the two directions approximately the same. This will enable us to keep the plate thicknesses to a minimum. The condition $m = n$ can be approached if the following equation is satisfied.

$$N \approx \sqrt{A_1} + \Delta$$

where
$$A_1 = \text{area of plate} = BN$$
$$\Delta = 0.5\,(0.95\, d - 0.80\, b_f)$$

$$B \approx \frac{A_1}{N}$$

7.6.2   Plate Thickness

To determine the required plate thickness, moments are taken in the two directions as though the plate is cantilevered out by the dimensions m and n. Reference is again made here to Fig. 7.14. These moments are respectively $(P_u/BN)(m)(m/2) = P_u m^2/2BN$ and $(P_u/BN)(n)(n/2) = P_u n^2/2BN$ both computed for a 1-in. width of plate.

The design moment strength of the plate per in. of width must at least equal the larger of these two moments. In the next few chapters we will learn to compute this value. For a plate it will equal $\phi F_y(t^2/4)$ where t is the plate thickness and $\phi = 0.9$. Equating this expression to the maximum computed moment the required value of t can be determined as follows:

$$\phi F_y \frac{t^2}{4} \geq \frac{P_u m^2}{2BN} \quad \text{or} \quad \frac{P_u n^2}{2BN}$$

$$t \geq m \sqrt{\frac{2P_u}{0.9F_y BN}} \quad \text{or} \quad n \sqrt{\frac{2P_u}{0.9F_y BN}}$$

If lightly loaded base plates, as for the columns of low-rise buildings and pre-engineered metal buildings, are designed by the procedure just described they will have quite small areas. They will, as a result, extend very little outside the edges of the columns and the computed moments, and the resulting plate thicknesses will be very small, perhaps so small as to be impractical.

Several procedures for handling this problem have been proposed. In 1990, W. A. Thornton[8] combined three of the methods into a single procedure applicable to either heavily loaded or lightly loaded base plates. This modified method is used for the example base plate problems in Part 14 of the LRFD Manual as well as for the example problems in this text.

Thornton proposed that the thickness of the plates be determined using the largest of m, n, or $\lambda n'$. He called this largest value ℓ

$$\ell = \max (m, n, \text{ or } \lambda n')$$

To determine $\lambda n'$ it is necessary to substitute into the following expressions which are developed in his paper.

$$\phi_c P_p = \phi_c \, 0.85 \, f_c' A_1 \quad \text{for plates covering the full area of the concrete support}$$

$$\phi_c P_p = \phi_c 0.85 \, f_c' A_1 \sqrt{\frac{A_2}{A_1}} \quad \text{where} \quad \sqrt{\frac{A_2}{A_1}} \text{ must be } \leq 2 \text{ for plates not covering the}$$

entire area of concrete support

$$X = \left[\frac{4db_f}{(d + b_f)^2} \right] \frac{P_u}{\phi_c P_p}$$

$$\lambda = \frac{2\sqrt{X}}{1 + \sqrt{1 - X}} \leq 1$$

$$\lambda n' = \frac{\lambda \sqrt{db_f}}{4}$$

Finally, the thickness of the plate is determined with

$$t = \ell \sqrt{\frac{2P_u}{0.9F_y \, BN}}$$

Three example base plate designs are presented in the next few pages. Example 7-5 illustrates the design of a base plate supported by a large reinforced concrete footing

[8]W. A. Thornton, "Design of Base Plates for Wide Flange Columns—A Concatenation of Methods," *Engineering Journal*, AISC, 27, no. 4 (4th quarter, 1990) pp. 173–174.

with A_2 many times as large as A_1. In Example 7-6, a base plate is designed which is supported by a concrete pedestal where the plate covers the entire concrete area.

Finally, in Example 7-7 a base plate is selected for a column that is to be supported on a pedestal 4 in wider on each side than the plate. This means A_2 cannot be determined until the plate area (A_1) is computed. This puts us into the following trial-and-error procedure:

1. Compute $A_1 \geq \dfrac{P_u}{\phi_c 0.85 f_c'}$

2. Then $A_2 = (B + 4)(N + 4)$

3. Now A_1 can be determined $\geq \dfrac{P_u}{\phi_c 0.85 f_c' \sqrt{\dfrac{A_2}{A_1}}} = \dfrac{\text{above } A_1}{\sqrt{\dfrac{A_2}{A_1}}}$

   and values selected for B and N.

4. The value of A_2 determined in Step 2 has now changed, so we continue through Steps 2 to 3 until A_1 changes very little.

Example 7-5

Design a base plate of A36 steel for a W12 $\times$ 65 column ($F_y = 50$ ksi), which supports a factored axial load $P_u = 720$ k. The concrete has a specified compressive strength $f_c' = 3$ ksi. Assume footing size is 9 ft $\times$ 9 ft.

Solution. Using a W12 $\times$ 65 column ($d = 12.10$ in, $b_f = 12.00$ in)

$$A_2 = (12 \times 9)(12 \times 9) = 11{,}664 \text{ in}^2$$

Determine required base plate area

Assuming the area of the supporting concrete will be far greater than the area of the base plate such that $\sqrt{A_2/A_1} \geq 2$

$$A_1 = \frac{P_u}{\phi_c(0.85 f_c')\sqrt{\dfrac{A_2}{A_1}}} = \frac{720}{(0.6)(0.85)(3)(2)} = 235.3 \text{ in}^2 \leftarrow$$

Checking

$$\sqrt{\frac{A_2}{A_1}} = \sqrt{\frac{11{,}664}{235.3}} = 7.04 > 2 \qquad \text{(OK)}$$

The base plate should be at least as large as the column.

$$A_1 = db_f = (12.10)(12.00) = 145.2 \text{ in}^2 < 235.3 \text{ in}^2 \qquad \text{(OK)}$$

Optimizing base-plate dimensions to make $n \sim m$

$$\Delta = \frac{0.95d - 0.8\,b_f}{2} = \frac{(0.95)(12.14) - (0.8)(12.00)}{2} = 0.947 \text{ in}$$

$$N = \sqrt{A_1} + \Delta = \sqrt{235.3} + 0.947 = 16.30 \text{ in Say 16 in}$$

$$B = \frac{A_1}{N} = \frac{235.3}{16} = 14.71 \text{ in Say 15 in}$$

Computing required base-plate thickness

$$m = \frac{N - 0.95\,d}{2} = \frac{16 - (0.95)(12.10)}{2} = 2.25 \text{ in}$$

$$n = \frac{B - 0.8\,b_f}{2} = \frac{15 - (0.8)(12.00)}{2} = 2.70 \text{ in}$$

$$\phi_c P_p = 0.6(0.85 f_c' A_1)\sqrt{\frac{A_2}{A_1}} = (0.6)(0.85)(3)(16)(15)(2) = 734.4 \text{ k}$$

$$X = \frac{4db_f}{(d + b_f)^2}\frac{P_u}{\phi_c P_p} = \frac{(4)(12.12)(12.00)}{(12.10 + 12.00)^2}\frac{720}{734.4} = 0.980$$

$$\lambda = \frac{2\sqrt{X}}{1 + \sqrt{1 - X}} = \frac{2\sqrt{0.980}}{1 + \sqrt{1 - 0.980}} = 1.73 > 1.0 \quad \text{Use 1.0}$$

$$\lambda n' = \frac{\lambda\sqrt{db_f}}{4} = \frac{1\sqrt{(12.10)(12.00)}}{4} = 3.01 \text{ in}$$

$$\ell = \max(m, n, \lambda n') = 3.01 \text{ in}$$

$$t_{\text{reqd}} = \ell\sqrt{\frac{2P_u}{0.9F_y BN}} = 3.01\sqrt{\frac{(2)(720)}{(0.9)(36)(15)(16)}} = 1.30 \text{ in}$$

Use PL$1\frac{1}{2}$ × 15 × 1 ft 4 in A36

Example 7-6

A base plate is to be designed for a W12 × 152 column($F_y = 50$ ksi) supporting a factored axial load $P_u = 960$ k. Determine the dimensions for the plate (A36 steel) and for a concrete pedestal ($f_c' = 3$ ksi) if the plate is to cover the entire concrete area.

Solution. Using a W12 × 152 ($d = 13.70$ in, $b_f = 12.5$ in) Determine required base-plate area

$$A_1 \geq \frac{P_u}{\phi_c(0.85 f_c')} = \frac{960}{(0.6)(0.85)(3)} = 627.5 \text{ in}^2 \leftarrow$$

$$A_1 \geq db_f = (13.70)(12.48) = 171.0 \text{ in}^2$$

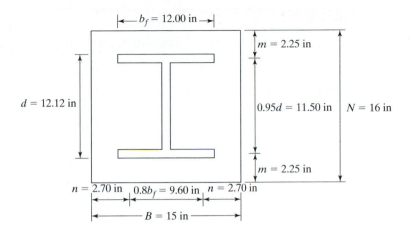

FIGURE 7.15

Optimizing base-plate dimensions to make $n \sim m$

$$\Delta = \frac{0.95\,d - 0.8\,b_f}{2} = \frac{(0.95)(13.70) - (0.8)(12.5)}{2} = 1.51 \text{ in}$$

$$N = \sqrt{A_1} + \Delta = \sqrt{627.5} + 1.51 = 26.56 \text{ in. Say 27 in}$$

$$B = \frac{A_1}{N} = \frac{627.5}{27} = 23.24 \text{ in Say 24 in}$$

Computing required base-plate thickness

$$m = \frac{N - 0.95d}{2} = \frac{27 - (0.95)(13.70)}{2} = 6.99 \text{ in}$$

$$n = \frac{B - 0.8\,b_f}{2} = \frac{24 - (0.8)(12.5)}{2} = 7.00 \text{ in}$$

$$\phi_c P_p = 0.6(0.85\,f'_c A_1) = (0.6)(0.85)(3)(27)(24) = 991.4 \text{ k} > P_u = 960 \text{ k}$$

$$X = \frac{4db_f}{(d + b_f)^2}\frac{P_u}{\phi_c P_p} = \frac{(4)(13.70)(12.5)}{(13.71 + 12.48)^2}\frac{960}{991.4} = 0.967$$

$$\lambda = \frac{2\sqrt{X}}{1 + \sqrt{1 - X}} = \frac{(2)\sqrt{0.967}}{1 + \sqrt{1 - 0.967}} = 1.66 > 1.0. \text{ Use 1.0.}$$

$$\lambda n' = \frac{\lambda\sqrt{db_f}}{4} = \frac{1.0\sqrt{(13.70)(12.5)}}{4} = 3.27 \text{ in}$$

$$\ell = \max{(m, n, \lambda n')} = 7.01 \text{ in}$$

$$t_{reqd} = \ell\sqrt{\frac{2P_u}{0.9F_y BN}} = 7.01\sqrt{\frac{(2)(960)}{(0.9)(36)(24)(27)}} = 2.12 \text{ in}$$

Use PL$2\frac{1}{4}$ × 24 × 2 ft 3 in. A36 with 24 × 27 pedestal

Example 7-7

Repeat Example 7-6 if the column is to be supported by a concrete pedestal 4 in. wider on each side than the base plate.

Solution. Using a W12 × 152 (d = 13.70 in, b_f = 12.5 in) Determining required base-plate area.

If base plate had same area as pedestal, A_1 could be determined as follows:

$$A_1 \geq \frac{P_u}{\phi_c(0.85f'_c)} = \frac{960}{(0.6)(0.85)(3)} = 627.5 \text{ in}^2$$

$$A_1 \geq db_f = (13.70)(12.5) = 171.2 \text{ in}^2$$

If we try a plate 25 × 26 (A_1 = 650 in^2) the pedestal area A_2 will equal

$(25 + 4)(26 + 4) = 870$ in^2. Then

$$\sqrt{\frac{A_2}{A_1}} = \sqrt{\frac{870}{650}} = 1.16$$

And

$$A_1 \geq \frac{P_u}{\phi_c(0.85f'_c)\sqrt{\dfrac{A_2}{A_1}}} = \frac{960}{(0.6)(0.85)(3)(1.16)} = 540.9 \text{ in}^2$$

Trying a plate 23 × 24(552 in^2) the pedestal area will be $(23 + 4) \times (24 + 4) =$ 756 in^2 and $\sqrt{A_2/A_1} = \sqrt{756/540.9} = 1.18$. Thus no appreciable change in A_1.

Optimizing base-plate dimensions

$$\Delta = \frac{0.95\, d - 0.8\, b_f}{2} = \frac{(0.95)(13.70) - (0.8)(12.5)}{2} = 1.51 \text{ in}$$

$$N = \sqrt{A_1} + \Delta = \sqrt{540.9} + 1.51 = 24.77 \text{ in Say 25 in}$$

$$B = \frac{540.9}{25} = 21.6 \text{ Say 22 in}$$

Use pedestal 26 × 29

$$\sqrt{\frac{A_2}{A_1}} = \sqrt{\frac{(26)(29)}{(22)(25)}} = 1.17 \quad \therefore \text{ No change}$$

Computing required base-plate thickness

$$m = \frac{N - 0.95d}{2} = \frac{25 - (0.95)(13.70)}{2} = 5.99 \text{ in}$$

$$n = \frac{B - 0.8b_f}{2} = \frac{22 - (0.8)(12.5)}{2} = 6.00 \text{ in}$$

$$\phi_c P_p = 0.6(0.85f'_c A_1)\sqrt{\frac{A_2}{A_1}} = (0.6)(0.85)(3)(22 \times 25)\sqrt{\frac{26 \times 29}{22 \times 25}}$$

$$= 984.6 \text{ k}$$

$$X = \frac{4db_f}{(d + b_f)^2} \frac{P_u}{\phi_c P_p} = \frac{(4)(13.70)(12.5)}{(13.70 + 12.5)^2} \frac{960}{984.6} = 0.973$$

$$\lambda = \frac{2\sqrt{X}}{1 + \sqrt{1 - X}} = \frac{2\sqrt{0.973}}{1 + \sqrt{1 - 0.973}} = 1.69 > 1.0 \text{ Use } 1.0$$

$$\lambda n' = \frac{\lambda\sqrt{db_f}}{4} = \frac{1\sqrt{(13.70)(12.5)}}{4} = 3.27 \text{ in}$$

$$\ell = \max{(m, n, \lambda n')} = 6.01 \text{ in}$$

$$t_{\text{reqd}} = \ell\sqrt{\frac{2P_u}{0.9F_y BN}} = 6.01\sqrt{\frac{(2)(960)}{(0.9)(36)(22)(25)}} = 1.97 \text{ in}$$

Use base PL2 × 22 × 2 ft 1 in. A36 and 26 × 29 pedestal

7.6.3   Moment Resisting Column Bases

The designer will often be faced with the need for moment resisting column bases. Before such a topic is introduced, however, the student needs to be familiar with the design of welds (Chapter 14) and moment resisting connections between members (Chapters 14 and 15). As a result, the subject of moment resisting base plates has been placed in Appendix E.

7.7   COMPUTER EXAMPLE

In Example 7-8 a column base plate is designed with INSTEP 32 for the information used in Example 7-5. To use the program it is necessary to input trial dimensions for the plate. Any reasonable values can be specified. The computer will then calculate the dimensions required based on those trial dimensions. The user will then see if the initial assumed dimensions are satisfactory. If not, some very accurate trial dimensions can be assumed from the results provided in the first solution.

Jackson and Mooreland Alignment Charts
Problem 2) problem 7-1 (page 211, McCormac and Nelson, 3rd Ed.)

Repeat Example 7-5 using INSTEP32.

Jackson and Mooreland Alignment Charts
Problem 2)

frame is subjected to sidesway

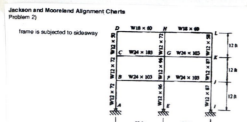

ind estimating the base plate area equal to

$$\frac{720}{)(0.85 \times 3)(2)}$$

.29 in^2 Try 15 in $\times$ 16 in $= 240$ in^2

for Column EF

$$G_{top\_EF} := \frac{\dfrac{833 \cdot in^4}{12 \cdot ft} + \dfrac{833 \cdot in^4}{12 \cdot ft}}{\dfrac{3000 \cdot in^4}{30 \cdot ft} + \dfrac{3000 \cdot in^4}{36 \cdot ft}}$$

<= two W12x16, I$_x$ from [LRFD table 1-1]

<= two W24x103, I$_x$ from [LRFD table 1-1]

$\boxed{G_{top\_EF} = 0.757}$

$\boxed{G_{bottom\_EF} := 10}$ for pinned base

rete footing parameters

$k_{EF} := 1.84$ from [text Figure 7-2 (b) pg 184]

3000 psi

108 in

108 in

for Column FG

$$G_{top\_FG} := \frac{\dfrac{833 \cdot in^4}{12 \cdot ft} + \dfrac{597 \cdot in^4}{12 \cdot ft}}{\dfrac{3000 \cdot in^4}{30 \cdot ft} + \dfrac{3000 \cdot in^4}{36 \cdot ft}}$$

<= W12x16 and W12x72, I$_x$ from [LRFD table 1-1] base plate

<= two W24x103, I$_x$ from [LRFD table 1-1]

5 in

$\boxed{G_{top\_FG} = 0.65}$

6 in

$G_{bottom\_FG} := G_{top\_EF}$

in

$\boxed{G_{bottom\_FG} = 0.757}$

$k_{FG} := 1.26$ from [text Figure 7-2 (b) pg 184]

Output:

for Column KL

$$G_{top\_KL} := \frac{\dfrac{475 \cdot in^4}{12 \cdot ft}}{\dfrac{984 \cdot in^4}{36 \cdot ft}}$$

<= W12x58, I$_x$ from [LRFD table 1-1]

<= two W18x60, I$_x$ from [LRFD table 1-1]

$\boxed{G_{top\_KL} = 1.448}$

4)

$= 734.4$ kips

$$G_{bottom\_KL} := \frac{\dfrac{740 \cdot in^4}{12 \cdot ft} + \dfrac{475 \cdot in^4}{12 \cdot ft}}{\dfrac{3000 \cdot in^4}{36 \cdot ft}}$$

<= W12x87 and W12x58, I$_x$ from [LRFD table 1-1]

<= W24x103, I$_x$ from [LRFD table 1-1]

$\boxed{G_{bottom\_KL} = 1.215}$

$k_{KL} := 1.44$ from [text Figure 7-2 (b) pg 184]

Lambda $= 1.73695$

L $= 3.01247$ in

t required $= 1.29636$ in

PROBLEMS

7-1. Using the Jackson and Moreland charts, determine the effective length factors for columns EF, FG, and KL of the frame shown in the accompanying illustration, assuming the frame is subject to sidesway and that all of the assumptions on which the alignment charts were developed are met. (*Ans.* 1.84, 1.26, and 1.44)

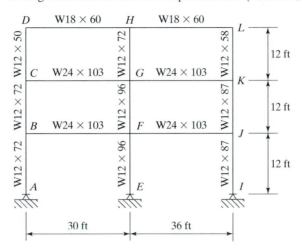

FIGURE P7-1

7-2. Determine the effective length factors for all the columns of the frame shown in the accompanying illustration. Note that the columns on the upper level are subject to sidesway while the ones on the lower level are braced against sidesway. Assume all of the assumptions on which the alignment charts were developed are met.

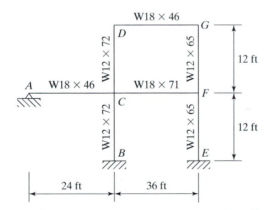

FIGURE P7-2

7-3. a. Select a W14 section for column AB in the frame shown if $P_u = 1100$ k, $F_y = 50$ ksi, and only in-plane behavior is considered. Furthermore, assume that the columns immediately above and below AB are approximately the same size as AB, and also that all the other assumptions on which the alignment charts were developed are met. (*Ans.* W14 × 109)

b. Repeat part (a) if inelastic behavior is considered. (*Ans.* W14 × 99)

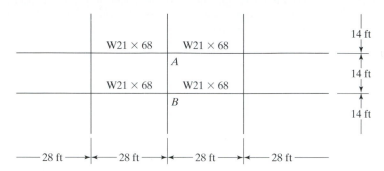

FIGURE P7-3

7-4. Repeat Prob. 7-3 if the beams are W24 × 62 and $P_u = 1200$ k.

7-5. Repeat Prob. 7-3 if a W12 is used. (*Ans.* W × 106, W12 × 96)

7-6. We desire to select a W14 section for column CD in the frame shown for which $P_u = 1300$ k and $F_y = 50$ ksi. Otherwise the conditions are exactly as those described for Prob. 7-3.

a. Assume elastic behavior.

b. Assume inelastic behavior.

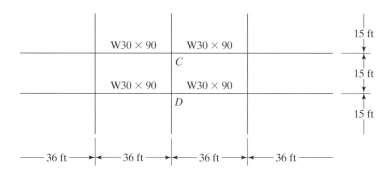

FIGURE P7-6

7-7. Design W14 columns for the bent shown in the accompanying illustration with 50 ksi steel using the inelastic K-factor procedure. The columns are braced top and bottom against sidesway out of the plane of the frame so that $K = 1.0$ in that direction. Sidesway is possible in the plane of the frame. Design the right-hand column using $K = 1.0$ and the left-hand column with K as determined from the alignment chart and $P_u = 1600$ k. The beam is rigidly connected to the left column but hsas only a simple connection to the right column. (*Ans.* W14 × 90, W14 × 176)

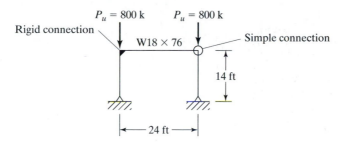

FIGURE P7-7

7-8.   Repeat Prob. 7-7 if the P_u loads are 1200 k each and A36 steel is used.

7-9.   The columns for the frames shown in the accompanying illustration are braced top and bottom against sidesway out of the plane of the frame so that $K = 1.0$ in that direction. Sidesway is possible in the plane of the frame. Design the interior column assuming $K = 1.0$ and the exterior columns with K as determined from the alignment chart and $P_u = 2000$ k. Use $F_y = 50$ ksi and a W14 section. (*Ans.* W14 × 193, W14 × 211)

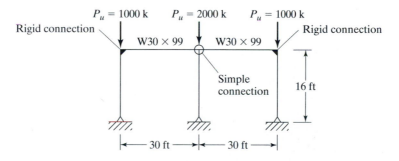

FIGURE P7-9

7-10.   Repeat Prob. 7-9 assuming outside column bases are fixed.

7-11.   For the frame shown in the accompanying illustration, the beams are rigidly connected to the exterior columns while all other connections are simple. The columns are braced top and bottom against sidesway out of the plane of the frame so that $K = 1.0$ in that direction. Sidesway is possible in the plane of the frame. Design W14 interior columns of 50 ksi steel assuming $K = 1.0$ and W14 exterior columns with K as determined from the alignment chart and $P_u = 1625$ k. (*Ans.* W14 $\times$ 90, W14 $\times$ 176)

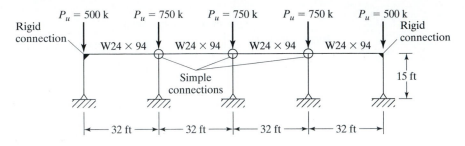

FIGURE P7-11

7-12. Repeat Prob. 7-11 using $F_y = 36$ ksi.

7-13. Design a base plate with A36 steel for a W14 $\times$ 82 column with a dead working load of 120 k and a working live load of 460 k. The concrete 28-day strength is 3 ksi. Footing size is 11 by 11 ft. (*Ans.* PL1$\frac{1}{2}$ $\times$ 15 $\times$ 1 ft 8 in)

7-14. Design a base plate with A36 steel for a W12 $\times$ 106 column with a dead working load of 100 k and a live working load of 420 k. The concrete 28-day strength is 4 ksi. Footing size is 12 by 12 ft.

7-15. Repeat Prob. 7-14 if the column is supported by a 28- by 28-in pedestal.
(*Ans.* PL1$\frac{1}{2}$ $\times$ 13 $\times$ 1 ft 4 in)

7-16. Design a column base plate of A36 steel for a W14 $\times$ 120 column supporting an axial load $P_u = 960$ k. Footing size is 10 by 10 ft and f'_c is 3 ksi.

7-17. Design a column base plate for a W14 $\times$ 90 column with $P_u = 650$ k if A36 steel is used and $f'_c = 3$ ksi for the concrete. Assume the column is to be supported by a pedestal 6 in wider on each side than the plate.
(*Ans.* PL1$\frac{1}{4}$ $\times$ 18 $\times$ 1 ft 6 in with 24 $\times$ 24 in pedestal)

7-18 to 7-19. *Repeat the problems indicated using INSTEP32.*

7-18. Prob. 7-14.

7-19. Prob. 7-16. (*Ans.* PL1$\frac{3}{4}$ $\times$ 18 $\times$ 1 ft 6 in)

CHAPTER 8

Introduction to Beams

8.1 TYPES OF BEAMS

Beams are usually said to be members that support transverse loads. They are probably thought of as being used in horizontal positions and subjected to gravity or vertical loads, but there are frequent exceptions—rafters, for example.

Among the many types of beams are joists, lintels, spandrels, stringers, and floor beams. *Joists* are the closely spaced beams supporting the floors and roofs of buildings, while *lintels* are the beams over openings in masonry walls such as windows and doors. A *spandrel beam* supports the exterior walls of buildings and perhaps part of the floor and hallway loads. The discovery that steel beams as a part of a structural frame could support masonry walls (together with the development of passenger elevators) is said to have permitted the construction of today's "skyscrapers." *Stringers* are the beams in bridge floors running parallel to the roadway, whereas *floor beams* are the larger beams in many bridge floors which are perpendicular to the roadway of the bridge and which are used to transfer the floor loads from the stringers to the supporting girders or trusses. The term *girder* is rather loosely used but usually indicates a large beam and perhaps one into which smaller beams are framed. These and other types of beams are discussed in the sections to follow.

8.2 SECTIONS USED AS BEAMS

The W shapes will normally prove to be the most economical beam sections and they have largely replaced channels and S sections for beam usage. Channels are sometimes used for beams subjected to light loads, such as purlins, and in places where clearances available require narrow flanges. They have very little resistance to lateral forces and need to be braced as illustrated by the sag rod problem in Chapter 4. The W shapes have more steel concentrated in their flanges than do S beams and thus have larger moments of inertia and resisting moments for the same weights. They are relatively wide and have appreciable lateral stiffness. (The small amount of space devoted to S beams in the LRFD Manual clearly shows how much their use has decreased from former years. Today, they are used primarily for special situations, such as when narrow

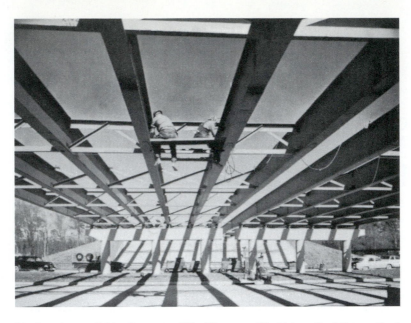

Harrison Avenue Bridge, Beaumont, TX.   (Courtesy of Bethlehem Steel Corporation.)

flange widths are desirable, where shearing forces are very high, or where the greater flange thickness next to the web may be desirable where lateral bending occurs, as perhaps with crane rails.)

Another common type of beam section is the open web joist or bar joist, which is discussed at length in Chapter 19. This type of section, which is commonly used to support floor and roof slabs, is actually a light shop-fabricated parallel chord truss. It is particularly economical for long spans and light loads.

8.3   BENDING STRESSES

For an introduction to bending stresses, the rectangular beam and stress diagrams of Fig. 8.1 are considered. (For this initial discussion, the beam's compression flange is assumed to be fully braced against lateral buckling. Lateral buckling is discussed at length in Chapter 9). If the beam is subjected to some bending moment, the stress at any point may be computed with the usual flexure formula $f_b = Mc/I$. *It is to be remembered, however, that this expression is only applicable when the maximum computed stress in the beam is below the elastic limit.* The formula is based on the usual elastic assumptions: stress is proportional to strain, a plane section before bending remains a plane section after bending, etc. The value of I/c is a constant for a particular section and is known as the *section modulus* (*S*). The flexure formula may then be written as follows:

$$f_b = \frac{M}{S}.$$

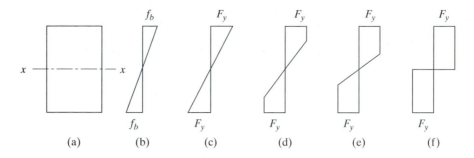

FIGURE 8.1

Variations in bending stresses due to increasing moment about x-axis.

Initially, when the moment is applied to the beam, the stress will vary linearly from the neutral axis to the extreme fibers. This situation is shown in part (b) of Fig. 8.1. If the moment is increased, there will continue to be a linear variation of stress until the yield stress is reached in the outermost fibers, as shown in part (c) of the figure. The *yield moment* of a cross section is defined as the moment that will just produce the yield stress in the outermost fiber of the section.

If the moment in a ductile steel beam is increased beyond the yield moment, the outermost fibers that had previously been stressed to their yield stress will continue to have the same stress but will yield, and the duty of providing the necessary additional resisting moment will fall on the fibers nearer to the neutral axis. This process will continue with more and more parts of the beam cross section stressed to the yield stress as shown by the stress diagrams of parts (d) and (e) of the figure, until finally a full plastic distribution is approached as shown in part (f). Note that the variation of strain from the neutral axis to the outer fibers remains linear for all of these cases. When the stress distribution has reached this stage, a *plastic hinge* is said to have formed because no additional moment can be resisted at the section. Any additional moment applied at the section will cause the beam to rotate with little increase in stress.

The *plastic moment* is the moment that will produce full plasticity in a member cross section and create a plastic hinge. The ratio of the plastic moment M_p to the yield moment M_y is called the *shape factor*. The shape factor equals 1.50 for rectangular sections and varies from about 1.10 to 1.20 for standard rolled-beam sections.

8.4 PLASTIC HINGES

This section is devoted to a description of the development of a plastic hinge in the simple beam shown in Fig. 8.2. The load shown is applied to the beam and increased in magnitude until the yield moment is reached and the outermost fiber is stressed to the yield stress. The magnitude of the load is further increased with the result that the outer fibers begin to yield. The yielding spreads out to the other fibers away from the section of maximum moment as indicated in the figure. The length in which this yielding occurs away from the section in question is dependent on the loading conditions and the member cross section. For a concentrated load applied at the center line of a

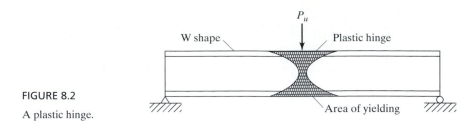

FIGURE 8.2

A plastic hinge.

simple beam with a rectangular cross section, yielding in the extreme fibers at the time the plastic hinge is formed will extend for one-third of the span. For a W shape in similar circumstances, yielding will extend for approximately one-eighth of the span. During this same period, the interior fibers at the section of maximum moment yield gradually until nearly all of them have yielded and a plastic hinge is formed, as shown in Fig. 8.2.

Although the effect of a plastic hinge may extend for some distance along the beam, it is assumed to be concentrated at one section for analysis purposes. For the calculation of deflections and for the design of bracing, the length over which yielding extends is quite important.

When steel frames are loaded to failure, the points where rotation is concentrated (plastic hinges) become quite visible to the observer before collapse occurs.

8.5   ELASTIC DESIGN

Until recent years, almost all steel beams were designed on the basis of the elastic theory. The maximum load that a structure could support was assumed to equal the load that first caused a stress somewhere in the structure to equal the yield stress of the material. The members were designed so that computed bending stresses for service loads did not exceed the yield stress divided by a safety factor (e.g., 1.5 to 2.0). Engineering structures have been designed for many decades by this method with satisfactory results. The design profession, however, has long been aware that ductile members do not fail until a great deal of yielding occurs after the yield stress is first reached. This means that such members have greater margins of safety against collapse than the elastic theory would seem to indicate.

8.6   THE PLASTIC MODULUS

The yield moment M_y equals the yield stress times the elastic modulus. The elastic modulus equals I/c or $bd^2/6$ for a rectangular section and the yield moment equals $F_y bd^2/6$. This same value can be obtained by considering the resisting internal couple shown in Fig. 8.3.

The resisting moment equals T or C times the lever arm between them, as follows:

$$M_y = \left(\frac{F_y bd}{4}\right)\left(\frac{2}{3}d\right) = \frac{F_y bd^2}{6}.$$

The elastic section modulus can again be seen to equal $bd^2/6$ for a rectangular beam.

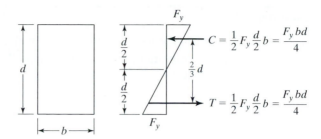

FIGURE 8.3

The resisting moment at full plasticity can be determined in a similar manner. The result is the so-called plastic moment, M_p. *It is also the nominal moment of the section, M_n.* This plastic or nominal moment equals T or C times the lever arm between them. For the rectangular beam of Fig. 8.4, we have

$$M_p = M_n = T\frac{d}{2} = C\frac{d}{2} = \left(F_y\frac{bd}{2}\right)\left(\frac{d}{2}\right) = F_y\frac{bd^2}{4}.$$

The plastic moment is said to equal the yield stress times the plastic modulus. From the foregoing expression for a rectangular section, the plastic modulus Z can be seen to equal $bd^2/4$. The shape factor, which equals $M_p/M_y = F_yZ/F_yS$, or Z/S, is $(bd^2/4)/(bd^2/6) = 1.50$ for a rectangular section. *A study of the plastic modulus determined here shows that it equals the statical moment of the tension and compression areas about the plastic neutral axis.* Unless the section is symmetrical, the neutral axis for the plastic condition will not be in the same location as for the elastic condition. The total internal compression must equal the total internal tension. As all fibers are considered to have the same stress (F_y) in the plastic condition, the areas above and below the plastic neutral axis must be equal. This situation does not hold for unsymmetrical sections in the elastic condition. Example 8-1 illustrates the calculations necessary to determine the shape factor for a tee beam and the nominal uniform load w_n that the beam can theoretically support.

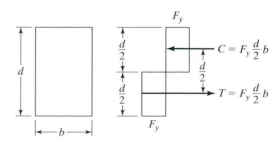

FIGURE 8.4

Example 8-1

Determine M_y, M_n, and Z for the steel tee beam shown in Fig. 8.5. Also calculate the shape factor and the nominal load (w_n) which can be placed on the beam for a 12-ft simple span. $F_y = 50$ ksi.

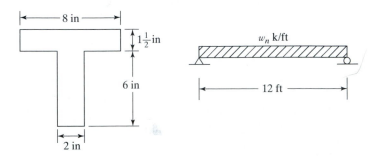

FIGURE 8.5

Solution. Elastic calculations:

$$A = (8)\left(1\frac{1}{2}\right) + (6)(2) = 24 \text{ in}^2$$

$$\bar{y} = \frac{(12)(0.75) + (12)(4.5)}{24} = 2.625 \text{ in from top of flange}$$

$$I = \left(\frac{1}{3}\right)(2)(1.125^3 + 4.875^3) + \left(\frac{1}{12}\right)(8)\left(1\frac{1}{2}\right)^3 + (12)(1.875)^2$$

$$= 122.4 \text{ in}^4$$

$$S = \frac{I}{c} = \frac{122.4}{4.875} = 25.1 \text{ in}^3$$

$$M_y = F_y S = \frac{(50)(25.1)}{12} = 104.6 \text{ ft-k}$$

Plastic calculations (neutral axis is at base of flange):

$$Z = (12)(0.75) + (12)(3) = 45 \text{ in}^3$$

$$M_n = M_p = F_y Z = \frac{(50)(45)}{12} = 187.5 \text{ ft-k}$$

$$\text{Shape factor} = \frac{M_p}{M_y} \quad \text{or} \quad \frac{Z}{S} = \frac{45}{25.1} = 1.79$$

$$M_n = \frac{w_n L^2}{8}$$

$$w_n = \frac{(8)(187.5)}{(12)^2} = 10.4 \text{ k/ft}$$

The values of the plastic moduli for the standard steel beam sections are tabulated in Table 5.3 in Part 5 of the LRFD Manual in the "Load Factor Design Selection Table for Shapes Used as Beams" as well as being listed for each shape in the "Dimensions and Properties" section of the Handbook (Part 1). These Z values will be used continuously throughout the text.

8.7   THEORY OF PLASTIC ANALYSIS

The basic plastic theory has been shown to be a major change in the distribution of stresses after the stresses at certain points in a structure reach the yield stress. The theory is that those parts of the structure that have been stressed to the yield stress cannot resist additional stresses. They instead will yield the amount required to permit the extra load or stresses to be transferred to other parts of the structure where the stresses are below the yield stress and thus in the elastic range and able to resist increased stress. Plasticity can be said to serve the purpose of equalizing stresses in cases of overload.

As early as 1914, Dr. Gabor Kazinczy, a Hungarian, recognized that the ductility of steel permitted a redistribution of stresses in an overloaded statically indeterminate structure.[1] In the United States, Prof. J. A. Van den Broek introduced his plastic theory which he called limit design. This theory was published in a paper entitled "Theory of Limit Design" in February 1939 in the *Proceedings of the ASCE*.

For this discussion, the stress-strain diagram is assumed to have the idealized shape shown in Fig. 8.6. The yield stress and the proportional limit are assumed to occur at the same point for this steel, and the stress-strain diagram is assumed to be a perfectly straight line in the plastic range. Beyond the plastic range there is a range of strain hardening. This latter range could theoretically permit steel members to withstand additional stress, but from a practical standpoint the strains occuring are so large

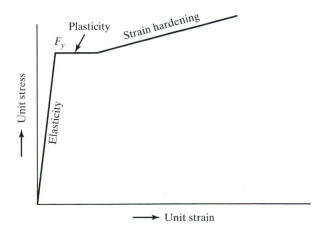

FIGURE 8.6

[1]Lynn S. Beedle, *Plastic Design of Steel Frames* (New York: Wiley, 1958), p. 3.

Ignore these two equations. For graphing purpose only. But these two equations can be determined by using the knowledge of homework 01. x := 0·ft, 0.25·ft .. L

$$V_{u2}(x) := \begin{cases} \left[W_{DL}\left(\dfrac{L}{2} - x\right) + \dfrac{P_{LL}\cdot(12\cdot ft)}{L} \right] & \text{if } 0\cdot ft \le x < 24\cdot ft \\[2mm] \left[W_{DL}\left(\dfrac{L}{2} - x\right) - \dfrac{P_{LL}\cdot(24\cdot ft)}{L} \right] & \text{if } 24\cdot ft \le x \le 36\cdot ft \end{cases}$$

$$M_{u2}(x) := \begin{cases} \left[\dfrac{(W_{DL})\cdot x}{2}\cdot(L - x) + \dfrac{P_{LL}\cdot(12\cdot ft)\cdot x}{L} \right] & \text{if } 0\cdot ft \le x < 24\cdot ft \\[2mm] \left[\dfrac{(W_{DL})\cdot x}{2}\cdot(L - x) + \dfrac{P_{LL}\cdot(12\cdot ft)\cdot x}{L} - P_{LL}\cdot(x - 24\cdot ft) \right] & \text{if } 24\cdot ft \le x \le 36\cdot ft \end{cases}$$

Shear Diagram for Case 2

Factored Shear (kip) vs Distance from left end of beam (ft)

+66.348

+6·33

1·6 L ≈ 64ᵏ

−57.67

−87.68

$V_{u2}(0\cdot ft) = 66.348\,kip$

$V_{u2}(24\cdot ft) = -57.671\,kip$

$V_{u2}(36\cdot ft) = -87.681\,kip$

$V_{u2\_max} := \left| V_{u2}(36\cdot ft) \right|$

use absolute value for <u>maximum shear</u> for design

$\boxed{V_{u2\_max} = 87.681\,kip}$

maximum reaction equal to maximum shear for this case

$R_{u2\_max} := V_{u2\_max}$

Moment Diagram for Case 2

Factored Moment (ft·kip) vs Distance from left end of beam (ft)

$M_{u2\_max} := M_{u2}(24\cdot ft)$

maximum moment $\boxed{M_{u2\_max} = 872.115\,ft\cdot kip}$

elastic buckling will limit the ability of even if strain hardening is significant.

astic hinge develops. To illustrate this loaded with a concentrated load at ld the load be increased until a plastic nent (underneath the load in this case) nown in part (b) of the figure. Any fur- esents the nominal or theoretical max-

fail, it is necessary for more than one nges required for failure of statically from structure to structure, but may Fig. 8.8 cannot fail unless the three d.

in a statically indeterminate structure, ilure if the geometry of the structure ge insofar as increased loading is con- ribution of moment because the plas- re plastic hinges are formed in the mber of them to cause collapse. Actu-

large to be permissible.

The propped beam of Fig. 8 plastic hinges develop. Three hing on the right end. In this beam the trated load is at the fixed end. As will form at that point.

The load may be further inc will be at the concentrated load

Real hinge

FIGURE 8.7

d) Design moment capacity for a <u>fully braced and compact beam</u> (10)

$\phi_b := 0.9$ for bending

Design criteria for "zone 1":

$\phi_b\cdot M_n = \phi_b\cdot M_p \le 1.5\cdot\phi_b\cdot M_y$

Plastic moment capacity:

$Z_x := 244\cdot in^3$ plastic section modulus of W27x84 from [LRFD Table 5-3 pg 5-46]

$F_y := 50\,ksi$ A992 steel

$M_p := Z_x\cdot F_y$

$M_p = 1016.67\,ft\cdot kip$

$\boxed{\phi_b\cdot M_p = 915\,ft\cdot kip}$ compare to value given in [LRFD Table 5-3 pg 5-46] $\phi_b\cdot M_{px} = 915\cdot ft\cdot kip$

Check the 1.5 yield moment limit:

$I_x = 2850\,in^4$ $d = 26.7\,in$ <= [LRFD Table 5-3 pg 5-46]

section modulus

$S_x := \dfrac{2\cdot I_x}{d}$

$S_x = 213.483\,in^3$ notice that section modulus is smaller than plastic section modulus.

Yield moment capacity:

$M_y := S_x\cdot F_y$

$M_y = 889.513\,ft\cdot kip$

$\phi_b\cdot M_y = 800.562\,ft\cdot kip$

$\phi_b\cdot M_p = 915\,ft\cdot kip$ < $\boxed{1.5\cdot\phi_b\cdot M_y = 1200.843\,ft\cdot kip}$

Therefore , the final moment design capacity is

$\phi M_n := \phi_b\,M_p$

$\boxed{\phi M_n = 915\,ft\cdot kip}$ > $M_{u1\_max} = 472.651\,ft\cdot kip$ Moment capacity is adequate.

$M_{u2\_max} = 872.115\,ft\cdot kip$

e) Design shear capacity

$\phi_v := 0.9$ for shear

Design criteria

$\phi_v \cdot V_n = \phi_v \cdot A_w \cdot (0.6 \cdot F_y)$

find area of web (we assume shear only carried by the web of the w-shape)

$A_w := t_w \cdot d$ $t_w = 0.46\,in$ $d = 26.7\,in$ [LRFD Table 1-1 page 1-14]

$A_w = 12.282\,in^2$

$V_n := A_w \cdot (0.6 \cdot F_y)$ $F_y = 50\,ksi$

$V_n = 368.46\,kip$

$\phi_v \cdot V_n = 331.614\,kip$ $>$ $\begin{array}{l} V_{u1\_max} = 52.517\,kip \\ V_{u2\_max} = 87.681\,kip \end{array}$ Shear capacity is adequate.

f) Check bearing capacity using 3" bearing length

Web yielding: [text page 288]

$\phi_{wy} := 1.0$

$F_{yw} := F_y$ note : yield strength of web equal to the yield strength of the W-shape for rolled section.
Might be different for built-up section, when different steel types are used for web and
$F_{yw} = 50\,ksi$ flanges.

$N := 3 \cdot in$ bearing length

$k = 1.24\,in$ web toes of fillets from [LRFD Table 1-1 page 1-14]

$t_w = 0.46\,in$ thickness of web

$R_n := (2.5 \cdot k + N) \cdot F_{yw} \cdot t_w$ use equation for concentrated load at the end of beam. (< distance d)

$R_n = 140.3\,kip$

$\phi_{wy} \cdot R_n = 140.3\,kip$ $>$ $\begin{array}{l} R_{u1\_max} = 52.517\,kip \\ R_{u2\_max} = 87.681\,kip \end{array}$ Resistance against web yielding is
enough.

FIGURE 8.9

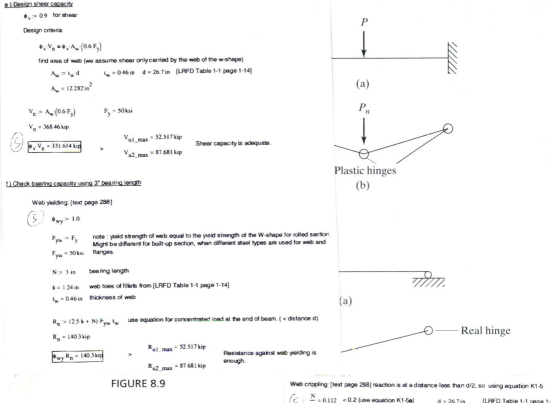

(a)

P

(b)

P_n

Plastic hinges

(a)

Real hinge

Web crippling: [text page 288] reaction is at a distance less than d/2, so using equation K1-5

$\dfrac{N}{d} = 0.112$ < 0.2 (use equation K1-5a) $d = 26.7\,in$ [LRFD Table 1-1 page 1-14]

$\phi_{wc} := 0.75$

$t_w = 0.46\,in$

$t_f = 0.64\,in$

$E = 29000\,ksi$

$F_{yw} = 50\,ksi$

$R_n := 0.4 \cdot t_w^2 \left[1 + 3 \cdot \left(\dfrac{N}{d} \right) \left(\dfrac{t_w}{t_f} \right)^{1.5} \right] \sqrt{\dfrac{E \cdot F_{yw} \cdot t_f}{t_w}}$

$R_n = 144.911\,kip$

$\phi_{wc} \cdot R_n = 108.683\,kip$ $>$ $\begin{array}{l} R_{u1\_max} = 52.517\,kip \\ R_{u2\_max} = 87.681\,kip \end{array}$ Resistance against web crippling is
enough.

cause the beam to collapse. The arrang
which permit collapse in a structure is
and 8.9 show mechanisms for various

After observing the large numbe
tration in this text, the student may
quently encounter such beams in e
difficult to find in actual structures bu
ples. They are particularly convenient
ous beams and frames are considered.

8.9   THE VIRTUAL-WORK METHOD

One very satisfactory method used for the plastic analysis of structures is the *virtual-work method*. The structure in question is assumed to be loaded to its nominal capacity, M_n, and is then assumed to deflect through a small additional displacement after the ultimate load is reached. The work performed by the external loads during this displacement is equated to the internal work absorbed by the hinges. For this discussion the *small-angle theory* is used. By this theory, the sine of a small angle equals the tangent of that angle and also equals the same angle expressed in radians. In the pages to follow, the author uses these values interchangeably because the small displacements considered here produce extremely small rotations or angles.

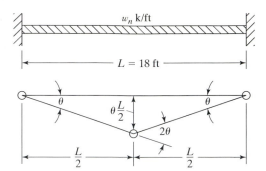

FIGURE 8.10

As a first illustration, the uniformly loaded fixed-ended beam of Fig. 8.10 is considered. This beam and its collapse mechanism are shown. Owing to symmetry, the rotations at the end plastic hinges are equal, and they are represented by θ in the figure; thus, the rotation at the middle plastic hinge will be 2θ.

The work performed by the total external load (w_nL) is equal to w_nL times the average deflection of the mechanism. The average deflection equals one-half the deflection at the center plastic hinge $(1/2 \times \theta \times L/2)$. The external work is equated to the internal work absorbed by the hinges or to the sum of M_n at each plastic hinge times the angle through which it works. The resulting expression can be solved for M_n and w_n as follows:

$$M_n(\theta + 2\theta + \theta) = w_nL\left(\frac{1}{2} \times \theta \times \frac{L}{2}\right)$$

$$M_n = \frac{w_nL^2}{16}$$

$$w_n = \frac{16M_n}{L^2}.$$

For the 18-ft span used in Fig. 8.10, these values become

$$M_n = \frac{(w_n)(18)^2}{16} = 20.25\,w_n$$

$$w_n = \frac{M_n}{20.25}.$$

Plastic analysis can be handled in a similar manner for the propped beam of Fig. 8.11. There the collapse mechanism is shown, and the end rotations (which are equal to each other) are assumed to equal θ.

The work performed by the external load P_n as it moves through the distance $\theta \times L/2$ is equated to the internal work performed by the plastic moments at the hinges, noting that there is no moment at the real hinge on the right end of the beam.

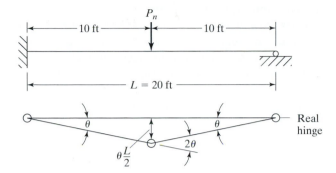

FIGURE 8.11

$$M_n(\theta + 2\theta) = P_n\left(\theta\,\frac{L}{2}\right)$$

$$M_n = \frac{P_n L}{6} \quad \text{(or 3.33 } P_n \text{ for the 20-ft beam shown)}$$

$$P_n = \frac{6M_n}{L} \quad \text{(or 0.3 } M_n \text{ for the 20-ft beam shown)}$$

The fixed-end beam of Fig. 8.12 together with its collapse mechanism and assumed angle rotations is next considered. From this figure, the values of M_n and P_n can be determined by virtual work as follows:

$$M_n(2\theta + 3\theta + \theta) = P_n\left(2\theta \times \frac{L}{3}\right)$$

$$M_n = \frac{P_n L}{9} \quad \text{(or 3.33 } P_n \text{ for this beam)}$$

$$P_n = \frac{9M_n}{L} \quad \text{(or 0.3 } M_n \text{ for this beam).}$$

A person beginning the study of plastic analysis needs to learn to think of all the possible ways in which a particular structure might collapse. Such a habit is of the

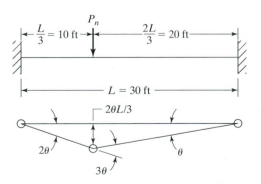

FIGURE 8.12

greatest importance when one begins to analyze more complex structures. In this light, the plastic analysis of the propped beam of Fig. 8.13 is considered by the virtual-work method. The beam with its two concentrated loads is shown together with four possible collapse mechanisms and the necessary calculations. It is true that the mechanisms of parts (b), (d), and (e) of the figure do not control, but such a fact is not obvious to the average student until he or she makes the virtual-work calculations for each case. Actually, the mechanism of part (e) is based on the assumption that the plastic moment is reached at both the concentrated loads simultaneously (a situation that might very well occur).

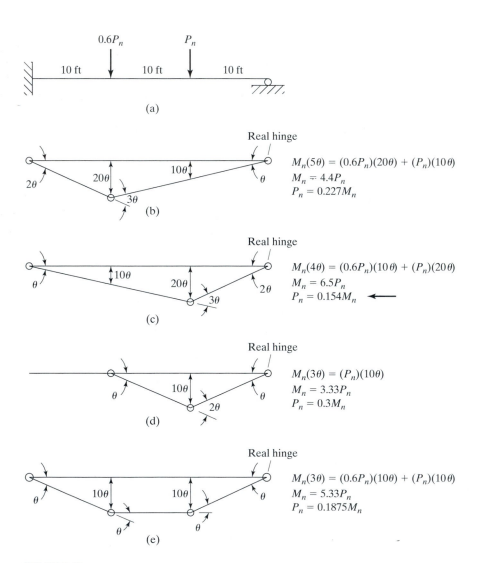

$M_n(5\theta) = (0.6P_n)(20\theta) + (P_n)(10\theta)$
$M_n = 4.4P_n$
$P_n = 0.227M_n$

$M_n(4\theta) = (0.6P_n)(10\theta) + (P_n)(20\theta)$
$M_n = 6.5P_n$
$P_n = 0.154M_n$ ⟵

$M_n(3\theta) = (P_n)(10\theta)$
$M_n = 3.33P_n$
$P_n = 0.3M_n$

$M_n(3\theta) = (0.6P_n)(10\theta) + (P_n)(10\theta)$
$M_n = 5.33P_n$
$P_n = 0.1875M_n$

FIGURE 8.13

The value for which the collapse load P_n is the smallest in terms of M_n is the correct value (or the value where M_n is the greatest in terms of P_n). For this beam, the second plastic hinge forms at the P_n concentrated load, and P_n equals $0.154\,M_n$.

8.10   LOCATION OF PLASTIC HINGE FOR UNIFORM LOADINGS

There was no difficulty in locating the plastic hinge for the unformly loaded fixed-end beam, but for other beams with uniform loads, such as propped or continuous beams, the problem may be rather difficult. For this discussion, the uniformly loaded propped beam of Fig. 8.14(a) is considered.

The elastic moment diagram for this beam is shown as the solid line in part (b) of the figure. As the uniform load is increased in magnitude, a plastic hinge will first form at the fixed end. At this time the beam will, in effect, be a "simple" beam with a plastic hinge on one end and a real hinge on the other. Subsequent increases in the load will cause the moment to change as represented by the dashed line in part (b) of the figure. This process will continue until the moment at some other point (a distance x from the right support in the figure) reaches M_n and creates another plastic hinge.

The virtual-work expression for the collapse mechanism of the beam shown in part (c) of Fig. 8.14 is written as follows:

$$M_n\left(\theta + \theta + \frac{L-x}{x}\theta\right) = (w_nL)(\theta)(L-x)\left(\frac{1}{2}\right).$$

Solving this equation for M_n, taking $dM_n/dx = 0$, the value of x can be calculated to equal $0.414L$. This value is also applicable to uniformly loaded end spans of continuous beams with simple end supports, as will be illustrated in the next section.

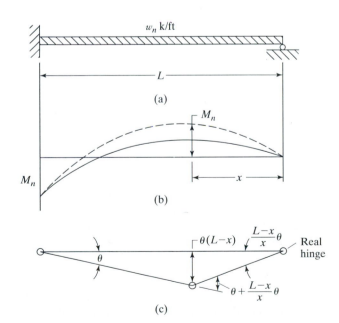

FIGURE 8.14

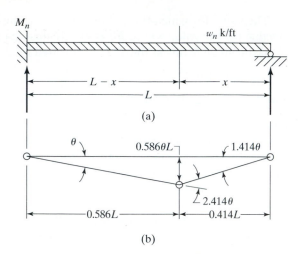

FIGURE 8.15

The beam and its collapse mechanism are redrawn in Fig. 8.15 and the following expression for the plastic moment and uniform load are written using the virtual-work procedure.

$$M_n(\theta + 2.414\theta) = (w_nL)(0.586\theta L)\left(\frac{1}{2}\right)$$

$$M_n = 0.0858w_nL^2$$

$$w_n = 11.65\frac{M_n}{L^2}$$

8.11   CONTINUOUS BEAMS

Continuous beams are very common in engineering structures. Their continuity causes analysis to be rather complicated in the elastic theory, and even though one of the complex "exact" methods is used for analysis, the resulting stress distribution is not nearly so accurate as is usually assumed.

Plastic analysis is applicable to continuous structures as it is to one-span structures. The resulting values definitely give a more realistic picture of the limiting strength of a structure than can be obtained by elastic analysis. Continuous statically indeterminate beams can be handled by the virtual-work procedure as they were for the single-span statically indeterminate beams. As an introduction to continuous beams, Examples 8-2 and 8-3 are presented to illustrate two of the more elementary cases.

Here it is assumed that if any or all of a structure collapses failure has occurred. Thus, in the continuous beams to follow virtual-work expressions are written separately for each span. From the resulting expressions, it is possible to determine the limiting or maximum loads that the beams can support.

Example 8-2

A W18 × 55(Z_x = 112 in³) has been selected for the beam shown in Fig. 8.16. Using 50 ksi steel and assuming full lateral support, determine the value of w_n.

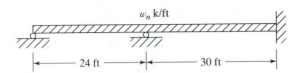

FIGURE 8.16

Solution

$$M_n = F_y Z = \frac{(50)(112)}{12} = 466.7 \text{ ft-k}$$

Drawing the (collapse) mechanisms for the two spans:

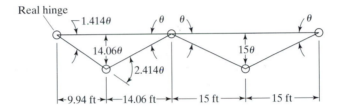

Left-hand span:

$$(M_n)(3.414\theta) = (24w_n)\left(\frac{1}{2}\right)(14.06\theta)$$

$$w_n = 0.0202 \, M_n = (0.0202)(466.7) = 9.43 \text{ klf}$$

Right-hand span:

$$(M_n)(4\theta) = (30w_n)\left(\frac{1}{2}\right)(15\theta)$$

$$w_n = 0.0178 \, M_n = (0.0178)(466.7) = 8.31 \text{ klf} \leftarrow$$

Additional spans have little effect on the amount of work involved in the plastic analysis procedure. The same cannot be said for elastic analysis. Example 8-3 illustrates the analysis of a three-span beam that is loaded with a concentrated load on each span. The student from his or her knowledge of elastic analysis can see that plastic hinges will initially form at the first interior supports and then at the center lines of the end spans, at which time each end span will have a collapse mechanism.

Example 8-3

Using a W21 × 44 ($Z_x = 95.4 \text{ in}^3$) consisting of A992 steel determine the value of P_n for the beam of Fig. 8.17.

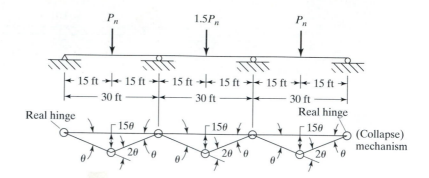

FIGURE 8.17

Solution

$$M_n = F_y Z = \frac{(50)(95.4)}{12} = 397.5 \text{ ft-k}$$

For first and third spans:

$$M_m(3\theta) = (P_n)(15\theta)$$
$$P_n = 0.2 M_n = (0.2)(397.5) = 79.5 \text{ k}$$

For center span:

$$(M_n)(4\theta) = (1.5\,P_n)(15\theta)$$
$$P_n = 0.178\,M_n = (0.178)(397.5) = 70.8 \text{ k} \leftarrow$$

8.12   BUILDING FRAMES

In this section, plastic analysis is applied to a small building frame. It is not the authors' purpose to consider frames at length in this chapter. Rather, they want to show the reader that the virtual-work method is applicable to frames as well as to beams, and that there are other types of mechanisms besides the beam types.

For the frame considered, it is assumed that the same W section is used for both the beam and the columns. If these members differed in size, it would be necessary to take that into account in the analysis.

The pin-supported frame of Fig. 8.18 is statically indeterminate to the first degree. The development of one plastic hinge will cause it to become statically determinate, while the forming of a second hinge can create a mechanism. There are, however, several types of mechanisms that might feasibly occur in this frame. A possible beam mechanism is shown in part (b), a sidesway mechanism is shown in part (c) and a combined beam and sidesway mechanism is shown in part (d). The critical condition is the one that will result in the smallest value of P_n.

Example 8-4 presents the plastic analysis of the frame of Fig. 8.18. The distances through which the loads tend to move in the various mechanisms should be carefully studied. The solution of this problem brings out a point of major significance: *Superposition*

Beam Design Moments (ϕ_b=0.9, C_b=1.0, F_y=50 ksi)

Beam Design Moments (ϕ_b=0.9, C_b=1.0, F_y=50 ksi)

$\phi_b M_n$, Design Moment (4 kip-ft increments)

L_b, Unbraced Length (0.5 ft increments)

(d) Combined beam and sidesway mechanism

d) check the bending capacity against factored moment, if only braced at ends and mid span.

Design Strength   "Load"

$\phi_b M_n = 698$ ft·kip   > $M_{max} = 551.61$ kip

The design is safe.

e) if the braced at mid span is removed.

unbraced length equal to the total span

$L_b := L$ $L_b = 30$ ft

determine factored moments at every 1/4 point of unbraced length (M_A, M_B and M_C).

$M_A := M_u\left(\dfrac{L_b}{4}\right)$ $\boxed{M_A = 413.707 \text{ ft·kip}}$ at $\dfrac{L_b}{4} = 7.5$ ft

$M_B := M_u\left(\dfrac{2 L_b}{4}\right)$ $\boxed{M_B = 551.61 \text{ ft·kip}}$ at $\dfrac{2 L_b}{4} = 15$ ft

$M_C := M_u\left(\dfrac{3 \cdot L_b}{4}\right)$ $\boxed{M_C = 413.707 \text{ ft·kip}}$ at $\dfrac{3 L_b}{4} = 22.5$ ft

$M_{max} := M_u\left(\dfrac{L}{2}\right)$ $\boxed{M_{max} = 551.61 \text{ ft·kip}}$ at $\dfrac{L}{2} = 15$ ft

Calculate C_b using LRFD equation F1-3 [text pg 256]

$C_b := \dfrac{12.5 \cdot M_{max}}{2.5 \cdot M_{max} + 3 \cdot M_A + 4 \cdot M_B + 3 \cdot M_C}$ $\boxed{C_b = 1.136}$ check with the value given in textbook figure 9.9 pg 257 $C_b = 1.14$ (top, left picture)

Design capacity :

$L_b = 30$ ft   > $L_r = 26$ ft

unbraced length is greater than L_r now, so use "zone 3" or elastic buckling equation.

[LRFD Equation F1-13] or textbook page 260

$C_w := 13600 \text{ in}^6$ warping constant [LRFD Table 1-25] $G := 11200$ ksi shear modulus of steel [text, page 260]

$J := 4.1 \cdot \text{in}^4$ torsional constant [LRFD Table 1-25] $E = 29000$ ksi   modulus of elasticity of steel

$I_y := 175 \cdot \text{in}^4$ [LRFD Table 1-1]

$C_b = 1.136$ $\phi_b = 0.9$

$M_{cr} := C_b \dfrac{\pi}{L_b} \sqrt{E \cdot I_y \cdot G \cdot J + \left(\dfrac{\pi \cdot E}{L_b}\right)^2 \cdot I_y \cdot C_w}$

$M_{cr} = 513.075$ ft·kip

$\phi_b \cdot M_{cr} = 461.767$ ft·kip   < $\phi_b M_{px} = 698$ ft·kip

so lateral torsional buckling controls

$\phi_b M_n := \phi_b \cdot M_{cr}$

$\boxed{\phi_b M_n = 461.767 \text{ ft·kip}}$ < $M_{max} = 551.61$ ft·kip

The design is no longer safe if the lateral bracing at mid span is removed. Design capacity is lower than factored moment

this by studying the virtual work ex-
e values of P_n obtained for the sepa-
up to the value obtained for the

he situation in which we have the
ause collapse. If you look at one of
becomes smaller as the number of
part (d) of Fig. 8.18. The frame can
a plastic hinge at the top of the left
e sufficient for collapse to occur.

and columns of the frame shown in

or parts (b), (c), and (d) of Fig. 8.18
The combined beam and sidesway
value of P_n is determined as follows:

= 56.25 k

PROBLEMS

8-1 to 8-10. *Find the values of S, Z, and the shape factor about the horizontal x-axes for the sections shown in the accompanying illustrations.*

8-1.

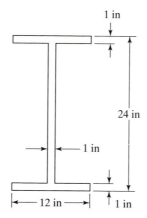

FIGURE P8-1 (*Ans.* 338.6, 397, 1.17)

8-2.

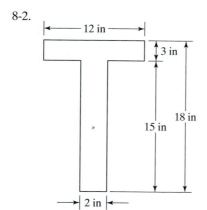

FIGURE P8-2

8-3.

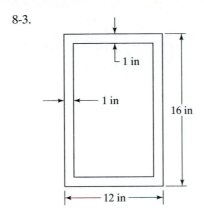

1 in

1 in

16 in

12 in

FIGURE P8-3 (*Ans.* 226.2, 278, 1.23)

8-4.

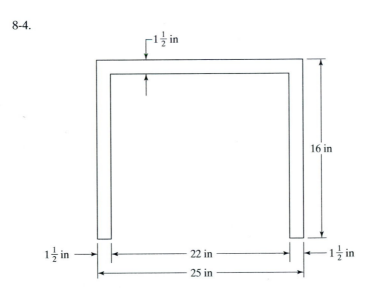

$1\frac{1}{2}$ in

16 in

$1\frac{1}{2}$ in

22 in

$1\frac{1}{2}$ in

25 in

FIGURE P8-4

8-5.

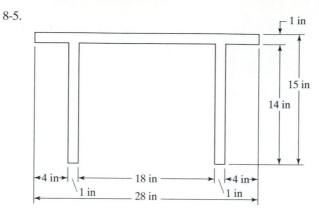

FIGURE P8-5 (*Ans.* 116, 210, 1.81)

8-6.

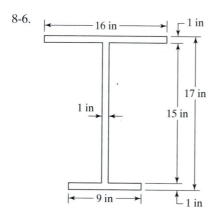

FIGURE P8-6

8-7.

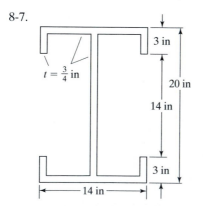

FIGURE P8-7 (*Ans.* 279.1, 321.1, 1.15)

8-8.

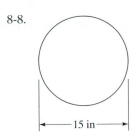

←——15 in——→

FIGURE P8-8

8-9.

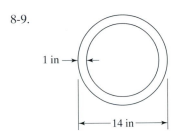

1 in →

←——14 in——→

FIGURE P8-9 (*Ans.* 124, 169.3, 1.37)

8-10.

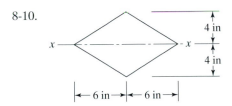

4 in

4 in

←—6 in—→←—6 in—→

FIGURE P8-10

8-11 to 8-20. *Determine the values of S, Z, and the shape factor about the horizontal x axes unless otherwise directed. Use the web and flange dimensions given in the LRFD Manual for making these calculations.*

8-11. A W27 × 178 (*Ans.* 499, 563.6, 1.13)

8-12. A W24 × 162 with one $\frac{3}{4}$ × 16 in PL on each flange.

8-13. Two 8 × 6 × $\frac{5}{8}$-in Ls long legs vertical and back to back. (*Ans.* 19.7, 35.7, 1.81)

8-14. Two C12 × 30 s back-to-back.

8-15. Four $8 \times 8 \times \frac{3}{4}$ in Ls arranged as shown in the accompanying illustration. (*Ans.* 64.5, 104.2, 1.62)

FIGURE P8-15

8-16. The section of Prob. 5-9(b).

8-17. The section of Prob. 8-1 considering the *y* axis. (*Ans.* 48.3, 77.5, 1.60)

8-18. Rework Prob. 8-2 considering the *y* axis.

8-19. Rework Prob. 8-12 considering the *y* axis. (*Ans.* 119.4, 201, 1.68)

8-20. Rework Prob. 8-14 considering the *y* axis.

8-21 to 8-40. *Using the given sections all of A992 steel and the plastic theory, determine the values of P_n and w_n as indicated.*

8-21.

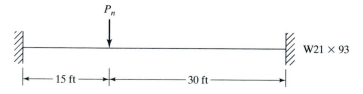

FIGURE P8-21 (*Ans.* $P_n = 184.2\,k$)

8-22.

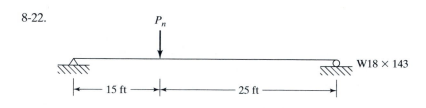

FIGURE P8-22

8-23.

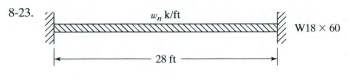

w_n k/ft

W18 × 60

28 ft

FIGURE P8-23 (*Ans.* $w_n = 10.46\ k/$ft)

8-24.

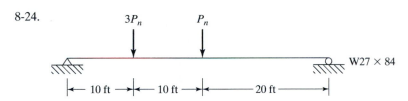

$3P_n$ P_n

W27 × 84

10 ft   10 ft   20 ft

FIGURE P8-24

8-25.

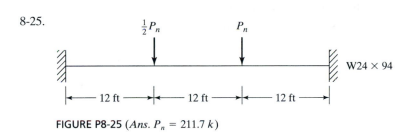

$\frac{1}{2}P_n$ P_n

W24 × 94

12 ft   12 ft   12 ft

FIGURE P8-25 (*Ans.* $P_n = 211.7\ k$)

8-26.

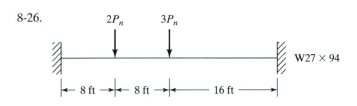

$2P_n$ $3P_n$

W27 × 94

8 ft   8 ft   16 ft

FIGURE P8-26

8-27.

$2P_n$ $3P_n$ $4P_n$

W24 × 176

12 ft   12 ft   12 ft   12 ft

FIGURE P8-27 (*Ans.* $P_n = 59.14\ k$)

8-28.

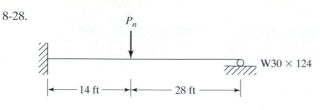

FIGURE P8-28

8-29.

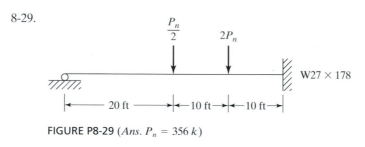

FIGURE P8-29 (*Ans. $P_n = 356\,k$*)

8-30.

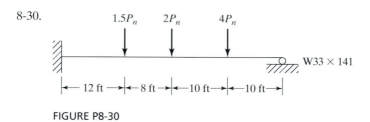

FIGURE P8-30

8-31.

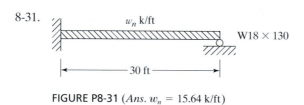

FIGURE P8-31 (*Ans. $w_n = 15.64$ k/ft*)

8-32.

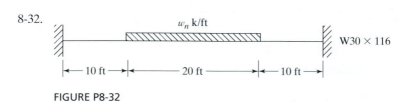

FIGURE P8-32

8-33.

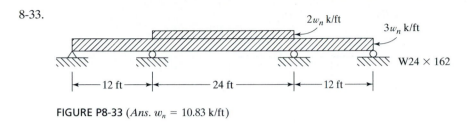

FIGURE P8-33 (*Ans. w_n = 10.83 k/ft*)

8-34.

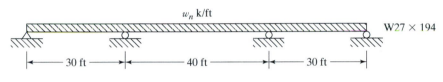

FIGURE P8-34

8-35.

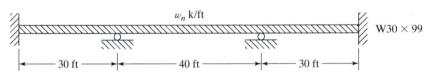

FIGURE P8-35 (*Ans. w_n = 26.29 k/ft*)

8-36.

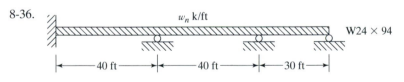

FIGURE P8-36

8-37.

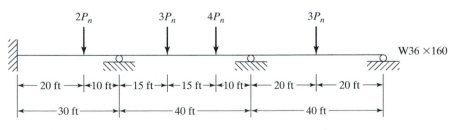

FIGURE P8-37 (*Ans. P_n = 120.6 k*)

8-38.

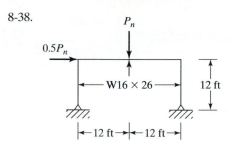

FIGURE P8-38

8-39.

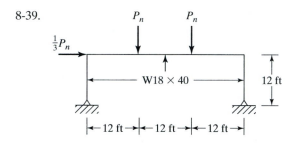

FIGURE P8-39 (*Ans. P_n = 40.8 k*)

8-40. Repeat Prob. 8-38 if the column bases are fixed.

CHAPTER 9

Design of Beams for Moments

9.1 INTRODUCTION

If gravity loads are applied to a fairly long, simply supported beam, the beam will bend downward, and its upper part will be placed in compression and will act as a compression member. The cross section of this "column" will consist of the portion of the beam cross section above the neutral axis. For the usual beam the "column" will have a much smaller moment of inertia about its y or vertical axis than about its x axis. If nothing is done to brace it perpendicular to the y axis, it will buckle laterally at a much smaller load than would otherwise have been required to produce a vertical failure. (You can verify these statements by trying to bend vertically a magazine held in a vertical position. The magazine will, just as a steel beam, always tend to buckle laterally unless it is braced in that direction.)

Lateral buckling will not occur if the compression flange of a member is braced laterally or if twisting of the beam is prevented at frequent intervals. In this chapter, the buckling moments of a series of compact ductile steel beams with different lateral or torsional bracing situations are considered. (As previously defined, a *compact section* is one that has a sufficiently stocky profile so that it is capable of developing a fully plastic stress distribution before buckling.)

In this chapter, we will look at beams as follows:

1. First, the beams will be assumed to have continuous lateral bracing for their compression flanges.
2. Next, the beams will be assumed to be braced laterally at short intervals.
3. Finally, the beams will be assumed to be braced laterally at larger and larger intervals.

In Fig. 9.1, a typical curve showing the nominal resisting or buckling moments of one of these beams with varying unbraced lengths is presented.

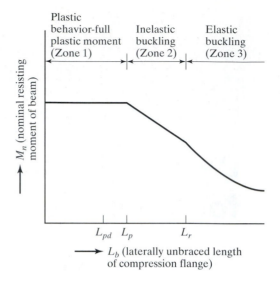

FIGURE 9.1

Nominal moment as function of unbraced length of compression flange.

An examination of Fig. 9.1 will show that beams have three distinct ranges or zones of behavior depending on their lateral bracing situation. If we have continuous or closely spaced lateral bracing, the beams will experience extreme yielding of the entire cross section and fall into what is classified as Zone 1. As the distance between lateral bracing is increased further, the beams will begin to fail inelastically at smaller moments and fall into Zone 2. Finally, with even larger unbraced lengths, the beams will fail elastically and fall into Zone 3. A brief discussion of these three types of behavior is presented in this section, while the remainder of the chapter is devoted to a detailed discussion of each type, together with a series of numerical examples.

9.1.1   Plastic Behavior (Zone 1)

If we were to take a compact beam whose compression flange is continuously braced laterally, we would find that we could load it until its full plastic moment M_p is reached at some point or points; further loading then produces a redistribution of moments as was described in Chapter 8. In other words, the moments in these beams can reach M_p and then develop a rotation capacity sufficient for moment redistribution.

If we now take one of these compact beams and provide closely spaced intermittent lateral bracing for its compression flanges, we will find that we can still load it until the plastic moment plus moment redistribution is achieved if the spacing between the bracing does not exceed a certain value called L_p herein. (The value of L_p is dependent on the dimensions of the beam cross section and on its yield stress.) *Most beams fall in Zone 1.*

9.1.2   Inelastic Buckling (Zone 2)

If we now further increase the spacing between points of lateral or torsional bracing, the section may be loaded until some, but not all, of the compression fibers are stressed to F_y. The section will have insufficient rotation capacity to permit full moment redistribution

Bridge over Allegheny River at Kittanning, PA.   (Courtesy American Bridge Company.)

and thus will not permit plastic analysis. In other words, in this zone we can bend the member until the yield strain is reached in some, but not all, of its compression elements before buckling occurs. This is referred to as *inelastic buckling*.

As we increase the unbraced length, we will find that the moment the section resists will decrease until finally it will buckle before the yield stress is reached anywhere in the cross section. The maximum unbraced length at which we can still reach F_y at one point is the end of the inelastic range. It's shown as L_r in Fig. 9.1; its value is dependent upon the properties of the beam cross section as well as on the yield and residual stresses of the beam. At this point, as soon as we have a moment which theoretically causes the yield stress to be reached at some point in the cross section (actually it's less than F_y because of residual stresses), the section will buckle.

9.1.3   Elastic Buckling (Zone 3)

If the unbraced length is greater than L_r, the section will buckle elastically before the yield stress is reached anywhere. As the unbraced length is further increased, the buckling moment becomes smaller and smaller. As the moment is increased in such a beam, the beam will deflect more and more transversely until a critical moment value M_{cr} is reached. At this time, the beam cross section will twist and the compression flange will move laterally. The moment M_{cr} is provided by the torsional resistance and the warping resistance of the beam, as will be discussed in Section 9.7.

150 Federal Street, Boston, MA.   (Courtesy Owen Steel Company, Inc.)

9.2   YIELDING BEHAVIOR—FULL PLASTIC MOMENT, ZONE 1

In this section and the next two, beam formulas for yielding behavior (Zone 1) are presented, while in Sections 9.5 through 9.7 formulas are presented for inelastic buckling (Zone 2) and elastic buckling (Zone 3). After seeing some of these latter expressions, the reader may become quite concerned that he or she is going to spend an enormous amount of time in formula substitution. This is not generally true, however, as the values needed are tabulated and graphed in simple form in Part 5 of the LRFD Manual.

When a steel section has a large shape factor, appreciable inelastic deformations may occur under service loads if the section in designed so that M_p is reached at the factored load condition. As a result, LRFD Specification F1.1 limits the amount of such deformation for sections with shape factors larger than 1.5. This is done by limiting M_p to a maximum value of 1.5 M_y.

If the unbraced length L_b of the compression flange of a compact I- or C-shaped section, including hybrid members, does not exceed L_p (if elastic analysis is being used) or L_{pd} (if plastic analysis is being used), then the member's bending strength about its major axis may be determined as follows:

$$M_n = M_p = F_y Z \leq 1.5\ M_y \qquad \text{(LRFD Equation F1-1)}$$
$$M_u = \phi_b M_n \text{ with } \phi_b = 0.90.$$

This part of the specification limiting M_n to 1.5 M_y for high shape factor sections such as WTs does not apply to hybrid sections with web yield stresses less than their flange yield stresses. Web yielding for such members does not result in significant inelastic deformations. For hybrid members, the yield moment $M_y = F_{yf} S$ (LRFD Commentary F1.1).

When a conventional elastic analysis approach is used to establish member forces, L_b may not exceed the value L_p to follow if M_n is to equal $F_y Z$.

$$L_p = 1.76\ r_y \sqrt{\frac{E}{F_{yf}}} \qquad \text{(LRFD Equation F1-4)}$$

For solid rectangular bars and box beams with A = cross-sectional area (in^2) and J = torsional constant (in^4),

$$L_p = \frac{0.13\ r_y E}{M_p}\sqrt{JA}. \qquad \text{(LRFD Equation F1-5)}$$

When a plastic analysis approach is used to establish member forces for doubly symmetric and singly symmetric I-shaped members with the compression flanges larger than their tension flanges (including hybrid members) loaded in the plane of the web, L_b (which is defined as the laterally unbraced length of the compression flange at plastic hinge locations associated with failure mechanisms) may not exceed the value L_{pd} to follow if M_n is to equal $F_y Z$.

$$L_{pd} = \left[0.12 + 0.076\left(\frac{M_1}{M_2}\right)\right]\left(\frac{E}{F_y}\right)r_y \qquad \text{(LRFD Equation F1-17)}$$

In this expression, M_1 is the smaller moment at the end of the unbraced length of the beam and M_2 is the larger moment at the end of the unbraced length, and the ratio M_1/M_2 is positive when the moments cause the member to be bent in double curvature (⁀‿) and negative if they bend it in single curvature (‿‿). Only steels with F_y values (F_y is the specified minimum yield stress of the compression flange) of 65 ksi or less may be considered. Higher-strength steels may not be ductile.

TABLE 9.1   Limiting Width-Thickness Ratios for Compression Elements

Description of Element	Width-Thickness Ratio	Limiting Width-Thickness Ratios λ_p	
		Nonseismic	Seismic
Flanges of I-shaped sections (including hybrid sections) and channels in flexure [a]	b/t	$0.38\sqrt{E/F_y}$	$0.31\sqrt{E/F_y}$
Webs in combined flexural and axial compression	h/t_w	For $P_u/\phi_b P_y \leq 0.125$	
		$3.76\sqrt{\dfrac{E}{F_y}}\left(1 - \dfrac{2.75P_u}{\phi_b P_y}\right)$	$3.05\sqrt{\dfrac{E}{F_y}}\left(1 - \dfrac{1.54P_u}{\phi_b P_y}\right)$
		For $P_u/\phi_b P_y > 0.125$	
		$1.12\sqrt{\dfrac{E}{F_y}}\left(2.33 - \dfrac{P_u}{\phi_b P_y}\right) \geq 1.49\sqrt{\dfrac{E}{F_y}}$	

[a] For hybrid beams use F_{yf} in place of F_y.

There is no limit of the unbraced length for circular or square cross sections or for I-shaped beams bent about their minor axes. (If I-shaped sections are bent about their minor or y axes, they will not buckle before the full plastic moment M_p about the y axis is developed, as long as the flange element is compact.) Equation F1-18 of the LRFD Specification also provides a value of L_{pd} for solid rectangular bars and symmetrical box beams.

For these sections to be compact, the width-thickness ratios of the flanges and webs of I- and C-shaped sections are limited to the maximum values shown in Table 9.1 (which is Table C-B5.1 in the Commentary to the LRFD Specification).

In these formulas, h is the distance from the web toe of the fillet in the top of the web to the web toe of the fillet in the bottom of the web (that is, twice the distance from the neutral axis to the inside face of the compression flange less the fillet or corner radius).

9.3   DESIGN OF BEAMS, ZONE 1

Included in the items that need to be considered in beam design are moments, shears, deflections, crippling, lateral bracing for the compression flanges, fatigue, and others. Beams will probably be selected that provide sufficient design moment capacities ($\phi_b M_n$) and then checked to see if any of the other items are critical. The factored moment will be computed, and a section having that much design moment capacity will be initially selected from the LRFD Manual. (See Table 5.3, entitled "W-Shapes Selection by Z_x," presented in Part 5 of the Manual.) From this table, steel shapes having sufficient plastic moduli to resist certain moments can quickly be selected. Two important items should be remembered in selecting shapes. These are:

1. These steel sections cost so many cents per pound and it is therefore desirable to select the lightest possible shape having the required plastic modulus (assuming that the resulting section is one that will reasonably fit into the structure). The

table has the sections arranged in various groups having certain ranges of plastic moduli. The heavily typed section at the top of each group is the lightest section in that group, and the others are arranged successively in the order of their plastic moduli. Normally, the deeper sections will have the lightest weights giving the required plastic moduli, and they will be generally selected unless their depth causes a problem in obtaining the desired headroom, in which case a shallower but heavier section will be selected.

2. The plastic moduli values in the table are given about the horizontal axes for beams in their upright positions. If a beam is to be turned on its side, the proper plastic modulus can be found in the aforementioned Table 5.3 or in the tables giving dimensions and properties of shapes in Part 1 of the LRFD Manual. A W shape turned on its side may only be from 10 to 30 percent as strong as one in the upright position when subjected to gravity loads. In the same manner, the strength of a wood joist with the actual dimensions 2 × 10 in turned on its side would be only 20 percent as strong as in the upright position.

The examples to follow illustrate the design of compact steel beams whose compression flanges have full lateral support or bracing, thus permitting plastic analysis. For the selection of such sections, the designer may enter the tables either with the required plastic modulus or with the factored design moment (if $F_y = 50$ ksi).

9.3.1   Beam Weight Estimates

In each of the examples to follow, the weight of the beam is included in the calculation of the bending moment to be resisted, as the beam must support itself as well as the external loads. The estimates of beam weight are very close here because the author was able to perform a little preliminary paperwork before making his estimates. The beginning student is not expected to be able to glance at a problem and estimate exactly the weight of the beam required. A very simple method is available, however, with which the student can quickly and accurately estimate beam weights. He or she can calculate the maximum factored bending moment, not counting the effect of the beam weight, and pick a section from LRFD Table 5.3. Then the weight of that section or a little bit more (since the beam's weight will increase the moment somewhat) can be used as the estimated beam weight. The results will almost always be very close to the weight of the member selected in the final design.

Example 9-1

Select a beam section for the span and loading shown in Fig. 9.2, assuming full lateral support is provided for the compression flange by the floor slab above (that is $L_b = 0$) and $F_y = 50$ ksi.

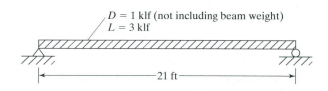

FIGURE 9.2

Solution

Beam weight estimate:

$$w_u \text{ not including beam weight} = (1.2)(1.0) + (1.6)(3.0) = 6.0 \text{ klf}$$

$$M_u = \frac{(6.0)(21)^2}{8} = 330.75 \text{ ft-k}$$

$$Z \text{ required} = \frac{M_u}{\phi_b F_y} = \frac{(12)(330.75)}{(0.9)(50)} = 88.2 \text{ in}^3$$

Referring to Table 5.3 in Part 5 of the Manual, a W21 × 44 is the lightest section available. Assume beam weight = 44 lb/ft, as its value will not increase moment significantly.

$$w_u = (1.2)(1.044) + (1.6)(3) = 6.05 \text{ klf}$$

$$M_u = \frac{(6.05)(21)^2}{8} = 333.5 \text{ ft-k}$$

$$Z \text{ required} \geq \frac{(12)(333.5)}{(0.9)(50)} = 88.9 \text{ in}^3$$

Use W21 × 44 F_y = 50 ksi.

Example 9-2

The 5-in reinforced-concrete slab shown in Fig. 9.3 is to be supported with steel W sections 8 ft 0 in on centers. The beams, which will span 20 ft, are assumed to be simply supported. If the concrete slab is designed to support a live load of 100 psf, determine the lightest steel section required to support the slab. It is assumed that the compression flange of the beam will be fully supported laterally by the concrete slab. The concrete weighs 150 lb/ft³. F_y = 50 ksi.

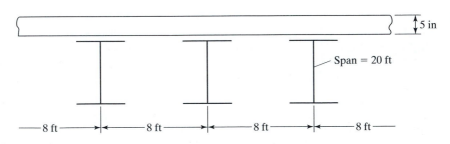

FIGURE 9.3

Solution

$$\text{Dead loads:} \quad \text{Slab} = \left(\tfrac{5}{12}\right)(150)(8) = 500 \text{ lb/ft}$$

$$\text{Estimated beam wt} = 22$$

$$\text{Total} = 522 \text{ lb/ft}$$

$$w_u = (1.2)(522) + (1.6)(8 \times 100) = 1906 \text{ lb/ft} = 1.906 \text{ klf}$$

$$M_u = \frac{(1.906)(20)^2}{8} = 95.3 \text{ ft-k}$$

$$Z \text{ required} \geq \frac{(12)(95.3)}{(0.9)(50)} = 25.4 \text{ in}^3$$

Use W10 × 22, $F_y = 50$ ksi.

9.3.2   Holes in Beams

It is often necessary to have holes in steel beams. They are obviously required for the installation of bolts and sometimes for pipes, conduits, ducts, etc. If at all possible, these latter types of holes should be completely avoided. When absolutely necessary, they should be placed through the web if the shear is small and through the flange if the moment is small and the shear is large. Cutting a hole through the web of a beam does not reduce its section modulus greatly or its resisting moment; but, as will be described in Section 10.2, a large hole in the web tremendously reduces the shearing strength of a steel section. When large holes are put in beam webs, extra plates are sometimes connected to the webs around the holes to serve as reinforcing against possible web buckling.

When large holes are placed in beam webs, the strength limit states of the beams such as local buckling of the compression flange, the web, or the tee-shaped compression zone above or below the opening—or the moment-shear interaction or serviceability limit states may control the size of the member. A general procedure for estimating these effects and the design of any required reinforcement is available for both steel and composite beams.[1,2]

The presence of holes of any type in a beam certainly does not make it stronger and in all probability weakens it somewhat. The effect of holes has been a subject which has been argued back and forth for many years. The questions, "Is the neutral axis affected by the presence of holes?" and "Is it necessary to subtract holes from the compression flange that are going to be plugged with bolts?" are frequently asked.

The theory that the neutral axis might move from its normal position to the theoretical position of its net section when bolt holes are present is rather questionable. Tests seem to show that flange holes for bolts do not appreciably change the location of the neutral axis. It is logical to assume that the location of the neutral axis will not follow the exact theoretical variation with its abrupt changes in position at bolt holes as

[1]D. Darwin, "*Steel and Composite Beams with Web Openings*," AISC Design Guide Series No. 2 (Chicago: American Institute of Steel Construction, 1990).

[2]ASCE Task Committee on Design Criteria for Composite Structures in Steel and Concrete "Proposed Specification for Structural Steel Beams with Web Openings," D. Darwin, Chairman, *Journal of Structural Engineering*, ASCE, Vol. 118 (New York: ASCE December, 1992).

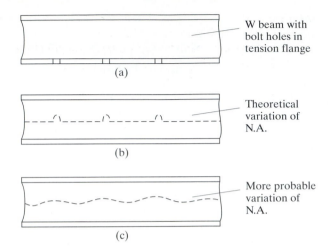

W beam with
bolt holes in
tension flange

(a)

Theoretical
variation of
N.A.

(b)

More probable
variation of
N.A.

FIGURE 9.4 (c)

shown in part (b) of Fig. 9.4. A more reasonable change in neutral axis location is shown
in part (c) of this figure where it is assumed to have a more gradual variation in position.

It is interesting to note that flexure tests of steel beams seem to show that their
failure is based on the strength of the compression flange even though there may be
bolt holes in the tension flange. The presence of these holes does not seem to be as serious as might be thought, particularly as compared with holes in a pure tension member. These tests show little difference in the strengths of beams with no holes and in
beams with an appreciable number of bolt holes in either flange.

The LRFD Specification does not require a reduction in flange area for bolts as
long as the following expression is satisfied:

$$0.75 F_u A_{fn} \geq 0.9 F_y A_{fg} \qquad \text{(LRFD Equation B10-1)}$$

In this equation, A_{fn} is the net flange area and A_{fg} is the gross flange area. Substituting into this expression, we find that no deduction is necessary if the net flange
area is equal to or greater than 74 percent of the gross flange area for A36 steel or 92
percent for A992 steel. These values are shown in Table 9.2.

Should $0.75 F_u A_{fn}$ be less than $0.9 F_y A_{fg}$, the LRFD Specification requires that
the flexural properties of the section be based on an effective tension flange area A_{fe},
determined as follows:

$$A_{fe} = \frac{5}{6} \frac{F_u}{F_y} A_{fn} \qquad \text{(LRFD Equation B10-3)}$$

TABLE 9.2 When to Ignore Bolt Holes in Beam and Girder Flanges

Steel	No Reduction if $A_{fn}/A_{fg} \geq$
A36 ($F_u = 58$ ksi)	0.74
A992 ($F_u = 65$ ksi)	0.92
A588 ($F_y = 50$ ksi, $F_u = 70$ ksi)	0.86

Bolt holes in the webs of beams are generally considered to be insignificant as they have almost no effect on Z calculations.

Some specifications, notably the bridge ones, and some designers have not adopted the idea of ignoring the presence of all or part of the holes in tension flanges. As a result, they follow the more conservative practice of deducting 100 percent of all holes. For such a case, the reduction in Z_x will equal the statical moment of the holes (in both flanges) taken about the neutral axis.

The usual practice is to assume that we have the same holes in both flanges whether or not they are actually present in the compression flange. (If we have bolt holes filled with bolts in the compression flange only, we forget the whole thing. This is because it is felt that the fasteners can adequately transmit compression through the holes by means of the bolts.) For a section with holes in the tension flange only, where a reduction in the tension flange area is required by the specification or is considered necessary, we make the same deduction from both flanges.

Should the application of LRFD Equation B10-1 show that $0.75F_uA_{fn} < 0.9F_yA_{fg}$, we will need to reduce Z_x. Its value will equal Z_x from the Manual tables minus the statical moment of $A_{fg} - A_{fe}$ for each flange taken about the neutral axis. The review of a beam with flange bolt holes is considered in Example 9-3.

Example 9-3

Determine ϕM_n for the fully braced A36 W24 × 176 shown in Fig. 9.5 for the following situations:

  a. Using the LRFD Specification and assuming two lines of 1-in bolts in standard holes in each flange.
  b. Using the LRFD Specification and assuming four lines of 1-in bolts in standard holes in each flange.
  c. Same as (b) above, but conservatively assuming we must deduct all holes for computing beam properties.

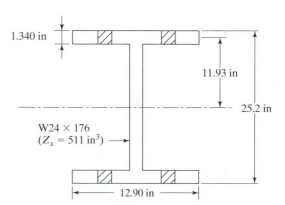

FIGURE 9.5

Solution

a. With two holes in each flange (LRFD Specification)

$A_{fg} = (12.9)(1.340) = 17.29 \text{ in}^2$

$A_{fn} = 17.29 - (2)(1\frac{1}{8})(1.340) = 14.27 \text{ in}^2$

For no reduction $0.75F_u A_{fn} \geq 0.9F_y A_{fg}$

$(0.75)(58)(14.27) = 620.7 \text{ k} > (0.9)(36)(17.29) = 560.2 \text{ k}$

$\therefore$ No reduction in flange area is required.

$M_u \leq \phi_b M_n = \dfrac{(0.9)(36)(511)}{12} = 1380 \text{ ft-k}$

b. With four holes in each flange (LRFD Specification)

$A_{fg} = 17.29 \text{ in}^2$

$A_{fn} = 17.29 - (4)(1\frac{1}{8})(1.340) = 11.26 \text{ in}^2$

For no reduction $0.75F_u A_{fn} \geq 0.9F_y A_{fg}$

$(0.75)(58)(11.24) = 488.9 \text{ k} < (0.9)(36)(17.27) = 559.5 \text{ k}$

$\therefore$ Flange area must be reduced

$A_{fe} = \left(\dfrac{5}{6}\right)\left(\dfrac{58}{36}\right)(11.26) = 15.12 \text{ in}^2$

Reduced $Z_x = 511 - (17.29 - 15.12)(11.93)(2) = 459.2 \text{ in}^3$

$M_u \leq \phi_b M_n = \dfrac{(0.9)(36)(459.2)}{12} = 1240 \text{ ft-k}$

c. Reducing each flange area by four holes (not LRFD Specification)

Area of four holes $= (4)(1\frac{1}{8})(1.340) = 6.03 \text{ in}^2$

Reduced $Z_x = 511 - (6.03)(11.93)(2) = 367.1 \text{ in}^3$

$M_y \leq \phi_b M_n = \dfrac{(0.9)(36)(367.1)}{12} = 991 \text{ ft-k}$

These LRFD requirements apply to the design of hybrid beams and girders (but not plate girders) whose flanges consist of a stronger grade of steel than their webs. This is true as long as these members are not required to resist an axial force larger than $\phi_b(0.15\,F_{yf})A_g$ of the flanges times their gross areas.

Should a hole be present in only one side of a flange of a W section, there will be no axis of symmetry for the net section of the shape. A correct theoretical solution of the problem would be very complex. Rather than going through such a lengthy process over a fairly minor point, it seems logical to assume holes in both sides of the flange. The results obtained will probably be just as satisfactory as those obtained by a laborious theoretical method.

A steel beam in place ready for installation of erection bolts. (Courtesy of the American Institute of Steel Construction, Inc.)

9.4 LATERAL SUPPORT OF BEAMS

Probably the large majority of steel beams are used in such a manner that their compression flanges are restrained against lateral buckling. (Unfortunately, however, the percentage has not been quite as high as the design profession has assumed.) The upper flanges of beams used to support concrete building and bridge floors are often incorporated in these concrete floors. For situations of this type, where the compression flanges are restrained against lateral buckling, the beams will fall into Zone 1.

Should the compression flange of a beam be without lateral support for some distance, it will have a stress situation similar to that existing in columns. As is well known, the longer and slenderer a column becomes, the greater becomes the danger of its buckling for the same loading condition. When the compression flange of a beam is long enough and slender enough, it may quite possibly buckle unless lateral support is provided.

There are many factors affecting the amount of stress that will cause buckling in the compression flange of a beam. Some of these factors are properties of the material, the spacing and types of lateral support provided, residual stresses in the sections, the types of end support or restraints, the loading conditions, etc.

The tension in the other flange of a beam tends to keep that flange straight and restrain the compression flange from buckling; but as the bending moment is increased, the tendency of the compression flange to buckle may become large enough to overcome the tensile restraint. When the compression flange does begin to buckle, twisting or torsion will occur, and the smaller the torsional strength of the beam the more rapid will be the failure. The W, S, and channel shapes so frequently used for beam sections do not have a great deal of resistance to lateral buckling and the resulting torsion. Some other shapes, notably the built-up box shapes, are tremendously stronger. These types of members have a great deal more torsional resistance than the

W, S, and plate girder sections. Tests have shown that they will not buckle laterally until the strains developed are well in the plastic range.

Some judgment needs to be used in deciding what does and what does not constitute satisfactory lateral support for a steel beam. Perhaps the most common question asked by practicing steel designers is "What is lateral support?" A beam that is wholly encased in concrete or that has its compression flange incorporated in a concrete slab is certainly well supported laterally. When a concrete slab rests on the top flange of a beam, the engineer must study the situation carefully before he or she counts on friction to provide full lateral support. Perhaps if the loads on the slab are fairly well fixed in position, they will contribute to the friction and it may be reasonable to assume full lateral support. If, on the other hand, there is much movement of the loads and appreciable vibration, the friction may well be reduced and full lateral support cannot be assumed. Such situations occur in bridges due to traffic, and in buildings with vibrating machinery such as printing presses.

Should lateral support of the compression flange not be provided by a floor slab, it is possible that such support may be provided with connecting beams or with special members inserted for that purpose. Beams that frame into the sides of the beam or girder in question and are connected to the compression flange can usually be counted on to provide full lateral support at the connection. If the connection is made primarily to the tensile flange, little lateral support is provided to the compression flange. Before support is assumed from the these beams, the designer should note whether they themselves are prevented from moving. The series of beams represented with horizontal dashed lines in Fig. 9.6 provide questionable lateral support for the main beams between columns. For a situation of this type, some system of *x*-bracing may be desirable in one of the bays. Such a system is shown in Fig. 9.6. This one system will provide sufficient lateral support for the beams for several bays.

The intermittent welding of metal roof or floor decks to the compression flanges of beams will probably provide sufficient lateral bracing. The corrugated sheet-metal roofs that are usually connected to the purlins with metal straps probably furnish only partial lateral support. A similar situation exists when wood flooring is bolted to supporting steel beams. At this time the student quite naturally asks, "If only partial support is

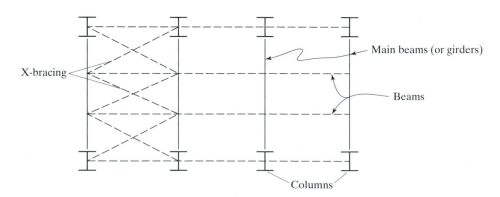

FIGURE 9.6

available, what am I to consider to be the distance between points of lateral support?" The answer to this question is for him or her to use his or her judgment. As an illustration, a wood floor is assumed to be bolted every 4 ft to the supporting steel beams in such a manner that it is thought only partial lateral support is provided at those points. After studying the situation, the engineer might well decide that the equivalent of full lateral support at 8-ft intervals is provided. Such a decision seems to be within the spirit of the specification.

Should there be doubt in the designer's mind as to the degree of lateral support provided, he or she should probably be advised to assume there is none.

9.5 INTRODUCTION TO INELASTIC BUCKLING, ZONE 2

If intermittent lateral bracing is supplied for the compression flange of a beam section, or if intermittent torsional bracing is supplied to prevent twisting of the cross section at the bracing points such that the member can be bent until the yield strain is reached in some but not all of its compression elements before lateral buckling occurs, we have inelastic buckling. In other words, the bracing is insufficient to permit the member to reach a full plastic strain distribution before buckling occurs.

Because of the presence of residual stresses (discussed in Section 5.2), yielding will begin in a section at applied stresses equal to $F_y - F_r$, where F_y is the yield stress of the web and F_r equals the compressive residual stress, assumed equal to 10 ksi for rolled shapes and 16.5 ksi for welded shapes (LRFD Specification F1.2a). It should be noted that the definition of plastic moment $F_y Z$ in Zone 1 is not affected by residual stresses because the sum of the compressive residual stresses equals the sum of the tensile residual stresses in the section and the net effect is theoretically zero.

When a constant moment occurs along the unbraced length, L_b, of a compact I- or C-shaped section and L_b, is larger than L_p the beam will fail inelastically unless L_b is greater than a distance L_r (to be discussed) beyond which the beam will fail elastically before F_y is reached (thus falling into Zone 3).

9.5.1 Bending Coefficients

In the formulas presented in these next few sections for inelastic and elastic buckling, we will use a term C_b. This is a moment coefficient that is included in the formulas to account for the effect of different moment gradients on lateral-torsional buckling. In other words, lateral buckling may be appreciably affected by the end restraint and loading conditions of the member.

As an illustration, the reader can see that the moment in the unbraced beam of part (a) of Fig. 9.7 causes a worse compression flange situation than does the moment in the unbraced beam of part (b). For one reason, the upper flange of the beam in part (a) is in compression for its entire length, while in (b) the length of the "column," that is, the length of the upper flange that is in compression is much less (thus a much shorter "column").

For the simply supported beam of part (a) of the figure, we will find that C_b is 1.14, while for the beam of part (b) it is 2.38. The basic moment capacity equations for Zones 2 and 3 were developed for laterally unbraced beams subject to single curvature with $C_b = 1.0$. Frequently, beams are not bent in single curvature with the result that

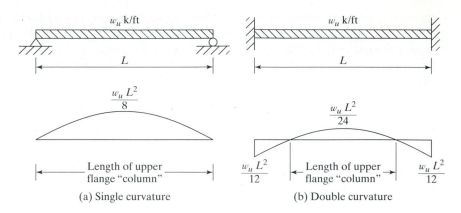

FIGURE 9.7

they can resist more moment. We have seen this in Fig. 9.7. To handle this situation, the LRFD Specification provides moment or C_b coefficients larger than 1.0 which are to be multiplied by the computed M_n values. The results are higher moment capacities. The designer who conservatively says, "I'll always use $C_b = 1.0$" is missing out on the possibility of significant savings in steel weight for some situations. *When using C_b values, the designer should clearly understand that the moment capacity obtained by multiplying M_n by C_b may not be larger than the plastic M_n of Zone I which is equal to F_yZ.* This situation is illustrated in Fig. 9.8.

The value of C_b is determined from the expression to follow in which M_{max} is the largest moment in an unbraced segment of a beam, while M_A, M_B, and M_c are, respectively, the moments at the 1/4 point, 1/2 point, and 3/4 point in the segment.

$$C_b = \frac{12.5M_{max}}{2.5M_{max} + 3M_A + 4M_B + 3M_c} \qquad \text{(LRFD Equation F1-3)}$$

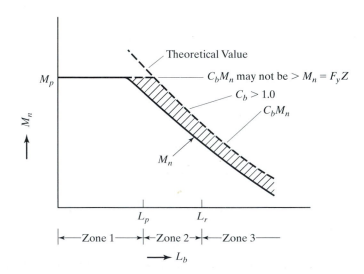

FIGURE 9.8

C_b is equal to 1.0 for cantilevers or overhangs where the free end is unbraced. Some typical values of C_b calculated with the above equation are shown in Fig. 9.9 for various beam and moment situations. Most of these values are also given in Table 5.1 of the Manual.

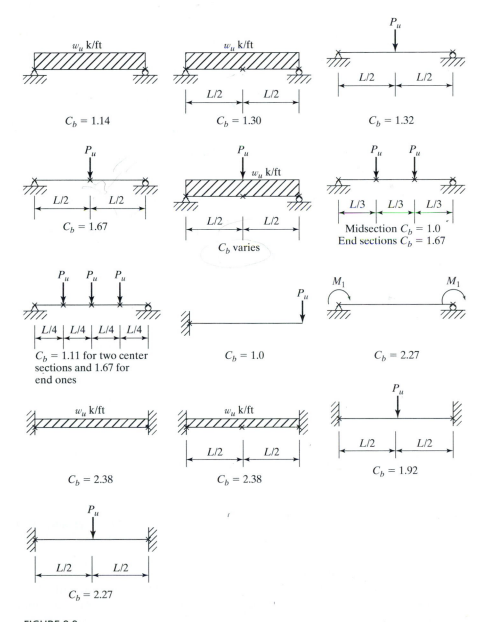

FIGURE 9.9

Sample C_b values. (The X marks represent points of lateral bracing.)

9.6   MOMENT CAPACITIES, ZONE 2

When constant moment occurs along the unbraced length, or as the unbraced length of the compression flange of a beam or the distance between points of torsional bracing is increased beyond L_p, the moment capacity of the section will become smaller and smaller. Finally, at an unbraced length L_r, the section will buckle elastically as soon as the yield stress is reached. Owing to the rolling operation, however, there is a residual stress in the section equal to F_r. Thus, the elastically computed stress caused by bending can only reach $F_y - F_r$. Assuming $C_b = 1.0$, the design moment capacity for a compact I- or C-shaped section bent about its x axis may be determined as follows if $L_b = L_r$.

$$\phi_b M_r = \phi_b S_x (F_y - F_r)$$

L_r is a function of several of the section's properties such as its cross-sectional area, modulus of elasticity, yield stress, and warping and torsional properties. The very complex formulas needed for its computation are given in the LRFD Specification ($F1$) and space is not taken to show them here. Fortunately, numerical values have been determined for sections normally used as beams and are given in LRFD Manual Table 5.3, entitled "W shapes selected by Z_x."

Going backward from an unbraced length of L_r toward an unbraced length L_p we can see that buckling does not occur when the yield stress is first reached. We are in the inelastic range (Zone 2) where there is some penetration of the yield stress into the section from the extreme fibers. For these cases when the unbraced length falls between L_p and L_r, the design moment strength will fall approximately on a straight line between $\phi M_n = \phi_b F_y Z$ at L_p and $\phi_b S_x (F_y - F_r)$ at L_r. For intermediate values of the unbraced length, the moment capacity may be determined by proportions or by substituting into the expression at the end of this paragraph. If C_b is larger than 1.0, the section will resist additional moment but not more than $\phi_b F_y Z = \phi_b M_p$.

$$\phi_b M_{nx} = C_b [\phi_b M_{px} - BF(L_b - L_p)] \leq \phi_b M_{px}$$

in which BF is a factor given in LRFD Table 5.3 for each section, which enables us to do the proportioning with a simple formula.

Alternatively, the value of M_n can be determined from the equation

$$M_n = C_b \left[M_p - (M_p - M_r) \left(\frac{L_b - L_p}{L_r - L_p} \right) \right] \leq M_p \qquad \text{(LRFD Equation F1-2)}$$

and then multiplied by ϕ_b to obtain $\phi_b M_n$.

Example 9-4 illustrates the determination of the moment capacities of a section with L_b between L_p and L_r, while Example 9-5 demonstrates the design of a beam in the same range.

Example 9-4

Determine the moment capacity of a W24 × 62 with $F_y = 50$ ksi if $L_b = 8.0$ ft and $C_b = 1.0$.

Solution if $F_y = 50$ ksi. From LRFD Table 5.3

$$L_p = 4.84$$
$$L_r = 13.3 \text{ ft}$$
$$\phi_b M_r = 396 \text{ ft-k}$$
$$\phi_b M_p = 578 \text{ ft-k}$$
$$BF = 21.6 \text{ k}$$

Since $L_b > L_p < L_r$

$$\phi_b M_n = 1.0[576 - (21.6)(8.0 - 4.84)] = 508 \text{ ft-k}$$

Example 9-5

Select the lightest available section for a factored moment of 290 ft-k if $L_b = 10.0$ ft. Use 50 ksi steel and assume $C_b = 1.0$.

Solution. Enter LRFD Table 5.3 and notice that $\phi_b M_p$ for a W18 × 40 is 294 ft-k, but L_p is 4.49 ft $< L_b$ of 10.0 ft. Also, $\phi_b M_r = 205$ ft-k and $BF = 11.7$ k.

$$\therefore \phi_b M_n = 1.0[294 - (11.7)(10.0 - 4.49)] \qquad \text{(NG)}$$
$$= 229.5 \text{ ft-k} < 290 \text{ ft-k}$$

Moving up in the table and after several trials, try a W21 × 48 (with $L_p = 6.09$ ft, $L_r = 15.4$ ft, $\phi_b M_r = 279$ ft-k, $\phi_b M_p = 401$ ft-k, and $BF = 13.2$ k).

$$\phi_b M_n = 1.0[401 - (13.2)(10.0 - 6.09)]$$
$$= 349 \text{ ft-k} > 290 \text{ ft-k} \qquad \text{(OK)}$$

Use W21 × 48

Note: A much easier solution is presented in Section 9.8.

9.7   ELASTIC BUCKLING, ZONE 3

When a beam is not fully braced, it may fail due to buckling of the compression portion of the cross section laterally about the weaker axis with twisting of the entire cross section about the beam's longitudinal axis between the points of lateral bracing. This will occur even though the beam is loaded so that it supposedly will bend about

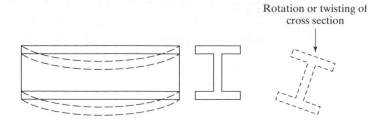

FIGURE 9.10

Lateral-torsional buckling of a simply supported beam.

the stronger axis. The beam will bend initially about the stronger axis until a certain critical moment M_{cr} is reached. At that time it will buckle laterally about its weaker axis. As it bends laterally, the tension in the other flange will try to keep the beam straight. As a result, the buckling of the beam will be a combination of lateral bending and a twisting (or torsion) of the beam cross section. A sketch of this situation is shown in Fig. 9.10.

The critical moment or flexural-torsional moment M_{cr} in a beam will be made up of the torsional resistance (commonly called *St-Venant torsion*) and the warping resistance of the section. These are combined as follows:

$$M_{cr} = \sqrt{\left(\begin{array}{c}\text{torsional}\\\text{resistance}\end{array}\right)^2 + \left(\begin{array}{c}\text{warping}\\\text{resistance}\end{array}\right)^2}$$

Returning to the LRFD Specification, if the unbraced length of the compression flange of a beam section or the distance between points that prevent twisting of the entire cross section is greater than L_r, the section will buckle elastically before the yield stress is reached anywhere in the section. In Section F1.1.2b of the LRFD Specification, the classic equation for determining this flexural-torsional buckling moment called M_{cr} is presented. This expression follows

$$M_{cr} = C_b \frac{\pi}{L_b} \sqrt{EI_y GJ + \left(\frac{\pi E}{L_b}\right)^2 I_y C_w} \qquad \text{(LRFD Equation F1-13)}$$

In this equation, G is the shear modulus of elasticity of the steel = 11,200 ksi, J is a torsional constant (in^4), while C_w is the warping constant (in^6). The values of J and C_w are provided for rolled sections in Tables 1.25 to 1.35 of the LRFD Manual.

This expression is applicable to compact doubly symmetric I-shaped members, channel sections loaded in the plane of their webs, and I-shaped singly symmetric sections with their compression flanges larger than their tension ones ($\perp$). Expressions also are given in Sections F1.1.2b and F1.1.2c of the LRFD Specification for M_{cr} in the elastic range for other sections such as solid rectangular bars, symmetric box sections, tees, and double angles.

It is not possible for lateral-torsional buckling to occur if the moment of inertia of the section about the bending axis is equal to or less than the moment of inertia out *of plane. As a result, the limit state of lateral-torsional buckling is not applicable for shapes bent about their minor axes, for shapes with $I_x \le I_y$, or for circular or square shapes. Furthermore, yielding controls if the section is noncompact.*

Example 9-6 illustrates the computation of $\phi_b M_{cr}$ for an elastic buckling situation.

Example 9-6

Compute $\phi_b M_{cr}$ for a W18 × 97 consisting of 50 ksi steel if the unbraced length L_b is 38 ft. Assume $C_b = 1.0$.

Solution. Noting $L_b = 44$ ft $> L_r = 27.5$ ft from Table 5.3 in the Manual.

The following values for the W18 × 97 are obtained from the W shapes tables and the torsion properties for W shapes tables in Part 1 of the Manual:

$$I_y = 201 \text{ in}^4, \ J = 5.86 \text{ in}^4, \ \text{and } C_w = 15{,}800 \text{ in}^6$$

$$\phi_b M_n = \phi M_{cr} = M_u = (0.9)(1.0)\left(\frac{\pi}{12 \times 38}\right)$$

$$\times \sqrt{(29 \times 10^3)(201)(11{,}200)(5.86) + \left(\frac{\pi 29 \times 10^3}{38 \times 12}\right)^2 (201)(15{,}800)}$$

$$= 4425 \text{ in k} = 369 \text{ ft-k}$$

LRFD Specification (F1.2b) also presents the elastic buckling equation in an alternate form as follows:

$$M_{cr} = \frac{C_b S_x X_1 \sqrt{2}}{L_b / r_y} \sqrt{1 + \frac{X_1^2 X_2}{2(L_b / r_y)^2}}$$

(LRFD Equation F1-13 Alternate Form)

In which

$$X_1 = \frac{\pi}{S_x} \sqrt{\frac{EGJA}{2}} \qquad \text{(LRFD Equation F1-8)}$$

$$X_2 = 4 \frac{C_w}{I_y} \left(\frac{S_x}{GJ}\right)^2 \qquad \text{(LRFD Equation F1-9)}$$

Values of X_1 and X_2 are shown for W shapes in the "W-Shapes Dimensions" section of Part 1 of the Manual.

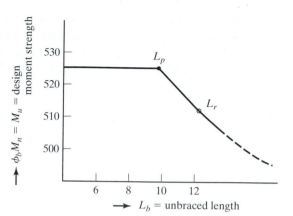

FIGURE 9.11

9.8   DESIGN CHARTS

Fortunately, the values of $\phi_b M_{cr} = \phi_b M_n$ for the sections normally used as beams are computed for different unbraced lengths and the results plotted as curves in Part 5 of the LRFD Manual. The values not only cover unbraced lengths in the elastic range but also the inelastic range, enabling us to very easily handle the problems presented in the last section as well as the ones in this section which fall in Zone 3. The moments are plotted for F_y equal to 50 ksi and for $C_b = 1.0$.

The curve for a typical W section is shown in Fig. 9.11. For each of the shapes L_p is indicated with a solid circle (●) while L_r is shown with a hollow circle (○).

The charts were developed without regard to such things as shear, deflection, etc.—items that may occasionally control the design as described in Chapter 10. They cover almost all of the unbraced lengths encountered in practice. If C_b is greater than 1.0, the values given will be magnified somewhat, as illustrated in Fig. 9.8.

To select a member it is only necessary to enter the chart with the unbraced length L_b and the factored design moment M_u. For an illustration, let's assume that $C_b = 1.0$, $F_y = 50$ ksi and that we wish to select a beam with $L_b = 30.0$ ft for a moment $M_u = 550$ ft-k. We enter the charts in Part 5 of the Manual entitled "Beam Design Moments" and go through the pages where $F_y = 50$ ksi until we find in the left-hand column $\phi_b M_n$ equal to 550 ft-k. We proceed up from the bottom of the chart for an unbraced length = 30.0 ft until we intersect a horizontal line from 550 k. Any section to the right and above this intersection point (↗) will have a greater unbraced length and a greater moment capacity. The appropriate page from the handbook is shown in Fig. 9.12 with the permission of the AISC.

Moving up and to the right we first encounter a W18 × 106. In this area of the charts this section is shown with a dashed line. This section will provide the necessary moment capacity but the dashed line indicates that it is in an uneconomical range. If we proceed farther upward and to the right, the first solid line that we encounter will represent the lightest satisfactory section. In this case, it's a W21 × 101. Other illustrations of the use of these charts are presented in Examples 9-7 and 9-8.

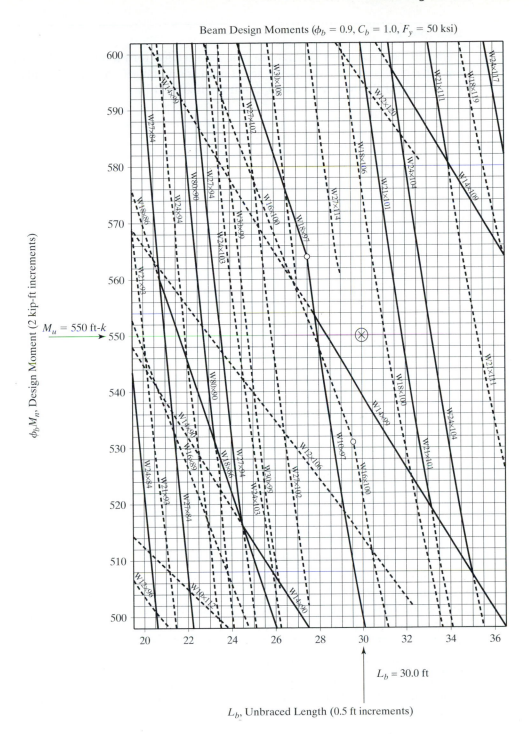

FIGURE 9.12

Design moments for beams with different unbraced lengths (from LRFD Manual)

Example 9-7

Using 50 ksi steel, select the lightest available section for the beam of Fig. 9.13 which has lateral bracing provided for its compression flange only at its ends. Assume $C_b = 1.00$ for this first chart example. (It's actually 1.14.)

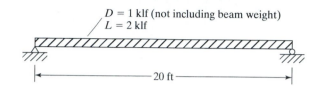

$D = 1$ klf (not including beam weight)
$L = 2$ klf

FIGURE 9.13 |← —————————— 20 ft —————————— →|

Solution

Assume beam weight $= 60$ lb/ft

$$w_u = (1.2)(1.060) + (1.6)(2) = 4.472 \text{ klf}$$

$$M_u = \frac{(4.472)(20)^2}{8} = 223.6 \text{ ft-k}$$

Entering the Beam Design Moments charts with $L_b = 20$ ft and $M_u = 223.6$ ft-k. Use W14 × 53, $F_y = 50$ ksi.

For the example problem that follows, C_b is greater than 1.0. For such a situation the reader should look back to Fig. 9.8. There he or she will see that the design moment strength of a section can go to $\phi_b C_b M_n$ when $C_b > 1.0$ but may under no circumstances exceed $\phi_b M_p = \phi_b F_y Z$.

To handle such a problem we calculate an effective moment as follows (the numbers being taken from Example 9-8):

$$M_{\text{effective}} = \frac{M_u}{C_b} = \frac{864.6}{1.67} = 517.7 \text{ ft-k}$$

Then we enter the charts with our unbraced length of 17 ft and with $M_{\text{effective}} = 517.7$ ft-k and there select a W18 × 76. We must, however, check to see that our $M_u(864.6)$ does not exceed $\phi_b F_y Z$ for the section. In this case, it does and we must keep going until we find the lightest section that has a $\phi_b M_n \geq 517.7$ ft-k at $L_b = 17$ ft and yet which has a $\phi_b F_y Z \geq 864.6$ ft-k.

Example 9-8

Using 50 ksi steel, select the lightest available section for the situation shown in Fig. 9.14. Bracing is provided only at the ends and center line of the member, and thus, $C_b = 1.67$.

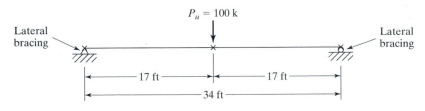

FIGURE 9.14

Solution

Assume beam weight $= 84$ lb/ft

$$w_u = (1.2)(0.084) = 0.1008 \text{ klf}$$

$$M_u = \frac{(100)(34)}{4} + \frac{(0.1008)(34)^2}{8} = 864.6 \text{ ft-k}$$

Entering the Beam Design Moments charts with $M_{u \text{ effective}} = 864.6/1.67 = 517.7$ ft-k and $L_b = 17$ ft we find a W18 × 76.

But $\phi_b M_n = 864.6$ ft-k may not exceed $\phi_b F_y Z$ of the section, and it does for a W18 × 76, as can be seen in the Load Factor Design Selection Table. Therefore, we continue looking in the table until we find the lightest section that has a $\phi_b F_y Z \geq 864.6$ ft-k.

Use W27 × 84, $F_y = 50$ ksi.

9.9   NONCOMPACT SECTIONS

You will remember from Section 5.7 of this textbook that a compact section is one that has a sufficiently stocky profile so that it is capable of developing a fully plastic stress distribution (assuming its compression flange has sufficient lateral bracing) before something buckles (web or flange). For a section to be compact the width thickness ratio of the flanges of W- or other I-shaped rolled sections must not exceed a b/t value $\lambda_p = 0.38\sqrt{E/F_y}$. Similarly, the webs in flexural compression must not exceed an h/t value $\lambda_p = 3.76\sqrt{E/F_y}$. The values of b, t, h, and t_f are shown in Fig. 9.15.

A noncompact section is one for which the yield stress can be reached in some but not all of its compression elements before buckling occurs. It is not capable of reaching a fully plastic stress distribution. The noncompact sections are those that have web thickness ratios greater than λ_p but not greater than λ_r. The λ_r values are provided in Table B5.1 of the LRFD Specification. For the noncompact range, the width thickness ratios of the flanges or W or other I-shaped rolled sections must not exceed $\lambda_r = 0.83 - \sqrt{E/F_L}$, while those for the webs must not exceed $\lambda_r = 5.70\sqrt{E/F_y}$. In the first of these two expressions for the noncompact range, F_L is the smaller of $F_{yf} - F_r$ or F_{yw}. Other values are provided in LRFD Table B5.1 in Part 16 of the Manual for λ_p and λ_r for other shapes and for shapes subject to both axial load and bending.

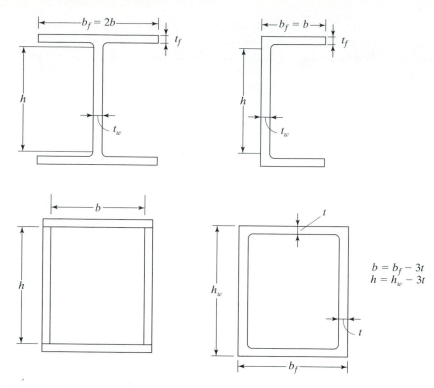

FIGURE 9.15

Values of h, b, and t to be used for computing λ = width-thickness ratios.

For noncompact beams, the nominal flexural strength M_n is the lowest of the lateral-torsional buckling strength, the flange local buckling strength, or the web local buckling strength.

If we have a noncompact section, that is one with $\lambda_p < \lambda \leq \lambda_r$, the value of M_n may be obtained by linear interpolation between M_p and M_r as represented by the equations to follow:

For lateral-torsional buckling

$$M_n = C_b\left[M_p - (M_p - M_r)\left(\frac{\lambda - \lambda_p}{\lambda_r - \lambda_p}\right)\right] \leq M_p$$

(LRFD Appendix Equation A-F1-2)

For flange and web local buckling

$$M_n = M_p - (M_p - M_r)\left(\frac{\lambda - \lambda_p}{\lambda_r - \lambda_p}\right)$$ (LRFD Appendix Equation A-F1-3)

Should $\lambda > \lambda_r$, the limit state of lateral-torsional buckling and flange local buckling is to be determined with the LRFD Appendix formula to follow in which S is the section modulus of the member and F_{cr} is the critical design stress for compression members as previously determined in Chapter 5 with LRFD Appendix Formulas A-E2-2 or A-E2-3, as applicable.

$$M_n = M_{cr} = SF_{cr} \leq M_p \qquad \text{(LRFD Appendix Equation A-F1-4)}$$

There are five sections with $F_y = 50$ ksi whose flanges are noncompact that are listed in "W-shapes Selection by Z_x" tables. These are the W21 × 48, W14 × 99, W14 × 90, W12 × 65 and the W10 × 12. They are indicated in the Manual with the a double dagger sign ††.

The equations listed here were used to obtain the values shown in the LRFD Manual "Load Factor Design Selection Table for shapes used as beams" for noncompact sections. The designer will have little trouble with noncompact sections when F_y is 36 or 50 ksi. He or she, however, will have to use the formulas presented in this section for shapes with F_y greater than 50 ksi.

9.10   COMPUTER EXAMPLE

In Example 9-9 the INSTEP program is used to obtain the design moment strength of a beam section for a certain unbraced length L_b. The strength for any other unbraced length can be quickly obtained by changing L_b.

Example 9-9

With the enclosed computer program determine the design moment capacity of an A992 W36 × 300 with $L_b = 48$ ft. Assume $C_b = 1.0$.

Input:

Factored design loads	Bracing parameters
Mu 0 kip-in	Lb 576 in
Vu 0 kips	Cb 1

Output:

Beam Flexural Design Summary

Section: W36 × 300

Calculated Capacity

Mu = 0.9Mn = 0.9[34530.1] = 31077.1 kip-in (2589.76 kip-ft)

Vu = 0.9Vn = 0.9[1040.44] = 936.4 kips

This section is adequate

PROBLEMS

9-1 to 9-8. *Select the most economical sections using F_y = 50 ksi unless otherwise specified and assuming full lateral bracing for the compression flanges. Working or service loads are given for each case, and beam weights are not included.*

9-1.

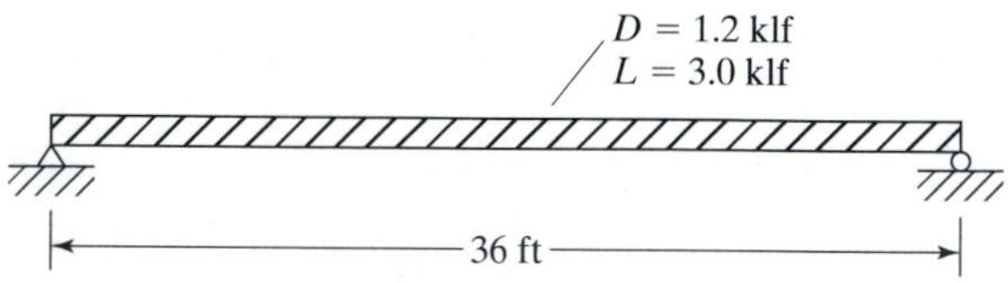

D = 1.2 klf
L = 3.0 klf

36 ft

FIGURE P9-1 (*Ans.* W30 × 90)

9-2.

P_L = 24 k P_L = 24 k

D = 2 klf

10 ft   10 ft   10 ft

FIGURE P9-2

9-3. Repeat Prob. 9-2 using P_L = 36 k. (*Ans.* W27 × 84)

9-4.

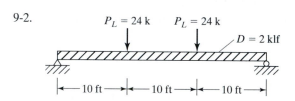

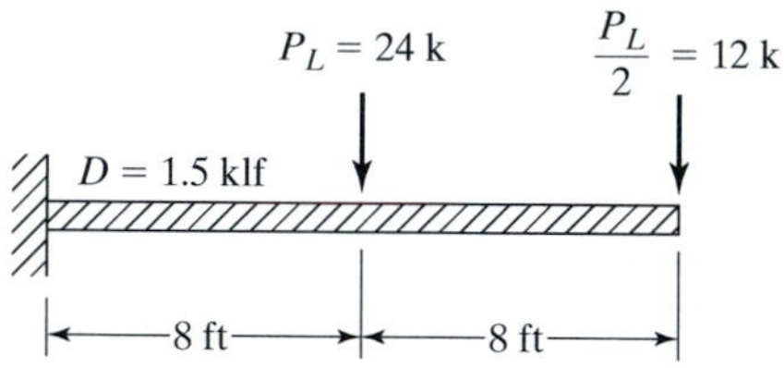

P_L = 24 k $\frac{P_L}{2}$ = 12 k

D = 1.5 klf

8 ft   8 ft

FIGURE P9-4

9-5.

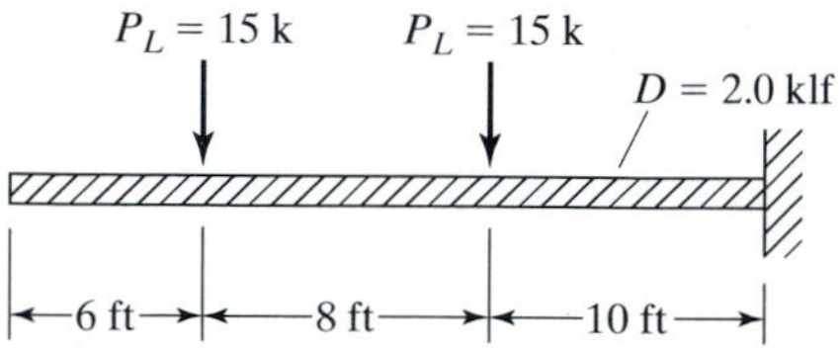

P_L = 15 k P_L = 15 k

D = 2.0 klf

6 ft   8 ft   10 ft

FIGURE P9-5 (*Ans.* W30 × 116)

9-6.

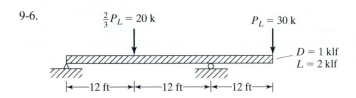

$\frac{2}{3}P_L = 20$ k $P_L = 30$ k

$D = 1$ klf
$L = 2$ klf

|←—12 ft—→|←—12 ft—→|←—12 ft—→|

FIGURE P9-6

9-7.

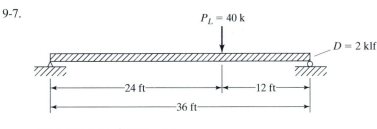

$P_L = 40$ k

$D = 2$ klf

|←————24 ft————→|←———12 ft———→|

|←————————36 ft————————→|

FIGURE P9-7 (*Ans.* W27 × 84)

9-8.   The accompanying illustration shows the arrangement of beams and girders that are used to support a 6-in. reinforced-concrete floor for a small industrial building. Design the beams and girders assuming they are simply supported. Assume full lateral support and a live load of 120 psf. Concrete weight is 150 lb/ft³.

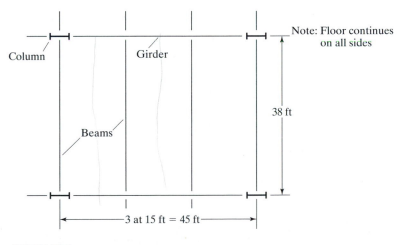

Column

Girder

Note: Floor continues
on all sides

38 ft

Beams

|←——————3 at 15 ft = 45 ft——————→|

FIGURE P9-8

9-9.   A beam consists of a W18 × 130 with a 1/2 × 14 in cover plate welded to each flange. Determine the design uniform load w_u the member can support in addition to its own weight for a 40-ft simple span. (*Ans.* 7.8 k/ft)

9-10. The member shown is made with 50 ksi steel. Determine the maximum service live load that can be placed on the beam if in addition to its own weight it is supporting a service dead load of 1.0 klf. The member is used for a 30-ft simple span.

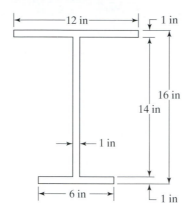

FIGURE P9-10

9-11 to 9-14. *Full lateral bracing is provided for these problems.*

9-11. Select a section for a 36-ft simple span to support a service dead uniform load of 2.5 klf and a live service uniform load of 2 klf if four holes for 1-in bolts are assumed present in each flange at the section of maximum moment. Use LRFD Specification and A36 steel. (*Ans.* W33 × 130)

9-12. Rework Prob. 9-11 assuming four holes for 1-in bolts pass through each flange at the point of maximum moment. Use LRFD Specification and A992 steel.

9-13. The section shown in the accompanying illustration has two 1-in bolts passing through each flange. Find the design or factored load w_u which the section can support in addition to its own weight for a 30-ft simple span if it consists of a steel with $F_y = 50$ ksi. Deduct all holes for calculating section properties. (*Ans.* 12.21 k/ft)

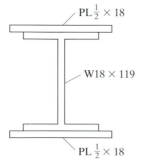

FIGURE P9-13

9-14. A 40-ft simple span beam is to support two movable service 30-k loads a distance of 15 ft apart. Assuming a dead load of 1.2 klf in addition to the beam weight, select a 50 ksi steel section to resist the largest possible moment.

9-15 to 9-28. *For these problems different values of L_b are given. Dead loads do not include beam weights.*

9-15. Determine the design moment strength $\phi_b M_n$ of a W21 × 73 for simple spans of 6, 12, and 22 ft if lateral bracing for the compression flanges is provided at the ends only. Use A992 steel and $C_b = 1.0$. (*Ans.* 645, 549, 324 ft-k)

9-16. Using $F_y = 50$ ksi select the lightest satisfactory section if bracing is provided at the ends only. Obtain C_b from Fig. 9.9 in text.

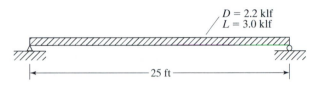

FIGURE P9-16

9-17. Select the lightest available section if $F_y = 50$ ksi. Lateral bracing is provided at the ends only. $C_b = 1.0$. (*Ans.* W21 × 111)

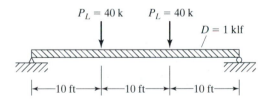

FIGURE P9-17

9-18. Repeat Prob. 9-17 if lateral bracing is supplied at each of the concentrated loads as well as at the ends of the span. Obtain C_b from Fig. 9.9 in text.

9-19. A W30 × 148 consisting of 50 ksi steel is used for a simple span of 27 ft and has lateral bracing provided at its ends only. If the only dead load present is the beam weight, what is the largest service concentrated live load that can be placed at the beam center line? (*Ans.* 122.7 k)

9-20. Repeat Prob. 9-19 if lateral bracing is supplied at beam ends and at the concentrated load.

9-21. If F_y is 65 ksi select the lightest section for the situation shown in the accompanying illustration if lateral bracing is supplied at the fixed end only. (*Ans.* W27 × 84)

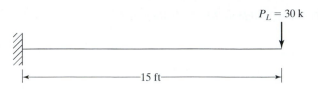

$P_L = 30$ k

15 ft

FIGURE P9-21

9-22. Using a steel with $F_y = 50$ ksi, what uniformly distributed service live load can a simply supported W33 × 118 carry for a span of 40 ft when (a) the compression flange is braced laterally for its full length and when (b) lateral bracing is supplied at the ends only.

9-23. What service live uniform load can be placed on a W18 × 130($F_y = 50$ ksi) which has lateral bracing provided at its ends only? The beam is simply supported and has a span of 30 ft. The dead uniform service load is 1 klf plus the beam weight. (*Ans.* 4.28 k/ft)

9-24. A W36 × 160 is used to support the loads shown in the accompanying illustration. Using $F_y = 50$ ksi and neglecting the dead weight of the beam, find if the beam is overloaded.

   a. If lateral support is provided for the entire span.

   b. If lateral support is provided at the beam ends only.

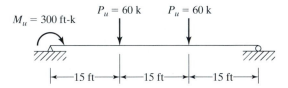

$M_u = 300$ ft-k $P_u = 60$ k $P_u = 60$ k

15 ft   15 ft   15 ft

FIGURE P9-24

9-25. Redesign the beam of Prob. 9-24 if lateral support is provided at the ends and at the concentrated loads. (*Ans.* W30 × 90)

9-26. The W30 × 99 shown has lateral support supplied at its center line as well as at its ends. If F_y is 50 ksi, determine the maximum permissible service live load P_L. Neglect beam weight.

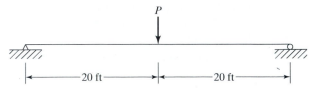

P

20 ft   20 ft

FIGURE P9-26

9-27.   Repeat Prob. 9-26 if a dead uniform service load of 1.2 klf including the beam weight is placed on the beam for its full length and if lateral bracing is provided at the ends and center line of the beam. (*Ans.* 55.1 k)

9-28.   Using 50 ksi steel, select the lightest available section for the situation shown in the accompanying illustration if bracing is provided only at the ends and center line.

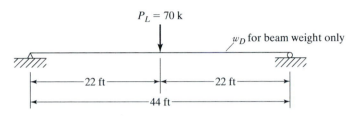

FIGURE P9-28

9-29.   A W30 × 148 is required for a certain span but, because of a strike in the steel mills, cannot be obtained on time; however, an extra W30 × 108 of sufficient length is available along with a good supply of plates. Select 7/8-in-thick cover plates to be welded to the flanges of the W30 × 108 to provide a satisfactory substitute for the original beam. Use F_y = 50 ksi and assume full lateral bracing is to be provided for the compression flange. (*Ans.* PL 7/8 × 6)

9-30.   A W33 × 152 has been specified for use on a certain job. By mistake, a W33 × 130 was shipped to the field. This beam must be erected today. Assuming that 1-in-thick plates are obtainable immediately, select cover plates to be welded to the flanges to obtain the necessary section. Use F_y = 50 ksi and assume full lateral bracing is supplied for the compression flange.

9-31.   Repeat Prob. 9-29 if the original beam was to have been of a steel with F_v = 65 ksi and the substitute material is available only in 50 ksi steel. Assume also that only 3/4-in plates are available. (*Ans.* PL 3/4 × 14 in)

9-32.   Design a beam of 50 ksi steel with a depth no greater than 12.00 in to support a uniform load of w_u = 15.0 klf (includes beam wt) for a 22-ft simple span. The member is to have full lateral bracing for its compression flange.

9-33 to 9-35.   *Using the enclosed computer diskette, repeat the following problems.*

9-33.   Prob. 9-1 (*Ans.* W30 × 90)

9-34.   Prob. 9-15

9-35.   Prob. 9-28 (*Ans.* W30 × 108)

Design of Beams— Miscellaneous Topics

10.1   DESIGN OF CONTINUOUS BEAMS

Section A5 of the LRFD Specification permits the design of beams analyzed either on the basis of elastic analysis with factored loads or by plastic analysis with the same factored loads. Design based on plastic analysis is permitted only for sections with yield stresses no greater than 65 ksi and is subject to some special requirements in Sections B5.2, C1.1, C2.1a, C2.2a, E1.2, F1.3, H1, and I1 of the LRFD Specification.

Both theory and tests show clearly that continuous ductile steel members meeting the requirements for compact sections with sufficient lateral bracing supplied for their compression flanges have the desirable ability of being able to redistribute moments caused by overloads. If plastic analysis is used this advantage is automatically included in the analysis.

If elastic analysis is used the LRFD handles the redistribution by a rule of thumb that approximates the real plastic behavior. The LRFD Specification (A5) states that for continuous compact sections the design *may* be made on the basis of nine-tenths of the maximum negative moments caused by gravity loads which are maximum at points of support if the positive moments are increased by one-tenth of the average negative moments at the adjacent supports. *(The 0.9 factor is applicable only to gravity loads and not to lateral loads such as those caused by wind and earthquake.* The factor can also be applied to columns that have axial stresses less than $0.15F_y$.) This moment reduction does not apply to members consisting of A514 steel, hybrid girders, or to moments produced by loading on cantilevers.

Example 10-1 illustrates the design of a three-span continuous beam analyzed by (a) plastic analysis and (b) elastic analysis.

Example 10-1

The beam shown in Fig. 10.1 is assumed to consist of 50 ksi steel. (a) Select the lightest W section available using plastic analysis and assuming full lateral support is provided for its compression flanges. (b) Design the beam using elastic analysis with the factored loads and the 0.9 rule and assuming full lateral support is provided for flanges. (c) Design the beam using elastic analysis with the factored loads and the 0.9 rule assuming full lateral support is provided for the upper flange *but* only for the lower flange at support points.

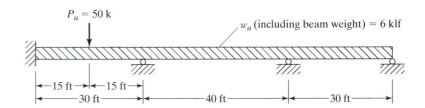

FIGURE 10.1

Solution

    a. Plastic analysis and design

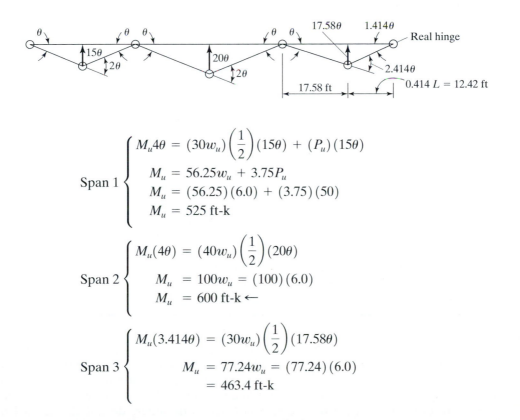

Span 1
$$
\begin{cases}
M_u 4\theta = (30w_u)\left(\dfrac{1}{2}\right)(15\theta) + (P_u)(15\theta) \\[2mm]
M_u = 56.25w_u + 3.75P_u \\
M_u = (56.25)(6.0) + (3.75)(50) \\
M_u = 525 \text{ ft-k}
\end{cases}
$$

Span 2
$$
\begin{cases}
M_u(4\theta) = (40w_u)\left(\dfrac{1}{2}\right)(20\theta) \\[2mm]
M_u = 100w_u = (100)(6.0) \\
M_u = 600 \text{ ft-k} \leftarrow
\end{cases}
$$

Span 3
$$
\begin{cases}
M_u(3.414\theta) = (30w_u)\left(\dfrac{1}{2}\right)(17.58\theta) \\[2mm]
M_u = 77.24w_u = (77.24)(6.0) \\
= 463.4 \text{ ft-k}
\end{cases}
$$

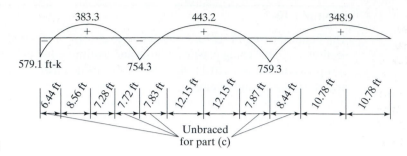

FIGURE 10.2

$$Z \text{ required} = \frac{M_u}{\phi_b F_y} = \frac{(12)(600)}{(0.9)(50)} = 160 \text{ in}^3$$

Use W21 × 68, F_y = 50 ksi. Section is Compact

b. Elastic analysis and design—full lateral bracing both flanges

Maximum negative moment for design

$$= (0.9)(759.3) = -683.4 \text{ ft-k} \leftarrow$$

Maximum positive moment for design

$$= 443.2 + \frac{1}{10}\left(\frac{754.3 + 759.3}{2}\right) = +518.9 \text{ ft-k}$$

$$Z \text{ required} = \frac{(12)(683.4)}{(0.9)(50)} = 182.2 \text{ in}^3$$

Use W24 × 76, F_y = 50 ksi. Section is Compact.

c. Elastic design—full lateral support supplied for top flange but only at support points for bottom flange

Referring to the moment diagram for the part (b) solution (Fig. 10.2) we see that the greatest unbraced length in the negative moment region is 8.44 ft and the reduced M_u is 683.4 ft-k.

From the Beam Design Charts in Section 5 of the Manual with M_u = 683.4 ft-k we pick a W24 × 76 Its L_p = 8.9 ft > 8.44 ft.

Use W24 × 76 F_y = 50 ksi. Section is Compact.

10.2   SHEAR

For this discussion the beam of Fig. 10.3(a) is considered. As the member bends, shear stresses occur because of the changes in length of its longitudinal fibers. For positive bending the lower fibers are stretched and the upper fibers are shortened while somewhere in between there is a natural axis where the fibers do not change in length.

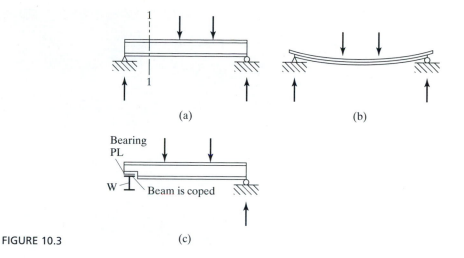

FIGURE 10.3

Owing to these varying deformations a particular fiber has a tendency to slip on the fiber above or below.

If a wooden beam was made by stacking boards on top of each other and not connecting them, they would obviously tend to take the shape shown in part (b) of the figure. The student may have observed short heavily loaded timber beams with large transverse shears that split along horizontal planes.

This presentation may be entirely misleading in seeming to completely separate horizontal and vertical shears. In reality, horizontal and vertical shear at any point are the same, as long as the critical section at which the shear stress is evaluated is taken parallel to the axis of symmetry, and it may not be separated. Furthermore, one cannot occur without the other.

Generally, shear is not a problem in steel beams because the webs of rolled shapes are capable of resisting rather large shearing forces. Perhaps it is well, however, to list here the most common situations where shear might be excessive. These are as follows.

1. Should large concentrated loads be placed near beam supports they will cause large internal forces without corresponding increases in bending moments. A fairly common example of this type of loading occurs in tall buildings where on a particular floor the upper columns are offset with respect to the columns below. The loads from the upper columns applied to the beams on the floor level in question will be quite large if there are many stories above.

2. Probably the most common shear problem occurs where two members (as a beam and a column) are rigidly connected together so their webs lie in a common plane. This situation frequently occurs at the junction of columns and beams (or rafters) in rigid frame structures.

3. Where beams are notched or coped as shown in Fig. 10.3(c) shear can be a problem. For this case shear forces must be calculated for the remaining beam depth.

Combined welded and bolted joint, Transamerica Pyramid, San Francisco, CA.
(Courtesy of Kaiser Steel Corporation.)

A similar discussion can be made where holes are cut in beam webs for duct work or other items.

4. Theoretically, very heavily loaded short beams can have excessive shears but practically this does not occur too often unless it is like Case 1.

5. Shear may very well be a problem even for ordinary loadings when very thin webs are used as in plate girders or in light gage cold formed steel members.

From his or her study of mechanics of materials the student is familiar with the horizontal shear stress formula $f_v = VQ/Ib$ where V is the external shear, Q is the statical moment of that portion of the section lying outside (either above or below) the line on which f_v is desired taken about the neutral axis, and b is the width of the section where the unit shearing stress is desired.

Figure 10.4(a) shows the variation in shear stresses across the cross section of an I-shaped member, while part (b) of the same figure shows the shear stress variation in a member with a rectangular cross section. It can be seen in part (a) of the figure that the shear in I-shaped sections is primarily resisted by the web.

If the load is increased on an I-shaped section until the bending yield stress is reached in the flange, the flange will be unable to resist shear stress and it will be carried in the web. If the moment is further increased the bending yield stress will penetrate farther down into the web and the area of web that can resist shear will be further reduced. Rather than assuming the nominal shear stress is resisted by part of the web, the LRFD Specification assumes that a reduced shear stress is resisted by the entire

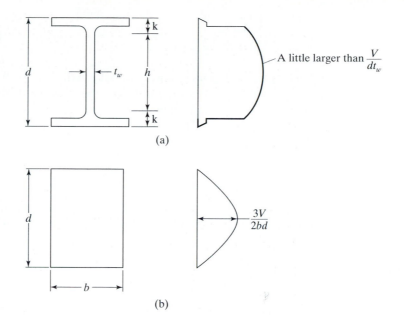

FIGURE 10.4

(a)

(b)

web area. This web area, A_w, is equal to the overall depth of the member, d, times the web thickness, t_w.

Shear strength expressions are given in LRFD Specification F2. In the expressions that follow, F_{yw} is the specified minimum yield stress of the web; h is the clear distance between the web toes of the fillets for rolled shapes, while for built-up welded sections it is the clear distance between flanges. For bolted built-up sections, h is the distance between adjacent lines of bolts in the web. Different expressions are given for different h/t_w ratios, depending on whether shear failures would be plastic, inelastic, or elastic.

1. Web yielding. Almost all rolled beam sections in the Manual fall into this classification.

   If $\dfrac{h}{t_w} \leq 2.45\sqrt{E/F_{yw}} = 59$ for $F_y = 50$ ksi and 52 for $F_y = 65$ ksi

   $V_n = 0.6\,F_{yw}A_w$ (LRFD Equation F2-1)

2. Inelastic buckling of web

   If $2.45\sqrt{E/F_{yw}} < \dfrac{h}{t_w} \leq 3.07\sqrt{E/F_{yw}} = 74$ for $F_y = 50$ ksi and 65 for $F_y = 65$ ksi

   $V_n = 0.6\,F_{yw}A_w(2.45\sqrt{E/F_{yw}})/(h/t_w)$ (LRFD Equation F2-2)

3. Elastic buckling of web

   If $3.07\sqrt{E/F_{yw}} < \dfrac{h}{t_w} \leq 260$

$$V_n = A_w \left[\frac{4.25E}{(h/t_w)^2} \right]$$ (LRFD Equation F2-3)

For each of the situations given $V_u \le \phi_v V_x$ with $\phi_v = 0.90$. The LRFD Appendix F2.2 gives expressions for the general design shear strength of webs with or without stiffeners. Appendix G3 also provides an alternative method for plate girders that utilizes tension field action (see Chapter 18).

In Example 10-2 the shear strength of a beam is computed.

Example 10-2

A W24 × 55($d = 23.6$ in, $t_w = 0.395$ in, $k =$ distance from extreme flange fiber to web toe of fillet $= 1.11$ in consisting of 50 ksi steel is used for the beam and loading of Fig. 10-5. Check its adequacy in shear.

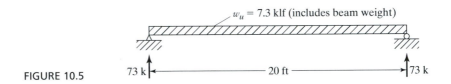

$w_u = 7.3$ klf (includes beam weight)

FIGURE 10.5   73 k |←——————— 20 ft ———————→| 73 k

Solution

$$h = 23.6 - (2)(1.11) = 21.38 \text{ in}$$

$$\frac{h}{t_w} = \frac{21.38}{0.395} = 54.13 < 2.45\sqrt{\frac{29.000}{50}} = 59$$

$$\phi_v V_n = (0.90)(0.6)(F_{yw})(A_w)$$

$$= (0.90)(0.6)(50)(23.6 \times 0.395) = 251.7 \text{ k} > 73 \text{ k}$$

(LRFD Equation F2-1)
(OK)

Note 1: Two values are given in the LRFD Manual for k. One is given in decimal form and is to be used for design calculations while the other one is given in fractions and is to be used for detailing. These two values are based respectively on the minimum and maximum radii of the fillets and will usually be quite different from each other.

Note 2: In Part 5 of the LRFD Manual there is a set of tables numbered 5-4 and entitled "Beams W-Shapes Maximum Factored Uniform Load." This title seems to the author to be quite inadequate because so much other information can be obtained from these tables, such as the design shear strengths or $\phi_v V_n$ values, and information concerning web yielding, web crippling, and other items to be discussed later.

Should V_u for a particular beam exceed the LRFD specified shear strength of the member, the usual procedure will be to select a little heavier section. If it is necessary, however, to use a much heavier section than required for moment, doubler plates (Fig. 10.6) may be welded to the beam web, or stiffeners may be connected to the webs in

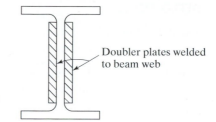

FIGURE 10.6

Increasing shear strength of beam by using doubler plates.

zones of high shear. Doubler plates must meet the width-thickness requirements for compact stiffened elements as prescribed in Section B.5 of the LRFD Specification. In addition, they must be welded sufficiently to the member webs to develop their proportionate share of the load.

The LRFD specified shear strength of a beam or girder is based on the entire area of the web. Sometimes, however, a connection is made to only a small portion or depth of the web. For such a case the designer may decide to assume that the shear is spread over only part of the web depth for purposes of computing shear strength. Thus he or she may compute A_w as being equal to t_w times the smaller depth for use in the shear strength expression.

When beams that have their top flanges at the same elevations (the usual situation) are connected to each other, it is frequently necessary to cope one of them, as shown in Fig. 10-7. For such cases there is a distinct possibility of a block shear failure along the broken lines shown. This subject was previously discussed in Section 3-7 and is continued in Chapter 15.

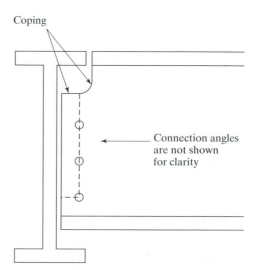

FIGURE 10.7

Block shear failure possible along dashed line.

10.3   DEFLECTIONS

The deflections of steel beams are usually limited to certain maximum values. Among the several excellent reasons for deflection limitations are the following:

1. Excessive deflections may damage other materials attached to or supported by the beam in question. Plaster cracks caused by large ceiling joist deflections are one example.
2. The appearance of structures is often damaged by excessive deflections.
3. Extreme deflections do not inspire confidence in the persons using a structure, although it may be completely safe from a strength standpoint.
4. It may be necessary for several different beams supporting the same loads to deflect equal amounts.

Standard American practice for buildings has been to limit service live-load deflections to approximately 1/360 of the span length. This deflection is supposedly the largest value that ceiling joists can deflect without causing cracks in underlying plaster. The 1/360 deflection is only one of many maximum deflection values in use because of different loading situations, different engineers, and different specifications. For situations where precise and delicate machinery is supported, maximum deflections may be limited to 1/1500 or 1/2000 of the span lengths. The 1996 AASHTO Specifications limit deflections in steel beams and girders due to live load and impact to 1/800 of the span. (For bridges in urban areas which are partly used by pedestrians, the AASHTO *recommends* a maximum value equal to 1/1000 of the span lengths.)

The LRFD Specification does not specify exact maximum permissible deflections. There are so many different materials, types of structures, and loadings that no one single set of deflection limitations is acceptable for all cases. Thus limitations must be set by the individual designer on the basis of his or her experience and judgment.

Before substituting blindly into a formula that will give the deflection of a beam for a certain loading condition, the student should thoroughly understand the theoretical methods of calculating deflections. These methods include the moment area, conjugate beam, and virtual work procedures. From these methods various expressions can be determined such as the common one given at the end of this paragraph for the center line deflection of a uniformly loaded simple beam.

$$\Delta_\ell = \frac{5wL^4}{384EI}$$

To use deflection expressions such as this one the reader must be very careful to use consistent units. Example 10-3 illustrates the application of the preceding expression. The author has changed all units to pounds and inches. Thus the uniform load given in the problem as so many kips per foot is changed to so many lb/in.

Example 10-3

A W24 × 55(I_x = 1360 in) has been selected for a 21-ft simple span to support a total service live load of 3 klf (including beam weight). Is the center line deflection of this section satisfactory for the service live load if the maximum permissible value is 1/360 of the span?

Solution

$$\Delta_\ell = \frac{5wL^4}{384EI} = \frac{(5)(3000/12)(12 \times 21)^4}{(384)(29 \times 10^6)(1360)} = 0.333 \text{ in total load deflection}$$

$$< \left(\frac{1}{360}\right)(12 \times 21) = 0.70 \text{ in} > 0.333 \text{ in} \qquad\qquad \text{(OK)}$$

On page 5-11 in the LRFD Manual the following simple formula for determining maximum beam deflections for *W, M, HP, S, C,* and *MC* sections for several different loading conditions is presented.

$$\Delta = \frac{ML^2}{C_1 I_x}$$

In this expression M is the service load moment in ft-k based on a uniformly distributed load, C_1 is a constant whose value can be determined from Fig. 10.8, L is the span length (ft), and I_x is the moment of inertia (in^4).

If we want to use this expression for the beam of Example 10-3, C_1 from part (a) of Fig. 10.6 would be 161, the center line bending moment is $wL^2/8 = (3)(21)^2/8 = 165.375$ ft-k, and the center line deflection for the live service load would be

$$\Delta_\ell = \frac{(165.375)(21)^2}{(161)(1360)} = 0.333 \text{ in}$$

Some specifications handle the deflection problem by requiring certain minimum depth-span ratios. For example, the AASHTO suggests the depth-span ratio be limited

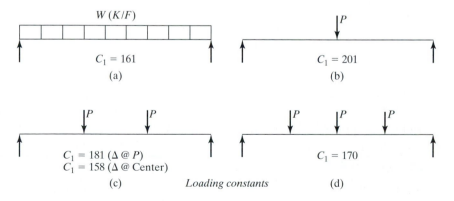

FIGURE 10.8

Values of constant C_1 for use in deflection expression. (Figure 5-2 in LRFD Manual)

to a minimum value of 1/25. A shallower section is permitted, but it should have sufficient stiffness to prevent a deflection greater than would have occurred if the 1/25 ratio had been used.

A steel beam can be cold-bent or cambered an amount equal to the deflection caused by dead load or the deflection caused by dead load plus some percentage of the live load. Approximately 25 percent of the camber so produced is elastic and will disappear when the cambering operation is completed. Detailed information for particular shapes is given in Tables 1-54 to 1-58 of the LRFD Manual in the section entitled "Standard Mill Practice." It should be remembered that a beam which is bent upward looks much stronger and safer than one which sags downward (even a very small distance).

A camber requirement is quite common for longer steel beams. In fact, a rather large percentage of the beams used in composite construction today (see Chapter 16) are cambered. On many occasions, however, it is more economical to select heavier beams with their larger moments of inertia so as to reduce deflections and avoid the labor costs involved in cambering. One commonly used rule of thumb is that it takes about one extra man-work-hour to camber each beam.

Cambering is something of a nuisance to many fabricators and it may very well introduce some additional problems. For instance, when beams are cambered it may be necessary to adjust the connection detail so that proper fitting of members is achieved. The end of a cambered beam will be rotated and thus it may be necessary to rotate the connection details by the same amount to insure proper fitting.

If we can move up one section in weight, thus reducing deflections so cambering is not needed, we may have a very desirable solution. Similarly, if a higher-strength steel is being used, it may be be desirable to switch the beams that need cambering to a lower yield stress steel. The results will be larger beams but smaller deflections and perhaps some economy if cambering can be eliminated.

The most economical way to camber beams is with a mechanical press. However, beams less than about 24 ft in length may not fit into a standard press. Such a situation may require the use of heat to carry out the cambering, but expenses will increase by 2 or 3 or more times. As a result, it is almost always wise for such short members to increase their sizes until cambering is not needed.

Deflections may very well control the sizes of beams for longer spans or for short ones where deflection limitations are severe. To assist the designer in selecting sections where deflections may control, the LRFD Manual includes a set of tables numbered 5-2 and entitled "W-shapes Selection by I_x," in which the I_x values are give in numerically descending order for the sections normally used as beams. In this table the sections are arranged in groups, with the lightest section in each group printed in bold type. Example 10-4 presents the design of a beam where deflections control the design.

Example 10-4

Select the lightest available section with $F_y = 50$ ksi to support a service dead load of 1.2 klf and a service live load of 3 klf for a 30-ft simple span. The section is to have full lateral bracing for its compression flange and the maximum total service load deflection is not to exceed 1/1500 the span length.

Solution. After some scratch work assume beam weight = 167 lb/ft

$$w_u = (1.2)(1.367) + (1.6)(3) = 6.44 \text{ klf}$$

$$M_u = \frac{(6.44)(30)^2}{8} = 724.5 \text{ ft-k}$$

$$Z \text{ required} \geq \frac{(12)(724.5)}{(0.9)(50)} = 193.2 \text{ in}^3$$

Try W24 × 76 $(I_x = 2100 \text{ in}^4)$

$$\text{Maximum permissible } \Delta = \left(\frac{1}{1500}\right)(12 \times 30) = 0.24 \text{ in}$$

$$\text{Actual } \Delta_\ell = \frac{ML^2}{C_1 I_x} = \frac{\dfrac{(4.367)(30)^2}{8}(30)^2}{(161)(2100)}$$

$$= 1.308 \text{ in} > 0.24 \text{ in} \qquad\qquad\text{(NG)}$$

Maximum I_x required to limit deflection to 0.24 in

$$\left(\frac{1.308}{0.24}\right)(2100) = 11.445 \text{ in}^4$$

From Moment of Inertia Selection Table
Use W40 × 167 or A36 (Its compact).

10.3.1   Vibrations

Though steel members may be selected that are satisfactory as to moment, shear, deflections, and so on, some very annoying floor vibrations may still occur. This is perhaps the most common serviceability problem faced by designers. Objectionable vibrations will frequently occur where long spans and large open floors without partitions or other items which might provide suitable damping are used. The reader may often have noticed this situation in the floors of large malls.

Damping of vibrations may be achieved by using framed-in-place partitions each attached to the floor system in at least 3 places or by installing "false" sheetrock partitions between ceilings and the underside of floor slabs. Further damping may be achieved by thickening the floor slabs, installing partitions, and by the office furniture and perhaps the equipment used in the building. A better procedure is to control the stiffness of the structural system.[1, 2] (A rather common practice in the past has been to try to limit vibrations by selecting beams no shallower than 1/20th times span lengths.)

---

[1] T.M. Murray, "Controlling Floor Movement." *Modern Steel Construction* (AISC, Chicago, IL, June 1991) pp. 17–19.
[2] T.M. Murray, "Acceptability Criterion for Occupant-Induced Floor Vibrations," *Engineering Journal*, AISC, 18, 2 (2nd quarter, 1981), pp. 62–70.

If the occupants of a building feel uneasy or are annoyed by vibrations the design is unsuccessful. It is rather difficult to correct a situation of this type in an existing structure. On the other hand, the situation can be easily predicted and corrected in the design stage. Several good procedures have been developed that enable the structural designer to estimate the acceptability of a given system by its users.

F. J. Hatfield[3] has prepared a useful chart that is used for estimating the perceptibility of vibrations of steel beams and concrete slabs for both office and residential buildings.

10.3   Ponding

If water on a flat roof accumulates faster than it runs off, the increased load causes the roof to deflect into a dish shape that can hold more water, which causes greater deflections and so on. This process of *ponding* continues until equilibrium is reached or until collapse occurs. Ponding is a serious matter, as illustrated by the large annual number of flat roof failures in the United States.

Ponding will occur on almost any flat roof to a certain degree, even though roof drains are present. Drains may be inadequate during severe storms, or they may become stopped up. Furthermore, they are often placed along the beam lines, which are actually the high points of the roof. The best method of preventing ponding is to have an appreciable slope on the roof (1/4 in/ft or more) together with good drainage facilities. It has been estimated that probably two thirds of the flat roofs in the United States have slopes less than this value, which is the minimum recommended by the National Roofing Contractors Association (NRCA). It costs approximately 3 to 6 percent more to construct a roof with this desired slope than building with no slope.[4]

When a very large flat roof (perhaps an acre or more) is being considered, the effect of wind on water depth may be quite important. A heavy rainstorm will frequently be accompanied by heavy winds. When a large quantity of water is present on the roof, a strong wind may very well push a great deal of water to one end, creating a dangerous depth of water as regards the load in pounds per square foot applied to the roof. For such situations *scuppers* are sometimes used. These are large holes or tubes in the walls or parapets which enable water above a certain depth to quickly drain off the roof.

Ponding failures will be prevented if the roof system (consisting of the roof deck and supporting beams and girders) has sufficient stiffness. The LRFD Specification (K2) describes a minimum stiffness to be achieved if ponding failures are to be prevented. If this minimum stiffness is not provided, it is necessary to make other investigations to be sure that a ponding failure is not possible.

---

[3] F. J. Hatfield, "Design Chart for Vibration of Office and Residential Floors," *Engineering Journal*, AISC, 29, 4 (4th quarter, 1992), pp. 141–144.
[4] Gary Van Ryzin, "Roof Design: Avoid Ponding by Sloping to Drain," *Civil Engineering* (New York: ASCE, January 1980), pp. 77–81.

Theoretical calculations for ponding are very complicated. The LRFD requirements are based on work by F. J. Marino[5] in which he considered the interaction of a two-way system of secondary or submembers supported by a primary system of main members or girders. A numerical ponding example is presented in Appendix F of this book.

10.4   WEBS AND FLANGES WITH CONCENTRATED LOADS

When steel members have concentrated loads applied that are perpendicular to one flange and symmetric to the web, their flanges and webs must have sufficient flange and web design strength in the areas of flange bending, web yielding, web crippling, and sidesway web buckling. Should a member have concentrated loads applied to both flanges, it must have a sufficient web design strength in the areas of web yielding, web crippling, and column web buckling. In this section formulas for determining strengths in these areas are presented.

If flange and web strengths do not satisfy the requirements of LRFD Specification Section K1.2 to K1.6, it will be necessary to use transverse stiffeners at the concentrated loads. Should web design strengths not satisfy the requirements of LRFD Section K1.7, it will be necessary to use doubler plates or diagonal stiffeners. These situations are discussed in the paragraphs to follow.

10.4.1   Local Flange Bending

The flange must be sufficiently rigid so that it will not deform and cause a zone of high stress concentrated in the weld in line with the web. The nominal tensile load that may be applied through a plate welded to the flange of a W section is to be determined by the expression to follow in which F_{yf} is the specified minimum yield stress of the flange (ksi) and t_f is the flange thickness (in.)

$$\phi = 0.90$$
$$R_n = 6.25\, t_f^2 F_{yf} \qquad \text{(LRFD Equation K1-1)}$$

It is not necessary to check this formula if the length of loading across the beam flange is less than 0.15 times the flange width b_f or if a pair of half-depth or deeper web stiffeners are provided. Fig. 10.10(a) shows a beam with local flange bending.

10.4.2   Local Web Yielding

The subject of local web yielding applies to all concentrated forces, tensile or compressive. Here we will try to limit the stress in the web of a member in which a force is being transmitted. Local web yielding is illustrated in part (b) of Fig. 10.10.

The nominal strength of the web of a beam at the web toe of the fillet when a concentrated load or reaction is applied is to be determined by one of the following two expressions in which k is the distance from the outer edge of the flange to the web

---

[5]F. J. Marino, "Ponding of Two-Way Roof System," *Engineering Journal* (New York: AISC, July 1966), pp. 93–100.

This beam is to be subjected to a heavy concentrated load from above. Partial web stiffeners are used to prevent web crippling and to keep the top flange from twisting or warping at the load. (Courtesy of American Institute of Steel Construction Inc.)

toe of the fillet, N is the bearing length (in) of the force parallel to the plane of the web, F_{yw} is the specified minimum yield stress (ksi) of the web, and t_w is the thickness of the web.

If the force is a concentrated load or reaction that causes tension or compression and is applied at a distance greater than the member depth from the end of the member

$$\phi = 1.0$$
$$R_n = (5k + N)F_{yw}t_w \qquad \text{(LRFD Equation K1-2)}$$

If the force is a concentrated load or reaction applied at a distance d or less from the member end

$$\phi = 1.0$$
$$R_n = (2.5k + N)F_{yw}t_w \qquad \text{(LRFD Equation K1-3)}$$

Reference to Fig. 10.9 clearly shows where these expressions were obtained. The nominal strength R_n equals the length over which the force is assumed to be spread when it reaches the web toe of the fillet times the web thickness times the yield stress of the web. Should a stiffener extending for at least half the member depth be provided on each side of the web at the concentrated force, it is not necessary to check for web yielding.

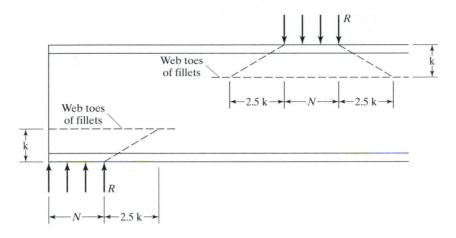

FIGURE 10.9

Local web yielding.

10.4.3 Web Crippling

Should concentrated compressive loads be applied to a member with an unstiffened web (the load being applied in the plane of the web), the nominal web crippling strength of the web is to be determined by the appropriate equation of the two that follow (in which d is the overall depth of the member). If one or two web stiffeners or one or two doubler plates are provided and extend for at least half of the web depth, web crippling will not have to be checked. Research has shown that when web crippling occurs it is located in the part of the web adjacent to the loaded flange. Thus it is thought that stiffening the web in this area for half its depth will prevent the problem. Web crippling is illustrated in part (c) of Fig. 10.10.

If the concentrated load is applied at a distance not less than $d/2$ from the end of the member

$$\phi = 0.75$$

$$R_n = 0.80t_w^2[1 + 3\left(\frac{N}{d}\right)\left(\frac{t_w}{t_f}\right)^{1.5}]\sqrt{\frac{EF_{yw}t_f}{t_w}} \qquad \text{(LRFD Equation K1-4)}$$

If the concentrated load is applied at a distance less than $d/2$ from the end of the member

For $\qquad \dfrac{N}{d} \leq 0.2$

$$\phi = 0.75$$

$$R_n = 0.40t_w^2\left[1 + 3\left(\frac{N}{d}\right)\left(\frac{t_w}{t_f}\right)^{1.5}\right]\sqrt{\frac{EF_{yw}t_f}{t_w}} \qquad \text{(LRFD Equation K1-5a)}$$

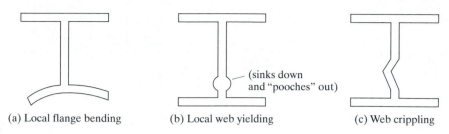

(a) Local flange bending   (b) Local web yielding   (c) Web crippling

(sinks down
and "pooches" out)

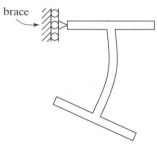

(d) Sidesway web buckling

FIGURE 10.10

For $\dfrac{N}{d} > 0.2$

$\phi = 0.75$

$$R_n = 0.40t_w^2\left[1 + \left(4\frac{N}{d} - 0.2\right)\left(\frac{t_w}{t_f}\right)^{1.5}\right]\sqrt{\frac{EF_{yw}t_f}{t_w}} \quad \text{(LRFD Equation K1-5b)}$$

10.4.4 Sidesway Web Buckling

Should compressive loads be applied to laterally braced compression flanges the web will be put in compression and the tension flange may buckle as shown in Fig. 10-10(d).

It has been found that sidesway web buckling will not occur if the compression flange is restrained against rotation with $(h/t_w)/(\ell/b_f) > 2.3$ or if $(h/t_w)/(\ell/b_f) > 1.7$ when the compression flange rotation is not restrained about its longitudinal axis. In these expressions h is the web depth between the web toes of the fillets, that is, $d - 2k$, and ℓ is the largest laterally unbraced length along either flange at the point of the load.

It is also possible to prevent sidesway web buckling with properly designed lateral bracing or stiffeners at the load point. The LRFD Commentary suggests that local bracing for *both* flanges be designed for 1 percent of the magnitude of the concentrated load applied at the point. If stiffeners are used they must extend from the point of load for at least one half of the member depth and should be designed to carry the full load. Flange rotation must be prevented if the stiffeners are to be effective.

Should members not be restrained against relative movement by stiffeners or lateral bracing and be subject to concentrated compressive loads, their strength may be determined as follows:

Girders for all-welded
Connecticut expressway bridge.
(Courtesy of the Lincoln
Electric Company.)

When the loaded flange is braced against rotation and $(h/t_w)/(\ell/b_f)$ is < 2.3

$$\phi = 0.85$$

$$R_n = \frac{C_r t_w^3 t_f}{h^2}\left[1 + 0.4\left(\frac{h/t_w}{\ell/b_f}\right)^3\right] \qquad \text{(LRFD Equation K1-6)}$$

When the loaded flange is not restrained against rotation and $\dfrac{h/t_w}{\ell/b_f} \le 1.7$

$$\phi = 0.85$$

$$R_n = \frac{C_r t_w^3 t_f}{h^2}\left[0.4\left(\frac{h/t_w}{\ell/b_f}\right)^3\right] \qquad \text{(LRFD Equation K1-7)}$$

It is not necessary to check Equations K1-6 and K1-7 if the webs are subject to distributed load. Furthermore, these equations were developed for bearing connections and *do not apply to moment connections*. In these expressions,

$C_r = 960{,}000$ ksi when $M_u < M_y$ at the location of the force, ksi.

$C_r = 480{,}000$ ksi when $M_u \ge M_y$ at the location of the force, ksi.

10.4.5 Compression Buckling of the Web

This limit state applies to concentrated compression loads applied to both flanges of a member as, for instance, moment connections applied to both ends of a column. For such a situation it is necessary to limit the slenderness ratio of the web to avoid the possibility of buckling. Should the concentrated loads be larger than the value of ϕR_n given below, it will be necessary to provide either one stiffener or a pair of them extending for the full depth of the web and meeting the requirements of LRFD Specification K1.9. (The equation to follow is applicable to moment connections, but not to bearing ones.)

$$\phi = 0.90$$

$$R_n = \frac{24\, t_w^3 \sqrt{EF_{yw}}}{h} \qquad \text{(LRFD Formula K1-8)}$$

If the concentrated forces to be resisted are applied at a distance from the member and less than $\frac{d}{2}$ the value of R_n shall be reduced by 50%.

Example 10-5, which follows, presents the selection of a beam for moment and its review for the applicable items discussed in this section.

Example 10-5

Select a W section for moment using 50 ksi steel for the span and service loads shown in Fig. 10-11. Lateral bracing is provided for both flanges at the ends and concentrated loads. If the end bearing length is 3.50 in and the concentrated load bearing length is 3.00 in, check the beam for web yielding, web crippling, and sidesway web buckling.

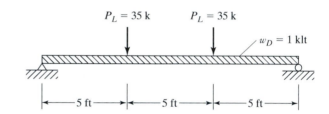

FIGURE 10.11

Solution

Assume beam weight = 44 lb/ft

$$M_u = \frac{(1.2)(1.044)(15)^2}{8} + (1.6)(35)(5) = 315.2 \text{ ft-k}$$

$$Z \text{ required} \geq \frac{(12)(315.2)}{(0.9)(50)} = 84.05 \text{ in}^3$$

Use W21 × 44 ($L_p = 4.45$ ft $< L_b$ of 5.0 ft, $d = 20.7$ in, $b_f = 6.500$ in, $t_w = 0.350$ in, $t_f = 0.450$ in, $k = 0.950$ in)

End reaction

$$R_u = (1.2)(1.044)\left(\frac{15}{2}\right) + (1.6)(35) = 65.4 \text{ k}$$

Concentrated load

$$P_u = (1.6)(35) = 56 \ k$$

Local web yielding

End reaction

$$\phi R_n = \phi(2.5 \ k + N)F_{yw}t_w$$
$$= 1.0(2.5 \times 0.950 + 3.50)(50)(0.350) \qquad \text{(LRFD Equation K1-3)}$$
$$= 102.8 \ k > 65.4 \ k \qquad \text{(OK)}$$

Concentrated load

$$\phi R_n = \phi(5k + N)F_{yw}t_w$$
$$= (1.0)(5 \times 0.950 + 3.0)(50)(0.350) \qquad \text{(LRFD Equation K1-2)}$$
$$= 135.6 \ k > 56 \ k \qquad \text{(OK)}$$

Web crippling

Concentrated loads

$$\phi R_n = \phi 0.80 \ t_w^2 \left[1 + 3\left(\frac{N}{d}\right)\left(\frac{t_w}{t_f}\right)^{1.5} \right] \sqrt{\frac{EF_{yw}t_f}{t_w}} \qquad \text{(LRFD Equation K1-4)}$$

$$= (0.75)(0.80)(0.350)^2 \left[1 + 3\left(\frac{3.0}{20.7}\right)\left(\frac{0.350}{0.450}\right)^{1.5} \right] \sqrt{\frac{(29 \times 10^3)(50)(0.450)}{0.350}}$$

$$= 130.3 \ k > 56 \ k \qquad \text{(OK)}$$

End reactions

$$\frac{N}{d} = \frac{3.5}{20.66} = 0.169 < 0.20$$

$$\phi R_n = \phi 0.40 \ t_w^2 \left[1 + 3\left(\frac{N}{d}\right)\left(\frac{t_w}{t_f}\right)^{1.5} \right] \sqrt{\frac{EF_{yw}t_f}{t_w}} \qquad \text{(LRFD Equation K1-5a)}$$

$$= (0.75)(0.40)(0.350)^2 \left[1 + 3\left(\frac{3.5}{20.7}\right)\left(\frac{0.350}{0.450}\right)^{1.5} \right] \sqrt{\frac{(29 \times 10^3)(50)(0.450)}{0.350}}$$

$$= 67.7 \ k > 65.4 \ k \qquad \text{(OK)}$$

Sidesway web buckling $\qquad\qquad$ (10.14)

$$\left(\frac{h}{t_w}\right) \Big/ \left(\frac{l}{b_f}\right) = \frac{(20.7 - 2 \times 0.950)}{0.350} \Big/ \left(\frac{12 \times 5}{6.50}\right)$$

$$= 5.82 > 2.3$$

$\therefore$ sidesways web buckling does not have to be checked.

To greatly simplify these kinds of calculations, a set of tables is presented in Part 9 of the Manual. These are numbered 9-5 and are entitled "Beam End Bearing Constants." The definition of the terms in these tables is given in the Manual together with several useful example problems.

For instance, for the sections normally used as beams and with $F_y = 50$ ksi, the values $\phi_b M_p$, $\phi_v V_n$, L_p, and L_u are provided along with the quantities ϕR_1, ϕR_2, $\phi_r R_3$, and $\phi_r R_4$. The latter values are useful in making the calculations necessary to check web yielding and web crippling. If we have a W21 × 44 ($t_w = 0.350$ in, $k = 0.95$ in; $F_{yw} = 50$ ksi) bearing on a length N of 3.5 in we can compute ϕR_n for web yielding as follows:

$$\phi R_n = \phi(2.5k + N)F_{yw}t_w$$
$$= (1.0)(2.5 \times 0.950 + 3.5)(50)(0.350) = 102.8 \text{ k}$$

Or we can use the table values for ϕR_1 and ϕR_2 where $k = 1\frac{1}{8}$ in.

$$\phi R_n = \phi R_1 + \phi R_2 N = 49.2 + (17.5)(3.5) = 110.5 \text{ k*}$$

If the requirements as to local flange bending, local web yielding, web crippling, or sidesway web buckling at concentrated loads are not met it is usually economical to take care of the situation by using a little larger section. If the problem cannot be handled by making a small increase in member size, we may decide to use web stiffeners as specified in detail in Section K1.9 of the LRFD Specification. Such stiffeners can be one-sided or two-sided. When stiffeners are required for local flange bending, sidesway web buckling or web yielding, they must be two sided. Should the load normal to the flange be tensile, the stiffeners need only be welded to the loaded flange. If, on the other hand, the load is compressive, the stiffeners must either bear against the compression flange or be welded to it.

10.5   UNSYMMETRICAL BENDING

From mechanics of materials, it should be remembered that each beam cross section has a pair of mutually perpendicular axes known as the principal axes for which the product of inertia is zero. Bending that occurs about any axis other than one of the principal axes is said to be unsymmetrical bending. When the external loads are not in a plane with either of the principal axes or when loads are simultaneously applied to the beam from two or more directions unsymmetrical bending results.

If a load is not perpendicular to one of the principal axes it may be broken into components which are perpendicular to those axes and the moments about each axis, M_{ux} and M_{uy}, determined as shown in Fig. 10.12.

When a section has one axis of symmetry, that axis is one of the principal axes and the calculations necessary for determining the moments are quite simple. For this reason unsymmetrical bending is not difficult to handle in the usual beam section, which is probably a W, S, M, or C. Each of these sections has at least one axis of

---

*The authors have consistently used design decimal dimensions throughout this text for their calculations. The writers of the Manual, however, used the detailing value of k here (11/8 in rather than the design value 0.950 in). As a result, the answers are not identical.

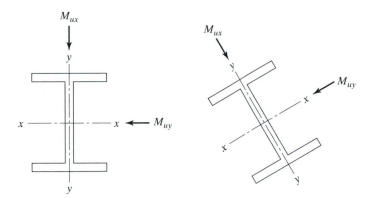

FIGURE 10.12

symmetry and the calculations are appreciably reduced. A further simplifying factor is that the loads are usually gravity loads and probably perpendicular to the x axis.

Among the beams that must resist unsymmetrical bending are crane girders in industrial buildings and purlins for ordinary roof trusses. The x axes of purlins are parallel to the sloping roof surface while the large percentage of their loads (roofing, snow, etc.) are gravity loads. These loads do not lie in a plane with either of the principal axes of the inclined purlins and the result is unsymmetrical bending. Wind loads are generally considered to act perpendicular to the roof surface and thus perpendicular to the x axes of the purlins with the result that they are not considered to cause unsymmetrical bending. The x axes of crane girders are usually horizontal, but the girders are subjected to lateral thrust loads from the moving cranes as well as to gravity loads.

To check the adequacy of members bent about both axes simultaneously, the LRFD provides an equation in Section H1 of their Specification. The equation to follow is for combined bending and axial tension or compression if $P_u/\phi P_n < 0.2$.

$$\frac{P_u}{2\phi P_n} + \left(\frac{M_{ux}}{\phi_b M_{nx}} + \frac{M_{uy}}{\phi_b M_{ny}} \right) \leq 1.0 \qquad \text{(LRFD Equation H1-1b)}$$

Since, for the problem discussed here, P_u is equal to zero, the formula reduces to

$$\frac{M_{ux}}{\phi_b M_{nx}} + \frac{M_{ny}}{\phi_b M_{ny}} \leq 1.0$$

This is an interaction or percentage equation. If M_{ux} is, say, 75 percent of the moment it could resist if bent about the x axis only ($\phi_b M_{nx}$), then M_{uy} can be no greater than 25 percent of what it could resist if bent about the y axis only ($\phi_b M_{ny}$).

Examples 10-6 and 10-7 illustrate the design of beams subjected to unsymmetrical bending. To illustrate the trial-and-error nature of the problem the author did not do quite as much scratchwork as in some of his earlier examples. The first design problems of this type which the student attempts may quite well take several trials. Consideration needs to be given to the question of lateral support for the compression flange. Should the lateral support be of questionable nature the engineer should reduce the design moment resistance by means of one of the expressions previously given for that purpose.

Example 10-6

A certain beam in its upright position is estimated to have a vertical bending moment $M_{ux} = 230$ ft-k and a lateral bending moment $M_{uy} = 56$ ft-k. These moments include the effect of the estimated beam weight. The loads are assumed to pass through the centroid of the section. Select a W24 shape of 50 ksi steel that can resist these moments assuming full lateral support for the compression flange.

Solution

Try W24 $\times$ 62 $(\phi_b M_p = \phi_b M_{nx} = 578$ ft-k, $Z_y = 15.8$ in^3, $S_y = 9.8$ in^3)

$$\phi_b M_{ny} = \frac{(0.9)(50)(15.7)}{12} = 58.88 \text{ ft-k} > \phi_b(1.5\,F_y S_y)$$

as per LRFD Section F 1.1 $= \dfrac{(0.90)(1.5)(50)(9.8)}{12} = 55.12$ ft-k $\leftarrow$

$$\frac{M_{ux}}{\phi_b M_{nx}} + \frac{M_{uy}}{\phi_b M_{ny}} = \frac{230}{578} + \frac{55}{55.12} = 1.40 > 1.00 \qquad \text{(NG)}$$

Try W24 $\times$ 68 $(\phi_b M_p = \phi_b M_{nx} = 664$ ft-k, $Z_y = 24.5$ in^3, $S_y = 15.7$ in^3)

$$\phi_b M_{ny} = \frac{(0.9)(50)(24.5)}{12} = 91.87 \text{ ft-k} \leq \phi_b(1.5\,F_y S_y)$$

$$= \frac{(0.90)(1.5)(50)(15.7)}{12} = 88.31 \text{ ft-k}$$

$$\frac{230}{664} + \frac{55}{88.31} = 0.969 < 1.00 \qquad \text{(OK)}$$

Use W24 $\times$ 68, $F_y = 50$ ksi

It should be noted in the solution for Example 10-6 that although the procedure used will yield a section that will adequately support the moments given the selection of the absolutely lightest section listed in the LRFD Manual could be quite lengthy because of the two variables Z_x and Z_y which affect the size. If we have a large M_{ux} and a small M_{uy} the most economical section will probably be quite deep and rather narrow, whereas if we have a large M_{uy} in proportion to M_{ux} the most economical section may be rather wide and shallow.

10.6 DESIGN OF PURLINS

To avoid bending in the top chords of roof trusses, it is theoretically desirable to place purlins only at panel points. For large trusses, however, it is more economical to space them at closer intervals. If this practice is not followed for large trusses the purlin sizes may become so large as to be impractical. When intermediate purlins are used the top chords of the truss should be designed for bending as well as axial stress as described in

Chapter 11. Purlins are usually spaced from 2 to 6 ft apart depending on loading conditions, while their most desirable depth-to-span ratios are probably in the neighborhood of 1/24. Channels or S sections are the most frequently used sections, but on some occasions other shapes may be convenient.

As described in Chapter 4, the channel and S sections are very weak about their web axes, and sag rods may be necessary to reduce the span lengths for bending about those axes. Sag rods, in effect, make the purlins continuous sections for their y axes and the moments about these axes are greatly reduced, as shown in Fig. 10.13. These moment diagrams were developed on the assumption that the changes in length of the sag rods are negligible. It is further assumed that the purlins are simply supported at the trusses. This assumption is on the conservative side since they are often continuous

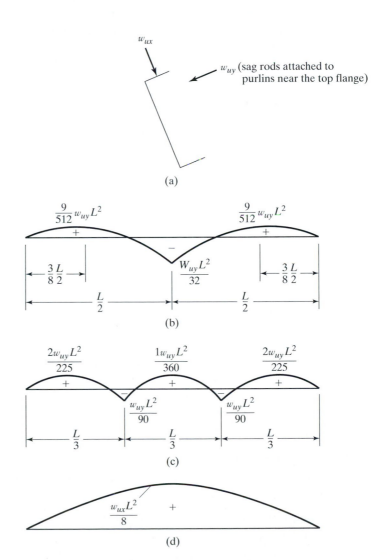

FIGURE 10.13

(a) Channel purlin, (b) Moment about web axis of purlins acting on the upper half of purlin—sag rods at midspan. (c) Moment about web axis of purlins acting on the upper half of purlin—sag rods at one-third points. (d) Moment about X axis of purlin.

over two or more trusses and appreciable continuity may be achieved at their splices. The student can easily reproduce these diagrams from his or her knowledge of moment distribution or other analysis methods. In the diagrams L is the distance between trusses, w_{uy} is the load component *perpendicular* to the web axis of the purlin, and w_{ux} is the load component *parallel* to the web axis.

If sag rods were not used the maximum moment about the web axis of a purlin would be $w_{uy}L^2/8$. When sag rods are used at midspan this moment is reduced to a maximum of $w_{uy}L^2/32$ (a 75 percent reduction), and when used at one-third points is reduced to a maximum of $w_{uy}L^2/90$ (a 91 percent reduction). In Example 10-7, sag rods are used at the midpoints and the purlins are designed for a moment of $w_{ux}L^2/8$ parallel to the web axis and $w_{uy}L^2/32$ perpendicular to the web axis.

In addition to being of advantage in reducing moments about the web axes of purlins, sag rods can serve other useful purposes. First, they can provide lateral support for the purlins; second, they are useful in keeping the purlins in proper alignment during erection until the roof deck is installed and connected to the purlins.

Example 10-7

Select a W6 purlin for the roof shown in Fig. 10.14. The trusses are 18 ft 6 in. on center and sag rods are used at the midpoints between trusses. Full lateral support is assumed to be supplied from the roof above. Use 50 ksi steel and the LRFD Specification. The full value of Z_y is to be used. Loads are as follows in terms of pounds per square foot of roof surface:

$$\text{Snow} = 30 \text{ psf}$$

$$\text{Roofing} = 6 \text{ psf}$$

$$\text{Estimated purlin weight} = 3 \text{ psf}$$

$$\text{Wind pressure} = 15 \text{ psf} \perp \text{ to roof surface}$$

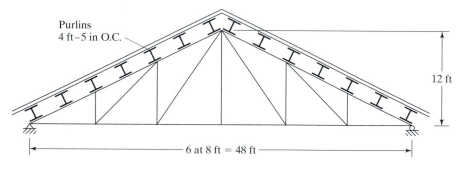

FIGURE 10.14

Solution

Computing the load w_{ux} to be supported per foot by the purlins (which are 4.42 ft on center).

$$w_u = 1.2D + 1.6S + 0.8W$$

$$= (1.2)(6 + 3)(4.42)\left(\frac{2}{\sqrt{5}}\right) + (1.6)(30)(4.42)\left(\frac{2}{\sqrt{5}}\right)$$

$$+ (0.8)(15)(4.42) = 285.6 \text{ lb/ft} \leftarrow$$

$$w_{ux} = 1.2D + 1.3W + 0.5S$$

$$= (1.2)(6 + 3)(4.42)\left(\frac{2}{\sqrt{5}}\right) + (1.3)(15)(4.42)$$

$$+ (0.5)(30)(4.42)\left(\frac{2}{\sqrt{5}}\right) = 188.2 \text{ lb/ft} \leftarrow$$

$$w_{uy} = (1.2D + 1.6S)\frac{1}{\sqrt{5}}$$

$$= [(1.2)(6 + 3)(4.42) + (1.6)(30)(4.42)]\frac{1}{\sqrt{5}}$$

$$= 116.5 \text{ lb/ft}$$

$$M_{ux} = \frac{(0.2856)(18.5)^2}{8} = 12.22 \text{ ft-k}$$

$$M_{uy} = \frac{(0.1165)(18.5)^2}{32} = 1.25 \text{ ft-k}$$

Try W6 × 9($Z_x = 6.23$ in^3, $Z_y = 1.72$ in^3, $S_y = 1.11$ in^3)

$$\phi_b M_{ny} = \frac{(0.9)(6.23)(6.23)}{12} = 23.4 \text{ ft-k}$$

$$\phi_b M_{ny} = \frac{1}{2}\phi_b F_y Z_y \text{ since sag rod attached to top of purlin}$$

$$= \frac{\left(\frac{1}{2}\right)(0.9)(50)(1.72)}{12} = 3.225 \text{ ft-k}$$

But $$\frac{\left(\frac{1}{2}\right)1.5\phi_b F_y S_y}{12} = \frac{\left(\frac{1}{2}\right)(0.9)(50)(1.11)}{12} = 3.12 \text{ ft-k}$$

$$\frac{M_{ux}}{\phi_b M_{nx}} + \frac{M_{uy}}{\phi_b M_{ny}} \leq 1.0 \text{ since } \frac{P_u}{\phi P_n} < 0.2$$

$$\frac{12.22}{23.4} + \frac{1.25}{3.12} = 0.922 < 1.0 \qquad\qquad \text{(OK)}$$

Use W6 × 9 $F_y = 50$ ksi.

10.7   THE SHEAR CENTER

The shear center is defined as the point on the cross section of a beam through which the resultant of the transverse loads must pass so that the stresses in the beam may be calculated only from the theories of pure bending and transverse shear. Should the resultant pass through this point; it is unnecessary to analyze the beam for torsional moments. For a beam with two axes of symmetry the shear center will fall at the intersection of the two axes, thus coinciding with the centroid of the section. For a beam with one axis of symmetry the shear center will fall somewhere on that axis but not necessarily at the centroid of the section. This surprising statement means that to avoid torsion in some beams the lines of action of the applied loads and beam reactions should not pass through the centroids of the sections.

The shear center is of particular importance for beams whose cross sections are composed of thin parts which provide considerable bending resistance but little resistance to torsion. Many common structural members such as the W, S, and C sections, angles, and various beams made up of thin plates (as in aircraft construction) fall into this class and the problem has wide application.

The location of shear centers for several open sections are shown with the solid dots (●) in Fig. 10-15. Sections such as these are relatively weak in torsion and for them and similar shapes the location of the resultant of the external loads can be a very serious matter. A previous discussion has indicated that the addition of one or more webs to these sections so they are changed into box shapes greatly increases their torsional resistance.

The average designer probably does not take the time to go through the sometimes tedious computations involved in locating the shear center and calculating the effect of twisting. He or she may instead simply ignore the situation, or may make a rough estimate of the effect of torsion on the bending stresses. One very poor but conservative estimate used occasionally is that of reducing $\phi_b M_{ny}$ by 50 percent for use in the interaction equation.

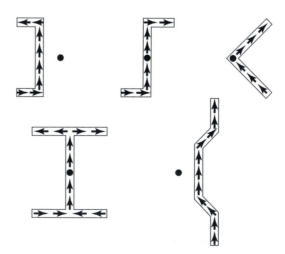

FIGURE 10.15

Shear centers can be located quickly for beams with open cross sections and relatively thin webs. For other beams the shear centers can probably be found, but only with considerable difficulty. The term *shear flow* is often used when speaking of thin-wall members, although there is really no flowing involved. It refers to the shear per inch of the cross section and equals the unit shearing stress times the thickness of the member. (The unit shearing stress F_v has been determined by the expression VQ/bI and the shear flow q_v can be determined by VQ/I if the shearing stress is assumed to be constant across the thickness of the section.) The shear flow acts parallel to the sides of each element of a member.

The channel section of Fig. 10.16(a) will be considered for this discussion. In this figure the shear flow is shown with the small arrows and in part (b) the values are totaled for each component of the shape and labeled H and V. The two H values are in equilibrium horizontally and the internal V value balances the external shear at the section. Although the horizontal and vertical forces are in equilibrium, the same cannot be said for the moment forces unless the lines of action of the resultant of the external forces passes through a certain point called the *shear center*. The horizontal H forces in part (b) of the figure can be seen to form a couple. The moment produced by this couple must be opposed by an equal and opposite moment which can only be produced by the two V values. The location of the shear center is a problem in equilibrium; therefore, moments should be taken about a point that eliminates the largest number of forces possible.

From this information the following equation can be written from which the shear center can be located.

$$Ve = Hh \text{ (moments taken about c.g. of web)}$$

Examples 10-8 and 10-9 illustrate the calculations involved in locating the shear center for two shapes. Note that the location of the shear center is independent of the value of the external shear. (*Shear center locations are provided for channels in* Table 1-5 *of Part I of the LRFD Manual.*)

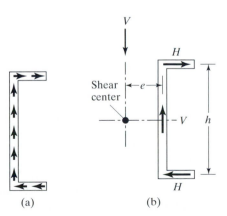

FIGURE 10.16   (a)   (b)

Example 10-8

The channel section shown in Fig. 10.17(a) is subjected to an external shear of V in the vertical plane. Locate the shear center.

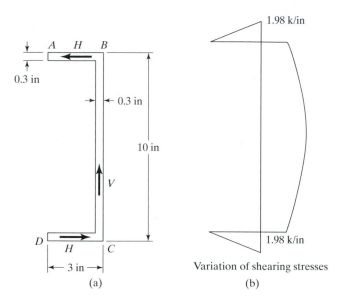

FIGURE 10.17   (a)   (b)

Solution

Properties of section:

$$I_x = \left(\frac{1}{12}\right)(3)(10)^3 - \left(\frac{1}{12}\right)(2.7)(9.4)^3 = 63 \text{ in}^4$$

$$q_v \text{ at } B = \frac{(V)(2.85 \times 0.3 \times 4.85)}{63} = 0.06582 \, V/\text{in}$$

$$\text{Total } H = \left(\frac{1}{2}\right)(2.85)(0.06582\,V) = 0.0938\,V$$

Location of shear center:

$$Ve = Hh$$
$$Ve = (0.0938\,V)(9.7)$$
$$e = 0.91 \text{ in from } \mathcal{L} \text{ of web}$$

The student should clearly understand that shear stress variation across the corners, where the webs and flanges join, cannot be determined correctly with the mechanics of materials expression (VQ/bI or VQ/I for shear flow) and cannot be determined

too well even after a complicated study with the theory of elasticity. As an approximation in the two examples presented here, we have assumed that shear flow continues up
to the middle of the corners on the same pattern (straight-line variation for horizontal
members and parabolic for others). The values of Q are computed for the corresponding dimensions. Other assumptions could have been made such as assuming a shear
flow variation continuing vertically for the full depth of webs and only for the protruding parts of flanges horizontally, or vice versa. It is rather disturbing to find that
whichever assumption is used the values do not check out perfectly. For instance in Example 10-9, which follows the sum of the vertical shear flow values, does not check out
very well with the external shear.

Example 10-9

The open section of Fig. 10.18 is subjected to an external shear of V in a vertical plane.
Locate the shear center.

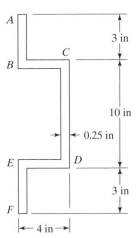

FIGURE 10.18

Solution

Properties of section:

$$I_x = \left(\frac{1}{12}\right)(0.25)(16)^3 + (2)(3.75 \times 0.25)(4.87)^2 = 129.8 \text{ in}^4$$

Values of shear flow (labeled q):

$$q_A = 0$$

$$q_B = \frac{(V)(3.12 \times 0.25 \times 6.44)}{129.8} = 0.0387 \, V/\text{in}$$

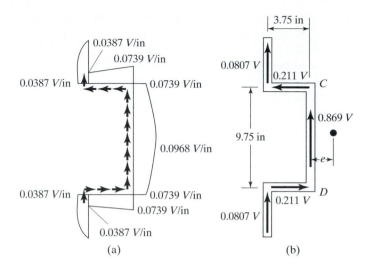

FIGURE 10.19

(a)

(b)

$$q_C = q_B + \frac{(V)(3.75 \times 0.25 \times 4.87)}{129.8} = 0.0739\,V/\text{in}$$

$$q_\ell = q_C + \frac{(V)(4.87 \times 0.25 \times 2.44)}{129.8} = 0.0968\,V/\text{in}$$

These shear flow values are shown in Fig. 10.19(a) and the summation for each part of the member is given in part (b) of the figure.

Taking moments about the center line of *CD*

$$-(0.211V)(9.75) + (2)(0.0807V)(3.75) + Ve = 0$$
$$e = 1.45 \text{ in}$$

The theory of the shear center is a very useful one in design, but it has certain limitations that should be clearly understood. For instance, the approximate analysis given in this section is valid only for thin sections. In addition, steel beams often have variable cross sections along their spans with the result that the loci of the shear centers are not straight lines along those spans. Thus, if the resultant of the loads passes through the shear center at one cross section, it might not do so at other cross sections.

When designers are faced with the application of loads to thin-walled sections such that twisting of those sections will be a problem, they will usually provide some means by which the twisting can be constrained. They may specify special bracing at close intervals, or attachments to flooring or roofing, or other similar devices. Should such solutions not be feasible they will probably consider selecting sections with greater torsional stiffnesses. Two references on this topic are given here.[6,7]

[6]C. G. Salmon and J. E. Johnson, *Steel Structures, Design and Behavior*, 2d ed. (New York: Harper & Row, 1980), pp. 672–675.
[7]AISC Engineering Staff, Torsional Analysis of Steel Members (Chicago: AISC, 1983).

10.8 BEAM-BEARING PLATES

When the ends of beams are supported by direct bearing on concrete or other masonry construction it is frequently necessary to distribute the beam reactions over the masonry by means of beam-bearing plates. The reaction is assumed to be spread uniformly through the bearing plate to the masonry and the masonry is assumed to push up against the plate with a uniform pressure equal to the factored reaction R_u over the area of the plate A_1. This pressure tends to curl up the plate and the bottom flange of the beam. The LRFD Manual recommends that the bearing plate be considered to take the entire bending moment produced and that the critical section for moment be assumed to be a distance k from the center line of the beam (see Fig. 10.20). The distance k is the same as the distance from the outer face of the flange to the web toe of the fillet given in the tables for each section (or it equals the flange thickness plus the fillet radius).

The determination of the true pressure distribution in a beam-bearing plate is a very formidable task, and the uniform pressure distribution assumption is usually made. This assumption is probably on the conservative side as the pressure is usually larger at the center of the beam than at the edges. The outer edges of the plate and flange tend to bend upward and the center of the beam tends to go down, concentrating the pressure there.

The required thickness of a 1-in-wide strip of plate can be determined as follows with reference being made to Fig. 10.20:

$$M_u = \frac{R_u}{A_1} n \frac{n}{2} = \frac{R_u n^2}{2A_1}$$

$$Z \text{ of a 1 in-wide piece of plate of } t \text{ thickness} = (1)\left(\frac{t}{2}\right)\left(\frac{t}{4}\right)(2) = \frac{t^2}{4}$$

$$\phi_b M_m \le \phi_b F_y Z \quad \text{with} \quad \phi_b = 0.9$$

$$\frac{R_u n^2}{2A_1} = (0.9)(F_y)\left(\frac{t^2}{4}\right)$$

$$t \ge \sqrt{\frac{2.22 R_u n^2}{A_1 F_y}}$$

In the absence of code regulations specifying different values, the design strength for bearing on concrete is to be taken equal to $\phi_c P_p$ according to LRFD Specification J9. This specification states that when a bearing plate extends for the full area of a concrete support, the bearing strength of the concrete can be determined as follows:

$$\phi_c = 0.60$$
$$P_p = 0.85 f'_c A_1 \qquad \text{(LRFD Equation J9-1)}$$

In this expression, f'_c is the compression strength of the concrete in psi and A_1 is the area of the plate (in^2) bearing concentrically on the concrete. For the design of such a plate, its required area A_1 can be determined by dividing the factored reaction R_u by $\phi_c P_p$:

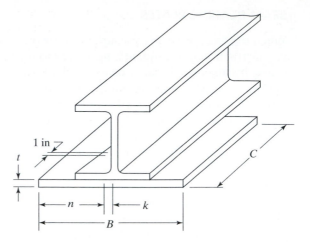

FIGURE 10.20

$$A_1 \geq \frac{R_u}{\phi_c P_p} = \frac{R_u}{\phi_c 0.85 f'_c}$$

(The bearing values provided are the same as those given in Section 10.17 of the 1999 *ACI Building Code Requirements for Structural Concrete*.)

Should the bearing load be applied to an area less than the full area of the concrete support, $\phi_c P_p$ is to be determined with the following equation in which A_2 is the maximum area of the supporting surface that is geometrically similar to and concentric with the loaded area with $\sqrt{A_2/A_1}$ having a maximum value of 2:

$$\phi_c = 0.60$$

$$P_p = 0.85 f'_c A_1 \sqrt{\frac{A_2}{A_1}} \qquad \text{(LRFD Equation J9-2)}$$

For this case in which the full area of the concrete support is not loaded, A_1 can be computed as follows:

$$A_1 \geq \frac{R_u}{\phi_c P_p} = \frac{R_u}{\phi_c f'_c A_1 \sqrt{\dfrac{A_2}{A_1}}}$$

$$= \frac{1}{A_2} \left(\frac{R_u}{\phi_c 0.85 f'_c} \right)^2 \quad \text{when} \quad A_2 < 4A_1$$

$$\text{or}$$

$$A_1 \geq \frac{R_u}{1.7 \, \phi_c f'_c} \quad \text{when} \quad A_2 \geq 4A_1.$$

After A_1 is determined, its length (parallel to the beam) and its width are selected. The length may not be less than the N required to prevent web yielding or web crippling of the beam, nor may it be less than about 3 1/2 or 4 in for practical construction reasons. It may not be greater than the thickness of the wall or other support, and actually it may have to be less than that thickness, particularly at exterior walls, to prevent the steel from being exposed.

Example 10-10 illustrates the calculations involved in designing a beam bearing plate. Notice that the width and the length of the plate are desirably taken to the nearest full inch.

Example 10-10

A W18 × 71 beam ($d = 18.5$ in, $t_w = 0.495$ in, $b_f = 7.64$ in, $t_f = 0.810$ in, $k = 1.21$ in) has one of its ends supported by a reinforced-concrete wall with $f'_c = 3$ ksi. Design a bearing plate for the beam with A36 steel. The factored end reaction is 120 k and the maximum length of end bearing ⊥ to the wall is the full wall thickness 8.0 in.

Solution

$$\text{Area required} = A_1 = \frac{R_u}{\phi_c 0.85 f'_c} = \frac{120}{(0.60)(0.85)(3)} = 78.43 \text{ in}^2$$

Minimum length of bearing ⊥ to wall required for local web yielding

$$R_u \le \phi(2.5\,k + N)F_{yw}t_w \qquad \text{(LRFD Equation K1-3)}$$

$$120 = 1.0(2.5 \times 1.21 + N)(50)(0.495)$$

$$N \ge 1.82 \text{ in}$$

Minimum length of bearing ⊥ to wall for web crippling

$$\text{Assume } \frac{N}{d} = \frac{8}{18.5} = 0.432 > 0.2$$

$$R_u \le \phi\,0.40\,t_w^2\left[1 + \left(4\frac{N}{d} - 0.2\right)\left(\frac{t_w}{t_f}\right)^{1.5}\right]\sqrt{\frac{EF_{yw}t_f}{t_w}} \qquad \text{(LRFD Equation K1-5b)}$$

$$120 = (0.75)(0.40)(0.495)^2\left[1 + \left(4\frac{N}{18.47} - 0.2\right)\left(\frac{0.495}{0.810}\right)^{1.5}\right] \times$$

$$\sqrt{\frac{(29 \times 10^3)(36)(0.810)}{0.495}}$$

$$N \ge 3.33 \text{ in}$$

Try 8×10 in PL

$$n = 5 - 1.21 = 3.79 \text{ in}$$

$$t \geq \sqrt{\frac{2.22 \, R_u n^2}{A_1 F_y}} = \sqrt{\frac{(2.22)(120)(3.79)^2}{(80.0)(36)}} = 1.15 \text{ in}$$

Use $1\frac{1}{4} \times 8 \times 10$ in, PL $F_y = 50$ ksi.

On some occasions, the beam flanges alone probably provide sufficient bearing area, but bearing plates are nevertheless recommended as they are useful in erection and ensure an even bearing surface for the beam. They can be placed separately from the beams and carefully leveled to the proper elevations. When the ends of steel beams are enclosed by the concrete or masonry walls, it is considered desirable to use some type of wall anchor to prevent the beam from moving longitudinally with respect to the wall. The usual anchor consists of a bent steel bar called a government anchor passing through the web of the beam and running parallel to the wall. Occasionally, clip angles attached to the web are used instead of government anchors. Should longitudinal loads of considerable size be anticipated, regular vertical anchor bolts may be used at the beam ends.

If we were to check to see if the flange thickness alone is sufficient, we would have $\left(\text{with } n = \dfrac{b_f}{2} - k\right) = \dfrac{7.64}{2} - 1.21 = 2.61 \text{ in}$

$$t = \sqrt{\frac{2.22 \, R_u n^2}{A_1 F_y}} = \sqrt{\frac{(2.22)(120)(2.61)^2}{(8 \times 7.64)(36)}} = 0.91 \text{ in} > t_f \text{ of W18} \times 71$$

$$\therefore \text{ Flange not sufficient alone}$$

10.9 COMPUTER EXAMPLE

A beam base plate is designed in Example 10-11 using the enclosed computer program. Here it is necessary as it was for column base plates to assume an initial plate size. The computer will print out the area and thickness required. You can then revise the initial dimensions and run the program again.

Example 10-11

Repeat Example 10-10 using the computer program INSTEP.

Input:

Factored bearing load	Concrete footing parameters
Ru = 120 kips	f′c = 3000 psi
Location on the span	b1 = 0 in
End of the span	
	Trial bearing plate
	B = 10 in
	N = 8 in
	t = 0 in

Output:

Beam Bearing Plate Design Summary
N = 8 in
B = 10 in

Area Provided = 80 sq. in
Required Thickness t = 1.15268 in

Bearing Capacity = 0.6Rn = 0.6(204) = 122.4 kips
Capacity-Local Web Yielding = 1.0Rn = 1.0(196.466) = 196.466 kips
Capacity-Web Crippling = 075Rn = 0.75(221.721) = 166.29 kips

PROBLEMS

10-1 to 10-5.   *Considering moment only and assuming full lateral support for the compression flanges, select the lightest sections available using 50 ksi steel. The loads shown include the effect of the beam weights. Use elastic analysis, factored loads, and the 0.9 rule.*

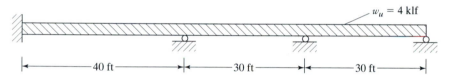

FIGURE P10-1 (*Ans.* W21 × 62)

10-2.

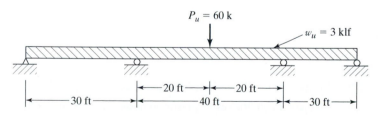

FIGURE P10-2

10-3.

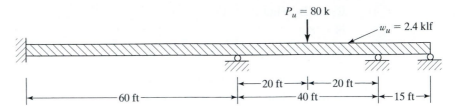

FIGURE P10-3 (*Ans.* W24 × 76)

10-4.

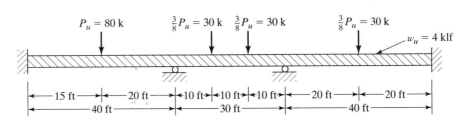

FIGURE P10-4

10-5.

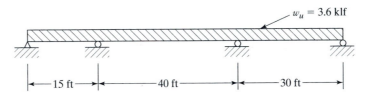

FIGURE P10-5 (*Ans.* W21 × 55)

10-6. Repeat Prob. 10-1 using plastic analysis.

10-7. Repeat Prob. 10-3 using plastic analysis. (*Ans.* W24 × 68)

10-8. Repeat Prob. 10-5 using plastic analysis.

10-9. Three methods of supporting a roof are shown in the accompanying illustration. Using an elastic analysis with factored loads, $F_y = 50$ ksi, and assuming full lateral support in each case, select the lightest section if a dead uniform service load (not including beam weight) of 1 klf and a live uniform service load of 1.2 klf is to be supported. Consider moment only.

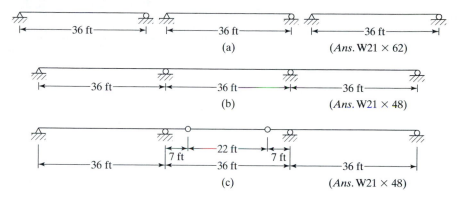

(a)   (*Ans.* W21 × 62)

(b)   (*Ans.* W21 × 48)

(c)   (*Ans.* W21 × 48)

FIGURE P10-9

10-10.   The welded section shown, made from 50 ksi steel, has full lateral support for its compression flange and is bent about its major axis. If C_b is 1.0, determine its design moment and shear strengths.

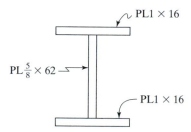

PL$\frac{5}{8}$ × 62

PL1 × 16

PL1 × 16

FIGURE P10-10

10-11 to 10-13.   *Using* $F_y = 50\,ksi$ *select the lightest available section for the span and loading shown. Consider moment and shear only and neglect beam weight in all calculations. The member is assumed to have full lateral bracing.*

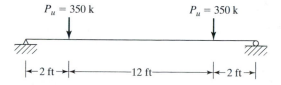

$P_u = 350$ k $P_u = 350$ k

├─2 ft─┤├────────12 ft────────┤├─2 ft─┤

FIGURE P10-11 (*Ans.* W30 × 90)

10-12.

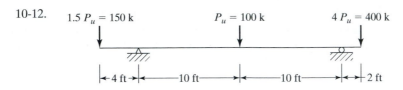

FIGURE P10-12

10-13. Repeat Prob. 10-11 using $F_y = 36$ ksi. (*Ans.* W33 × 118)

10-14. If a fully braced W14 × 34 section consisting of 50 ksi steel is used for a simple span of 6 ft 6 in., determine the maximum uniform load w_u that it can support in addition to its own factored weight. Use elastic analysis and consider shear and moment only.

10-15. A W16 × 40 consisting of 50 ksi steel is used as a simple beam for a span of 8 ft. If it has full lateral support determine the maximum uniform load w_u which it can support in addition to its own factored weight. Use an elastic analysis and consider shear and moment only. (*Ans.* 32.95 klf)

10-16. A fully braced W36 × 245 consisting of A992 steel is used as a simple beam for a span of 16 ft. Considering moment and shear only determine the maximum uniform load w_u it can support in addition to its own factored weight using an elastic analysis.

10-17. A 30-ft simply supported beam is to support a moving concentrated load $P_u = 80$ k. Using 50 ksi steel, select the most economical section considering moment and shear only. Use an elastic analysis and neglect beam weight. (*Ans.* W21 × 68)

10-18. A 40-ft simple beam that supports a service concentrated load $P_L = 30$ k at midspan is laterally unbraced except at its ends and center line. If the maximum permissible center line deflection under service loads equals 1/1000 of the span, select the most economical W section of 50 ksi steel considering moment, shear, and deflection. Neglect beam weight.

10-19. Design a beam for a 24-ft simple span to support the working uniform loads $w_D = 1.2$ klf (includes beam weight) and $w_L = 2.8$ klf. The maximum permissible deflection under working loads is 1/1200 of the span. Use 50 ksi steel and consider moment, shear, and deflection. The beam is to be braced laterally for its full length. (*Ans.* W30 × 108)

10-20. Select the lightest available W section of A992 steel for the span and service loads shown. The beam will have full lateral support for its compression flange. Its maximum service load center-line deflection may not exceed 1/1500 of the span under working loads. Consider moment, shear, and deflection only.

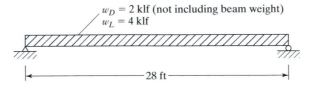

FIGURE P10-20

10-21.   Select the lightest section available if F_y = 50 ksi for the span and working loads shown if the section is to be fully braced laterally and have a maximum factored load deflection of 1/800 of its span length. Neglect beam-weight in all calculations. Consider moment, shear and deflection. (*Ans.* W44 × 230)

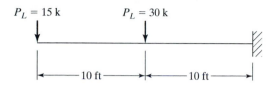

P_L = 15 k P_L = 30 k

10 ft   10 ft

FIGURE P10-21

10-22.   If the maximum permissible service load deflection for the fully braced beam shown is 1/1200 of the span, select the lightest available section using 50 ksi steel. Consider moment, shear, and deflections. Working loads are shown.

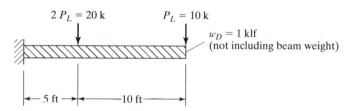

2 P_L = 20 k P_L = 10 k

w_D = 1 klf
(not including beam weight)

5 ft   10 ft

FIGURE P10-22

10-23.   Select the lightest available W section (F_y = 50 ksi) for the beams shown in the accompanying illustration. The floor slab is 4 in reinforced concrete (weight=150 lb/ft^3) and supports a 100 psf uniform live load. The maximum permissible deflection for the working loads is $\frac{1}{240} L$. Assume continuous lateral bracing is provided. (*Ans.* W21 × 48)

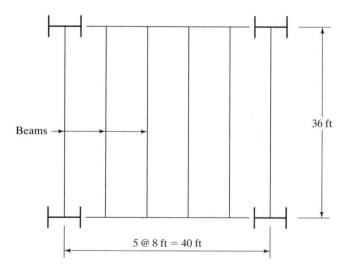

Beams

36 ft

5 @ 8 ft = 40 ft

FIGURE P10-23

10-24. Repeat Prob. 10-23 if the live load is 200 psf.

10-25. Repeat Prob. 10-23 if the span is 40 ft. and the maximum beam depth is 20 in. (*Ans.* W18 × 76)

10-26. Select the lightest available W section (50 ksi steel) for the working live loads and span shown in the accompanying illustration. Lateral support is provided only at the 12-ft points and a maximum deflection (under working loads) equal to 1/1500 of the 24-ft span is permitted. Neglect beam weight. Consider moment, shear, and deflections. (*Ans.* W30 × 108)

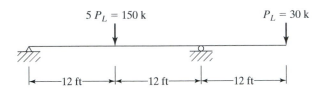

FIGURE P10-26

10-27. Repeat Prob. 10-26 if $F_y = 36$ ksi. (*Ans.* W30 × 108)

10-28. A simply supported W24 × 146 consisting of 50 ksi steel is supporting a 300-k service live load as shown in the figure. If the length of bearing at the left support is 8 in and at the concentrated load is 12 in, check the beam for shear, web yielding, and web crippling.

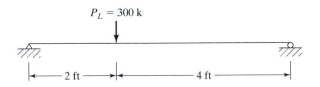

FIGURE P10-24

10-29. A 32-ft beam with full lateral support for its compression flange is supporting a moving service concentrated load of 60 k. Using 50 ksi steel, select the lightest section available for moment. Then check to see if the section is satisfactory in shear and compute the minimum length of bearing required at the supports from the standpoint of web yielding and web crippling. (*Ans.* W24 × 84)

10-30. Select the lightest available W30 section consisting of 50 ksi steel to resist a gravity moment $M_{ux} = 500$ ft-k and a lateral bending moment of $M_{uy} = 200$ ft-k. The section is assumed to have full lateral support.

10-31. The 20-ft simple beam shown has full lateral support for its compression flange and consists of 50 ksi steel. The beam supports a gravity service dead load of 1.2 klf (includes beam weight) and a gravity live load of 3.4 klf. The loads are assumed to act through the e.g. of the section. Select the lightest available W33 section. (*Ans.* W33 × 118)

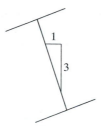

FIGURE P10-31

10-32. Repeat Prob. 10-31 if the gravity live load is 4.8 klf.

10-33. Design a steel bearing plate from A36 steel for a W21 × 68 beam with an end reaction $R_u = 100$ k. The beam will bear on a reinforced-concrete wall with $f'_c = 4$ ksi. In a direction perpendicular to the wall the bearing plate may not be longer than 8 in. (*One ans.* PL 1 × 8 × 0 ft 9 in)

10-34. Design a steel bearing plate of 50 ksi steel for a W30 × 116 beam supported by a reinforced-concrete wall with $f'_c = 3$ ksi. The maximum beam reaction R_u is 170 k. Assume the width of the plate perpendicular to the wall is 8 in.

10-35. Repeat Prob. 10-34, using a steel with $F_y = 36$ ksi. (*One ans.* $1\frac{3}{4}$ × 8 × 1 ft 2 in)

10-36 to 10-33. *Solve the problems shown using the computer program INSTEP*

10-36. Repeat Prob. 10-33.

10-37. Repeat Prob. 10-34 (One ans. PL $1\frac{1}{2}$ × 8 × 1 ft 2 in.)

Bending and Axial Force

11.1 OCCURRENCE

Structural members that are subjected to a combination of bending and axial force are far more common than the student may realize. This section is devoted to listing some of the more obvious cases. Columns that are part of a steel building frame must nearly always resist sizable bending moments in addition to the usual compressive loads. It is almost impossible to erect and center loads exactly on columns, even in a testing lab, and in an actual building, one can see that it is even more impossible. Even if building loads could be perfectly centered at one time they would not stay in one place. Furthermore, columns may be initially crooked or have other flaws resulting in lateral bending. The beams framing into columns are commonly supported with framing angles or brackets on the sides of the columns. These eccentrically applied loads produce moments. Wind and other lateral loads cause columns to bend laterally, and the columns in rigid frame buildings are subjected to moments even when the frame is supporting gravity loads alone. The members of bridge portals must resist combined forces, as do building columns. Among the causes of the combined forces are heavy lateral wind loads, vertical traffic loads—whether symmetrical or not—and the centrifugal effect of traffic on curved bridges.

The previous experience of the student has probably been to assume that truss members are only axially loaded. Purlins for roof trusses, however, are frequently placed between truss joints, causing the top chords to bend. Similarly, the bottom chords may be bent by the hanging of light fixtures, ductwork, and other items between the truss joints. All horizontal and inclined truss members have moments caused by their own weights, while all truss members—whether vertical or not—are subjected to secondary bending forces. Secondary forces are developed because the members are not connected with frictionless pins; as assumed in the usual analysis, the member centers of gravities or those of their connectors do not exactly coincide at the joints, etc.

Moments in tension members are not as serious as those in compression members because tension tends to reduce lateral deflections, while compression increases them. Increased lateral deflections in turn result in larger moments, which cause larger lateral deflections, etc. It is hoped that members in such situations are stiff enough to prevent the additional lateral deflections from becoming excessive.

11.2   MEMBERS SUBJECT TO BENDING AND AXIAL TENSION

A few types of members subject to both bending and axial tension are shown in Fig. 11.1.

The Alcoa Building under construction in San Francisco, California.
(Courtesy of Bethlehem Steel Corporation.)

In Section H1 of the LRFD Specification, the following interaction equations are given for symmetric shapes subjected simultaneously to bending and axial tensile forces. These equations are also applicable to members subjected to bending and axial compression forces as will be described in Sections 11-7 to 11-9.

If $\dfrac{P_u}{\phi_t P_n} \geq 0.2$

$$\frac{P_u}{\phi_t P_n} + \frac{8}{9}\left(\frac{M_{ux}}{\phi_b M_{nx}} + \frac{M_{uy}}{\phi_b M_{ny}}\right) \leq 1.0 \qquad \text{(LRFD Equation H1-1a)}$$

If $\dfrac{P_u}{\phi_t P_n} < 0.2$

$$\frac{P_u}{2\phi_t P_n} + \left(\frac{M_{ux}}{\phi_b M_{nx}} + \frac{M_{uy}}{\phi_b M_{ny}}\right) \leq 1.0 \qquad \text{(LRFD Equation H1-1b)}$$

The terms in these equations have previously been defined: P_u and M_u are the required tensile and flexural strengths, P_n and M_n are the nominal tensile and flexural strengths, and the resistance factors ϕ_t and ϕ_b are determined as in previous chapters. Usually only a

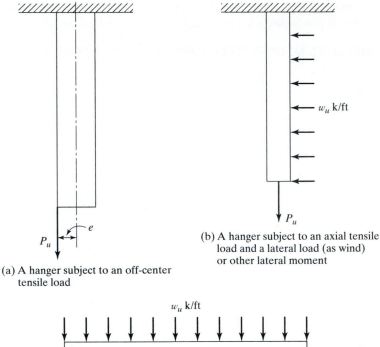

(a) A hanger subject to an off-center tensile load

(b) A hanger subject to an axial tensile load and a lateral load (as wind) or other lateral moment

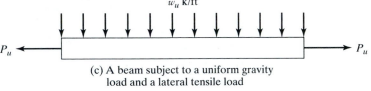

(c) A beam subject to a uniform gravity load and a lateral tensile load

FIGURE 11.1

Some members subject to bending and axial tension.

first-order analysis (that is, not including any secondary forces as described in the next section) is made for members subject to bending and axial tension. However, the analyst, may—in fact, is encouraged to—make second-order analyses for these members and use the results in his or her designs. Examples 11-1 and 11-2 illustrate the use of the interaction equations to review members subjected simultaneously to bending and axial tension, while Example 11-3 illustrates the design of such a member.

Example 11-1

A W12 × 35 tension member with no holes consisting of 50 ksi steel is subjected to a factored tensile force P_u of 80 k and a factored bending moment M_{uy} of 35 ft-k. Is the member satisfactory if $L_b < L_p$?

Solution

Using a W12 × 35 ($A = 10.3$ in^3, $Z_y = 11.5$ in^3)

$$\phi_t P_n = \phi_t F_y A_g = (0.9)(50)(10.3) = 463.5 \text{ k}$$

$$\frac{P_u}{\phi_t P_n} = \frac{80}{463.5} = 0.173 < 0.2$$

$$\therefore \ \text{Use LRFD Equation H1-1b}$$

$$\phi_b M_{ny} = \phi_b F_y Z_y = \frac{(0.9)(50)(11.5)}{12} = 43.12 \text{ ft-k}$$

$$\leq \phi_b (1.5) F_y S_y = \frac{(0.9)(1.5)(50)(7.47)}{12} = 42.02 \text{ ft-k}$$

$$\frac{P_u}{2\phi_t P_n} + \left(\frac{M_{ux}}{\phi_b M_{nx}} + \frac{M_{uy}}{\phi_b M_{ny}} \right) = \frac{80}{(2)(463.5)} + \left(0 + \frac{35}{42.02} \right)$$

$$= 0.919 < 1.0 \tag{OK}$$

Example 11-2

A W10 × 30 tensile member with no holes consisting of 50 ksi steel and with $L_b = 12.0$ ft is subjected to a factored tensile force $P_u = 120$ k and to the factored moments $M_{ux} = 90$ ft-k and $M_{uy} = 0$. If $C_b = 1.0$, is the member satisfactory?

Solution

Using a W10 × 30 ($A = 8.84 \text{ in}^2$, $L_p = 4.84$ ft, $L_r = 14.6$ ft)

$$\phi_t P_n = (0.9)(50)(8.84) = 397.8 \text{ k}$$

$$\frac{P_u}{\phi_t P_n} = \frac{120}{397.8} = 0.302 > 0.2$$

$$\therefore \ \text{Use LRFD Equation H1-1a}$$

Noting $L_p < L_b < L_r$ from Load Factor Design Selection Table

$$\phi_b M_p = 137 \text{ ft-k}$$

$$\phi_b M_r = 97.2 \text{ ft-k}$$

$$\text{BF} = 4.11$$

$$\phi_b M_n = C_b [\phi_b M_p - \text{BF}(L_b - L_p)] \leq \phi_b M_p$$

$$= 1.0[137 - 4.11(12.0 - 4.84)] = 107.6 \text{ ft-k} < 137 \text{ ft-k} \tag{OK}$$

$$\frac{P_u}{\phi_t P_n} + \frac{8}{9} \left(\frac{M_{ux}}{\phi_b M_{nx}} + \frac{M_{uy}}{\phi_b M_{ny}} \right) = \frac{120}{397.8} + \frac{8}{9} \left(\frac{90}{107.6} + 0 \right)$$

$$= 1.05 > 1.0 \tag{NG}$$

Example 11-3

Select an 8-ft long W10 to support a factored tensile load of 150 k applied with an eccentricity of 5 in. with respect to the x axis. The 50 ksi member is to be welded along all of its elements and braced laterally only at its supports. Assume C_b is equal to 1.0 and $A_e = 0.90A_g$ and $U = 1.0$.

Solution

Try a W10 × 26 ($A = 7.61$ in^2, $L_p = 4.8$ ft, $L_r = 13.6$ ft, $M_{ux} = (5)(150) = 750$ in-k $= 62.5$ ft-k, $\phi_t P_n = (0.9)(50)(7.61) = 342.4$ k and $\phi_t = \phi_t F_u U A_e = (0.75)(65)(1.0)(0.9)(7.61) = 333.9$ k $\leftarrow$

$$\frac{P_u}{\phi_t P_n} = \frac{150}{333.9} = 0.449 > 0.2$$

$$\therefore \text{ Use LRFD Equation H1-1a}$$

Noting that $L_b > L_p < L_r$, we determine $\phi_b M_n$ as follows, with reference to the Load Factor Design Selection Table in the Manual.

$$\phi_b M_p = 117 \text{ ft-k}$$
$$\phi_b M_r = 83.7 \text{ ft-k}$$
$$\text{BF} = 3.84$$
$$\phi_b M_{nx} = C_b[\phi_b M_p - \text{BF}(L_b - L_p)]$$
$$= 1.0[117 - (3.84)(8.0 - 4.8)] = 104.7 \text{ ft-k}$$
$$\frac{P_u}{\phi_t P_n} + \frac{8}{9}\left(\frac{M_{ux}}{\phi_b M_{nx}} + \frac{M_{uy}}{\phi_b M_{ny}}\right) = \frac{150}{342.4} + \frac{8}{9}\left(\frac{62.5}{104.7} + 0\right)$$
$$= 0.969 \leq 1.0 \qquad \text{(OK)}$$

Subsequent check of W10 × 22 shows it will not do.

Use W10 × 26 $F_y = 50$ ksi.

11.3   COMPUTER EXAMPLES FOR MEMBERS SUBJECT TO BENDING AND AXIAL TENSION

The computer program INSTEP 32 is used in Examples 11-4 and 11-5 to repeat Examples 11-2 and 11-3. For convenience the solutions of the problems use a few different but obvious abbreviations such as P_{na} for the design axial load strength $\phi_t P_n$ of a member, and M_{na} which is the design bending strength of the member $\phi_b M_n$.

Example 11-4

Repeat Example 11-2 using INSTEP 32.

Input:

Factored design loads	Design parameters
Pu = 120	L = 144
Mtx/u = 1080	Lb = 144
Mlt = 0	Cb = 1
	Cm = 1
	B2 = 0
	Ae = 8.84
	Kb = 1

Output:

Beam-Column Design Summary
Axial Pna = 397.8
Flexure Mna = 1294.06
Flexure Mue = 1080
Unity Check = 1.04351

Example 11-5

Repeat Example 11-3 using INSTEP 32.

Solution

Input:

Factored design loads	Design parameters
Pu = 150	L = 96
Mtx/u = 750	Lb = 96
Mlt = 0	Cb = 1
	Cm = 1
	B2 = 0
	Ae = 0
	Kb = 1

Output:

Beam-Column Design Summary
Axial Pna = 342.45
Flexure Mna = 1261.41
Flexure Mue = 750
Unity Check = 0.966531

11.4   FIRST-ORDER AND SECOND-ORDER MOMENTS FOR MEMBERS SUBJECT TO AXIAL COMPRESSION AND BENDING

When a beam column is subjected to moment along its unbraced length, it will be displaced laterally in the plane of bending. The result will be an increased or secondary moment equal to the axial compression load times the lateral displacement or eccentricity. In Fig. 11-2 we can see that the member moment is increased by an amount $P_u\delta$. This moment will cause additional lateral deflection, which will in turn cause a larger column moment, which will cause a larger lateral deflection, and so on until equilibrium is reached.

If a frame is subject to sidesway where the ends of the columns can move laterally with respect to each other, additional secondary moments will result. In Fig. 11.3 the secondary moment produced due to sidesway is equal to $P_u\Delta$.

The moment M_u is assumed by the LRFD Specification to equal M_{lt} (which is the moment due to the lateral loads) plus the moment due to $P_u\Delta$.

The required total flexural strength of a member must equal at least the sum of the first-order and second-order moments. Several methods are available for determining this required strength, ranging from very simple approximations to very rigorous procedures.

The LRFD Specification C.1 states that we can either (1) make a second-order analysis to determine the maximum factored load strength or (2) use a first-order elastic analysis and amplify the moments obtained with some amplification factors called B_1 and B_2 and described in the paragraphs to follow.

Should the designer make a second-order analysis, he or she should realize that it must account for the interaction of the factored load effects. That is, we must consider combinations of factored loads acting at the same time. We cannot correctly make separate analyses and superimpose the results.

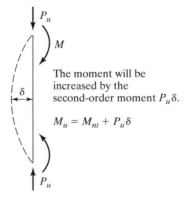

The moment will be increased by the second-order moment $P_u\delta$.

$$M_u = M_{nt} + P_u\delta$$

FIGURE 11.2

Moment amplification of a column that is braced against sidesway.

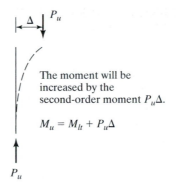

The moment will be increased by the second-order moment $P_u\Delta$.

$M_u = M_{lt} + P_u\Delta$

FIGURE 11.3

Column in an unbraced frame.

In this chapter the author presents the approximate method of analysis given in the LRFD Manual. We will make two first-order elastic analyses—one an analysis where the frame is assumed to be braced so that it cannot sway. We will call these moments M_{nt} and will multiply them by a magnification factor called B_1 to account for the P-δ effect (see Fig. 11-2). Then we will analyze the frame again, allowing it to sway. We will call these moments M_{lt} and will multiply them by a magnification factor called B_2 to account for the P-Δ effect (see Fig. 11-3). The final moment in a particular member will equal

$$M_u = B_1 M_{nt} + B_2 M_{lt} \qquad \text{(LRFD Equation C1-1)}$$

Instead of using the LRFD empirical procedure described here, the designer may and is encouraged to use a theoretical second-order elastic analysis, provided he or she meets the requirements of Sections C1 and C2 of the Specification. These requirements pertain to axial deformations, maximum axial forces permitted in members, bracing, K factors, and so on.

11.5   MAGNIFICATION FACTORS

The magnification factors are B_1 and B_2. With B_1, the analyst attempts to estimate the $P_u\delta$ effect for a column whether the frame is braced or unbraced against sidesway. With B_2 he or she attempts to estimate the $P_u\Delta$ effect in unbraced frames.

These factors are theoretically applicable when the connections are fully restrained or when they are completely unrestrained. The LRFD Manual indicates that the determination of secondary moments in between these two classifications for partially restrained moment connections is beyond the scope of their specification. The terms *fully restrained* and *partially restrained* are discussed at length in Chapter 15.

In the expression for B_1 that follows, C_m is a term that will be defined in the next section, P_u is the required axial strength of the member, and P_{e1} is the member's Euler buckling strength equal to $A_g F_y/\lambda_c^2$ for a braced frame. In this expression $\lambda_c = (KL/r\pi)\sqrt{F_y/E}$ as given by LRFD Equation E2-4. I and KL are both taken in the plane of bending determined in accordance with LRFD Specification C2.1 for a *braced frame*.

Substituting this value of λ_c into the expression for P_{e1} and replacing A_g with I/r^2, it becomes

$$P_{e1} = \frac{A_g F_y}{\lambda_c^2} = \frac{\dfrac{I}{r^2} F_y}{\left(\dfrac{KL}{2\pi} \sqrt{\dfrac{F_y}{E}} \right)^2}$$

$$P_{e1} = \frac{\pi^2 EI}{(KL)^2}$$

In a similar fashion, P_{e2} is the Euler buckling strength $\pi^2 EI/(KL)^2$ with K determined in the plane of bending for a column in the unbraced condition. P_{e1} and P_{e2} are the elastic buckling loads, respectively, for braced and unbraced frames. Their values can be obtained from Table 4-2 in Part 4 of the Manual. *Should we compute a B_1 or B_2 value that is less than 1.0, we must use 1.0.* (If we are supposedly magnifying moments, we don't want to multiply them by a number less than 1.0.)

The expression to follow for B_1 was derived for a member braced against sidesway. It will be used only to magnify the M_{nt} moments (*those moments computed assuming there is no lateral translation of the frame*).

$$B_1 = \frac{C_m}{1 - \dfrac{P_u}{P_{e1}}} \geq 1.0 \qquad \text{(LRFD Equation C1.2)}$$

The horizontal deflection of a multistory building due to wind or seismic load is called *drift*. It is represented by Δ in Figs. 11-3 and 11-4. In the formula to follow the term Δ_{oh} is used. It represents the lateral deflection of the story in question calculated with service loads with respect to the story below. Drift is measured with the so-called *drift index* Δ_{oh}/L where Δ_{oh} is the lateral inter-story deflection and L is the story height. For the comfort of occupants of a building the index usually is limited at working or service loads to a value between 0.0015 and 0.0030, and at factored loads to about 0.0040.

The designer may use either of the expressions that follow for B_2. In the first of these expressions, ΣP_u represents the total required axial strength of all the columns on the floor in question, Δ_{oh}/L represents the story drift index, and ΣH is the sum of all the story horizontal forces producing Δ_{oh}. The value of P_{e2} is defined as before for

FIGURE 11.4

Drift in a building frame.

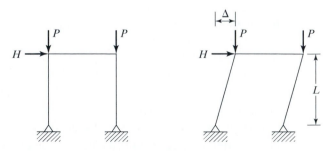

P_{e1}, except that the effective length factor K is determined in the plane of bending for *an unbraced frame*.

The values shown for ΣP_u and ΣP_{e2} are for all of the columns on the floor in question. This is considered to be necessary because the B_2 term is used to magnify column moments for sidesway. For sidesway to occur in a particular column, it is necessary for all of the columns on the floor to sway simultaneously. The ΣH value used in the first of the B_2 expressions represents the sum of the lateral loads acting above the floor being considered. To compute the ratio $\Delta_{oh}/\Sigma H$ we may use either factored or unfactored loads.

$$B_2 = \frac{1}{1 - \Sigma P_u(\Delta_{oh}/\Sigma HL)} \qquad \text{(LRFD Equation C1-4)}$$

or

$$B_2 = \frac{1}{1 - \Sigma P_u/\Sigma P_{e2}} \qquad \text{(LRFD Equation C1-5)}$$

We must remember that the amplification factor B_2 is only applicable to moments caused by forces that cause sidesway and is to be computed for an entire story. (Of course, if you want to be conservative, you can multiply B_2 times the sum of the no-sway and the sway moments—that is, M_{nt} and M_{lt}—but that's probably overdoing it.) To use the B_2 value given by LRFD Equation C1-5, we must select initial member sizes (that is so we can compute a value for P_{e2}). Should you be designing a building frame where you are limiting the drift index Δ_{oh}/L to a certain maximum value, you can compute B_2 with LRFD Equation C1-4 before you design the member. In this way you can set a drift limit in advance so that secondary bending is insignificant.

To calculate the values of ΣP_u and ΣP_e some designers will calculate the values for the columns in the one frame under consideration (or for that single line of columns perpendicular to the wind). This, however, is a rather bad practice unless all the other frames on that level are exactly the same as the one under study.

Of the two expressions given for B_2, the first one (C1-4), which involves a drift index, is better suited for design office practice. If we assume drifts as large as these values, we are probably being quite conservative. That is, the real structures may not drift this much.

11.6 MOMENT MODIFICATION OR C_m FACTORS

In Sections 11-4 and 11-5, the subject of moment magnification due to lateral deflections was introduced, and the factors B_1 and B_2 were presented with which the moment increases could be estimated. In the expression for B_1 a term C_m, called the *modification factor*, was included. The magnification factor B_1 was developed for the largest possible lateral displacement. On many occasions the displacement is not that large and B_1 overmagnifies the column moment. As a result, the moment may need to be reduced or modified with the C_m factor. In Fig. 11-5, we have a column bent in single curvature with equal end moments such that the column bends laterally by an amount δ

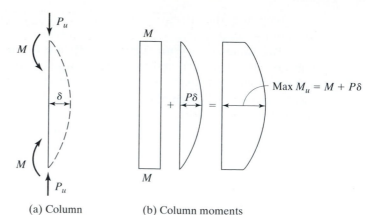

FIGURE 11.5

Moment magnification for column bent in single curvature.

(a) Column   (b) Column moments

at mid-depth. The maximum total moment occuring in the column clearly will equal M plus the increased moment $P_u\delta$. As a result, no modification is required and $C_m = 1.0$.

An entirely different situation is considered in Fig. 11-6, where the end moments tend to bend the member in reverse curvature. The initial maximum moment occurs at one of the ends, and we shouldn't increase it by a value $P_u\delta$ that occurs some distance out in the column because we will be overdoing the moment magnification. The purpose of the modification factor is to modify or reduce the magnified moment when the variation of the moments in the column is such that B_1 is made too large. If we didn't use a modification factor, we would end up with the same total moments in the columns of both Figs. 11-5 and 11-6, assuming the same dimensions and initial moments and load.

Modification factors are based on the rotational restraint at the member ends and on the moment gradients in the members. The LRFD Specification (C1) includes two categories of C_m as described in the next few paragraphs.

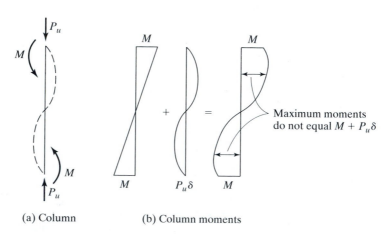

(a) Column   (b) Column moments

FIGURE 11.6

Moment magnification for column bent in double curvature.

TABLE 11.1 Modification Factors for Beam Columns Subject to Transverse Loads Between Joints.

Case	ψ	C_m
	0	1.0
	-0.4	$1 - 0.4\dfrac{P_u}{P_{e1}}$
	-0.4	$1 - 0.4\dfrac{P_u}{P_{e1}}$
	-0.2	$1 - 0.2\dfrac{P_u}{P_{e1}}$
$L/2$	-0.3	$1 - 0.3\dfrac{P_u}{P_{e1}}$
	-0.2	$1 - 0.2\dfrac{P_u}{P_{e1}}$

In Category 1 the members are prevented from joint translation or sidesway and they are not subject to transverse loading between their ends. For such members the modification factor is based on an elastic first-order analysis.

$$C_m = 0.6 - 0.4\frac{M_1}{M_2} \qquad \text{(LRFD Equation C1-3)}$$

In this expression M_1/M_2 is the ratio of the smaller moment to the larger moment at the ends of the unbraced length in the plane of bending under consideration. The ratio is negative if the moments cause the member to bend in single curvature and positive if they bend the members in reverse or double curvature. As previously described, a member in single curvature has larger lateral deflections than a member bent in reverse curvature. With larger lateral deflections the moments due to the axial loads will be larger.

Category 2 applies to members that are subjected to transverse loading between the joints in the plane of loading. The compression chord of a truss with a purlin load between its joints is a typical example of this category. The LRFD Specification states that the value of C_m is to be determined either by rational analysis or by using one of the values to follow.

a. For members with restrained ends $C_m = 0.85$.
b. For members with unrestrained ends $C_m = 1.0$.

If either one of these two values is used, or if we interpolate between them where we feel our conditions warrant it, the results probably will be quite reasonable.

Instead of using these values for transversely loaded members, the values of C_m for Category 2 may be determined for various end conditions and loads by the values given in Table 11-1, which is a reproduction of Table C-C1.1 of the Commentary on the LRFD Specification. In the expressions given in the table P_u is the factored column axial load and P_{e1} is the elastic buckling load for a braced column for the axis about which bending is being considered.

$$P_{e1} = \frac{\pi^2 EI}{(KL)^2}$$

In Table 11-1 note that some members have restrained ends and some do not. Sample values of C_m are calculated for four beam columns and shown in Fig. 11-7.

11.7   REVIEW OF BEAM–COLUMNS IN BRACED FRAMES

The same interaction equations are used for members subject to axial compression and bending as were used for members subject to axial tension and bending. However, some of the terms involved in the equations are defined somewhat differently. For instance, P_u and P_n refer to compressive forces rather than tensile forces, ϕ_c is 0.85 for axial compression, and ϕ_b is 0.9 for bending.

To analyze a particular beam column or a member subject to both bending and axial compression we need to make both a first-order and a second-order analysis to obtain the bending moments. The first-order moment is usually obtained by making an elastic analysis and consists of the moments M_{nt} (these are the moments in beam columns caused by gravity loads) and the moments M_{lt} (these are the moments in beam columns due to the lateral loads, that is, due to lateral translation).

Theoretically, if both the loads and frame are symmetrical, M_{lt} will be zero. Similarly, if the frame is braced, M_{lt} will be zero. For practical purposes you can have lateral deflections in taller buildings with symmetrical dimensions and loads.

Examples 11-6 to 11-8 illustrate the application of the interaction equations to beam-columns that are members of braced frames. Thus, only B_1 will be computed, as B_2 is not applicable. *It is to be remembered that C_m was developed for braced frames and thus must be used in these three examples for calculating B_1.*

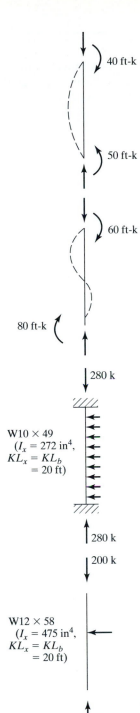

(a) No sidesway and no transverse loading.
Moments bend member in single curvature

$$C_m = 0.6 - (0.4)\left(-\frac{40}{50}\right) = 0.92$$

(b) No sidesway and no transverse loading.
Moments bend member in reverse curvature

$$C_m = 0.6 - 0.4\left(+\frac{60}{80}\right) = 0.30$$

(c) Member has restrained ends and transverse loading
and is bent about x axis.
$C_m = 0.85$ or can be determined from Table 11-1
(LRFD Table C-C1.1) as below

$$P_{e1} = \frac{\pi^2 EI}{(KL)_x^2} = \frac{(\pi^2)(29 \times 10^3)(272)}{(12 \times 20)^2}$$

$$= 1351 \text{ k}$$

$$C_m = 1 - 0.4\left(+\frac{280}{1351}\right) = 0.92$$

(d) Member has unrestrained ends and transverse loading
and is bent about x axis.
$C_m = 1.0$ or can be determined from Table 11-1
(LFRD Table C-C1.1)

$$P_{e1} = \frac{(\pi^2)(29 \times 10^3)(475)}{(12 \times 20)^2} = 2360 \text{ k}$$

$$C_m = 1 - 0.2\left(+\frac{200}{2360}\right) = 0.98$$

FIGURE 11.7

Example modification or C_m factors.

Example 11-6

A 12-ft W12 × 96 (50 ksi steel) is used as a beam-column in a braced frame. It is bent in single curvature with equal and opposite end moments and is not subjected to intermediate transverse loads. Is the section satisfactory if $P_u = 700$ k and first-order moment $M_{ntx} = 175$ ft-k?

Solution. Using a W12 × 96 ($A = 28.2$ in^2, $I_x = 833$ in^4, $\phi_b M_p = 551$ ft-k, $\phi_b M_r = 393$ ft-k, $L_p = 10.9$ ft, $L_r = 41.3$ ft, $BF = 5.20$),

For a braced frame let $K = 1.0$

$$\therefore K_x L_x = K_y L_y = (1.0)(12) = 12 \text{ ft}$$
$$\phi_c P_n = 1020 \text{ k from LRFD column tables}$$
$$\frac{P_u}{\phi_c P_n} = \frac{700}{1020} = 0.686 > 0.2$$
$$\therefore \text{ Must use LRFD Equation H1-1a}$$

As the only moment is M_{ntx}, there is no lateral translation of the frame—that is, $M_{lt} = 0$.

$$\therefore M_{ux} = B_1 M_{ntx}$$
$$C_m = 0.6 - 0.4\left(\frac{M_1}{M_2}\right) = 0.6 - 0.4\left(-\frac{175}{175}\right) = 1.0$$
$$P_{e1} = \frac{\pi^2 E I_x}{(K_x L_x)^2} = \frac{(\pi^2)(29 \times 10^3)(833)}{(12 \times 12)^2} = 11,498 \text{ k}$$

The values $\dfrac{P_{ex}(kL^2)}{104}$ and $\dfrac{P_{ey}(kL^2)}{104}$ can be obtained for W shapes from Table 4-2 in Part 4 of the Manual. For this shape

$$\frac{P_{ex}(kL^2)}{10^4} = 23,800$$
$$P_{ex} = \frac{(10^4)(23,800)}{(12 \times 12)^2} = 11,477 \text{ k}$$
$$B_1 = \frac{C_m}{1 - \dfrac{P_u}{P_{e1}}} = \frac{1.0}{1 - \dfrac{700}{11,477}} = 1.065 > 1.0 \qquad \text{(OK)}$$
$$M_{ux} = (1.065)(175) = 186.4 \text{ ft-k}$$

Applying LRFD Equation H1-1a. Since $L_b = 12$ ft $> L_p = 10.9$ ft $< L_r = 41.3$ ft,
$$\phi_b M_{nx} = 1.0\,[551 - (5.20)(12 - 10.9) = 545.3 \text{ ft-k}$$

$$\frac{P_u}{\phi_c P_n} + \frac{8}{9}\left(\frac{M_{ux}}{\phi_b M_{nx}} + \frac{M_{uy}}{\phi_b M_{ny}}\right)$$

$$= \frac{700}{1020} + \frac{8}{9}\left(\frac{186.4}{545.3} + 0\right) = 0.990 < 1.0$$

∴ Section is satisfactory.

Example 11-7

A 14-ft W14 × 120 (50 ksi steel) is used as a beam-column in a braced frame. It is bent in single curvature with equal and opposite moments. Its ends are rotationally restrained and it is not subjected to intermediate transverse loads. Is the section satisfactory if $P_u = 250\ k$ and if it has the first-order moments $M_{ntx} = 210$ ft-k and $M_{nty} = 140$ ft-k?

Solution. Using a W14 × 120 ($A = 35.3$ in^2, $I_x = 1380$ in^4, $I_y = 495$ in^4, $Z_x = 212$ in^3, $Z_y = 102$ in^3, $L_r = 46.3$ ft, $L_p = 13.2$ ft, $BF = 6.81$)

For a braced frame let $K_x L_x = K_y L_y = (1.0)(14) = 14$ ft

$$\phi_c P_n = 1290\,k \text{ from LRFD column tables}$$

$$\frac{P_u}{\phi_c P_n} = \frac{250}{1290} = 0.194 < 0.2$$

∴ Must use LRFD Equation H1-1b

As we have only the moments M_{ntx} and M_{nty} there is no lateral translation of the frame and thus $M_{ltx} = M_{lty} = 0$.

$$M_{ux} = B_{1x} M_{ntx} \quad \text{and} \quad M_{uy} = B_{1y} M_{nty}$$

$$C_m = 0.6 - 0.4\left(-\frac{210}{210}\right) = 1.0$$

$$P_{e1x} \text{ from column table} = \frac{(10)^4(39,500)}{(12 \times 14)^2} = 13,995\ k$$

$$B_{1x} = \frac{C_m}{1 - \dfrac{P_u}{P_{e1x}}} = \frac{1.0}{1 - \dfrac{250}{13,995}} = 1.018 > 1.0 \qquad \text{(OK)}$$

$$M_{ux} = (1.018)(210) = 214 \text{ ft-k}$$

$$P_{e1y} \text{ from column table} = \frac{(10)^4(14,200)}{(12 \times 14)^2} = 5031\ k$$

$$B_{1y} = \frac{C_m}{1 - \dfrac{P_u}{P_{e1y}}} = \frac{1.0}{1 - \dfrac{250}{5031}} = 1.053 > 1.0 \qquad \text{(OK)}$$

$$M_{uy} = (1,053)(140) = 147.4 \text{ ft-k}$$

Since $L_b = 14 \text{ ft} > L_p = 13.2 \text{ ft} < L_r = 46.2 \text{ ft}$

$$\phi_b M_x = 1.0[795 - 6.83(14 - 13.2)] = 789.5 \text{ ft-k}$$

$$\phi_b M_{ny} = \phi_b F_y Z = \frac{(0.9)(50)(102)}{12} = 382.5 \text{ ft-k} >$$

$$\phi_b 1.5 F_y S_y = \frac{(0.9)(1.5)(50)(67.5)}{12} = 379.7 \text{ ft-k} \leftarrow$$

Applying LRFD Equation H1-1b

$$\frac{P_u}{2\phi P_n} + \left(\frac{M_{ux}}{\phi_b M_{nx}} + \frac{M_{uy}}{\phi_b M_{ny}}\right)$$

$$= \frac{250}{(2)(1290)} + \left(\frac{261}{789.5} + \frac{147.4}{379.7}\right)$$

$$= 0.756 < 1.0 \qquad\qquad\qquad \text{(OK but overdesigned)}$$

Example 11-8

For the truss shown in Fig. 11-8(a), a W8 × 31 is used as a continuous top chord member from joint L_0 to joint U_3. If the member consists of 50 ksi steel, does it have sufficient strength to resist the factored loads shown in part (b) of the figure. Part (b) shows the portion of the chord from L_0 to U_1 and the 16-k load represents the effect of a purlin. It is assumed that lateral support is provided for this member at its ends and center line.

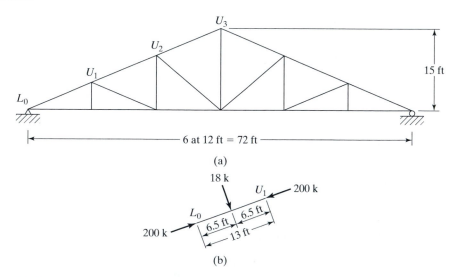

FIGURE 11.8

A truss whose top chord is subject to intermediate loads.

Solution. Using a W8 × 31 ($A = 9.14$ in², $I_x = 110$ in⁴, $r_x = 3.47$ in, $r_y = 2.02$ in, $L_p = 7.1$ ft as calculated with LRFD Equation F1-4, $Z_x = 30.4$ in³)

For a frame braced against sidesway $K = 1.0$

$$\left(\frac{KL}{r}\right)_x = \frac{(1.0)(12 \times 13)}{3.47} = 44.96 \leftarrow$$

$$\left(\frac{KL}{r}\right)_y = \frac{(1.0)(12 \times 6.5)}{2.02} = 38.61$$

$$\phi_c F_{cr} = 36.7 \text{ ksi (from LFRD Table 3.50)}$$

$$\phi_c P_n = (36.7)(9.13) = 335 \text{ k}$$

$$\frac{P_u}{\phi_c P_n} = \frac{200}{335} = \; > 0.2$$

Must use LRFD Equation H1-1a

For a braced frame M_{lt} (the moment due to lateral loads) $= 0$. Note that P_{e1x} is not given in Manual for a W8 × 31, but it can be computed by

$$\frac{\pi^2 EI}{(k_x L_x)^2} = \frac{(\pi^2)(29 \times 10^3)(110)}{(12 \times 13)^2} = 1294 \text{ k}$$

For ⤸⤸ $C_{mx} = 1 - 0.2 \dfrac{P_u}{P_{elx}} = 1 - 0.2 \left(\dfrac{200}{1294}\right) = 0.969$

For ⤸⤸ $C_{mx} = 1 - 0.3 \dfrac{P_u}{P_{elx}} = 1 - 0.3 \left(\dfrac{200}{1294}\right) = 0.954$

Average $C_m = 0.961$

Computing moment:

For ⤸⤸ $M_{mtx} = \dfrac{(18)(13)}{4} = 58.5 \text{ ft-k}$

For ⤸⤸ $M_{mtx} = \dfrac{(3)(18)(13)}{16} = 43.9 \text{ ft-k}$

Avg $M_{mtx} = \dfrac{58.5 + 43.9}{2} = 51.2 \text{ ft-k}$

Computing B_{1x} and M_{ux}:

$$B_{1x} = \frac{C_{mx}}{1 - \dfrac{P_u}{P_{e1x}}} = \frac{0.961}{1 - \dfrac{200}{1294}} = 1.18 > 1.0 \qquad \text{(OK)}$$

$$M_{ux} = (1.18)(51.2) = 60.4 \text{ ft-}k$$

Since $L_b = 6.5$ ft $< L_p = 7.1$ ft

$$\therefore \phi_b M_{nx} = \phi_b F_y Z_x = \frac{(0.9)(50)(30.4)}{12} = 114 \text{ ft-k}$$

Applying LRFD Equation H1-1a:

$$\frac{P_u}{\phi P_n} + \frac{8}{9}\left(\frac{M_{ux}}{\phi_b M_{nx}} + \frac{M_{uy}}{\phi_b M_{ny}}\right)$$

$$= \frac{200}{335} + \frac{8}{9}\left(\frac{60.4}{114} + 0\right) = 1.07 > 1.0 \qquad \text{(NG)}$$

11.8   REVIEW OF BEAM–COLUMNS IN UNBRACED FRAMES

The maximum primary moments in unbraced frames almost always occur at the column ends. As you can see in Fig. 11-3, the maximum sidesway moments always occur at the member ends and the total moment for a particular column is determined by adding its primary end moment to its sidesway moment. Therefore it is not necessary to use a modification factor and C_1 is not used in the B_2 expressions.

Example 11-9

A 10-ft W12 × 106 ($F_y = 50$ ksi) is used as a beam-column in an unbraced frame. It is bent in reverse curvature with equal and opposite moments and is not subject to intermediate transverse loads. Is the section satisfactory if $P_u = 250$ k, $M_{ntx} = 60$ ft-k, $M_{nty} = 40$ ft-k, $M_{ltx} = 100$ ft-k and $M_{lty} = 80$ ft-k? The total factored gravity load ΣP_u above this level has been calculated to equal 5000 k. Assume that $\Sigma P_{ex} = 40{,}000$ k and $\Sigma P_{ey} = 20{,}000$ k. $K_x = K_y = 1.2$.

Solution

Using a W12 × 106 ($A = 31.2$ in^2, $I_x = 933$ in^4, $I_y = 301$ in^4, $L_p = 11.0$ ft, $Z_x = 164$ in^3, $Z_y = 75.1$ in^3)

$$K_x L_x = K_y L_y = (1.2)(10) = 12 \text{ ft}$$
$$\phi_c P_u = 1130 \text{ k from LRFD column tables}$$

$$\frac{P_u}{\phi_c P_n} = \frac{250}{1130} = 0.221 > 0.2$$

∴ Must use LRFD Equation H1-1a

As we have the moments M_{utx} and M_{nty} as well as M_{ltx} and M_{lty}

$$M_{ux} = B_{1x} M_{ntx} + B_{2x} M_{ltx}$$
$$M_{uy} = B_{1y} M_{nty} + B_{2y} M_{lty}$$

For reverse curvature and equal end moments

$$C_{mx} = C_{my} = 0.6 - 0.4(+1.0) = 0.2$$

$$P_{e1x} \text{ from Manual Table 4-2} = \frac{(10)^4 (26,700)}{(12 \times 10)^2} = 18,542 \text{ k}$$

$$B_{1x} = \frac{C_m}{1 - \dfrac{P_u}{P_{e1x}}} = \frac{0.2}{1 - \dfrac{250}{18,542}} = 0.203 < 1.0 \qquad \text{Use } 1.0$$

$$P_{e1y} \text{ from Manual Table 4-2} = \frac{(10)^4 (8620)}{(12 \times 10)^2} = 5986 \text{ k}$$

$$B_{1y} = \frac{C_m}{1 - \dfrac{P_u}{P_{e1y}}} = \frac{0.2}{1 - \dfrac{250}{5986}} = 0.209 < 1.0 \qquad \text{Use } 1.0$$

$$B_{2x} = \frac{1}{1 - \dfrac{\Sigma P_u}{\Sigma P_{e2x}}} = \frac{1}{1 - \dfrac{5000}{40,000}} = 1.143$$

$$B_{2y} = \frac{1}{1 - \dfrac{\Sigma P_u}{\Sigma P_{e2y}}} = \frac{1}{1 - \dfrac{5000}{20,000}} = 1.333$$

$$M_{ux} = (1.0)(60) + (1.143)(100) = 174.3 \text{ ft-k}$$
$$M_{uy} = (1.0)(40) + (1.333)(80) = 146.6 \text{ ft-k}$$

Since $L_b = 10 \text{ ft} < L_p = 11.0 \text{ ft}$

$$\phi_b M_{nx} = \phi_b F_y Z_x = \frac{(0.9)(50)(164)}{12} = 615 \text{ ft-k}$$

$$\phi_b M_{ny} = \phi_b F_y Z_y = \frac{(0.9)(50)(75.1)}{12} = 281.6 \text{ ft-k} > \phi_b \, 1.5 \, F_y S_y$$

$$= \frac{(0.9)\,(1.5)\,(50)\,(49.3)}{12} = 277.3 \text{ ft-k} \leftarrow$$

Applying LRFD Equation H1-1a, we have

$$\frac{P_u}{\phi P_n} + \frac{8}{9}\left(\frac{M_{ux}}{\phi_b M_{nx}} + \frac{M_{uy}}{\phi_b M_{ny}}\right) = \frac{250}{1130} + \frac{8}{9}\left(\frac{174.3}{615} + \frac{146.6}{277.3}\right)$$

$$= 0.943 < 1.0 \qquad\qquad\qquad (\text{OK})$$

11.9 DESIGN OF BEAM–COLUMNS—BRACED OR UNBRACED

The design of beam columns involves a trial-and-error procedure. A trial section is selected by some process and is then checked with the appropriate interaction formula. If the section does not satisfy the equation, or if it's too much on the safe side (that is, if it's overdesigned), another section is selected, and the interaction equation is applied again.

Based on work by A. Aminmansour[1], a set of design aids for checking and selecting beam columns is presented in Part 6 of the LRFD Manual. These aids were developed initially assuming that LRFD Equation H1-1a is applicable—that is, where $P_u/\phi P_m \geq 0.2$. This equation is presented here in a little different fashion:

$$\left(\frac{1}{\phi_c P_m}\right)P_u + \left(\frac{8}{9\,\phi_b\,M_{nx}}\right)M_{ux} + \left(\frac{8}{9\phi_b\,M_{ny}}\right)M_{uy} \leq 1.0$$

If the values shown in parantheses are referred to in order as b, m, and n, this equation can be written as

$$bP_u + mM_{ux} + nM_{uy} \leq 1.0$$

This latter equation is exactly equivalent to the original one and can be used in its place. Thus, if a W-shape is selected (as described later in this section) using this equation, it will obviously satisfy the requirements of the LRFD Specification. Look back at the expressions for b, m, and n, and notice that their values are based on steel grades, cross sectional properties, and dimensions of the shapes, as well as values of KL, L_b and C_b. Their values have been computed for 50 ksi W-shapes with different KL values and recorded in Table 6-2 of the Manual.

To show that this new form of LRFD Equation H1-1a may be used to check a beam column, the W-shape of Example 11-6 is reviewed in Example 11-10. Notice that the values for b, m, and n in the tables have been multiplied by 1000 to avoid working with several decimal places.

---

[1] A Aminmansour, "A New Approach for Design of Steel Beam-Columns," Engineering Journal, AISC, *37*, no. 2 (2nd quarter, 2000), pp. 41–72.

Example 11-10

Apply the Aminmansour method to the 50 ksi W12 × 96 of Example 11-6, where KL was equal to 12 ft, P_u was 700 k and the magnified M_{ux} was 186.4 ft-k.

Solution. From Table 6-2 in the Manual the following values are selected:

$$b = 0.978 \times 10^{-3}$$
$$m = 1.63 \times 10^{-3}$$
$$n = 3.56 \times 10^{-3}$$

Substituting into the equation.

$$(0.978 \times 10^{-3})(700) + (1.63 \times 10^{-3})(186.4) + (3.56 \times 10^{-3})(0) = 0.988 < 1.0$$
OK

The revised LRFD interaction equations can be used with the tables in Part 6 of the Manual to make fairly good estimates of beam-column sizes needed for certain situations. To make an estimate in this manner, the following steps should be taken:

1. If it appears that axial loads dominate a median value of m can be selected from Table 6-1 in Part 6 of the Manual. This table is entitled "W-Shapes Median values of b, m and n for Beam-Columns." In addition, if there is bending about the y axis, a value of n is selected from the bottom of the table.

2. The value of m (and perhaps n) selected in step 1 is substituted into the Aminmansour equation, and the equation is then solved for b:

$$bP_u + mM_{ux} + nM_{uy} \leq 1.0$$

3. From Table 6-2, a shape is selected having values of b, m and n approximately equal to those obtained in the preceding two steps. This shape is then checked with the appropriate interaction equation. If its over- or underdesigned, another shape is tried and so on. Sometimes this may take several trials, but they can be made very quickly with the Aminmamsour equation.

Example 11-11 makes use of the Aminmansour procedure to design a beam column.

Example 11-11

Select a W14 beam column using A992 steel for the following: $KL = 12$ ft, $P_u = 350$ k, $M_{ux} = 270$ ft-k and $M_{uy} = 70$ ft-k. Use the Aminmamsour procedure.

Solution Assume axial load dominates and select values for m and n from Table 6.1.

$$m = 0.826 \times 10^{-3}$$
$$n = 1.64 \times 10^{-3}$$

Substituting into the Aminmansour equation

$$bP_u + mM_{ux} + nM_{uy} \leq 1.0$$

$$(b)(350) + (0.826 \times 10^{-3})(270) + (1.64 \times 10^{-3})(70) = 1.0$$

$$b = 1.89 \times 10^{-3}$$

From Table 6.2, a section is sought having the following approximate values: $b = 1.89 \times 10^{-3}$, $m = 0.826 \times 10^{-3}$ and $n = 1.64 \times 10^{-3}$. From the tables we can see that the desired section will fall somewhere between the W14 × 61 and the W14 × 159. The authors tried a section halfway in between. Try W14 × 99 $(b = 0.903 \times 10^{-3}$, $m = 1.38 \times 10^{-3}$ and $n = 2.88 \times 10^{-3})$

Noting that $\dfrac{P_u}{\phi_c P_n} = bP_u = (0.903 \times 10^{-3})(350) = 0.316 > 0.2$

$$(0.903 \times 10^{-3})(350) + (1.38 \times 10^{-3})(270) + (2.88 \times 10^{-3})(70) = 0.890 < 1.0$$
$$\text{(OK, but it may be overdesigned.)}$$

A subsequent check of the next lightest section, a W14 × 90 results in a value 0.987 < 1.0   (OK)

USE W14 × 90 A992 Steel

In step 1, if it appears that the moment rather than axial load dominates a median value of b (rather than m) is selected from Table 6-1 (along with n if there is bending about the y-axis) and the equation used to solve for m. Steps 2 and 3 are then repeated as before.

When $bP_u < 0.2$, the tablulated values of b, m and n are still applicable but it is necessary to use the second interaction equation H1-1b which is written by Aminmansour as

$$\frac{1}{2}bP_u + \frac{9}{8}(mM_{ux} + nM_{uy}) \leq 1.0$$

Because so many different loading situations exist, we present quite a few examples in the pages that follow. Trial sections are selected by the procedure illustrated in Example 11-11 and then checked with the appropriate Aminmansour equation. Often it was necessary to try 2 or 3 sections before the most economical one was found. To conserve space, only the final trial is shown in each of the examples.

Example 11-12

Select a 14-ft W14 A992 steel beam-column to support the following: $P_u = 600\,k$, first-order moment $M_{ux} = 200$ ft-k and $M_{uy} = 0$. The member is to be bent in single curvature with no transverse loading and the frame is to be braced. Use $K = 1.0$ and $C_m = 0.85$.

Solution

Try W14 × 90

$$\frac{(P_{e1x})(KL)^2}{10^4} = 28{,}600 \text{ (from Table 4-2 in Manual)}$$

$$P_{e1x} = \frac{(10^4)(28{,}600)}{(12 \times 14)^2} = 10{,}133 \text{ k}$$

$$B_1 = \frac{C_m}{1 - \dfrac{P_u}{P_{e1x}}} = \frac{0.85}{1 - \dfrac{600}{10{,}133}} < 1.0 \;\therefore\; \text{Use 1.0}$$

$$\therefore\; M_{ux} = (1.0)(200) = 200 \text{ ft-}k$$

Checking W14 × 90 $(b = 1.03 \times 10^{-3}, m = 1.54 \times 10^{-3}, n = 3.25 \times 10^{-3})$

$$bP_u = (1.03 \times 10^{-3})(600) = 0.618 > 0.2$$

$$\therefore\; \text{Use } bP_u + mM_{ux} + nM_{uy} \le 1.0$$

$$(1.03 \times 10^{-3})(600) + (1.54 \times 10^{-3})(200) + 0 = 0.926 < 1.0 \qquad \text{(OK)}$$

Use W14 × 90 A992.

Example 11-13

A 12-ft beam-column of A992 steel is to be used in a braced frame for the following: $P_u = 200$ k; first-order moment $M_{ux} = 150$ ft-k and $M_{uy} = 100$ ft-k. Assuming it is to have restrained ends and to be subject to intermediate transverse loads, select a W12 section. Assume $K = 1.0$.

Solution

Try W12 × 79

$C_{mx} = C_{my}$ for members with rotationally restrained ends subject to transverse loads between ends = 0.85

$$\frac{(P_{e1x})(KL)^2}{10^4} = 18{,}900 \text{ (from Table 4-2 in Manual)}$$

$$P_{e1x} = \frac{(10)^4(18,900)}{(12 \times 12)^4} = 9115 \text{ k}$$

$$B_{1x} = \frac{C_{mx}}{1 - \dfrac{P_u}{P_{e1x}}} = \frac{0.85}{1 - \dfrac{200}{9115}} = 0.869 < 1.0 \ \therefore \ \text{Use } 1.0$$

$$M_{ux} = (1.0)(150) = 150 \text{ ft-k}$$

$$P_{ey1} = \frac{(10)^4(6180)}{(12 \times 12)^4} = 2980 \text{ k}$$

$$B_{1y} = \frac{C_{my}}{1 - \dfrac{P_u}{P_{e1y}}} = \frac{0.85}{1 - \dfrac{200}{2980}} = 0.911 < 1.0 \ \therefore \ \text{Use } 1.0$$

$$M_{uy} = (1.0)(100) = 100 \text{ ft-}k$$

Checking W12 × 79 ($b = 1.19 \times 10^{-3}$, $m = 2.02 \times 10^{-3}$, $n = 4.41 \times 10^{-3}$)

$$bP_u = (1.19 \times 10^{-3})(200) = 0.238 > 0.2$$

$$\therefore \ \text{Use } bP_u + mM_{ux} + nM_{uy} \leq 1.0$$

$$0.238 + (2.02 \times 10^{-3})(150) + (4.41 \times 10^{-3})(100) = 0.982 < 1.0.$$

Use W12 × 79 A992.

Example 11-14

Select a 10-ft W12 beam-column (A992 steel) for an unbraced symmetrical frame with $P_u = 200$ k; first-order moments $M_{ltx} = 160$ ft-k, $M_{lty} = 110$ ft-k, $M_{ntx} = M_{nty} = 0$. The computed ΣP_u for all of the columns on this level is 3600 k while ΣP_{ex2} has been calculated to equal 50,000 k and ΣP_{ey2} to equal 25,000 k. Assume $K_x = K_y = 1.2$. The column is to be bent in single curvature with no intermediate loads.

Solution. Determining M_{ux} and M_{uy} from LRFD Equation C1-2.

$$M_u = B_1 M_{nt} + B_2 M_{lt}$$

$$B_{2x} = \frac{1.0}{1 - \dfrac{\Sigma P_u}{\Sigma P_{ex2}}} = \frac{1.0}{1 - \dfrac{3600}{50,000}} = 1.078$$

$$B_{2y} = \frac{1.0}{1 - \dfrac{\Sigma P_u}{\Sigma P_{ey2}}} = \frac{1.0}{1 - \dfrac{3600}{25,000}} = 1.168$$

The second order moment strengths are

$$M_{ux} = (1.078)(160) = 172.5 \text{ ft-k}$$
$$M_{uy} = (1.168)(110) = 128.5 \text{ ft-k}$$
$$KL = (1.2)(10) = 12 \text{ ft}$$

Try w12 × 96 ($b = 0.978 \times 10^{-3}$, $m = 1.63 \times 10^{-3}$, $x = 3.56 \times 10^{-3}$):

$$bP_u = (0.978 \times 10^{-3})(200) = 0.1956 < 0.2$$

$$\therefore \text{ Use } \frac{1}{2} bP_u + \frac{9}{8}(mM_{ux} + nM_{uy}) \leq 1.0$$

$$\left(\frac{1}{2}\right)(0.1956) + \left(\frac{9}{8}\right)\left[(1.63 \times 10^{-3})(172.5) + (3.56 \times 10^{-3})(128.5)\right]$$

$$= 0.929 < 1.0$$

Use w12 × 96 A992.

Example 11-15

Select a W14 beam-column of A922 steel with $K_x = 1.2 = K_y$ for the following: $P_u = 500$ k, first order moment $M_{uy} = 100$ ft-k due to wind, with all other moments equal to 0. The 14-ft member is to be used in a symmetrical unbraced frame with an allowable story drift index $\Delta_{oh}/L = 0.0020$ as a result of a total service or unfactored load H of 100 k. The total factored gravity load ΣP_u above has been calculated to equal 5000 k.

Solution

$$K_x L_x = K_y L_y = (1.2)(14) = 16.8 \text{ ft}$$

Having set a drift index $= 0.002$ we can backfigure B_2 with LRFD Formula C1-4.

$$B_2 = \frac{1}{1 - \Sigma P_u \left(\dfrac{\Delta_{oh}}{\Sigma HL}\right)} = \frac{1}{1 - 5000\left(\dfrac{0.0020}{100}\right)} = 1.111$$

$$\therefore B_2 M_{uy} = (1.111)(100) = 111.1 \text{ ft-k}$$

Try w14 × 90 ($b = 1.10 \times 10^{-3}$, $m = 1.54 \times 10^{-3}$, $n = 3.25 \times 10^{-3}$):

$$bP_u = (1.10 \times 10^{-3})(500) = 0.550 > 0.2$$

$$\therefore \text{ Use } bP_u + mM_{ux} + nM_{uy} \leq 1.0$$

$$0.550 + (1.54 \times 10^{-3})(0) + (3.25 \times 10^{-3})(111.1) = 0.911 < 1.0$$

Use W14 × 90 A992.

Example 11-16

Select a 12-ft W12 A992 steel for a symmetrical unbraced frame to support the following: $P_u = 360$ k, $M_{ntx} = 70$ ft-k, $M_{nty} = 50$ ft-k, $M_{ltx} = 140$ ft-k, and $M_{lty} = 100$ ft-k. The member is to be subjected to intermediate transverse loads and its ends are rotationally restrained. It is to have an allowable drift index = 0.0020 due to the total horizontal service or unfactored loads of 120 k perpendicular to the x axis and 80 k perpendicular to the y axis. Assume $K_x = K_y = 1.2$ and $\Sigma P_u = 5000$ k.

Solution

We do not know the value of P_{e2x}, P_{e2y}, B_{1x}, or B_{1y}. Assume $B_{1x} = B_{1y} = 1.0$ and compute the values of B_{2x} and B_{2y} as follows using LRFD Equation C1-4.

$$B_{2x} = \cfrac{1}{1 - \Sigma P_u \left(\cfrac{\Delta_{oh}}{\Sigma HL}\right)} = \cfrac{1}{1 - 5000 \left(\cfrac{0.002}{120}\right)} = 1.09$$

$$B_{2y} = \cfrac{1}{1 - 5000\left(\cfrac{0.002}{80}\right)} = 1.14$$

$$M_{ax} = (1.00)(70) + (1.09)(140) = 222.6 \text{ ft-k}$$
$$M_{uy} = (1.00)(50) + (1.14)(100) = 164 \text{ ft-k}$$
$$KL = (1.2)(12) = 14.4 \text{ ft}$$

Try W12 × 136($b = 0.734 \times 10^{-3}$, $m = 1.13 \times 10^{-3}$, $n = 2.46 \times 10^{-3}$):

$$bP_u = (0.734 \times 10^{-3})(360) = 0.264 > 0.2$$
$$\therefore \text{ Use } bP_u + mM_{ux} + nM_{uy} \leq 1.0$$
$$0.264 + (1.13 \times 10^{-3})(222.6) + (2.46 \times 10^{-3})(164) = 0.919 < 1.0$$

Use W12 × 136 A992 steel.

Example 11-17

Select the lightest satisfactory W12 section (A992 steel) for a 12-ft column in a building frame for the situation to be described. It is to be unbraced in the plane of the frame but is to be braced at each story out of plane so that $K_y = 1.0$. In the plane of the frame K_x has been estimated to equal 1.50.

A first-order analysis has been made with the factored loads and the results are given in Fig. 11.9. It is assumed that the 1.2D + 0.5L + 1.3W load combination controls. Assume that ΣP_u for all of the columns on this floor has been calculated to be 7200 k, while ΣP_{e2} has been estimated to equal 114,000 k.

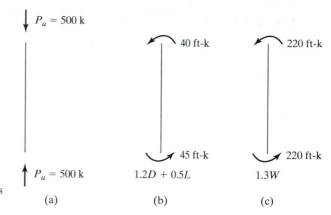

FIGURE 11.9

Column factored load and moments considered in Example 11-13.

(a)   (b)   (c)

Solution

$$K_x L_x = (1.5)(12) = 18 \text{ ft}$$

$$C_{mx} = 0.6 - 0.4\left(+\frac{40}{45}\right) = 0.244$$

$$P_{exl} = \frac{(10)^4(26{,}700)}{(12 \times 18)^2} = 8241 \text{ k}$$

$$B_1 = \frac{C_m}{1 - \dfrac{P_u}{P_{ex1}}} = \frac{0.244}{1 - \dfrac{500}{8241}} = \; < 1.0 \quad \text{Use 1.0}$$

$$B_2 = \frac{1}{1 - \left(\dfrac{\Sigma P_u}{\Sigma P_{el}}\right)} = \frac{1}{1 - \dfrac{7200}{114{,}000}} = 1.07$$

$$M_{ux} = (1.0)(45) + (1.07)(220) = 280.4 \text{ ft-k}$$

Try W12 × 106 ($b = 1.07 \times 10^{-3}$, $m = 1.54 \times 10^{-3}$, $n = 3.21 \times 10^{-3}$):

$$bP_u = (1.07 \times 10^{-3})(500) = 0.535 > 0.2$$

$$\therefore \; \text{Use } bP_u + mM_{ux} + nM_{uy} \le 1.0$$

$$0.535 + (1.54 \times 10^{-3})(280.4) + 0 = 0.967 < 1.0$$

Use W12 × 106 A992.

11.10 COMPUTER EXAMPLES FOR MEMBERS SUBJECT TO BENDING AND AXIAL COMPRESSION

Example 11-18 illustrates the use of INSTEP 32 to check a member subject to bending and axial compression. Then in Example 11-17 INSTEP 32 is used to select the lightest W14 shape for given moment and axial compression values. You can quickly repeat this design for the lightest W10, W12, etc. These two examples are repeats of Examples 11-6 and 11-10. *Notice that a sign must be given to P_u (+ for tension, − for compression).*

Example 11-18

Repeat Example 11-6 using INSTEP 32.

Solution

Input:

Factored design loads	Design parameters
Pu = −700	L = 144
Mtx/u = 2100	Lb = 144
Mlt = 0	Cb = 1
	Cm = 0.85
	B2 = 0
	Ae = 0
	Kb = 1

Output:

Beam-Column Design Summary
Axial Pna = 1022.53
Flexure Mna = 6547.29
Flexure Mue = 2235.87
Unity Check = 0.988129

Example 11-19

Repeat Example 11-12 using INSTEP 32.

Solution

Instep32-Structural Steel Design with AISC LRFD Procedures Clemson University Department of Civil Engineering

Client:	Designed by:
Project:	Checked by:
Topic:	Date:

Design of a Member in Axial Tension

Properties for Specified Material

$F_y = 50$ ksi

$F_u = 65$ ksi

Lightest Section. W14 × 90 in the W14 series

Factored Design Loads

$$M_{nt/u} = 2400 \text{ kip-in}$$
$$M_{lt} = 0 \text{ kip-in}$$
$$P_u = -600 \text{ kips}$$

Beam/Column Design Parameters

$$L = 168 \text{ in}$$
$$L_b = 168 \text{ in}$$
$$C_b = 1$$
$$C_m = 0.85$$
$$A_e = 28.2 \text{ sq. in}$$
$$K_b = 1$$

Calculated Design Data

Axial Pna = 968.653 kips
Flexure Mna = 6912.71 kip-in
Flexure Mue = 2400 kip-in
Unity Check = 0.928027

This member is acceptable.

PROBLEMS

Bending and Axial Tension

11-1. A W14 × 30 tension member with no holes consisting of 50 ksi steel is subjected to a factored tensile load $P_u = 250$ k and a factored bending moment M_{ux} of 30-ft-k. Is the member satisfactory if $L_b = 4.0$ ft and if $C_b = 1.0$? (*Ans.* 0.779 < 1.00 OK)

11-2. A W12 × 58 tension member with no holes and $F_y = 50$ ksi is subjected to $P_u = 125$ k and $M_{ux} = 140$ ft-k. Is the member satisfactory if $L_b = 8.0$ ft and $C_b = 1.0$?

11-3. Repeat Prob. 11-1 if the member is also subject to a factored bending moment $M_{uy} = 20$ ft-k. (*Ans.* 1.32 > 1.00 NG)

11-4. Select the lightest W12 available to support a factored tensile load $P_u = 80$ k and a factored bending moment $M_{uy} = 60$ ft-k. Assume the member consists of 50 ksi steel, has no holes, and is braced laterally for its full length.

11-5. Select the lightest W12 section available to support a factored tensile load $P_u = 300$ k using 50 ksi steel to support a factored bending moment $M_{ux} = 75$ ft-k, $L_b = 10$ ft, and $C_b = 1.0$. The member is assumed to have no holes. (*Ans.* W12 × 40)

11-6. Select the lightest available W12 section (50 ksi steel) to support an axial factored tensile load equal to 240 k placed with an eccentricity of 5 in with respect to the x axis. The member which is to be welded is to be 8 ft long and is to be braced laterally only at its supports.

11-7. Repeat Prob. 11-6 if a W14 section is to be used. (*Ans.* W14 × 38)

11-8. A W12 × 72 member consisting of 50 ksi steel has continuous lateral bracing. If M_{ux} is 100 ft-k and M_{uy} is 0 determine the maximum axial tensile force P_u that it can support. It is assumed to have no holes.

11-9. Repeat Prob. 11-8 if F_y = 36 ksi (*Ans.* 475.2 k)

Bending and Axial Compression

11-10. A W10 × 54 pin-connected beam-column that is not subjected to sidesway is 15 ft long. A load P_u = 300 k is applied to the column at its upper end with an eccentricity of 2 in. so as to cause bending about the major axis of the section. Check the adequacy of the member if it consists of 50 ksi steel and is used in a braced frame so that $M_{ltx} = M_{lty} = 0$. K_y = 1.0 while K_x = 1.00. Assume C_b = 1.0 and $C_{mx} = C_{my}$ = 0.85.

11-11. Sidesway is prevented for the beam-column shown in the accompanying illustration. If the first-order moments shown are about the major axis, is the member satisfactory if it consists of 50 ksi steel? Assume $K_x = K_y$ = 1.0. (*Ans.* 0.726 < 1.00 OK)

11-12. A pin-connected W10 × 49 consisting of 50 ksi steel is used as a beam-column. If sidesway is prevented and a load P_u = 300 k is applied 1.50 in. off center at both ends so as to cause single curvature bending about the x axis, is the member satisfactory? Length = 12 ft. K_x = 1.4, K_y = 1.0.

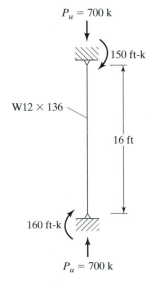

P_u = 700 k

150 ft-k

W12 × 136

16 ft

160 ft-k

P_u = 700 k

FIGURE P11-11

11-13. The W12 × 40 truss member shown in the accompanying illustration consists of 50 ksi steel and is fixed at its ends in the x and y directions. The purlin shown is assumed to provide pinned support in the y direction at the midpoint of the member. Is the member satisfactory to support the factored loads shown? (*Ans.* 0.395 < 1.0 but way overdesigned)

11-14. Repeat Prob. 11-13 if the ends of the member are assumed to be pinned and braced against lateral movement.

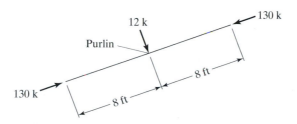

FIGURE 11-13

11-15. Select a W14 section consisting of 50 ksi steel to support a factored load $P_u = 800$ k and the factored moments $M_{ntx} = 150$ ft-k and $M_{nty} = 40$ ft-k. The 12-ft member is to be used in a braced frame with M_{ltx} and $M_{lty} = 0$. Assume $K_x = K_y = 1.0$ and $C_m = 0.85$. (*Ans.* W14 × 109)

11-16. Repeat Prob. 11-15 if a W12 section consisting of 50 ksi steel is to be used.

11-17. A W14 × 109 section (50 ksi steel) is used for a 15-ft column in a building frame. It is unbraced in the plane of the frame but is braced at each story out of plane so that $K_y = 1.00$. K_x has been estimated to equal 1.40. A first-order analysis has been made with factored loads, and the results are shown in the accompanying illustration. Is the member satisfactory? The total factored load above this level has been calculated to equal 6000 k and ΣP_{ex} is estimated to equal 36,000 k. (*Ans.* 0.610 < 1.00 OK)

11-18. Repeat Prob. 11-17 if the moments bend the member in single curvature.

11-19. Select a W12 with $F_y = 50$ ksi for the load and moments of Prob. 11-17. (*Ans.* W14 × 132)

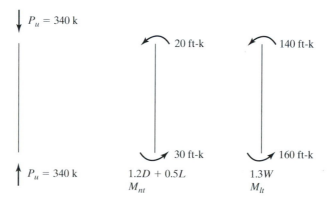

FIGURE P11-17

11-20. Select a 14-ft 50 ksi W12 beam-column for $P_u = 900\,k$ and the first-order moments $M_{ntx} = 300$ ft-k and $M_{nty} = 0$. This pinned-ended member is to be located in a braced frame and is to be bent in single curvature with no intermediate transverse loads.

11-21. Select a 12-ft A572 Grade 50 W12 beam-column; for $P_u = 260$ k and the first-order moments $M_{ntx} = 180$ ft-k and $M_{nty} = 80$ ft-k. The member is to be pinned at its ends and is subjected to transverse loads. Assume the frame is braced $C_b = 1.0$ and $C_m = 0.85$. (*Ans.* W12 × 79)

11-22. Repeat Prob. 11-21 if M_{ntx} is 260 ft-k and $F_y = 50$ ksi.

11-23. Select a 14-ft W14 (50 ksi steel) with $K = 1.2$, $P_u = 400$ k and a first-order moment $M_{ltx} = 140$ ft-k due to wind. There are no other calculated moments and the maximum drift index is 0.0020 due to an unfactored wind load of 75 k. Assume the total factored gravity load above this story is 5000 k. The frame is symmetrical and unbraced. (*Ans.* W14 × 90)

11-24. Repeat Prob. 11-23 using a W12 section with $F_y = 50$ ksi.

11-25. Select a 15-ft W14 (50 ksi steel) with $K_x = K_y = 1.2$, $P_u = 300$ k; the first-order moments $M_{ltx} = 150$ ft-k, $M_{lty} = 100$ ft-k; and $M_{ntx} = M_{nty} = 0$. The frame is symmetrical and unbraced and ΣP_u for all of the loads above this story is 2500 k, while $\Sigma P_{e2x} = 50{,}000$ k for bending about the x axis and $\Sigma P_{e2y} = 30{,}000$ k for bending about the y axis. Column is bent in single curvature with no intermediate loads. (*Ans.* W14 × 90)

11-26. Repeat Prob. 11-25 if a 14-ft W12 with $F_y = 65$ ksi is to be used.

11-27 to 11-34. *Repeat the problems listed using the enclosed diskette.*

11-27. Prob. 11-1 (*Ans.* 0.778 < 1.0 OK)

11-28. Prob. 11-10

11-29. Prob. 11-12 (*Ans.* 0.803 < 1.0 OK)

11-30. Prob. 11-13

11-31. Prob. 11-17 (*Ans.* 0.575 < 1.0 OK)

11-32. Prob. 11-19

11-33. Prob. 11-20 (*Ans.* W12 × 152)

CHAPTER 12

Bolted Connections

12.1 INTRODUCTION

For many years, riveting was the accepted method used for connecting the members of steel structures. For the last few decades, however, bolting and welding have been the methods used for making structural steel connections, and riveting is almost never used. This chapter and the next are almost entirely devoted to bolted connections, although some brief remarks are presented at the end of Chapter 13 concerning rivets.

Bolting of steel structures is a very rapid field erection process that requires less skilled labor than does riveting or welding. This gives bolting a distinct economic advantage over the other connection methods in the United States where labor costs are so very high. Even though the purchase price of a high-strength bolt is several times that of a rivet, the overall cost of bolted construction is cheaper than that for riveted construction because of reduced labor and equipment costs and the smaller number of bolts required to resist the same loads.

12.2 TYPES OF BOLTS

There are several types of bolts that can be used for connecting steel members. They are described in the paragraphs that follow.

Unfinished bolts are also called *ordinary* or *common bolts*. They are classified by the ASTM as A307 bolts and are made from carbon steels with stress-strain characteristics very similar to those of A36 steel. They are available in diameters from 5/8 to 1 1/2 in in 1/8-in increments.

A307 bolts generally have square heads and nuts to reduce costs, but hexagonal heads are sometimes used because they have a little more attractive appearance, are easier to turn and easier to hold with the wrenches, and require less turning space. As they have relatively large tolerances in shank and thread dimensions, their design strengths are appreciably smaller than those for high-strength bolts. They are primarily used in light structures subjected to static loads and for secondary members (such as purlins, girts, bracing, platforms, small trusses, and so forth).

349

Designers often are guilty of specifying high-strength bolts for connections when common bolts would be satisfactory. *The strength and advantages of common bolts have usually been greatly underrated in the past.* The analysis and design of A307 bolted connections are handled exactly as are riveted connections in every way except that the design stresses are slightly different.

High-strength bolts are made from medium carbon heat-treated steel and from alloy steel and have tensile strengths two or more times those of ordinary bolts. There are two basic types, the A325 bolts (made from a heat-treated medium carbon steel) and the higher strength A490 bolts (also heat-treated but made from an alloy steel). High-strength bolts are used for all types of structures, from small buildings to "skyscrapers" and monumental bridges. These bolts were developed to overcome the weaknesses of rivets—primarily insufficient tension in their shanks after cooling. The resulting rivet tensions may not be large enough to hold them in place during the application of severe impactive and vibrating loads. The result is that they may become loose and vibrate and may eventually have to be replaced. High-strength bolts may be tightened until they have very high tensile stresses so that the connected parts are clamped tightly together between the bolt and nut heads, permitting loads to be transferred primarily by friction.

Sometimes high-strength bolts are made from A449 steel in sizes larger than the 1 1/2-in maximum diameter A325 and A490 bolts. These larger bolts may also be used as high-strength anchor bolts and for threaded rods of many different diameters.

12.3   HISTORY OF HIGH-STRENGTH BOLTS

The joints obtained using high-strength bolts are superior to riveted joints in performance and economy and they are the leading field method of fastening structural steel members. C. Batho and E. H. Bateman first claimed in 1934 that high-strength bolts could satisfactorily be used for the assembly of steel structures,[1] but it was not until 1947 that the Research Council on Riveted and Bolted Structural Joints of the Engineering Foundation was established. This group issued their first specifications in 1951, and high-strength bolts were adopted by building and bridge engineers for both static and dynamic loadings with amazing speed. They not only quickly became the leading method of making field connections, they also were found to have many applications for shop connections. The construction of the Mackinac Bridge in Michigan involved the use of more than one million high-strength bolts.

Connections that were formerly made with ordinary bolts and nuts were not too satisfactory when they were subjected to vibratory loads because the nuts frequently became loose. For many years this problem was dealt with by using some type of lock-nut, but the modern high-strength bolts furnish a far superior solution.

---

[1]C. Batho and E. H. Bateman, "Investigations on Bolts and Bolted Joints," H. M. Stationery Office (London, 1934).

High-strength bolt. (Courtesy of
Bethlehem Steel Corporation.)

12.4   ADVANTAGES OF HIGH-STRENGTH BOLTS

Among the many advantages of high-strength bolts, partly explaining their great success, are the following:

1. Smaller crews are involved compared with riveting. Two two-person bolting crews can easily turn out over twice as many bolts in a day as the number of rivets driven by the standard four-person riveting crew. The result is quicker steel erection.
2. Compared with rivets, fewer bolts are needed to provide the same strength.
3. Good bolted joints can be made by people with a great deal less training and experience than is necessary to produce welded and riveted connections of equal quality. The proper installation of high-strength bolts can be learned in a matter of hours.
4. No erection bolts are required that may have to be later removed (depending on specifications) as in welded joints.
5. Though quite noisy, bolting is not nearly as loud as riveting.
6. Cheaper equipment is used to make bolted connections.
7. No fire hazard is present, nor danger from the tossing of hot rivets.
8. Tests on riveted joints and fully tensioned bolted joints under identical conditions show that bolted joints have a higher fatigue strength. Their fatigue strength is also equal to or greater than that obtained with equivalent welded joints.
9. Where structures are to be later altered or disassembled, changes in connections are quite simple because of the ease of bolt removal.

12.5   SNUG-TIGHT, PRETENSIONED, AND SLIP-CRITICAL BOLTS

High-strength bolted joints are said to be *snug-tight, pretensioned* or *slip-critical*. These terms are defined in the paragraphs to follow. The type of joint used is dependent on the type of load that the fasteners will have to carry.

a. **Snug-tight bolts**
   For most connections, bolts are tightened only to what is called a *snug-tight* condition. Snug-tight is the situation existing when all the plies of a connection are in

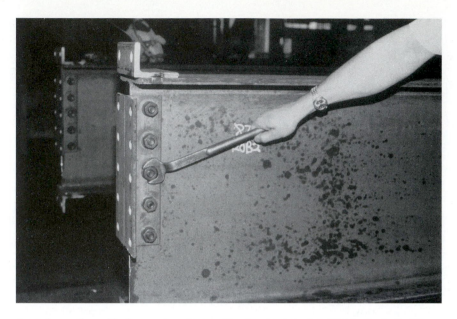

A spud wrench used by ironworkers for erecting structural steel and tightening bolts. One end of the wrench is sized for the hexagonal ends of bolts and nuts, while the other end is tapered to a rounded point and is used to align bolt holes between different connection pieces. (Courtesy of the AISC.)

firm contact with each other. It usually means the tightness obtained by the full effort of a person using a spud wrench, or the tightness obtained after a few impacts of the pneumatic wrench. Obviously there is some variation in the degree of tightness obtained under these conditions. Snug-tight bolts must be clearly identified on both design and erection drawings.

Snug-tight bolts are permitted for all situations in which pretensioned or slip-critical bolts are not required. In this type of connection, the plies of steel being connected must be brought together so they are solidly seated against each other, but they do not have to be in continuous contact. The installed bolts do not have to be inspected to determine their actual pretensioned stresses.

b. **Pretensioned joints**

The bolts in a pretensioned joint are brought to very high tensile stresses equal to approximately 70% of their minimum tensile stresses. To properly tighten them it is necessary to first bring them to a snug-tight condition. Then they are further tightened by one of the four methods described in Section 12-6 of this chapter.

Pretensioned joints are required for connections subjected to appreciable load reversals where nearly full or full design loads are applied to them in one direction, after which the nearly full or full design loads are applied in the other direction. Such a condition is typical of seismic loadings, but not for wind loads. Pretensioned bolts are also required for joints subject to fatigue loads where there is no reversal of the load direction. In addition, they are used where the bolts are subjected to tensile fatigue stresses. A490 bolts should be pretensioned if they are subjected to tension or if they are subjected to combined shear and tension, whether or not there is fatigue. Pretensioned bolts are permitted when slip resistance is of no concern.

c. **Slip-critical joints**

The installation of slip-critical bolts is identical with that for pretensioned joints. The only difference between the two is in the treatment of the faying surfaces. Their inspection is the same except that the inspector needs to check the faying surface for slip-critical joints.

Slip-critical joints are required only for situations involving shear or combined shear and tension. They are not required for situations involving only tension. In addition, they are to be used for joints with oversized holes and for joints with slotted holes where the load is applied approximately normal (within $80°$ to $100°$) to the long direction of the slot.

When loads are applied to snug-tight bolts there may be a little slippage, as the holes are a little larger in diameter than the shanks of the bolts. As a result the parts of the connection may bear against the bolts. You can see that if we have a fatigue situation with constantly changing loads, this is not a desirable situation.

For fatigue situations, and for connections subject to direct tension, it is desirable to use connections that will not slip. These are referred to as *slip-critical connections*. To achieve this situation the bolts must be tightened until they reach a fully tensioned condition in which they are subject to extremely large tensile forces.

Fully tensioning bolts is an expensive process, as is the inspection necessary to see that they are fully tensioned. Thus, they should be used only where absolutely necessary, as where the working loads cause large numbers of stress changes resulting in fatigue problems. Section J1.11 of the LRFD Specification gives a detailed list of connections that must be made with fully tensioned bolts. Included in this list are connections for supports of running machinery or for live loads producing impact and stress reversal; column splices in all tier structures 200 ft or more in height; connections of all beams and girders to columns and other beams or girders on which the bracing of the columns is dependent for structures over 125 ft in height; and so on.

Snug-tight bolts have several advantages over fully tensioned ones. One worker can properly tighten bolts to a snug-tight condition with an ordinary spud wrench or with only a few impacts of an impact wrench. The installation is quick and only a visual inspection of the work is needed. (Such is not the case for fully tensioned bolts.) Furthermore, snug-tight bolts may be installed with electric wrenches, thus eliminating the need for air compression on the site. As a result the use of snug-tight bolts saves time and money and is safer than the procedure needed for fully tensioned bolts. *Therefore, for most situations snug-tight bolts should be used.*

Tables 12.1 and 12.1M provide the minimum fastener tensions required for slip-resistant connections and for connections subject to direct tension. These are, respectively, reproductions of Tables J3.1 and J3.1M of the LRFD Specification.

The quality-control provisions specified in the manufacture of the A325 and A490 bolts are more stringent than those for the A449 bolts. As a result, despite the method of tightening, the A449 bolts may not be used in slip-resistant connections.

Although many engineers felt that there would be some slippage as compared with rivets (because of the fact that the hot driven rivets more nearly filled the holes),

TABLE 12.1 Minimum Bolt Pretension, kips*

Bolt Size, In.	A325 Bolts	A490 Bolts
$1/2$	12	15
$5/8$	19	24
$3/4$	28	35
$7/8$	39	49
1	51	64
$1^1/_8$	56	80
$1^1/_4$	71	102
$1^3/_8$	85	121
$1^1/_2$	103	148

*Equal to 0.70 of minimum tensile strength of bolts, rounded off to nearest kip, as specified in ASTM specifications for A325 and A490M bolts with UNC threads.

TABLE 12.1M Minimum Bolt Pretension, kN*

Bolt Size, mm	A325M Bolts	A490M Bolts
M16	91	114
M20	142	179
M22	176	221
M24	205	257
M27	267	334
M30	326	408
M36	475	595

*Equal to 0.70 of minimum tensile strength of bolts, rounded off to nearest kN, as specified in ASTM specifications for A325M and A490M bolts with UNC threads.

it was found that there is less slippage in fully tensioned high-strength bolted joints than in riveted joints under similar conditions.

It is interesting to note that the nuts used for fully tensioned high-strength bolts need no special provisions for locking. Once these bolts are installed and sufficiently tightened to produce the tension required, there is almost no tendency for the nuts to come loose. There are, however, a few situations where they will work loose under heavy vibrating loads. What do we do then? Some steel erectors have replaced the offending bolts with longer ones with two fully tightened nuts. Others have welded the nuts onto the bolts. Apparently the results have been somewhat successful.

12.6 METHODS FOR FULLY TENSIONING HIGH-STRENGTH BOLTS

We have already commented on the tightening required for snug-tight bolts. For fully tensioned bolts several methods of tightening are available. These methods, including the turn-of-the-nut method, the calibrated wrench method, and the use of alternate design bolts and direct tension indicators, are permitted without preference by the LRFD Specification.

12.6.1   Turn-of-the-Nut Method

The bolts are brought to a snug-tight condition and then, with an impact wrench, they are given from one-third to one full turn, depending on their length and the slope of the surfaces under their heads and nuts. Section 8 of Part 16 of the LRFD Manual presents the amounts of turn to be applied. (The amount of turn given to a particular bolt can easily be controlled by marking the snug-tight position with paint or crayon.)

12.6.2   Calibrated Wrench Method

With this method the bolts are tightened with an impact wrench that is adjusted to stall at that certain torque which is theoretically necessary to tension a bolt of that diameter and ASTM classification to the desired tension. It is necessary that wrenches be calibrated daily and that hardened washers be used. Particular care needs to be given to protecting the bolts from dirt and moisture at the job site. The reader should refer to the "Specification for Structural Joints Using ASTM A325 or A490 Bolts" in Part 16 of the Manual for additional tightening requirements.

12.6.3   Direct Tension Indicator

The direct tension indicator (which was originally a British device) consists of a hardened washer which has protrusions on one face in the form of small arches. The arches will be flattened as a bolt is tightened. The amount of gap at any one time is a measure of the bolt tension. For fully tensioned bolts the gaps should measure about 0.015 in or less.

12.6.4   Alternate Design Fasteners

In addition to the preceding methods there are some alternate design fasteners that can be tensioned quite satisfactorily. Bolts with splined ends that extend beyond the threaded portion of the bolts, called "twist-off bolts," are one example. Special wrench chucks are used to tighten the nuts until the splined ends shear off. This method of tightening bolts in quite satisfactory and will result in smaller labor costs.

A maximum bolt tension is not specified in any of the preceding tightening methods. This means that the bolt can be tightened to the highest load that will not break it and the bolt still will do the job. Should the bolt break, another one is put in with no damage done. It might be noted that the nuts are stronger than the bolt and the bolt will break before the nut strips. (The bolt specification mentioned above requires that a tension measuring device be available at the job site to ensure that specified tensions are achieved.)

For fatigue situations where members are subjected to constantly fluctuating loads, the slip-resistant connection is very desirable. If, however, the force to be carried is less than the frictional resistance and thus no forces are applied to the bolts, how could we ever have a fatigue failure of the bolts? The slip-resistant connection is a serviceability limit state in that it is based on working loads. For a slip-resistant joint the working loads are not permitted to exceed the permissible frictional resistance.

Other situations where slip-resistant connections are desirable include joints where bolts are used in oversized holes, joints where bolts are used in slotted holes where the loads are applied parallel or nearly so to the slots, joints that are subjected to

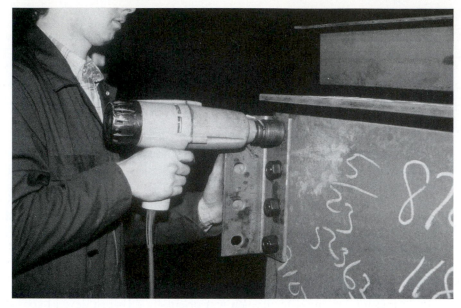

An impact wrench used to tighten bolts to either a snug-tight or a fully tensioned condition. It may be electric, such as this one, or it may be pneumatic. (Courtesy of the AISC.)

Torquing the nut for a high-strength bolt with an air-driven impact wrench. (Courtesy of Bethlehem Steel Corporation.)

significant force reversals, and joints where bolts and welds resist shear together on a common *faying surface*. (The faying surface is the contact or shear area between the members.)

A "twist-off bolt," also called a "load-indicator bolt" or a "tension control bolt." Notice the spline at the end of the bolt shank. (Courtesy of AISC.)

12.7   SLIP-RESISTANT CONNECTIONS AND BEARING-TYPE CONNECTIONS

When high-strength bolts are fully tensioned they clamp the parts being connected tightly together. The result is a considerable resistance to slipping on the faying surface. This resistance is equal to the clamping force times the coefficient of friction.

If the shearing load is less than the permissible frictional resistance the connection is referred to as slip-resistant. If the load exceeds the frictional resistance, the members will slip on each other and will tend to shear off the bolts; at the same time, the connected parts will push or bear against the bolts as shown in Fig. 12.1.

The surfaces of joints, including the area adjacent to washers, need to be free of loose scale, dirt, burrs, and other defects that might prevent the parts from solid seating. It is necessary for the surface of the parts to be connected to have slopes of not more than 1 to 20 with respect to the bolt heads and nuts, unless beveled washers are used. For slip-resistant joints the faying surfaces must also be free from oil, paint, and lacquer. (Actually, paint may be used if it is proved to be satisfactory by test.)

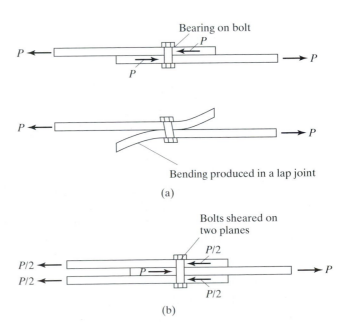

FIGURE 12.1

(a) Lap joint. (b) Butt joint.

If the faying surfaces are galvanized, the slip factor will be reduced to almost half of its value for clean mill scale surfaces. The slip factor, however, may be significantly improved if the surfaces are subjected to hand wire brushing or to "brush off" grit blasting. However, such treatments do not seem to provide increased slip resistance for sustained loadings where there seems to be a creeplike behavior.[2]

The 1996 AASHTO Specifications permit hot-dip galvanization if the coated surfaces are scored with wire brushes or sandblasted after galvanization and before steel erection.

The ASTM Specification permits the galvanization of the A325 bolts themselves, but not the A490 bolts. There is a danger of embrittlement of this higher-strength steel during galvanization due to the possibility that hydrogen may be introduced into the steel in the pickling operation of the galvanization process.

If special faying surface conditions (such as blast-cleaned surfaces or blast-cleaned surfaces with special slip-resistant coatings applied) are used to increase the slip resistance, the designer may increase the values used here to the ones given by the Research Council on Structural Joints in Part 16 of the LRFD Manual.

12.8   MIXED JOINTS

Bolts may on occasion be used in combination with welds and on other occasions with rivets (as where they are added to old riveted connections to enable them to carry increased loads). The LRFD Specification contains some specific rules for these situations.

12.8.1   Bolts in Combination with Welds

For new work neither A307 common bolts nor high-strength bolts designed for bearing or snug-tight connections may be considered to share the load with welds. (Before the connection's ultimate strength is reached the bolts will slip, with the result that the welds will carry a larger proportion of the load—the actual proportion being difficult to determine.) For such circumstances welds will have to be proportioned to resist the entire loads.

If high-strength bolts are designed for slip-critical conditions, they may be allowed to share the load with welds. For such situations the LRFD Commentary J1.9 states that it is necessary to fully tighten the bolts before the welds are made. If the weld is made first, the heat from the weld may very well distort the connection so that we will not get the slip-critical resistance desired from the bolts. If the bolts are placed and fully tightened before the welds are made, the heat of the welding will not change the mechanical properties of the bolts. Thus, for such a case the slip-critical bolts and the welds may be assumed to share the load.

If we are making alterations for an existing structure that is connected with bearing or snug-tight bolts or with rivets, we can assume that any slipping that is going to

---

[2]J. W. Fisher and J. H. A. Struik, *Guide to Design Criteria for Bolted and Riveted Joints* (New York: John Wiley & Sons, 1974) pp. 205–206.

occur has already taken place. Thus, if we are using welds in the alteration, we will design those welds neglecting the forces that would be produced by the existing dead load.

12.8.2 High-Strength Bolts in Combination with Rivets

High-strength bolts may be considered to share loads with rivets for new work or for alterations of existing connections that were designed as slip-critical. (The ductility of the rivets allows the capacity of both sets of fasteners to act together.)

12.9   SIZES OF BOLT HOLES

In addition to the standard size bolt holes (STD) which are 1/16 in larger in diameter than the bolts, there are three types of enlarged holes: oversized, short-slotted, and long-slotted. Oversized holes will on occasion be very useful in speeding up steel erection. In addition, they give some latitude for adjustments in plumbing frames during erection. The use of nonstandard holes requires the approval of the designer and is subject to the requirements of Section J3 of the LRFD Specification. Table 12.2 provides the nominal dimensions in inches for the various kinds of enlarged holes permitted by the LRFD, while Table 12.2M provides the same information in millimeters. (These tables are, respectively, Tables J3.3 and J3.3M of the LRFD Specification.)

The situations in which we may use the various types of enlarged holes are now described.

Oversized holes (OVS) may be used in all plies of connections as long as the applied load does not exceed the permissible slip resistance. They may not be used in bearing-type connections. It is necessary for hardened washers to be used over oversized holes that are located in outer plies. The use of oversized holes permits the use of larger construction tolerances.

Short-slotted holes (SSL) may be used regardless of the direction of the applied load for slip-critical or bearing-type connections. Should the load be applied in a direction approximately normal (between 80 and 100 degrees) to the slot, these holes may

TABLE 12.2   Nominal Hole Dimensions, Inches

		Hole Dimensions		
Bolt Diameter	Standard (Dia.)	Oversize (Dia.)	Short-slot (Width × Length)	Long-slot (Width × Length)
$\frac{1}{2}$	$\frac{9}{16}$	$\frac{5}{8}$	$\frac{9}{16} \times \frac{11}{16}$	$\frac{9}{16} \times 1\frac{1}{4}$
$\frac{5}{8}$	$\frac{11}{16}$	$\frac{13}{16}$	$\frac{11}{16} \times \frac{7}{8}$	$\frac{11}{16} \times 1\frac{9}{16}$
$\frac{3}{4}$	$\frac{13}{16}$	$\frac{15}{16}$	$\frac{13}{16} \times 1$	$\frac{13}{16} \times 1\frac{7}{8}$
$\frac{7}{8}$	$\frac{15}{16}$	$1\frac{1}{16}$	$\frac{15}{16} \times 1\frac{1}{8}$	$\frac{15}{16} \times 2\frac{3}{16}$
1	$1\frac{1}{16}$	$1\frac{1}{4}$	$1\frac{1}{16} \times 1\frac{5}{16}$	$1\frac{1}{16} \times 2\frac{1}{2}$
$\geq 1\frac{1}{8}$	$d + \frac{1}{16}$	$d + \frac{5}{16}$	$(d + \frac{1}{16}) \times (d + \frac{3}{8})$	$(d + \frac{1}{16}) \times (2.5 \times d)$

TABLE 12.2M   Nominal Hole Dimensions, mm

Bolt Diameter	Standard (Dia.)	Oversize (Dia.)	Short-slot (Width × Length)	Long-slot (Width × Length)
		Hole Dimensions		
M16	18	20	18 × 22	18 × 40
M20	22	24	22 × 26	22 × 50
M22	24	28	24 × 30	24 × 55
M24	27 [a]	30	27 × 32	27 × 60
M27	30	35	30 × 37	30 × 67
M30	33	38	33 × 40	33 × 75
≥ M36	$d + 3$	$d + 8$	$(d + 3) \times (d + 10)$	$(d + 3) \times 2.5d$

[a] Clearance provided allows the use of a 1 in. bolt if desirable.

Source: American Institute of Steel Construction, *Manual of Steel Construction Load & Resistance Factor Design*, 3rd ed. (Chicago: AISC, 2001), Table J3.3 and J3.3M, p. 16-62. Reprinted with the permission of the AISC.

be used in any or all plies of connections for bearing-type connections. It is necessary to use washers (hardened if high-strength bolts are being used) over short-slotted holes in an outer ply. The use of short-slotted holes provides for some mill and fabrication tolerances, but does not result in the necessity for slip-critical procedures.

Long-slotted holes (LSL) may be used in *only one* of the connected parts of slip-critical or bearing-type connections at any one faying surface. For slip-critical joints these holes may be used in any direction, but for bearing-type connections the loads must be normal (between 80 and 100 degrees) to the axes of the slotted holes. If long-slotted holes are used in an outer ply they will need to be covered with plate washers or a continuous bar. For high-strength bolted connections the washers or bar do not have to be hardened, but they must be made of structural-grade material and may not be less than $\frac{5}{16}$ in. thick. Long-slotted holes are usually used when connections are being made to existing structures where the exact positions of the members being connected are not known.

Generally, washers are used to prevent scoring or galling of members when bolts are tightened. Most persons think that they also serve the purpose of spreading out the clamping forces more uniformly to the connected members. Tests have shown, however, that standard size washers don't affect the pressure very much except when oversized or short-slotted holes are used. Section J3.1 of the specification provides detailed information concerning the use of washers.

12.10   LOAD TRANSFER AND TYPES OF JOINTS

The following paragraphs present a few of the elementary types of bolted joints subjected to axial forces (that is, the loads are assumed to pass through the centers of gravities of the groups of connectors). For each of these joint types, some comments are made about the methods of load transfer. Eccentrically loaded connections are discussed in Chapter 13.

For this initial discussion, the reader is referred to part (a) of Figure 12.1. It is assumed that the plates shown are connected with a group of snug-tight bolts. In other words, the bolts are not tightened sufficiently so as to significantly squeeze the plates together. If there is assumed to be little friction between the plates they will slip a little due to the applied loads shown. As a result the loads in the plates will tend to shear the connectors off on the plane between the plates and press or bear against the sides of the bolts as shown in the figure. These connectors are said to be in *single shear and bearing* (also called *unenclosed bearing*). They must have sufficient strength to satisfactorily resist these forces, and the members forming the joint must be sufficiently strong to prevent the connectors from tearing through.

When rivets were used instead of the snug-tight bolts, the situation was somewhat different because hot-driven rivets would cool and shrink and squeeze or clamp the connected pieces together with sizable forces that greatly increased the friction between the pieces. As a result a large portion of the loads being transferred between the members was transferred by friction. The clamping forces produced in riveted joints, however, were generally not considered to be dependable, and as a result specifications normally consider such connections to be snug-tight with no frictional resistance. The same assumption is made for A307 common bolts as they are not tightened to large dependable tensions.

Fully tensioned high-strength bolts are in a different class altogether. Using the tightening methods previously described, a very dependable tension is obtained in the bolts resulting in large clamping forces and large dependable amounts of frictional resistance to slipping. Unless the loads to be transferred are larger than the frictional resistance, the entire forces are resisted by friction and the bolts are not really placed in shear or bearing. If the load exceeds the frictional resistance there will be slippage, with the result that the bolts will be placed in shear and bearing.

12.10.1 The Lap Joint

The joint shown in part (a) of Figure 12.1 is referred to as a *lap joint*. This type of joint has a disadvantage in that the center of gravity of the force in one member is not in line with the center of gravity of the force in the other member. A couple is present that causes an undesirable bending in the connection, as shown in the figure. For this reason the lap joint, which is desirably used only for minor connections, should be designed with at least two fasteners in each line parallel to the length of the member to minimize the possibility of a bending failure.

12.10.2 The Butt Joint

A *butt joint* is formed when three members are connected as shown in Figure 12.1(b). If the slip resistance between the members is negligible the members will slip a little and tend to shear off the bolts simultaneously on the two planes of contact between the members. Again, the members are bearing against the bolts and the bolts are said to be in *double shear and bearing* (also called *enclosed bearing*). The butt joint is more desirable than the lap joint for two main reasons:

1. The members are arranged so that the total shearing force, P, is split into two parts, causing the force on each plane to be only about one-half of what it would

be on a single plane if a lap joint were used. From a shear standpoint, therefore, the load-carrying ability of a group of bolts in double shear is theoretically twice as great as the same number of bolts in single shear.

2. A more symmetrical loading condition is provided. (In fact, the butt joint does provide a symmetrical situation if the outside members are the same thickness and resist the same forces. The result is a reduction or elimination of the bending described for a lap joint.)

12.10.3 Double-Plane Connections

The double-plane connection is one in which the bolts are subjected to single shear and bearing, but in which bending moment is prevented. This type of connection, which is shown for a hanger in Fig. 12.2(a), subjects the bolts to single shear.

12.10.4 Miscellaneous

Bolted connections generally consist of lap or butt joints or some combination of them, but there are other cases. For instance, there are occasionally joints in which more than three members are being connected and the bolts are in multiple shear, as shown in Fig. 12.2(b). In this figure you can see how the loads are tending to shear this bolt on four separate planes (quadruple shear). Although the bolts in this connection are being sheared on more than two planes, the usual practice is to consider no more than double shear for strength calculations. It seems rather unlikely that shear failures can occur simultaneously on three or more planes. Several other types of bolted connections are discussed in this chapter and the next. These include bolts in tension, bolts in shear and tension, etc.

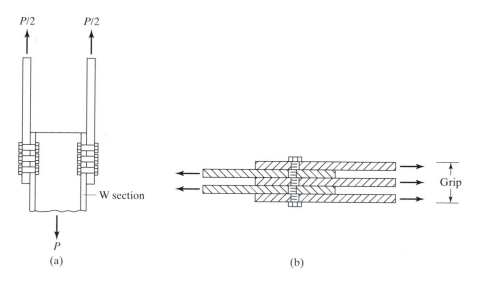

FIGURE 12.2

(a) Hanger connection. (b) Bolts in multiple shear.

12.11   FAILURE OF BOLTED JOINTS

Figure 12.3 shows several ways in which failure of bolted joints can occur. To design bolted joints satisfactorily it is necessary to understand these possibilities. These are described as follows.

1. The possibility of failure in a lap joint by shearing of the bolt on the plane between the members (single shear) is shown in part (a).
2. The possibility of a tension failure of one of the plates through a bolt hole is shown in part (b).
3. A possible failure of the bolts and/or plates by bearing between the two is given in (c).
4. The possibility of failure due to the shearing out of part of the member as shown in part (d).
5. The possibility of a shear failure of the bolts along two planes (double shear) as shown in part (e).

12.12   SPACING AND EDGE DISTANCES OF BOLTS

Before minimum spacings and edge distances can be discussed it is necessary for a few terms to be explained. The following definitions are given for a group of bolts in a connection and are shown in Fig. 12.4.

> *Pitch* is the center-to-center distance of bolts in a direction parallel to the axis of the member.
>
> *Gage* is the center-to-center distance of bolt lines perpendicular to the axis of the member.

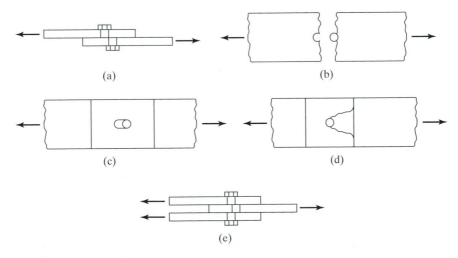

FIGURE 12.3

(a) Failure by single shearing of bolt. (b) Tension failure of plate. (c) Crushing failure of plate. (d) Shear failure of plate behind bolt. (e) Double shear failure of a butt joint.

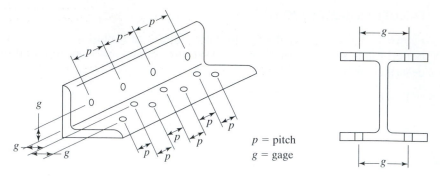

FIGURE 12.4

The *edge distance* is the distance from the center of a bolt to the adjacent edge of a member.

The *distance between bolts* is the shortest distance between fasteners on the same or different gage lines.

12.12.1 Minimum Spacings

Bolts should be placed a sufficient distance apart to permit efficient installation and to prevent bearing failures of the members between fasteners. The LRFD Specification (J3.3) provides a minimum center-to-center distance for standard, oversized, or slotted fastener holes. For standard oversized or slotted holes the minimum center-to-center distance should not be less than 2 2/3 diameters (with three diameters being preferred). Test results have clearly shown that bearing strengths are directly proportional to the 3d center-to-center value up to a maximum of 3d. No additional bearing strength is obtained when edge distances greater than 3d are used.

12.12.2 Minimum Edge Distances

Bolts should not be placed too near the edges of a member for two major reasons. First, the punching of holes too close to the edges may cause the steel opposite the hole to bulge out or even crack. The second reason applies to the ends of members where there is danger of the fastener tearing through the metal. The usual practice is to place the fastener a minimum distance from the edge of the plates equal to about 1.5 to 2.0 times the fastener diameter so the metal there will have a shearing strength at least equal to that of the fasteners. For more exact information it is necessary to refer to the specification being used. The LRFD Specification (J3.4) states that the distance from the center of a standard hole to the edge of a connected part may not be less than the applicable value given in Table 12.3 or 12.3M.

A reduced minimum edge distance is permitted (it's 1.25 in or 32 mm) by the LRFD Specification for end connections bolted to beam webs and designed for beam shear reactions only. This information is shown as a footnote in these tables.

The minimum edge distance from the center of an oversized hole or a slotted hole to the edge of a connected part must equal the minimum distance required for a standard hole plus an increment C_2, values of which are provided in Table 12.4 or

TABLE 12.3	Minimum Edge Distance,[a] Inches, From Center of Standard Hole [b] to Edge of Connected Part	
Nominal Rivet or Bolt Diameter (in.)	At Sheared Edges	At Rolled Edges of Plates Shapes or Bars, or Gas Cut Edges [c]
$1/2$	$7/8$	$3/4$
$5/8$	$1^1/8$	$7/8$
$3/4$	$1^1/4$	1
$7/8$	$1^1/2$[d]	$1^1/8$
1	$1^3/4$[d]	$1^1/4$
$1^1/8$	2	$1^1/2$
$1^1/4$	$2^1/4$	$1^6/8$
Over $1^1/4$	$1^3/4 \times$ Diameter	$1^1/4 \times$ Diameter

[a] Lesser edge distances are permitted to be used provided Equations from LRFD Section J3.10, as appropriate, are satisfied.

[b] For oversized or slotted holes, see Table 12.4 herein

[c] All edge distances in this column are permitted to be reduced $1/4$-in. when the hole is at a point where stress does not exceed 25 percent of the maximum design strength in the element.

[d] These are permitted to be $1^1/4$-in. at the ends of beam connection angles and shear end plates.

TABLE 12.3M	Minimum Edge Distance, [a] mm, From Center of Standard Hole [b] to Edge of Connected Part	
Nominal Rivet or Bolt Diameter (mm)	At Sheared Edges	At Rolled Edges of Plat Shapes or Bars, or Gas Cut Edges [c]
16	28	22
20	34	26
22	38 [d]	28
24	42 [d]	30
27	48	34
30	52	38
36	64	46
Over 36	1.75d	1.25d

[a] Lesser edge distances are permitted to be used provided Equations from LRFD Section J3.10, as appropriate, are satisfied.

[b] For oversized or slotted holes, see Table 12.4 M herein.

[c] All edge distances in this column are permitted to be reduced 3 mm when the hole is at a point where stress does not exceed 25 percent of the maximum design strength in the element.

[d] These are permitted to be 32 mm at the ends of beam connection angles and shear end plates.

Source: American Institute of Steel Construction, *Manual of Steel Construction Load & Resistance Factor Design*, 3rd ed. (Chicago: AISC, 2001), Table J3.4 and J3.4M, p. 16–63. Reprinted with the permission of the AISC.

TABLE 12.4 Values of Edge Distance Increment C_2, Inches

Nominal Diameter of Fastener (in.)	Oversized Holes	Slotted Holes		Long Axis Parallel to Edge
		Long Axis Perpendicular to Edge		
		Short Slots	Long Slots [a]	
$\leq {}^7/_8$	$^1/_{16}$	$^1/_8$		
1	$^1/_8$	$^1/_8$	$^3/_4 d$	0
$\geq 1^1/_6$	$^1/_8$	$^3/_{16}$		

[a] When length of slot is less than maximum allowable (see Table 12.5 herein), C_2 may be reduced by one-half the difference between the maximum and actual slot lengths.

TABLE 12.4M Values of Edge Distance Increment C_2, mm

Nominal Diameter of Fastener (mm)	Oversized Holes	Slotted Holes		Long Axis Parallel to Edge
		Long Axis Perpendicular to Edge		
		Short Slots	Long Slots [a]	
≤ 22	2	3		
24	3	3	0.75d	0
≥ 27	3	5		

[a] When length of slot is less than maximum allowable (see Table J3.5), C_2 are permitted to be reduced by one-half the difference between the maximum and actual slot lengths.

Source: American Institute of Steel Construction, *Manual of Steel Construction Load & Resistance Factor Design*, 3rd ed. (Chicago: AISC, 2001), Table J3.6 and J3.6M, p. 16-65. Reprinted with permission of AISC.

Table 12.4M. (These tables are respectively Tables J3.6 and J3.6M of the LRFD Specification.) As will be seen in the pages to follow, the computed bearing strengths of connections will have to be reduced if these requirements are not met.

12.12.3 Maximum Spacing and Edge Distances

Structural steel specifications provide maximum edge distances for bolted connections. The purpose of such requirements is to reduce the chances of moisture getting between the parts. When fasteners are too far from the edges of parts being connected, the edges may sometimes separate, thus permitting the entrance of moisture. When this happens and there is a failure of the paint, corrosion will develop and accumulate, causing increased separations between the parts. The LRFD maximum permissible edge distance (J3.5) is 12 times the thickness of the connected part, but not more than 6 in. (150 mm).

The maximum edge distances and spacings of bolts used for weathering steel are smaller than they are for regular painted steel subject to corrosion, or for regular unpainted steel not subject to corrosion. One of the requirements for using weathering steel is that it must not be allowed to be constantly in contact with water. As a result the LRFD Specification tries to insure that the parts of a built-up weathering steel member are connected tightly together at frequent intervals to prevent forming of pockets that might catch and hold water. The LRFD Specification (J3.5) states that the maximum spacing of bolts center-to-center for painted members or for unpainted members not subject to corrosion is 24 times the thickness of the thinner plate, not to exceed 12 in. (305 mm). For unpainted members consisting of weathering steel subject to atmospheric corrosion, the maximum is 14 times the thickness of the thinner plate, not to exceed 7 in (180 mm).

Holes cannot be punched very close to the web-flange junction of a beam or the junction of the legs of an angle. They can be drilled, but this rather expensive practice should not be followed unless there is an unusual situation. Even if the holes are drilled in these locations, there may be considerable difficulty in placing and tightening the bolts in the limited available space.

12.13 BEARING-TYPE CONNECTIONS—LOADS PASSING THROUGH CENTER OF GRAVITY OF CONNECTIONS

12.13.1 Shearing Strength

In bearing-type connections, it is assumed that the loads to be transferred are larger than the frictional resistance caused by tightening the bolts, with the result that the members slip a little on each other, putting the bolts in shear and bearing. The design strength of a bolt in single shear equals ϕ times the nominal shearing strength of the bolt in ksi times its cross-sectional area. The LRFD ϕ values for shear are 0.75 for high-strength bolts, rivets, and A307 common bolts.

The nominal shearing strengths of bolts and rivets are given in Table 12.5 (Table J3.2 in the LRFD Specification). For A325 bolts the values are 48 ksi if threads are not excluded from shear planes and 60 ksi if threads are excluded. (The values are 60 ksi and 75 ksi, respectively, for A490 bolts.) Should a bolt be in double shear, its shearing strength is considered to be twice its single shear value.

The student may very well wonder what is done in design practice concerning threads excluded or not excluded from the shear planes. If normal bolt and member sizes are used, the threads will almost always be excluded from the shear plane. It is true, however, that some extremely conservative individuals always assume the threads are not excluded from the shear plane.

Sometimes the designer needs to use high-strength bolts with diameters larger than those available with A325 and A490 bolts. One example is the use of very large bolts for fastening machine bases. For such situations, LRFD Specification A3.3 permits the use of the quenched and tempered A449 bolts. The LRFD Specification A3.3 says that such bolts can be used only when diameters larger than $1\frac{1}{2}$in. are required, and then only for bearing type connections.

12.13.2 Bearing Strength

The bearing strength of a bolted connection is not, as you might expect, determined from the strength of the bolts themselves; rather, it is based upon the strength of the parts being connected and the arrangement of the bolts. In detail, its computed strength is dependent upon the spacing of the bolts and their edge distances, the specified tensile strength F_u of the connected parts, and their thicknesses.

TABLE 12.5   Design Strength of Fasteners

Description of Fasteners	Tensile Strength		Shear Strength in Bearing-type Connections		
	Resistance Factor ϕ	Nominal Strength, ksi (MPa)	Resistance Factor ϕ	Nominal Strength, ksi (MPa)	
A307 bolts		45 (310) [a]		24 (165) [b,e]	
A325 or A325M bolts, when threads are not excluded from shear planes		90 (620) [d]		48 (330) [e]	
A325 or A325M bolts, when threads are excluded from shear planes		90 (620) [d]		60 (414) [e]	
A490 or A490M bolts, when threads are not excluded from shear planes		113 (780) [d]		60 (414) [e]	
A490 or A490M bolts, when threads are excluded from shear planes	0.75	113 (780) [d]	0.75	75 (520) [e]	
Threaded parts meeting the requirements of Section A3, wrn threads are not excluded from shear planes		$0.75F_u$[a,c]		$0.40F_u$	
Threaded parts meeting the requirements of Section A3, when threads are excluded from shear planes		$0.75F_u$[a,c]		$0.50F_u$[a,c]	
A502, Gr. 1, hot-driven Rivets		45 (310) [a]		25 (172) [e]	
A502, Gr. 2 & 3, hot-driven Rivets		60 (414) [a]		33 (228) [e]	

[a] Static loading only.

[b] Threads permitted in shear planes.

[c] The nominal tensile strength of the threaded portion of an upset rod, based upon the cross-sectional area at its major thread diameter, A_D shall be larger than the nominal body area of the rod before upsetting times F_y.

[d] For A325 or A325M and A490 or A490M bolts subject to tensile fatigue loading, see LRFD Appendix K3.

[e] When bearing-type connections used to splice tension members have a fastener pattern whose length, measured parallel to the line of force, exceeds 50 in. (1 270 mm), tabulated values shall be reduced by 20 percent.

Source: American Institute of Steel Construction, Manual of Steel Construction *Load & Resistance Factor Design*, 2nd ed. (Chicago: AISC, 2001), Table J3.2, p. 16–61. Reprinted with permission of AISC.

The bearing design strength of a bolt equals ϕ (which is 0.75) times the nominal bearing strength of the connected part (R_n). Expressions for R_n are provided in Section J3.10 of the LRFD Specification. The various expressions listed there include bolt diameters (d), the thicknesses of members bearing against the bolts (t), and the clear distances (L_c) between the edges of holes and the edges of the adjacent holes or edges of the material in the direction of the force. Finally, F_u is the specified minimum tensile strength of the connected material.

To be consistent throughout this text the authors have conservatively assumed that the diameter of a bolt hole equals the bolt diameter plus $\frac{1}{8}$ in. This dimension is used in computing the value of L_c for substituting into LRFD Equation J3-2a which is given below. The bearing values given in Table 7.13 of the LRFD Manual were computed with hole diameters equal to bolt diameters plus 1/16 in. As a result their bearing values are slightly higher than the ones computed in this text. It is perfectly permissible to use the table values for your calculations.

The expressions to follow are used to compute the nominal bearing strengths of bolts used in connections that have standard, oversized, or short-slotted holes, regardless of the direction of loading. They also are applicable to connections with long-slotted holes if the slots are perpendicular to the direction of the bearing forces.

a. If deformation around bolt holes is a design consideration (that is, if we want deformations to be ≤ 0.25 in), then

$$R_n = 1.2\, L_c t F_u \leq 2.4\, dt F_u \qquad \text{(LRFD Equation J3-2a)}$$

For the problems considered in this text we will normally assume that deformations around the bolt holes are important. Thus, unless specifically stated otherwise, Equation J3-2a will be used for bearing calculations.

If deformation around bolt holes is not a design consideration (that is, if deformations > 0.25 in are acceptable), then

$$R_n = 1.5\, L_c t F_u \leq 3.0\, dt F_u \qquad \text{(LRFD Equation J3-2b)}$$

b. For bolts used in connections with long-slotted holes, the slots being perpendicular to the forces

$$R_n = 1.0\, L_c t F_u \leq 2.0\, dt F_u \qquad \text{(LRFD Equation J3-2c)}$$

As described in Section 12-9 of this chapter, remember that oversized holes cannot be used in bearing connections, but that short- and long-slotted holes can be used in bearing type connections—if the loads are perpendicular to the long directions of the slots.

Tests of bolted joints have shown that neither the bolts nor the metal in contact with the bolts actually fail in bearing. However, these tests also have shown that the efficiency of the connected parts in tension and compression is affected by the magnitude of the bearing stress. Therefore, the nominal bearing strengths given by the LRFD Specification are values above which they feel the strength of the connected parts is impaired. In other words, these apparently very high design bearing stresses are not really bearing stresses at all, but, rather, indexes of the efficiencies of the connected parts. If bearing stresses larger than the values given are permitted, the holes seem to elongate more than about $\frac{1}{4}$ in. and impair the strength of the connections.

From the preceding we can see that the bearing strengths given are not specified to protect fasteners from bearing failures because they do not need such protection. Thus the same bearing values will be used for a particular joint regardless of the grades of bolts used and regardless of the presence or absence of bolt threads in the bearing area.

12.13.3 Minimum Connection Strength

The LRFD Specification (Section J1.7) states that except for lacing, sag rods, and girts, connections must be provided with design strengths sufficient to support factored loads of at least 10 kips.

Example 12-1 illustrates the calculations involved in determining the strength of a bearing-type connection. Using a similar procedure the number of bolts required for a certain loading condition is calculated in Example 12-2. In each case the bearing thickness to be used equals the smaller total thickness on one side or the other, since the steel grade for all plates is the same and since the edge distances are identical for all plates. For instance, in Fig. 12.6 it equals the smaller of $2 \times 1/2$ in on the left or 3/4 in on the right.

The values given for the strength of bolts in this chapter, whether bearing-type or slip-critical, can be obtained from Tables 7.10 to 7.16 in Part 7 of the Manual.

In the connection tables of the LRFD Manual and in various bolt literature we constantly see abbreviations used when referring to various types of bolts. For instance, we may see A325-SC, A325-N, A325-X, A490-SC, and so on. These are used to represent the following

A325-SC—slip-critical or fully tensioned A325 bolts

A325-N—snug-tight or bearing A325 bolts with threads *included* in the shear planes

A325-X—snug-tight or bearing A325 bolts with threads *excluded* from the shear planes

---

Example 12-1

Determine the design strength $\phi_c P_n$ of the bearing-type connection shown in Fig. 12.5. The steel is A572 Grade 50, the bolts are 7/8-in. A325, the holes are standard sizes and the threads are excluded from the shear plane.

Solution

Design strength of plates:

$$A_g = \left(\tfrac{1}{2}\right)(12) = 6.0 \text{ in}^2$$
$$A_n = 6.00 - (2)(1.0)\left(\tfrac{1}{2}\right) = 5.0 \text{ in}^2 = A_e$$

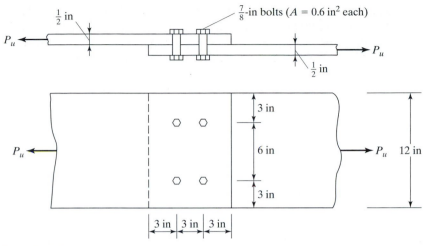

FIGURE 12.5

$$\phi_c P_n = \phi_t F_y A_g = (0.9)(50)(6.0) = 270 \text{ k}$$
$$\phi_t P_n = \phi_t F_u A_e = (0.75)(65)(5.0) = 243.7 \text{ k}$$

Bearing strength of bolts:

Noting n = number of bolts:

$$L_c = \text{lesser of } 3 - \tfrac{1}{2} = 2\tfrac{1}{2} \text{ in, or } 3 - 1 = 2.0 \text{ in}$$
$$\phi_v R_n = \phi\, 1.2 \ L_c t F_u n = (0.75)(1.2)(2.0)\left(\tfrac{1}{2}\right)(65)(4) = 234 \text{ k}$$

or

$$\phi_v R_n = \phi\, 2.4 \ dt F_u n = (0.75)(2.4)\left(\tfrac{7}{8}\right)\left(\tfrac{1}{2}\right)(65)(4) = 204.7 \text{ k}$$

Shearing strength of bolts:

Bolts in single shear:

$$P_u = \phi(0.6)(60)(4) = (0.75)(0.6)(60)(4) = 108 \text{ k} \leftarrow$$
Design $\phi P_n = 108 \text{ k}$

Example 12-2

How many 3/4-in A325 bolts in standard-size holes with threads excluded from the shear plane are required for the bearing-type connection shown in Fig. 12.6? Use $F_u = 65$ ksi and assume edge distances 2 m and the distance center-to-center of holes to be 3 in.

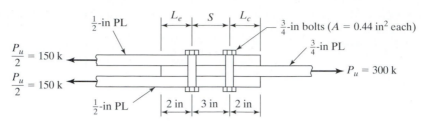

FIGURE 12.6

Solution. Bolts in double shear and bearing on 3/4 in.

Bearing strength per bolt:

$$L_c = \text{lesser of } 2 - \frac{0.875}{2} \text{ or } 3 - 0.875$$

$$\phi R_n = (0.75)(1.2)\left(2 - \tfrac{0.875}{2}\right)\left(\tfrac{3}{4}\right)(65) = 68.6 \text{ k}$$

or

$$\phi R_n = (0.75)(2.4)\left(\tfrac{3}{4}\right)\left(\tfrac{3}{4}\right)(65) = 65.8 \text{ k}$$

Shearing strength per bolt:

$$\phi R_n = (0.75)(2 \times 0.44)(60) = 39.6 \text{ k} \leftarrow$$

$$No \ of \ bolts \ required = \frac{300}{39.6} = 7^+$$

Use eight 3/4-in A325 bolts

Where cover plates are bolted to the flanges of W sections, the bolts must carry the longitudinal shear on the plane between the plates and the flanges. With reference to the cover-plated beam of Fig. 12.7, the unit longitudinal shearing stress to be resisted between a cover plate and the W flange can be determined with the expression $f_v = VQ/Ib$. The total shear force across the flange for a 1-in length of the beam equals $(b)(1.0)(VQ/Ib) = VQ/I$.

The LRFD Specification (E4) provides a maximum permissible spacing for bolts used in the outside plates of built-up members. It equals the thinner outside plate thickness times $0.75\sqrt{E/F_y}$ and may not be larger than 12 in.

The spacing of pairs of bolts in Fig. 12.7 can be determined by dividing the design strength of two bolts by the shear force to be taken per inch at a particular section. The theoretical spacings will vary as the external shear varies along the span. Example 12-3 illustrates the calculations involved in determining bolt spacing for a cover-plated beam.

The reader should note that LRFD Specification B.10 states that the total cross-sectional area of the cover plates of a bolted girder may not be greater than 70 percent of the total flange area.

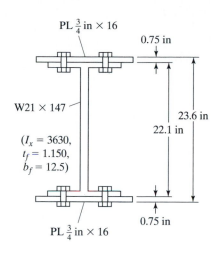

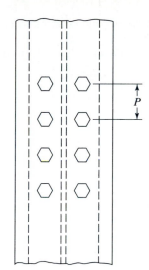

FIGURE 12.7

Example 12-3

At a certain section in the cover-plated beam of Fig. 12-7 the external factored shear V_u is 275 k. Determine the spacing required for 7/8-in A325 bolts used in a bearing-type connection. Assume the bolt threads are excluded from the shear plane and that the edge distance $=3.5$ in, $F_y = 50$ ksi. $F_u = 65$ ksi.

Solution

Checking LRFD Specification B.10:

A of 1 cover plate $= \left(\frac{3}{4}\right)(16) = 12.00$ in^2

A of 1 flange $= 12.00 + (12.5)(1.15) = 26.38$ in^2

Plate area $\div$ flange area $= \dfrac{12.00}{26.38} < 0.70$ \hfill (OK)

Computing shearing force to be taken:

$$I_g = 3630 + (2)\left(\tfrac{3}{4} \times 16\right)\left(\frac{22.1}{2} + \frac{0.75}{2}\right)^2 = 6760 \text{ in}^4$$

Factored shear per in. $= \dfrac{V_u Q}{I}$

$$= \frac{(275)\left(\tfrac{3}{4} \times 16 \times 11.425\right)}{6760} = 5.578 \text{ k/in}$$

Bolts in single shear and bearing on 0.75 in:

Design shear strength of 2 bolts $= \phi R_n$

$= N_b \phi A_b F_{nv} = (2)(0.75)(0.6)(60) = 54 \text{ k} \leftarrow$

Design bearing strength of 2 bolts $= \phi R_n$

$= N_b \phi 2.4 \ dt F_u = (2)(0.75)(2.4)\left(\frac{7}{8}\right)\left(\frac{3}{4}\right)(65) = 153.6 \text{ k}$

Spacing of pairs of bolts:

$$\rho = \frac{54}{5.578} = 9.68 \text{ in. } \textit{Say 9 in on center}$$

Next, we check the other bearing strength expression now that we have a bolt spacing and can calculate L_c:

$$L_c = \text{lesser of } 9 - 1 \quad \text{or} \quad 3.5 - \frac{1.0}{2}$$

$$\phi R_n = (0.75)(2)(1.2)\left(3.5 - \frac{1.0}{2}\right)(0.75)(65) = 263.2 \text{ k} > 153.6 \text{ k} \qquad \text{(OK)}$$

Maximum spacing for bolts in outside:

$$\text{plates} = 0.75\sqrt{\frac{29 \times 10^3}{50}} = 18.06 \text{ in nor } 12 \text{ in} > 9 \text{ in} \qquad \text{(OK)}$$

Place pairs of bolts 9 in O.C.

The assumption has been made that the loads applied to a bearing type connection are equally divided between the bolts if edge distances and spacings are satisfactory. For this distribution to be correct, the plates must be perfectly rigid and the bolts perfectly elastic, but, actually, the plates being connected are elastic, too, and have deformations that decidedly affect the bolt stresses. The effect of these deformations is to cause a very complex distribution of load in the elastic range.

Should the plates be assumed to be completely rigid and nondeforming, all bolts would be deformed equally and have equal stresses. This situation is shown in part (a) of Fig. 12.8. Actually, the loads resisted by the bolts of a group are probably never equal (in the elastic range) when there are more than two bolts in a line. Should the plates be deformable, the plate stresses and thus the deformations will decrease from the ends of the connection to the middle, as shown in part (b) of Fig. 12.8. The result is that the highest stressed elements of the top plate will be over the lowest stressed elements of the lower plate, and vice versa. The slip will be greatest at the end bolts and smallest at the middle bolts. The bolts at the ends will then have stresses much greater than those in the inside bolts.

The greater the spacing of bolts in a connection the greater will be the variation in bolt stresses due to plate deformation; therefore, the use of compact joints is very

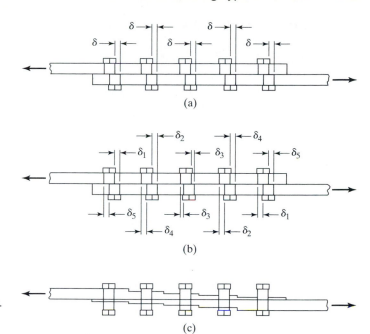

FIGURE 12.8

(a) Assuming nondeforming plates.
(b) Assuming deformable plates.
(c) Stepped joint (impractical).

Three high-strength bolted structures in Constitution Plaza Complex, Hartford, Conn., using approximately 195,000 bolts. (Courtesy of Bethlehem Steel Corporation.)

desirable as they will tend to reduce the variation in bolt stresses. It might be interesting to consider a theoretical (although not practical) method of roughly equalizing bolt stresses. The theory would involve the reduction of the thickness of the plate toward its end in proportion to the reduced stresses by stepping. This procedure, which is shown in Fig. 12.8(c), would tend to equalize the deformations of the plate and thus the bolt stresses. A similar procedure would be to scarf the overlapping plates.

The calculation of the theoretically correct elastic stresses in a bolted group based on plate deformations is a tedious problem and is rarely handled in the design office. On the other hand, the analysis of a bolted joint based on the plastic theory is a very simple problem. In this theory, the end bolts are assumed to be stressed to their yield point. Should the total load on the connection be increased, the end bolts will deform without resisting additional load, the next bolts in the line will have their stresses increased until they too are at the yield point, and so forth. Plastic analysis seems to justify to a certain extent the assumption of rigid plates and equal bolt stresses that is usually made in design practice. This assumption is used in the example problems of this chapter.

When there are only a few bolts in a line, the plastic theory of equal stresses seems to be borne out very well, but when there are a large number of bolts in a line the situation changes. Tests have clearly shown that the end bolts will fail before the full redistribution takes place.[3]

The LRFD Specification requires that the design shearing strength of a bearing-type connection used to splice a tension member be reduced if it has a fastener pattern longer than 50 in. parallel to the line of force. As a footnote to their Table J3.2 (it's Table 12.5 herein) it is stated that tabulated values for such situations must be reduced by 20 percent.

For load-carrying bolted joints, it is common for specifications to require a minimum of two or three fasteners. The feeling is that a single connector may fail to live up to its specified strength because of improper installation, material weakness, etc., but if several fasteners are used the effects of one bad fastener in the group will be overcome.

12.14 SLIP-CRITICAL CONNECTIONS—LOADS PASSING THROUGH CENTER OF GRAVITY OF CONNECTIONS

Slip critical connections may be designed using factored loads as described in Section J3.8a of the LRFD Specification or by using service loads as described in Appendix J3.8b of the same specification. Though the number of bolts obtained by the two methods will be approximately the same, there can be a little variation with different ratios of live load to dead load. We present the factored load approach here.

If bolts are tightened to their required tensions for slip-critical connections (see Tables 12.1 and 12.1M), there is very little chance of their bearing against the plates that they are connecting. In fact, tests show that there is very little chance of slip occuring unless there is a calculated shear of at least 50 percent of the total bolt tension. As

---

[3] *Trans.* ASCE 105 (1940), p. 1193.

we have said all along, this means that slip-critical bolts are not stressed in shear; however, the LRFD Specification J3.8a provides "design shear strengths" (they are really design friction values on the faying surfaces) so the designer can handle the connections in just about the same manner he or she uses for bearing-type connections.

Although there is little or no bearing on the bolts used in slip critical connections, the LRFD in its Section J3.10 states that bearing strength is to be checked for both bearing type and slip critical connections.

Using factored loads, the design slip resistance of a high strength bolt is determined by

$$\phi r_{str} = \phi 1.13\,\mu T_b N_s \qquad \text{(LRFD Equation J3-1)}$$

where

T_b = minimum fastener tension as given in Table 12.1 or 12.1M

N_s = number of slip planes

μ = mean slip coefficient

The specification provides three slip coefficients with the different values depending on the surface conditions between the connected parts. These values were determined by testing and are given in Table 12.6. They were taken from LRFD Specification J3.8a.

In applying the slip resistance equation, it is necessary to use different values of the resistance factor ϕ, depending on the type of holes used:

ϕ = 1.0 for standard holes

ϕ = 0.85 for oversized and short slotted holes

ϕ = 0.70 for long slotted holes transverse to the direction of the loads

ϕ = 0.60 for long slotted holes parallel to the direction of the loads

It is permissible to introduce finger shims up to 1/4 in thick into slip-critical connections with standard holes without the necessity of reducing the bolt design strength values to those specified for slotted holes (LRFD J3.8a).

The preceding discussion concerning slip-critical joints does not present the whole story because during erection the joints may be assembled with bolts, and as the

TABLE 12.6   Slip Coefficients

Surface	μ
Class A (unpainted clean mill scale or surfaces with Class A coating on blast-cleaned steel)	0.33
Class B (unpainted blast-cleaned surfaces or surfaces with Class B coating on blast-cleaned steel)	0.50
Class C (hot-dip galvanized and roughened surfaces)	0.35

members are erected their weights will often push the bolts against the side of the holes before they are tightened and put them in some bearing and shear.

Example 12-4 presents the design of a slip-critical connection for a lap joint. First, the number of bolts required for the load limit state of no slippage is determined. Then the number of bolts required for the factored load limit state is computed assuming the slip resistance is overcome and the bolts subjected to bearing.

Example 12-4

We want to design a slip-critical connection for the plates shown in Fig. 12.9 to resist the axial service loads $P_D = 30$ k and $P_L = 50$ k using 1-in. A325 high-strength bolts with standard-size holes. Edge distance $= 1\ 3/4$ in and c. to c. spacing of bolts $= 3$ in steel $F_y = 50$ ksi, $F_u = 65$ ksi. Surfaces are unpainted mill scale.

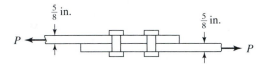

FIGURE 12.9

Solution. $P_u = (1.2)(30) + (1.6)(50) = 116$ k

Slip critical design:

$$\phi = 1, 0 \text{ for standard holes}$$

$$\mu = 0, 33 \text{ for class A surfaces from Table 12-6 in this chapter}$$

$$T_b = 51 \text{ from Table 12.1}$$

$$\phi r_{str} = \phi 1.13\,\mu T_b N_s = (1.0)(1.13)(0.33)(51)(1)$$

$$= 19.02 \text{ k}$$

$$\text{No of bolts reqd} = \frac{116}{19.02} = 6.10 \text{ say 7 or 8}$$

Checking bearing:

Bolts bearing on 5/8 in

Bearing strength of 1 bolt where $L_c = $ lesser of $3 - 1.125$ or $1.75 - 1.125/2$

$$\phi R_n = \phi 1.2\,L_c t F_u n = (0.75)(1.2)\left(1\tfrac{3}{4} - \tfrac{1.125}{2}\right)\left(\tfrac{5}{8}\right)(65)(1) = 43.5 \text{ k} \leftarrow$$

or

$$\phi R_n = \phi 2.4\,dt F_u n = (0.75)(2.4)(1.0)\left(\tfrac{5}{8}\right)(65)(1) = 73.12 \text{ k}$$

$$\text{No of bolts reqd} = \frac{116}{43.5} = 2.67$$

Use seven or eight 1-in A325-SC bolts.

Bridge over Allegheny River at Kittaning, PA. (Courtesy American Bridge Company.)

In Example 12-5, bearing-type bolts are used to connect a beam to a pair of gusset plates. The tensile strengths of the W section, the plates, and the bolts are determined. Finally, the block shear strength of the W section is calculated.

Example 12-5

The connection shown in Fig. 12.10 is made with 7/8-in A325 bearing-type bolts in standard-size holes with the threads excluded from the shear planes. The beam and gusset plates consist of 50 ksi steel ($F_u = 65$ ksi). Check the following items: (a) the tensile strengths of the W section and the gusset plates, (b) the strength of the bolts in single shear and bearing, (c) the block shear strength of the W section crosshatched areas shown in part (b) of the figure, and (d) the block shear strength for the gusset plates.

Solution

a. *Tensile design strength of W section*:

$$\phi_t P_n = \phi_t F_y A_g = (0.9)(50)(11.2) = 504 \text{ k} > P_u \, 330 \text{ k} \qquad \text{(OK)}$$
$$A_n = 11.2 - (4)(1)(0.515) = 9.14 \text{ in}^2$$

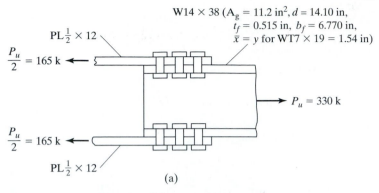

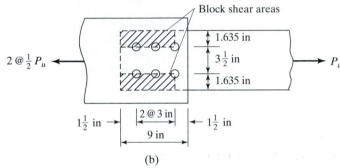

FIGURE 12.10

$$U \text{ from Chapter 3} = \left(1 - \frac{\overline{x}}{L}\right) \le 0.9 = \left(1 - \frac{1.54}{6}\right) = 0.74$$

$$A_e = (0.74)(9.14) = 6.76 \text{ in}^2$$

$$\phi_t P_n = \phi_t F_u A_e = (0.75)(65)(6.76) = 329.6 \text{ k} \approx P_u = 330 \text{ k} \quad \text{(OK)}$$

$$P_u = \phi_t F_y A_g = (0.9)(50)\left(\tfrac{1}{2} \times 12\right)(2) = 540 \text{ k} > 330 \text{ k}$$

$$A_n \text{ of 2 plates} = [\left(\tfrac{1}{2}\right)(12) - (2)(1)\left(\tfrac{1}{2}\right)]2 = 10 \text{ in}^2 \leftarrow$$

$$0.85 \, A_g = (0.85)\left(\tfrac{1}{2}\right)(1.2)(2) = 10.2 \text{ in}^2$$

$$\phi_t F_u A_n = (0.75)(65)(10) = 487.5 \text{ k} > 330 \text{ k} \quad \text{(OK)}$$

b. *Bolts in ss and bearing on $\frac{1}{2}$ in:*

ss design strength of bolts $= (0.75)(0.6)(60)(12)$

$$= 324 \text{ k} < 330 \text{ k} \quad \text{(NG)}$$

Bearing design strength of bolts:

$$L_c = \text{lesser of } 3 - 1.0 \quad \text{or } 1.635 - \tfrac{1.0}{2}$$

$$\phi_t R_n = (0.75)(1.2)(1.635 - \tfrac{1.0}{2})\left(\tfrac{1}{2}\right)(65)(12) = 332 \text{ k} > 330 \text{ k} \quad \text{(NG)}$$

or

$$\phi_t R_m = (0.75)(2.4)\left(\tfrac{7}{8}\right)\left(\tfrac{1}{2}\right)(65)(12) = 614.2 \text{ k}$$

c. *Block shearing strength for W section*:

Total areas for top and bottom

A_{gv} = gross area subject to shear

$$= (4)(0.515)\left(7\tfrac{1}{2}\right) = 15.45 \text{ in}^2$$

A_{gt} = gross area subject to tension

$$= (4)(0.515)(1.635) = 3.37 \text{ in}^2$$

A_{nv} = net area subject to shear

$$= (4)(7.50 - 2.5 \times 1.0)(0.515) = 10.30 \text{ in}^2$$

A_{nt} = net area subject to tension

$$= (4)\left(1.635 - \tfrac{1}{2} \times 1\right)(0.515) = 2.34 \text{ in}^2$$

Checking to see which equation is applicable:

$$F_u A_{nt} = (65)(2.34) = 152.1 \text{ k}$$
$$< 0.6 F_u A_{nv} = (0.6)(65)(10.30) = 401.7 \text{ k}$$

$\therefore$ Must use LRFD Equation J4-3b

$$\phi R_n = \phi[0.6 F_u A_{nv} + F_y A_{gt}]$$
$$= 0.75[(0.6)(65)(10.30) + (50)(3.37)]$$
$$= 427.6 \text{ k} > 330 \text{ k} \qquad \text{(OK)}$$

d. *Block shear strength for gusset plates*:

Total areas for both plates

$$A_{gv} = 2[(2)(7.50)(0.50)] = 15.00 \text{ in}^2$$
$$A_{gt} = 2[(3.50)(0.50)] = 3.50 \text{ in}^2$$
$$A_{nv} = 2\{2[7.50 - 2\tfrac{1}{2}(1)](0.50)\} = 10.00 \text{ in}^2$$
$$A_{nt} = 2\{[3.50 - 1(1)](0.50)\} = 2.50 \text{ in}^2$$
$$F_u A_{nt} = (65)(2.50) = 162.5 \text{ k}$$
$$< 0.60 F_u A_{nv} = (0.60)(65)(10.00) = 390 \text{ k}$$

∴ Use LRFD Equation J4-36

$$\phi R_n = \phi[0.6\,F_u A_{nv} + F_y A_{gt}] = 0.75[390 + (50)\,(2.50)]$$
$$= 386\,\text{k} > \text{P} = 330\,\text{k} \tag{OK}$$

The calculations for riveted connections and for connections made with A307 common bolts are made almost exactly as are the ones for high-strength bolts in bearing-type connections. The only difference is that the shear strength values for these connections are much smaller. The LRFD Specification does not permit the design of slip-critical joints using rivets or common bolts. Connections made with rivets or with common bolts are discussed in Chapter 13.

12.15   COMPUTER EXAMPLE

In Example 12-6 INSTEP 32 is used to determine the strength of a bolted joint.

Example 12-6

Repeat Example 12-1 using INSTEP.

Input:

Joint parameters

Bolt type A325

Bolt size = 0.875 inches

Threads SS bearing, excluded

Joint style Shear w/wo tension

t1 = 0.5 in.

t2 = 0.5 in.

Output:

Bolted Joint Design Summary

Vu = 108.2 kips

Pu = 0.0 kips

(handwritten annotations)

Bearing-type   12-7 a) calc design tensile strength of middle Plate

b) calc total shear strength of bolts

c) check bearing strength of connection (J3-2, p369)

d) check block shear capacity

e) $P_u = ?$

a) Yield $= .9(F_y)(A_g) = .9(36)(12) = 388.8$ kip

Fracture $= .75(A_e)(F_u) = .9(8.625)(58) = \boxed{375.2 kip}$

b) $\phi R_n = \phi(m)(A_b)(F_v)n_b$ F_v from Table

$= .75(2)\left(\frac{\pi(1)^2}{4}\right)(48)(9) = \boxed{508.9 kip}$

c) $\phi R_n = \phi(1.2)(F_u)(L_c)(t)(n_b)$

F_u from table $= 58$ ksi

$L_c = (L_{c\,end}) = 2 - \left(\frac{1\frac{1}{8}}{2}\right) = 1.438$

$(L_{c\,int}) = 3 - (1\frac{1}{8}) = 1.875$

min $\phi R_n = .75(1.2)(58)(1.438)(1)(9) = \boxed{675.3 kip}$

d) Block Shear → $= \boxed{398.8 kip}$ front on p83

e) $P_u = \boxed{375.2 kip}$

PROBLEMS

For each of the problems listed, the following information is to be used unless other-wise indicated: (a) LRFD Specification; (b) standard-size holes; (c) members have clean mill-scale surfaces (class A); (d) $F_y = 36$ ksi and $F_u = 58$ ksi. Do not consider block shear.

12-1 to 12-5. *Determine the design tensile strength P_u for the member and the connection shown assuming a bearing-type connection.*

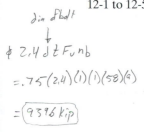

d in of bolt
$\downarrow$
$\phi\; 2.4\; d\; t\; F_u\, n_b$

$= .75(2.4)(1)(1)(58)(9)$

$= \boxed{9396\; kip}$

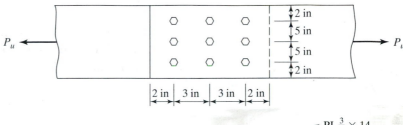

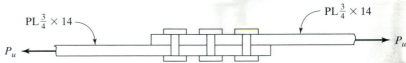

FIGURE P12-1 to 12-5

12-1. A325 $\frac{7}{8}$-in bolts, threads excluded from shear plane. (*Ans.* 243 k)

12-2. A325 1-in bolts, threads excluded from shear plane.

12-3. A325 $\frac{3}{4}$-in bolts, threads not excluded from shear plane. (*Ans.* 142.6 k)

12-4. 1-in A490 bolts, threads excluded from shear plane.

12-5. 7/8-in A490 bolts, threads excluded from shear plane. (*Ans.* 243 k)

12-6 to 12-10. *Determine the design tensile strength P_u for the member and the bearing-type con-nection shown.*

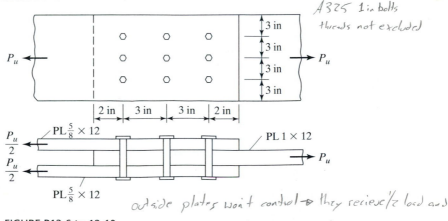

A325 1 in bolts
threads not excluded

FIGURE P12-6 to 12-10

outside plates won't control → they recieve $\frac{1}{2}$ load and
are > than $\frac{1}{2}$ thickness of middle plate

for part c)
$\phi R_n = \phi 1.2 (F_u)\, L_c (t)\, n_b \leq \phi 2.4 (d)\, t (F_u)\, n_b$

12-6.   A325 $\frac{7}{8}$ in bolts, threads excluded from shear planes.

12-7.   A325 1-in bolts, threads not excluded from shear planes. (*Ans.* 375.2 k)

Bearing-type Connection:
Problem 1) Problem 12-7 (page 383-384, McCormac and Nelson, 3rd Ed.)
(a) Calculate the design tensile strength of the middle plate. (Yielding of A_g and fracture of A_e, $U = 1$ for this connection). Note: The outer plates will not control because they receive half the factored load, P_u, but the thickness is more than half of the middle plate.
(b) Calculate the total shear strength of the bolts. This is a double shear connection.
(c) Check the bearing strength of the connection using LRFD equation J3-2a [text, page 369]
(d) Check block shear capacity.
(e) What is the final design tensile strength, P_u?

cluded from shear plane.

70 ksi, 1-in A490 bolts, threads excluded from shear

= 70 ksi, $\frac{3}{4}$-in A490 bolts, threads excluded from

12-11 to 12-13.   *How many bolts are required for the bearing-type connection shown?*

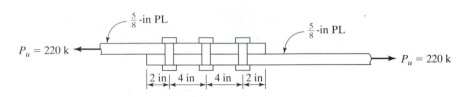

FIGURE P12-11 to 12-13

12-11.   A325 $\frac{3}{4}$-in bolts, threads excluded from shear plane. (*Ans.* 12 bolts)

12-12. $F_y = 50$ ksi, $F_u = 70$ ksi, $\frac{7}{8}$-in A325 bolts, threads excluded from shear plane.

12-13.   A325-1 in bolts, threads not excluded from shear plane. (*Ans.* 8 bolts)

12-14 to 12-16.   *How many bolts are required for the bearing type connection shown?*

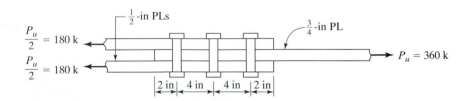

FIGURE P12-14 to 12-16

12-14.   A325 $\frac{7}{8}$-in bolts, threads excluded from shear planes.

12-15.   A325 $\frac{3}{4}$-in bolts, threads not excluded from shear planes. (*Ans.* 12 bolts)

12-16.   A325 = $1\frac{1}{8}$-in bolts, threads excluded from shear planes.

12-17.   The truss member shown in the accompanying illustration consists of two C12 × 25s (A36 steel) connected to a 1-in gusset plate. How many $\frac{7}{8}$-in A325 bolts (threads excluded from shear plane) are required to develop the full design tensile capacity of the member if it is used as a bearing-type connection? Assume $U = 0.85$. (*Ans.* 9 bolts)

b) Total shear strength of the bolts

$$\phi_{vb} \cdot R_n = \phi_{vb} \cdot m \cdot (A_b \cdot F_v) \cdot n_b$$

$\phi_{vb} := 0.75$

$m := 2$ two shear planes

$n_b := 9$ number of bolts

$F_v := 48 \, ksi$ shear strength of 1" dia. A325N (threads included) [Table 12.5 pg 368]

$d_{bolt} := 1 \, in$

$A_b := \dfrac{\pi \cdot d_{bolt}^2}{4}$ $A_b = 0.785 \, in^2$

$\phi R_n := \phi_{vb} \cdot m \cdot (A_b \cdot F_v) \cdot n_b$

$\boxed{\phi R_n = 508.9 \, kip}$

c) Bearing strength of the connection

shear failure of plate

$$\phi_{vpl} \cdot R_n = \phi_{vpl} \cdot 1.2 \cdot F_u \cdot L_c \cdot t \cdot n_b$$

$\phi_{vpl} := 0.75$

$F_u := 58 \, ksi$

$t := 1 \, in$

$n_b = 9$

$L_{c\_end} := 2 \cdot in - \dfrac{d_{hole}}{2}$ $L_{c\_end} = 1.438 \, in$ <= controls

$L_{c\_int} := 3 \cdot in - d_{hole}$ $L_{c\_int} = 1.875 \, in$

$L_c := L_{c\_end}$

$\phi R_n := \phi_{vpl} \cdot 1.2 \cdot F_u \cdot L_c \cdot t \cdot n_b$

$\boxed{\phi R_n = 675.3 \, kip}$

bearing failure of plate around bolt holes

$$\phi_{bpl} \cdot R_n = \phi_{bpl} \cdot 2.4 \cdot d_{bolt} \cdot t \cdot F_u \cdot n_b$$

$\phi_{bpl} := 0.75$

$d_{bolt} = 1 \, in$

$t = 1 \, in$

$F_u = 58 \, ksi$ A36 steel

$n_b = 9$

$\phi R_n := \phi_{bpl} \cdot 2.4 \cdot d_{bolt} \cdot t \cdot F_u \cdot n_b$

$\boxed{\phi R_n = 939.6 \, kip}$

d) Block shear

$A_{gv} := (3 \cdot in + 3 \cdot in + 2 \cdot in) \cdot t$ $A_{gv} = 8 \, in^2$ gross shear area

$A_{gt} := (3 \cdot in + 3 \cdot in + 3 \cdot in) \cdot t$ $A_{gt} = 9 \, in^2$ gross tension area

$A_{nv} := A_{gv} - 2.5 \cdot (d_{hole} \cdot t)$ $A_{nv} = 5.188 \, in^2$ net shear area

$A_{nt} := A_{gt} - 2.5 \cdot (d_{hole} \cdot t)$ $A_{nt} = 6.187 \, in^2$ net tension area

$F_y = 36 \, ksi$ A36 steel

$F_u = 58 \, ksi$

$F_u \cdot A_{nt} = 358.875 \, kip$

$0.6 \cdot F_u \cdot A_{nv} = 180.525 \, kip$

$\phi_{block} := 0.75$

$F_u \cdot A_{nt} > 0.6 \cdot F_u \cdot A_{nv}$

shear yield and tensile fracture, LRFD equation J4-3a

$\boxed{\phi_{block} \cdot (0.6 \cdot F_y \cdot A_{gv} + F_u \cdot A_{nt}) = 398.8 \, kip}$ <= controls block shear strength

all fracture mode

$\phi_{block} \cdot (0.6 \cdot F_u \cdot A_{nv} + F_u \cdot A_{nt}) = 404.5 \, kip$

e) final design tensile strength

Summary
Yield of gross section = 388.8 kip
Fracture of effective area = **375.2 kip**
Shear strength of bolts = 508.9 kip
Bearing strength of bolts
  (a) shear failure of plate = 675.3 kip
  (b) bearing failure around bolt holes = 939.6 kip
Block shear capacity = 398.8 kip

Final design tensile strength P_u = **375.2** kip

FIGURE P12-22

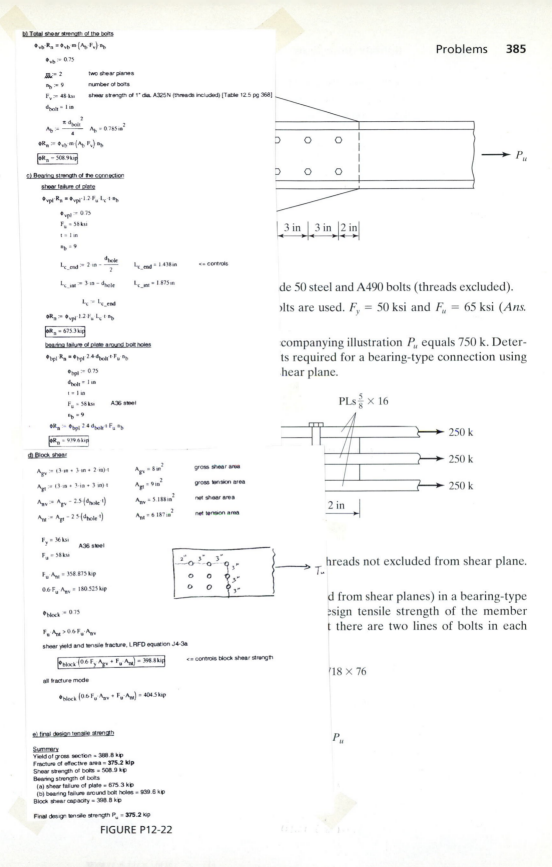

P_u

de 50 steel and A490 bolts (threads excluded).
olts are used. $F_y = 50$ ksi and $F_u = 65$ ksi (*Ans.*

companying illustration P_u equals 750 k. Deter-
ts required for a bearing-type connection using
hear plane.

$PLs\,\dfrac{5}{8} \times 16$

250 k

250 k

250 k

2 in

T_u

hreads not excluded from shear plane.

d from shear planes) in a bearing-type
esign tensile strength of the member
t there are two lines of bolts in each

W18 × 76

P_u

12-23. For the beam shown in the accompanying illustration, what is the required spacing of 3/4-in A325 bolts (threads excluded from shear plane) in a bearing-type connection at a section where the external shear V_u is 400 k? Assume L_c = 1.0 in (*Ans.* 4 ¼ in)

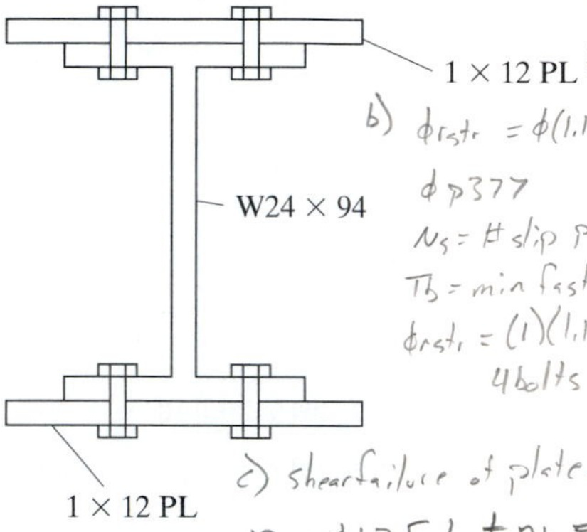

1 × 12 PL
W24 × 94
1 × 12 PL

FIGURE P12-23

a) yield .9(Ag)(Fy) = .9(3)(36) = 108 kip

Fract .75(Ae)(Fu) = .75(2.125)(58) = 92.4 kip

b) ϕrstr = ϕ(1.13)(μ)(Tb)(Ns)

ϕ p377

Ns = # slip planes

Tb = min fastener tension force p354

ϕrstr = (1)(1.13)(.33)(28)(1) = 10.441 kip/bolt

4 bolts → 10.441(4) = 41.8 kip

c) shear failure of plate ≡ bearing failure of plate around holes

ϕRn = ϕ1.2 Fu Lc t nb ≤ ϕ2.4 d t Fu nb

Lc = 1.5 − (3/4 + 1/8)/2 = 1.063

slip μ p377

ϕ p377

Tb p354

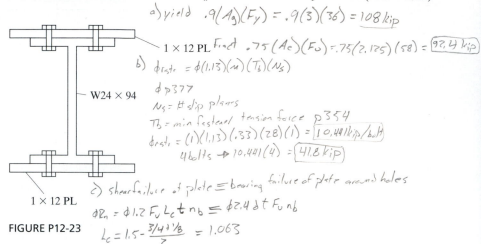

12-24. The cover-plated section shown in the accompanying illustration is used to support a uniform load w_D = 12 klf (includes beam weight effect and w_L = 15 klf for an 18-ft simple span). If $\frac{7}{8}$-in A325 bolts (threads excluded) are used in a bearing-type connection, work out a spacing diagram for the entire span.

Fu of plate .75(58)(1.063)(.5)(4) = 110.9 kip ≤ .75(2.4)(.75)(.5)(58)(4) = 156.6 kip

110.9 kip

d) Block shear → 123.7 kip

e) Tu = 41.8 kip

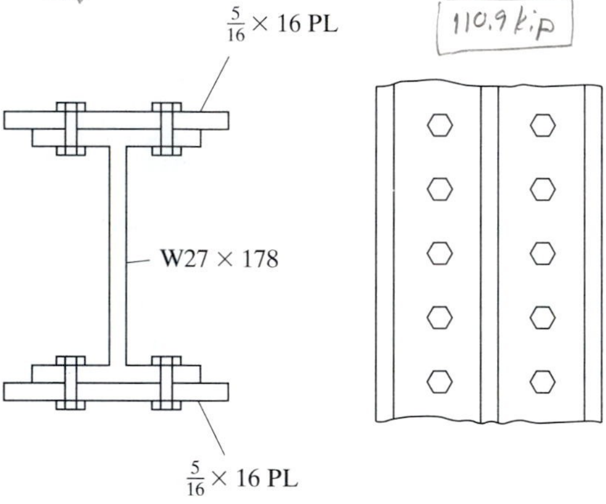

5/16 × 16 PL
W27 × 178
5/16 × 16 PL

FIGURE P12-24

12-25. For the section shown in the accompanying illustration determine the required spacing of 7/8-in A490 bolts (threads excluded) for a bearing-type connection if the member consists of A572 grade 60 steel (F_u = 75 ksi). V_D = 100 k and V_L = 140 k. Assume Class A surfaces and L_c = 1.0 in (*Ans.* 6 1/4 in)

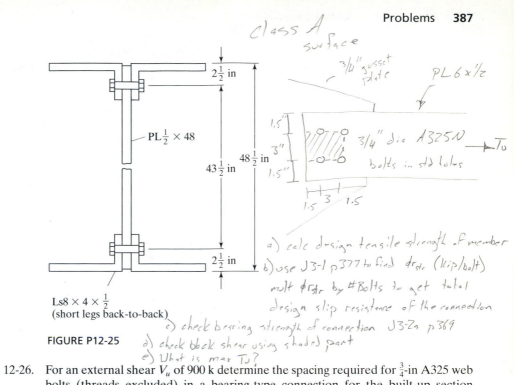

class A surface

3/4" gusset plate

PL 6 × 1/2

2½ in

PL ½ × 48

48½ in

43½ in

2½ in

1.5"
3"
1.5"

1.5 3 1.5

3/4" dia A325N bolts in std holes

$\longleftarrow T_u$

a) calc design tensile strength of member
b) use J3-1 p377 to find ϕr_{str} (kip/bolt) mult ϕr_{str} by #Bolts to get total design slip resistance of the connection
c) check bearing strength of connection J3-2a p369
d) check block shear using shaded part
e) What is max T_u?

Ls8 × 4 × ½
(short legs back-to-back)

FIGURE P12-25

12-26. For an external shear V_u of 900 k determine the spacing required for $\frac{3}{4}$-in A325 web bolts (threads excluded) in a bearing-type connection for the built-up section shown in the accompanying illustration. Assume $L_c = 1.5$ in.

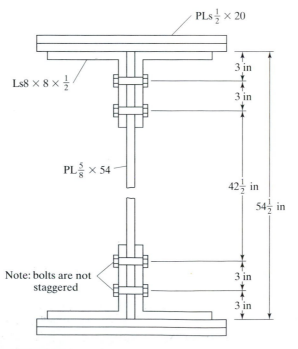

PLs ½ × 20

Ls8 × 8 × ½

PL ⅝ × 54

3 in

3 in

42½ in

54½ in

Note: bolts are not staggered

3 in

3 in

FIGURE P12-26

Slip-critical Connection:

Problem 2) Determine the maximum factored load, T_u, for the connection as shown in Figure 2 if no slip is permitted. Four ¾"-dia. A325N (threads included in the shear plane) bolts in standard holes are used. Both the tension member and the gusset plate are A36 steel. Assume Class A surfaces.

(a) Calculate the design tensile strength of the member. (Yielding of A_g and fracture of A_e, U = 1 for this connection).

(b) Use LRFD equation J3-1 [text, page 377] to determine ϕr_n, ($^{kips}/_{bolt}$). Multiply ϕr_n by number of bolts to obtain total design slip resistance of the connection.

(c) Check the bearing strength of the connection using LRFD equation J3-2a [text, page 369]

(d) Check block shear of the member using the shaded pattern as shown in Figure 2.

(e) What is the maximum factored load, T_u, allowed?

ngth P_u of the bearing-type connection shown if $\frac{7}{8}$ in d) are used. Use A36 steel. (*Ans.* 156.6 k)

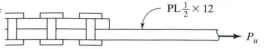

$PL\frac{1}{2} \times 12$

P_u

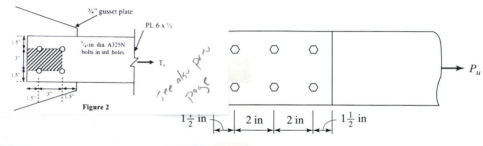

¾" gusset plate

PL 6 x ½

¾-in. dia. A325N bolts in std. holes

T_u

see also prob page

Figure 2

P_u

$1\frac{1}{2}$ in   2 in   2 in $1\frac{1}{2}$ in

Slip-critical Connection
Problem 2)

a) Tensile design strength of member (tension member controls because of thinner than gusset plate)

Yielding of gross area

$\phi_{t\_yield} := 0.90$

$F_y := 36 \cdot ksi$ [A36 steel]

$t := 0.5 \cdot in$ thickness of plate

$A_g := (6 \cdot in) \cdot t$ $A_g = 3 \, in^2$ gross area of PL 6 x ½

nominal strength

$P_n := F_y \cdot A_g$

$P_n = 108 \, kip$

design strength

$\boxed{\phi_{t\_yield} \cdot P_n = 97.2 \, kip}$

Fracture of the effective area

$\phi_{t\_fracture} := 0.75$

$F_u := 58 \cdot ksi$ [A36 steel]

$d_{bolt} := 0.75 \cdot in$ diameter of the bolts

$d_{hole} := d_{bolt} + \frac{1}{8} \cdot in$ for standard holes

$d_{hole} = 0.875 \, in$

Calculate effective area

$U := 1$ note: U = 1, no eccentricity

$A_e = A_n \cdot U$

$A_e := \left[A_g - 2 \cdot d_{hole} \cdot (t) \right] \cdot U$ 1/2" thick plate   t = 0.5 in

$A_e = 2.125 \, in^2$

nominal strength

$P_n := A_e \cdot F_u$

$P_n = 123.25 \, kip$

design strength

$\boxed{\phi_{t\_fracture} \cdot P_n = 92.4 \, kip}$

slip-critical connections, Class A surfaces and the service

90 k

= 100 k (*Ans.* 22 bolts)

= 160 k

= 80 k (*Ans.* 7 or 8 bolts)

= 100 k

= 360 k (*Ans.* 19 or 20 bolts)

slip-critical connections assuming Class A surfaces.

le strength P_u of the connection shown if eight $\frac{7}{8}$-in reads excluded from shear plane) are used in each n your calculations.

b) design slip resistance

$\phi r_{str} = \phi \cdot 1.13 \cdot \mu \cdot T_b \cdot N_s$

$\phi := 1.0$ for standard holes [text, page 377]

$\mu := 0.33$ slip coefficient for Class A surfaces [text, table 12.6, page 377]

$T_b := 28 \, kip$ minimum fastener tension force for 3/4" dia. A325 [text, table 12.1, page 354]

$N_s := 1$ number of slip plane

$\phi r_{str} := \phi \cdot 1.13 \cdot \mu \cdot T_b \cdot N_s$

$\phi r_{str} = 10.441 \, kip$ slip resistance per bolt

$\boxed{4 \cdot (\phi r_{str}) = 41.8 \, kip}$ total slip resistance of 4 bolts

c) Bearing strength of the connection

shear failure of plate

$\phi_{vpl} \cdot R_n = \phi_{vpl} \cdot 1.2 \cdot F_u \cdot L_c \cdot t \cdot n_b$

$\phi_{vpl} := 0.75$

$F_u = 58 \, ksi$ A36 steel

$t = 0.5 \, in$

$n_b := 4$

$L_{c\_end} := 1.5 \cdot in - \dfrac{d_{hole}}{2}$ $L_{c\_end} = 1.063 \, in$ <= controls

$L_{c\_int} := 3 \cdot in - d_{hole}$ $L_{c\_int} = 2.125 \, in$

$L_c := L_{c\_end}$

$\phi R_n := \phi_{vpl} \cdot 1.2 \cdot F_u \cdot L_c \cdot t \cdot n_b$

$\boxed{\phi R_n = 110.9 \, kip}$

bearing failure of plate around bolt holes

$\phi_{bpl} \cdot R_n = \phi_{bpl} \cdot 2.4 \cdot d_{bolt} \cdot t \cdot F_u \cdot n_b$

$\phi_{bpl} := 0.75$

$d_{bolt} = 0.75 \, in$

$t = 0.5 \, in$

$F_u = 58 \, ksi$ A36 steel

$n_b = 4$

$\phi R_n := \phi_{bpl} \cdot 2.4 \cdot d_{bolt} \cdot t \cdot F_u \cdot n_b$

$\boxed{\phi R_n = 156.6 \, kip}$

d) Block shear

$A_{gv} := 2 \cdot [(3 \cdot in + 1.5 \cdot in) \cdot t]$ $A_{gv} = 4.5 \, in^2$ gross shear area

$A_{gt} := (3 \cdot in) \cdot t$ $A_{gt} = 1.5 \, in^2$ gross tension area

$A_{nv} := A_{gv} - 2 \cdot [1.5 \cdot (d_{hole} \cdot t)]$ $A_{nv} = 3.187 \, in^2$ net shear area

$A_{nt} := A_{gt} - 1.0 \cdot (d_{hole} \cdot t)$ $A_{nt} = 1.062 \, in^2$ net tension area

$F_y = 36 \, ksi$ A36 steel

$F_u = 58 \, ksi$

$F_u \cdot A_{nt} = 61.625 \, kip$

$0.6 \cdot F_u \cdot A_{nv} = 110.925 \, kip$

$\phi_{block} := 0.75$

$F_u \cdot A_{nt} < 0.6 \cdot F_u \cdot A_{nv}$

shear fracture and tensile yield, LRFD equation J4-3b

$\boxed{\phi_{block} \cdot (0.6 \cdot F_u \cdot A_{nv} + F_y \cdot A_{gt}) = 123.7 \, kip}$ <= controls block shear strength

all fracture mode

$\phi_{block} \cdot (0.6 \cdot F_u \cdot A_{nv} + F_u \cdot A_{nt}) = 129.4 \, kip$

e) final design tensile strength

Summary
Yield of gross section = 97.2 kip
Fracture of effective area = 92.4 kip
Slip resistance of bolts = **41.8 kip**
Bearing strength of bolts
 (a) shear failure of plate = 110.9 kip
 (b) bearing failure around bolt holes = 156.6 kip
Block shear capacity = 123.7 kip

The maximum tension load $T_u = 41.8 \, kip$

$1\frac{1}{2}$ in

$\longrightarrow P_u$

W21 × 101

$\longrightarrow P_u$

25 bearing-type bolts. $F_y = 50$ ksi and

losed computer diskette.

75 k)

CHAPTER 13

Eccentrically Loaded Bolted Connections and Historical Notes on Rivets

13.1 BOLTS SUBJECTED TO ECCENTRIC SHEAR

Eccentrically loaded bolt groups are subjected to shears and bending moments. You might think that such situations are rare, but they are much more common than most people suspect. For instance, in a truss it is desirable to have the center of gravity of a member lined up exactly with the center of gravity of the bolts at its end connections. This feat is not quite as easy to accomplish as it may seem, and connections are often subjected to moments.

Eccentricity is quite obvious in Fig. 13.1(a), where a beam is connected to a column with a plate. In part (b) of the figure another beam is connected to a column with a pair of web angles. It is obvious that this connection must resist some moment because the center of gravity of the load from the beam does not coincide with the reaction from the column.

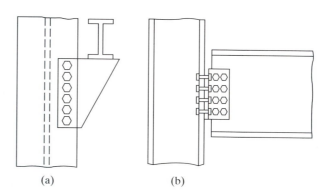

FIGURE 13.1

Eccentrically loaded bolt groups.   (a)   (b)

In general, specifications for bolts and welds clearly state that the center of gravity of the connection should coincide with the center of gravity of the member unless the eccentricity is accounted for in the calculations. However, Section J1.8 of the LRFD Specification provides some exceptions to this rule. It states that the rule is not applicable to the end connections of statically loaded single angles, double angles, and similar members. In other words, the eccentricities between the centers of gravity of these members and the centers of gravity of the connections may be ignored unless fatigue loadings are involved. *Furthermore, the eccentricity between the gravity axes and the gage lines of bolted members may be neglected for statically loaded members.*

The LRFD Specification presents values for computing the design strengths of individual bolts, but does not specify a method for computing the forces on these fasteners when they are eccentrically loaded. As a result the method of analysis to be used is left up to the designer.

Three general approaches for the analysis of eccentrically loaded connections have been developed through the years. The first of the methods is the very conservative *elastic method* in which friction or slip resistance between the connected parts is neglected. In addition these connected parts are assumed to be perfectly rigid. This type of analysis has been commonly used since at least 1870.[1,2]

Tests have shown that the elastic method usually provides very conservative results. As a consequence various *reduced* or *effective eccentricity methods* have been proposed.[3] The analysis is handled just as it is in the elastic method except that smaller eccentricities and thus smaller moments are used in the calculations.

The third method, called the *ultimate strength method*, provides the most realistic values as compared with test results, but is extremely tedious to apply, at least with hand-held calculators. Tables 7.17 to 7.24 in Part 7 of the Manual for eccentrically loaded connections are based on the ultimate strength method and enable us to solve most of these types of problems quite easily, as long as the bolt patterns are symmetrical. The remainder of this section is devoted to these three analysis methods.

13.1.1   Elastic Analysis

For this discussion the bolts of Fig. 13.2(a) are assumed to be subjected to a load P that has an eccentricity of e from the c.g. (center of gravity) of the bolt group. To consider the force situation in the bolts, an upward and downward force—each equal to P—is assumed to act at the c.g. of the bolt group. This situation, shown in part (b) of the figure, in no way changes the bolt forces. The force in a particular bolt will, therefore, equal P divided by the number of bolts in the group as seen in part (c), plus the force due to the moment caused by the couple shown in part (d) of the figure.

The magnitude of the forces in the bolts due to the moment Pe will now be considered. The distances of each bolt from the c.g. of the group are represented by the values d_1, d_2, etc., in Fig. 13.3. The moment produced by the couple is assumed to cause the plate to rotate about the c.g. of the bolt connection with the amount of rotation or

[1]W. McGuire, *Steel Structures* (Englewood Cliffs, N.J.: Prentice-Hall, 1968), p. 813.
[2]C. Reilly, "Studies of Iron Girder Bridges," *Proc. Inst. Civil Engrs.* 29 (London, 1870).
[3]T. R. Higgins, "New Formulas for Fasteners Loaded Off Center," *Engr. News Record* (May 21, 1964).

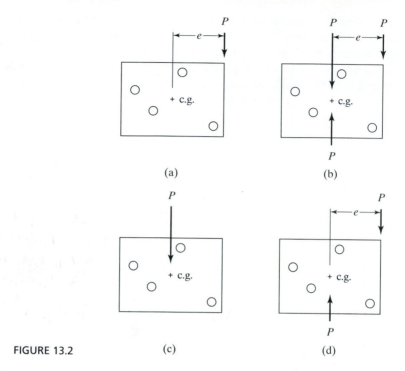

FIGURE 13.2

(a)   (b)   (c)   (d)

strain at a particular bolt being proportional to its distance from the c.g. (For this derivation the gusset plates are again assumed to be perfectly rigid and the bolts are assumed to be perfectly elastic.) Stress is greatest at the bolt that is the greatest distance from the c.g., because stress is proportional to strain in the elastic range.

The rotation is assumed to produce reactions of $r_1, r_2, r_3,$ and r_4, respectively, from the bolts in the figure. The moment transferred to the bolts must be balanced by resisting moments of the bolts as follows.

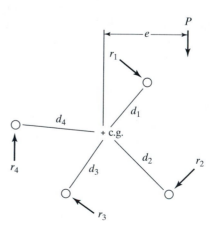

FIGURE 13.3

Bridge over New River Gorge in Fayette County, West VA, near Charleston. (Courtesy American Bridge Company.)

$$M_{c.g.} = Pe = r_1 d_1 + r_2 d_2 + r_3 d_3 + r_4 d_4 \qquad (1)$$

Fillet Weld:

Problem 3) Design the fillet welds with E70 electrodes to develop the full strength of the L6 x 4 x $^3/_8$ shown in Figure 3. Assume the gusset plate does not control the design strength.

(a) Determine the tensile strength of the angle. Check limitations due to yielding of the gross section and fracture of the effective area. Use the average value of U as given in the AISC Commentary, (U = 0.85 for this case).

(b) Determine the minimum, maximum and maximum effective size welds.

(c) Using the minimum weld size from part (b) and calculate the design strength of weld per inch length.

(d) Compute the total length of weld required using the controlling design tensile strength, from part (a). Round up your answer to the nearest inches.

(e) Minimize the effect of eccentricity of the welds, i.e. balance the welds by calculating L$_{top}$ and L$_{bottom}$. Round up your answers for L$_{top}$ and L$_{bottom}$. Sketch and label your final design.

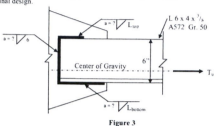

Figure 3

It is assumed to be directly proportional to the dis-

$$= \frac{r_2}{d_2} = \frac{r_3}{d_3} = \frac{r_4}{d_4}$$

r_1 and d_1, we get

$$= \frac{r_1 d_2}{d_1} \qquad r_3 = \frac{r_1 d_3}{d_1} \qquad r_4 = \frac{r_1 d_4}{d_1}$$

ion (1) and simplifying yields

$$\frac{_1 d_3^2}{d_1} + \frac{r_1 d_4^2}{d_1} = \frac{r_1}{d_1}(d_1^2 + d_2^2 + d_3^2 + d_4^2)$$

Therefore,

$$M = \frac{r_1 \Sigma d^2}{d_1}$$

w be written as

$$= \frac{d_2}{d_1} r_1 = \frac{M d_2}{\Sigma d^2} \qquad r_3 = \frac{M d_3}{\Sigma d^2} \qquad r_4 = \frac{M d_4}{\Sigma d^2}$$

ndicular to the line drawn from the c.g. to the particular ient to break these reactions down into vertical and hor- plishing this purpose reference is made to Fig. 13.4. ntal and vertical components of the distance d_1 are rep- ely, and the horizontal and vertical components of force

Fillet Weld
Problem 3)

a) Tensile design strength of member (Middle plate controls)

Yielding of gross area

$\phi_{t\_yield} := 0.90$

$F_y := 50 \cdot ksi$ [A572 Gr. 50 steel]

$A_g := 3.58 \cdot in^2$ gross area of L 6 x 4 x 3/8 [LRFD table 1-7 page 1-34]

nominal strength

$P_n := F_y \cdot A_g$

$P_n = 179 \, kip$

design strength

$\boxed{\phi_{t\_yield} \cdot P_n = 161.1 \, kip}$

Fracture of the effective area

$\phi_{t\_fracture} := 0.75$

$F_u := 65 \cdot ksi$ [A572 Gr. 50 steel]

Calculate effective area

$U := 0.85$ note: using average U value of commentary of AISC

$A_n := A_g$

$A_e := A_n \cdot U$

$A_e = 3.043 \, in^2$

nominal strength

$P_n := A_e \cdot F_u$

$P_n = 197.795 \, kip$

design strength

$\boxed{\phi_{t\_fracture} \cdot P_n = 148.3 \, kip}$

b) min, max and max effective weld sizes

$t := \frac{3}{8} \cdot in$ thickness of L 6 x 4 x 3/8

$\boxed{a_{min} := \frac{3}{16} \cdot in}$ 3/8" thick plate, [Table 14.2, page 448]

$a_{max} := t - \frac{1}{16} \cdot in$ $\boxed{a_{max} = \frac{5}{16} \, in}$ 3/8" thick plate, [Table 14.2, page 448]

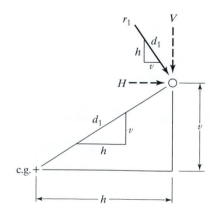

r_1 are represented by H and V, respectively. It is now possible to write the following ratio from which H can be obtained:

$F_{E70} := 70 \cdot ksi$ yield strength of the welds, E70 electrodes

$F_y = 50\,ksi$

$a_{max\_eff} := \dfrac{0.9 \cdot t \cdot (0.6 \cdot F_y)}{0.75 \cdot (0.707) \cdot (0.6 \cdot F_{E70})}$ $a_{max\_eff} = 0.455\,in$ greater than $a_{max} = 0.312\,in$ therefore a_{max} controls $\dfrac{v}{d_1}\Big)$

c) design strength of fillet weld per inch length

$a := a_{min}$

$a = \dfrac{3}{16}\,in$

Weld design strength

$\phi_{weld} = 0.75$

$\phi R_{n\_weld} := \phi_{weld} \cdot (0.707 \cdot a) \cdot (0.6 \cdot F_{E70})$

$\boxed{\phi R_{n\_weld} = 4.176\,\dfrac{kip}{in}}$ <= this controls the overall weld design strength

Strength of the base metal

$\phi_{base} = 0.90$

$t = 0.375\,in$

$F_y = 50\,ksi$

$\phi R_{n\_base} := \phi_{base} \cdot t \cdot (0.6 \cdot F_y)$

$\boxed{\phi R_{n\_base} = 10.125\,\dfrac{kip}{in}}$

d) Total length of the fillet welds required

from part (a) fracture of the effective area controls. 148.3 kip

$\dfrac{148.3 \cdot kip}{\phi R_{n\_weld}} = 35.515\,in$ <= from part (c) base metal strength controls

$\boxed{L_{total} := 36 \cdot in}$ greater than $4 \cdot a = 0.75\,in$

e) Balancing the welds

$y_{bar} := 1.94 \cdot in$ centroid measure from the shorter leg, L 6 x 4 x 3/8 [LRFD table 1-7 page 1-34]

$L_{tr} := 6 \cdot in$ use the full transverse length of L6 x 4 x 3/8

Matching the center of gravity (C.G.) of the welds group to the C.G. of the angle. C.G. of each component is measured from the bottom of the shorter leg.

$y_{bar} = \dfrac{L_{tr}\left(\dfrac{6 \cdot in}{2}\right) + L_{top} \cdot (6 \cdot in) + L_{bottom}(0 \cdot in)}{L_{total}}$

solve for L_{top}

$L_{top} = \dfrac{L_{total} \cdot (y_{bar}) - L_{tr}\left(\dfrac{6 \cdot in}{2}\right)}{6 \cdot in}$ $L_{top} = 8.64\,in$

use $\boxed{L_{top} := 9 \cdot in}$

$L_{bottom} := L_{total} - L_{tr} - L_{top}$ $\boxed{L_{bottom} = 21\,in}$

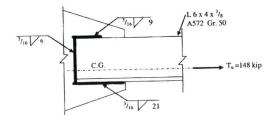

group shown in Fig. 13.5 using the

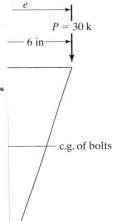

e

$P = 30\,k$

6 in

c.g. of bolts

ed to it by the direct load and the ketch, the student can see that the re the most stressed and that their

5 in

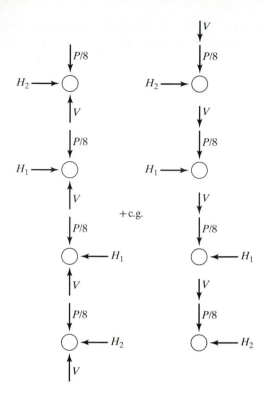

FIGURE 13.6

$$M = Pe = (30)(7.5) = 225 \text{ in-k}$$

$$\Sigma d^2 = \Sigma h^2 + \Sigma v^2$$

$$\Sigma d^2 = (8)(1.5)^2 + (4)(1.5^2 + 4.5^2) = 108$$

$$H = \frac{Mv}{\Sigma d^2} = \frac{(225)(4.5)}{108} = 9.38 \text{ k} \rightarrow$$

$$V = \frac{Mh}{\Sigma d^2} = \frac{(225)(1.5)}{108} = 3.13 \text{ k} \downarrow$$

$$\frac{P}{8} = \frac{30}{8} = 3.75 \text{ k} \downarrow$$

These components for the lower right-hand bolt are sketched as follows:

The resultant force applied to this bolt is

$$R = \sqrt{(3.13 + 3.75)^2 + (9.38)^2} = 11.63 \text{ k}$$

If the eccentric load is inclined, it can be broken down into vertical and horizontal components, and the moment of each about the c.g. of the bolt group can be determined. Various design formulas can be developed that will enable the engineer to directly design eccentric connections, but the process of assuming a certain number and arrangement of bolts, checking stresses, and redesigning probably is just as satisfactory.

The trouble with this inaccurate but very conservative method of analysis is that, in effect, we are assuming that there is a linear relation between loads and deformations in the fasteners; further, we assume that their yield stress is not exceeded when the ultimate load on the connection is reached. Various experiments have shown that these assumptions are incorrect.

13.1.2 Reduced Eccentricity Method

The elastic analysis method just described appreciably overestimates the moment forces applied to the connectors. As a result quite a few proposals have been made through the years that make use of an effective eccentricity, thus in effect taking into account the slip resistance on the faying or contact surfaces. One set of reduced eccentricity values that were fairly common at one time follow:

1. With one gage line of fasteners and where n is the number of fasteners in the line,

$$e_{effective} = e_{actual} - \frac{1 + 2n}{4}$$

2. With two or more gage lines of fasteners symmetrically placed and where n is the number of fasteners in each line,

$$e_{effective} = e_{actual} - \frac{1 + n}{2}$$

The reduced eccentricity values for two fastener arrangements are shown in Fig. 13.7.

To analyze a particular connection with the reduced eccentricity method, the value of $e_{effective}$ is computed as described above and used to compute the eccentric moment. Then the elastic procedure is used for the remainder of the calculations.

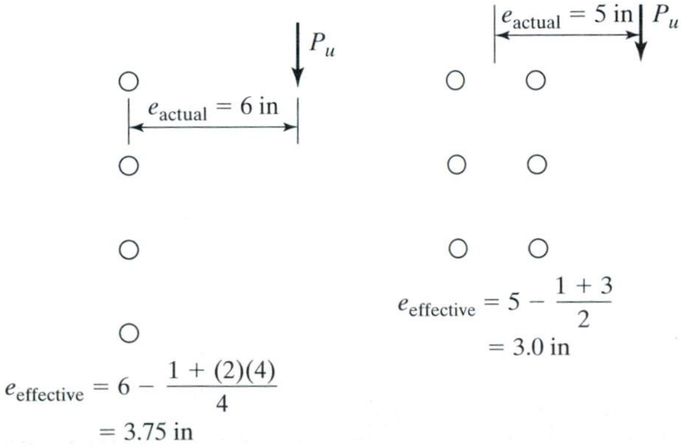

$$e_{effective} = 6 - \frac{1 + (2)(4)}{4}$$
$$= 3.75 \text{ in}$$

$$e_{effective} = 5 - \frac{1 + 3}{2}$$
$$= 3.0 \text{ in}$$

FIGURE 13.7

13.1.3 Ultimate Strength Method

Both the elastic and reduced eccentricity methods for analyzing eccentrically loaded fastener groups are based on the assumption that the behavior of the fasteners is elastic. A much more realistic method of analysis is the *ultimate strength method*, which is described in the next few paragraphs. The values given in the tables of Part 7 of the LRFD Manual for eccentrically loaded fastener groups were computed using this method.

If one of the outermost bolts in an eccentrically loaded connection begins to slip or yield, the connection will not fail. Instead, the magnitude of the eccentric load may be increased, the inner bolts will resist more load, and failure will not occur until all of the bolts slip or yield.

The eccentric load tends to cause both a relative rotation and translation of the connected material. In effect this is equivalent to pure rotation of the connection about a single point called the *instantaneous center of rotation*. An eccentrically loaded bolted connection is shown in Fig. 13.8, and the instantaneous center is represented by point 0. It is located a distance e' from the center of gravity of the bolt group.

The deformations of these bolts are assumed to vary in proportion to their distances from the instantaneous center. The ultimate shear force that one of them can resist is not equal to the pure shear force that a bolt can resist. Rather, it is dependent upon the load-deformation relationship in the bolt. Studies by Crawford and Kulak[4] have shown that this force may be closely estimated with the following expression:

$$R = R_{ult}(1 - e^{-10\Delta})^{0.55}$$

In this formula R_{ult} is the ultimate shear load for a single fastener equaling 74 k for an A325 bolt, e is the base of the natural logarithm (2.718), and Δ is the total deformation of a bolt. Its maximum value experimentally determined is 0.34 in. The Δ values for the other bolts are assumed to be in proportion to R as their d distances are to d for the bolt with the largest d. The coefficients 10.0 and 0.55 also were experimentally obtained. Figure 13.9 illustrates this load-deformation relationship.

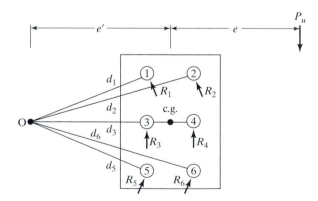

FIGURE 13.8

---

[4]S. F. Crawford and G. L. Kulak, "Eccentrically Loaded Bolt Connections," *Journal of Structural Division*, ASCE 97, ST3 (March 1971), pp. 765–783.

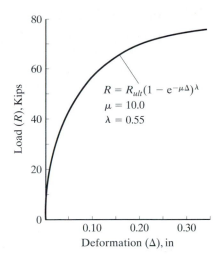

FIGURE 13.9

Ultimate shear force R in a single bolt at any given deformation.

This expression clearly shows that the ultimate shear load taken by a particular bolt in an eccentrically loaded connection is affected by its deformation. Thus, the load applied to a particular bolt is dependent upon its position in the connection with respect to the instantaneous center of rotation.

The resisting forces of the bolts of the connection of Fig. 13.8 are represented with the letters R_1, R_2, R_3, and so on. Each of these forces is assumed to act in a direction perpendicular to a line drawn from point 0 to the center of the bolt in question. For this symmetrical connection the instantaneous center of rotation will fall somewhere on a horizontal line through the center of gravity of the bolt group. This is the case because the sum of the horizontal components of the R forces must be zero as also must be the sum of the moments of the horizontal components about point 0. The position of point 0 on the horizontal line may be found by a tedious trial-and-error procedure to be described here.

With reference to Fig. 13.8, the moment of the eccentric load about point 0 must be equal to the summation of the moments of the R resisting forces about the same point. If we knew the location of the instantaneous center we could compute R values for the bolts with the Crawford-Kulak formula and determine P_u from the expression to follow in which e and e' are distances shown in Figs. 13-8 and 13-11.

$$P_u(e' + e) = \Sigma Rd$$
$$P_u = \frac{\Sigma Rd}{e' + e}$$

To determine the design strength of such a connection according to the LRFD Specification, we can replace R_{ult} in the Crawford-Kulak formula with the design shearing strength of one bolt in a connection where the load is not eccentric. For instance, if we have 7/8-in A325 bolts (threads excluded from shear plane) in single shear bearing on a sufficient thickness so bearing does not control, R_{ult} will equal $(0.75)(0.60)(60) = 27.0$ k.

The location of the instantaneous center is not known, however. Its position is estimated, the R values determined, and P_u calculated as described. It will be noted that P_u must be equal to the summation of the vertical components of the R resisting forces (ΣR_v).

If the value is computed and equals the P_u computed by the formula above we have the correct location for the instantaneous center. If not, we try another location, and so on.

In Example 13-2 the author demonstrates the very tedious trial-and-error calculations necessary to locate the instantaneous center of rotation for a symmetrical connection consisting of four bolts. In addition the design strength P_u of the connection is determined.

To solve such a problem it is very convenient to set the calculations up in a table similar to the one used in the solution to follow. In the table shown the h and v values given are the horizontal and vertical components of the d distances from point 0 to the center of gravities of the individual bolts. The bolt that is located at the greatest distance from point 0 is assumed to have a Δ value of 0.34 in. The Δ values for the other bolts are assumed to be proportionatal to their distances from point 0. The Δ values so determined are used in the R formula.

A set of tables entitled "Coefficients C for Eccentrically Loaded Bolt Groups" is presented in Part 7 of the LRFD Manual. The values in these tables were determined by the procedure described here. A large percentage of the practical cases that the designer will encounter are included in the tables. Should some other situation not covered be faced, the designer may very well decide to use the more conservative elastic procedure previously described.

Example 13-2

The bearing-type 7/8-in A325 bolts of the connection of Fig. 13.10 have a design shear strength $R_u = (0.75)\,(0.60)\,(60) = 27.0$ k. Locate the instantaneous center of rotation of the connection using the trial-and-error procedure and determine the value of P_u.

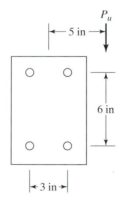

FIGURE 13.10

Solution. *By trial and error:* Try a value of $e' = 3$ in, reference being made to Fig. 13.11. In the following table Δ for bolt 1 equals $(3.3541/5.4083)\,(0.34) = 0.211$ in and R for the same bolt equals $27(1 - e^{-(10)\,(0.211)})^{0.55}$.

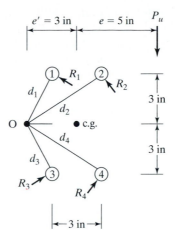

FIGURE 13.11

Bolt No.	h (in)	v (in)	d (in)	Δ (in)	R(kips)	R_v (kips)	Rd (in -kips)
1	1.5	3	3.3541	0.211	25.15	11.25	84.36
2	4.5	3	5.4083	0.34	26.50	22.05	143.32
3	1.5	3	3.3541	0.211	25.15	11.25	84.36
4	4.5	3	5.4083	0.34	26.50	22.05	143.32
						$\Sigma = 66.60$	$\Sigma = 455.36$

$$P_u = \frac{\Sigma Rd}{e' + e} = \frac{455.36}{3 + 5} = 56.92 \text{ k not } = 66.60 \text{ k} \qquad \text{(NG)}$$

After several trials assume $e' = 2.40$ in.

Bolt No.	h (in)	v (in)	d (in)	Δ (in)	R(kips)	R_v(kips)	Rd (in -kips)
1	0.90	3	3.1321	0.216	25.24	7.23	79.06
2	3.90	3	4.9204	0.34	26.50	21.00	130.39
3	0.90	3	3.1321	0.216	25.24	7.23	79.06
4	3.90	3	4.9204	0.34	26.50	21.00	130.39
						$\Sigma = 56.46$	$\Sigma = 418.90$

Then, we have

$$P_u = \frac{\Sigma Rd}{e' + e} = \frac{418.90}{2.4 + 5} = 56.61 \text{ k almost } = 56.46 \text{ k} \qquad \text{(OK)}$$

$$P_u = 56.6 \text{ k}$$

Although the development of this method of analysis was actually based on bearing-type connections where slip may occur, both theory and load tests have shown that it may conservatively be applied to slip-critical connections.[5]

The ultimate-strength method may be expanded to include inclined loads and unsymmetrical bolt arrangements, but the trial-and-error calculations are extraordinarily long for such situations.

Examples 13-3 and 13-4 provide illustrations of the use of the ultimate strength tables in Part 7 of the LRFD Manual for both analysis and design.

Example 13-3

Repeat Example 13-2 using ultimate strength tables in Part 7 of the Manual. These tables are entitled "Coefficients C for Eccentrically Loaded Bolt Groups."

Solution. Enter Manual Table 7.18 with angle $= 0$, $s = 6$ in., $e_x = 5$ in. and $n = 2$ vertical rows.

$$C = 2.10$$
$$\phi R_n = C \times \phi r_n$$
$$\phi R_n = (2.10)(27.0) = 56.7 \text{ k} \qquad\qquad \text{(OK)}$$

Example 13-4

Determine the number of $\frac{7}{8}$-in. A325 bolts in standard size holes required for the connection shown in Fig. 13.12. Use A36 steel and assume the connection is to be a bearing type with threads excluded from the shear plane. Further assume that the bolts are in single shear and bearing on $\frac{1}{2}$ in. Use the ultimate strength analysis method presented in Part 7 of the LRFD Manual.

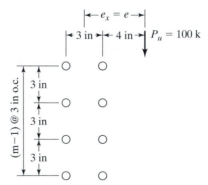

FIGURE 13.12

---

[5]G. L. Kulak, "Eccentrically Loaded Slip-Resistant Connections," *Engineering Journal*, AISC, *12*, No. 2 (second quarter, 1975), pp. 52–55.

Solution. Using Table 7.18 in Part 7 of the LRFD Manual, we obtain

$$e_x = e = 5\tfrac{1}{2} \text{ in} = 5.5 \text{ in}$$

Bolts in single shear and bearing on 1/2 in:

$$\phi r = \text{shear design strength per fastener}$$
$$= (0.75)(0.6)(60) = 27.0 \text{ k} \leftarrow$$
$$\phi r = \text{bearing design strength per fastener} = \phi(2.4\ dt\ F_u)$$
$$= (0.75)(2.4)\left(\tfrac{7}{8}\right)\left(\tfrac{1}{2}\right)(58) = 45.7 \text{ k}$$

With reference to Table 7.18 the necessary value of *C* required to provide a sufficient number of bolts can be determined as follows:

$$C_{\text{reqd}} = \frac{P_u}{\phi r_n} = \frac{100}{27.0} = 3.70$$

Then with $e_x = 5\tfrac{1}{2}$ in and a vertical spacing *s* of 3 in, we move horizontally in the table until we find the number of bolts in each vertical row so as to provide a *C* of 3.70 or more. With $e_x = 5$ in and $n = 4$. We find $C = 4.21$. Then with $e_x = 6$ in and $n = 4$. We find $C = 3.69$. Interpolating for $e_x = 5\tfrac{1}{2}$ in. We find $C = 3.95 > 3.70$.   (OK)

Use four 7/8 in A325 $\times$ bolts in each row.

Note: If the situation faced by the designer does not fit the ultimate strength tables given in Part 7 of the LRFD Manual, it is recommended that he or she use the conservative elastic procedure to handle the problem, whether analysis or design.

13.2   BOLTS SUBJECTED TO SHEAR AND TENSION

The bolts used for a good many structural steel connections are subjected to a combination of shear and tension. One obvious case is shown in Fig. 13.13, where a diagonal brace is attached to a column. The 111.8-k component shown in the figure is trying to shear the bolts off at the face of the column, while the 223.6 k component is tending to pull off their heads.

Tests on bearing-type bolts subject to combined shear and tension show that their ultimate strengths can be represented with an elliptical interaction curve as shown in Fig. 13.14. Particularly note the values F_t and F_v. F_t is the limiting tensile stress if there is no shear, and F_v is the limiting shearing stress if there is no externally applied tension.

The three straight dashed lines shown in Fig. 13.14 very closely represent the test result interaction curve. Equations for these lines are given in Table 13.1 (Table J3.5 of the LRFD Specification). In these expressions f_v and f_t are, respectively, the computed shear and tension stresses in the bolts due to the factored loads.

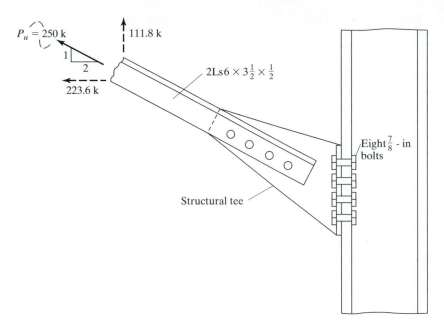

FIGURE 13.13

Combined shear and tension connection.

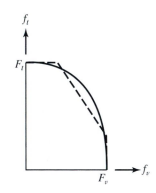

FIGURE 13.14

Bolts in a bearing-type connection
subject to combined shear and tension.

The design tensile stress for a bolt subjected to combined shear and tension
equals ϕF_t, while its total design strength is $\phi F_t A_b$ with $\phi = 0.75$. The nominal tensile
stress F_t is to be computed with the appropriate equation given in Table 13.1.

Example 13-5 illustrates the calculations involved in checking a high-strength
bearing-type bolted connection for shear and tension. (There is another stress situation that may affect this connection called *prying action*, which we discuss in
Section 13-4.)

TABLE 13.1   Nominal Tension Stress (F_t), ksi (MPa) Fasteners in Bearing-type Connections

Description of Fasteners	Threads included in the Shear Plane	Threads Excluded from the Shear Plane
A307 bolts	$59 - 2.5f_v \leq 45$ $(171 - 2.5f_v \leq 310)$	
A325 bolts A325M bolts	$117 - 2.5f_v \leq 90$ $(807 - 2.5f_v \leq 621)$	$117 - 2.0f_v \leq 90$ $(807 - 2.0f_v \leq 621)$
A490 bolts A490M bolts	$147 - 2.5f_v \leq 113$ $(1010 - 2.5f_v \leq 779)$	$147 - 2.0f_v \leq 113$ $(1010 - 2.0f_v \leq 779)$
Threaded parts A449 bolts over $1\frac{1}{2}$ diameter	$0.98F_u - 2.5f_v \leq 0.75F_u$	$0.98F_u - 2.0f_v \leq 0.75F_u$
A502 Gr. 1 rivets	$59 - 2.4f_v \leq 45$ $(407 - 2.4f_v \leq 310)$	
A502 Gr. 2 rivets	$78 - 2.4f_v \leq 60$ $(538 - 2.4f_v \leq 414)$	

Example 13-5

The tension member shown in Fig. 13.13 is connected to the column shown with eight 7/8-in A325 high-strength bolts in a bearing-type connection with the threads excluded from the shear plane and standard-size holes. Is this a sufficient number of bolts to resist the applied load according to the LRFD Specification neglecting prying action?

Solution

Shearing stress:

$$f_v = \frac{V_u}{N_b A_b} = \frac{111.8}{(8)(0.6)} = 23.29 \text{ ksi}$$

$$\phi F_{nv} = (0.75)(60) = 45 \text{ ksi} > f_v = 23.29 \text{ ksi} \qquad \text{(OK)}$$

Tension stress:

$$f_t = \frac{T_u}{N_b A_b} = \frac{223.6}{(8)(0.6)} = 46.58 \text{ ksi}$$

Limiting tension: stress $\phi F_t = \phi[(117 - 1.5f_v) \leq 90]$

$$= 0.75[(117 - 2.0 \times 23.29) \leq 90] = 52.81 \text{ ksi} > 46.58 \text{ ksi} \qquad \text{(OK)}$$

Connection is satisfactory

When an axial tension force is applied to a slip-critical connection, the clamping force will be reduced, and the design shear strength must be decreased in some proportion to the loss in clamping or prestress. This is accomplished in the LRFD Specification (Section J3.9) by requiring that the nominal slip-critical shear strengths given in Section 12-14 of Chapter 12 (Section J3.9a of the LRFD Specification) be multiplied by the reduction factor $1 - T_u/1.13T_bN_b$. In this expression, T_b is the minimum bolt pre-tension given in Table 12.1 of this text (Table J3.1 in the LRFD Specification), while N_b is the number of bolts carrying the factored tension load T_u. For such a situation ϕ is equal to 1.0 per LRFD Specification J3.8a.

Example 13-6

A group of twelve 7/8 in- A325 high strength bolts is used with standard holes for a slip critical connection to resist a factored shearing force $V_u = 130$ k and a factored tensile force $T_u = 150$ k. Determine the reduced slip resistance due to the tensile force assuming Class B surfaces with $\mu = 0.50$.

Solution

$$\text{Reduction factor} = 1 - \frac{T_u}{1.13 \, T_b N_b}$$

$$= 1 - \frac{150}{(1.13)(39)(12)} = 0.716$$

$$\text{Reduced slip resistance} = 0.716\phi r_{str}$$

$$= (0.716)(\phi 1.13\mu T_b N_s)$$

$$= (0.716)(1.0)(1.13)(0.50)(39)(1)$$

$$= 15.78 \text{ k} < \frac{130}{(12)(0.6)} = 18.06 \text{ ksi (N.G.)}$$

13.3   TENSION LOADS ON BOLTED JOINTS

Bolted and riveted connections subjected to pure tensile loads have been avoided as much as possible in the past by designers. The use of tensile connections was probably used more often for wind-bracing systems in tall buildings than for any other situation. Other locations exist, however, where they have been used, such as hanger connections for bridges, flange connection for piping systems, etc. Figure 13.15 shows a hanger-type connection with an applied tensile load.

Hot-driven rivets and fully tensioned high-strength bolts are not free to shorten, with the result that large tensile forces are produced in them during their installation. These initial tensions are actually close to their yield points. There has always been considerable reluctance among designers to apply tensile loads to connectors of this type for fear that the external loads might easily increase their already present tensile stresses and cause them to fail. The truth of the matter, however, is that when external tensile loads are applied to connections of this type, the connectors probably will experience little if any change in stress.

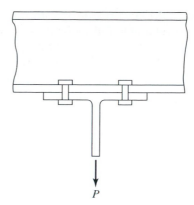

FIGURE 13.15

Hanger connection.

P

Hot-driven rivets that have cooled and shrunk or fully tensioned high-strength bolts actually prestress the joints in which they are used against tensile loads. (Think of a prestressed concrete beam that has external compressive loads applied at each end.) The tensile stresses in the connectors squeeze together the members being connected. If a tensile load is applied to this connection at the contact surface, it cannot exert any additional load on the bolts or rivets until the members are pulled apart and additional strains put on the bolts or rivets. The members cannot be pulled apart until a load is applied that is larger than the total tension in the connectors of the connection. This statement means that the joint is prestressed against tensile forces by the amount of stress initially put in the shanks of the connectors.

Another way of saying this is that if a tensile load P is applied at the contact surface it tends to reduce the thickness of the plates somewhat, but at the same time the contact pressure between the plates will be correspondingly reduced and the plates will tend to expand by the same amount. The theoretical result, then, is no change in plate thickness and no change in connector tension. This situation continues until P equals the connector tension. At this time an increase in P will result in separation of the plates and thereafter the tension in the connector will equal P.

Should the load be applied to the outer surfaces there will be some immediate strain increase in the connector. This increase will be accompanied by an expansion of the plates even though the load does not exceed the prestress, but the increase will be very slight because the load will go to the plate and connectors roughly in proportion to their stiffness. As the plate is stiffer it will receive most of the load. An expression can be developed for the elongation of the bolt based on the bolt area and the assumed contact area between the plates. Depending on the contact area assumed, it will be found that unless P is greater than the bolt tension, its stress increase will be in the range of 10 percent. Should the load exceed the prestress, the bolt stress will rise appreciably.

The preceding rather lengthy discussion is approximate, but should explain why an ordinary tensile load applied to a riveted or bolted joint will not change the stress situation very much.

The LRFD tensile design strength of a single bolt, rivet, or threaded part is given by the expression to follow, which is independent of any initial tightening force.

$$\phi P_n = \phi \, 0.75 \, A_b F_u$$

When fasteners are loaded in tension there is usually some bending due to the deformation of the connected parts. As a result the value of ϕ in this expression is a rather small 0.75. Table 12.5 of this text (Table J3.2 of the LRFD Specification) gives values of F_t the nominal tensile strength (ksi) for the different kinds of connectors with the values of rivets and threaded parts being quite conservative.

In this expression A_b is the nominal body area of a rivet, or the unthreaded portion of a bolt, or its threaded part not including upset rods. An upset rod has its ends made larger than the regular rod and the threads are placed in the enlarged section so that the area at the root of the thread is larger than that of the regular rod. An upset rod was shown in Fig. 4.3. The use of upset rods is not usually economical and should be avoided unless a large order is being made.

If an upset rod is used the nominal tensile strength of the threaded portion is set equal to $0.75 \, F_u$ times the cross-sectional area at its major thread diameter. This value must be larger than F_y times the nominal body area of the rod before upsetting.

Example 13.7 illustrates the determination of the strength of a tension connection.

Example 13-7

Determine the design tensile strength of the bolts of the hanger connection of Fig. 13.15 if eight 7/8-in A490 high-strength bolts are used. Neglect prying action.

Solution

$$\phi P_n = \phi(0.75 \, A_b F_u) \, N_b = (0.75)(0.75)(0.6)(113)(8) = 305.1 \text{ k}$$

The load applied to a tensile connection shall be the sum of the factored external loads plus any tension forces that result from prying action as described in the next section.

13.4   PRYING ACTION

A further consideration that should be given to tensile connections is the possibility of prying action. A tensile connection is shown in Fig. 13.16(a) that is subjected to prying action as illustrated in part (b) of the same figure. Should the flanges of the connection be quite thick and stiff or have stiffener plates like those in Fig. 13.16(c), the prying action will probably be negligible, but this is not the case if they are thin and flexible and have no stiffeners.

It is usually desirable to limit the number of rows of bolts in a tensile connection because a large percentage of the load is carried by the inner rows of multirow connections even at ultimate load. The tensile connection shown in Fig. 13.17 illustrates this point as the prying action will throw a large part of the load to the inner connectors, particularly if the plates are thin and flexible. For connections subjected to pure tensile loads, estimates should be made of possible prying action and its magnitude.

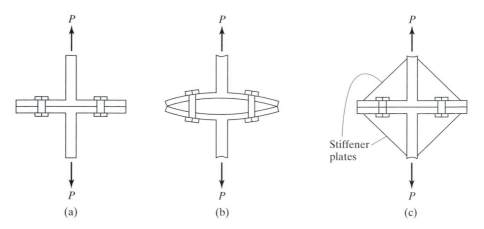

FIGURE 13.16

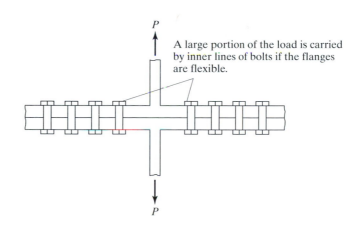

FIGURE 13.17

The additional force in the bolts resulting from prying action should be added to the tensile force resulting directly from the applied forces. The actual determination of prying forces is quite complex, and research on the subject is still being conducted. Several empirical formulas have been developed that approximate test results. Among these are the LRFD expressions included in this section. The reader should realize that we don't know very much about prying action and our formulas keep changing almost yearly.

Only fully tensioned bolts should be used for connections for which the applied loads subject the bolts to axial tension. This is true whether the connections are classified as slip-critical and whether they are subject to fatigue loads and whether there is prying action. If snug-tight bolts are used for any of these situations, the tensile loads will immediately start increasing bolt tensions.

Hanger and other tension connections should be so designed as to prevent significant deformations. The most important item in such designs is the use of rigid flanges. Rigidity is more important than bending resistance. To achieve this goal the distance b

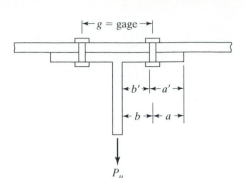

FIGURE 13.18

shown in Fig. 13.18 should be made as small as possible with a minimum value equal to the space required to use a wrench for tightening the bolts. Information concerning wrench clearance dimensions is presented in a table entitled "Entering and Tightening Clearances" in Table 7.3 in Part 7 of the LRFD Manual.

In Part 9 of the LRFD Manual a detailed procedure is presented for designing hanger connections and computing prying forces. This empirical method is for factored forces only. If it is desired to consider service loads for the purposes of investigating fatigue, deflections, or drift limitations, the reader may refer to the service load formulas for prying that were presented in the ninth edition of the *AISC Manual of Steel Construction Allowable Stress Design.*

For reasons of space limitations only one numerical prying action example is presented here. In this example the author checks to see if the flange thickness of a WT hanger connection is satisfactory according to the LRFD procedure. For a much more detailed example the reader can refer to Section 11 of the Manual.

Many abbreviations and equations are used in the LRFD prying analysis procedure; they follow with reference made to Fig. 13.18.

$$\phi r_n = \text{design tensile strength of each bolt} = \phi F_t A_b$$

$b = \dfrac{g - t_w}{2}$, where g is the gage. It must be sufficient for wrench clearance as given in Table 7.3 in Part 7 of the Manual.

$$a = \text{distance from bolt center line to the edge of tee flange or}$$
$$\text{angle leg but not greater than } \frac{b_f - g}{2} \le 1.25b$$

$$b' = b - \frac{d}{2} \text{ where } d = \text{bolt diameter}$$

$$a' = a + \frac{d}{2}$$

$$\rho = \frac{b'}{a'}$$

$$r_{ut} = \frac{P_u}{\text{number of bolts}}$$

$$\beta = \frac{1}{\rho}\left(\frac{\phi r_n}{r_{ut}} - 1\right)$$

If $\beta \geq 1$, set $\alpha' = 1.0$

If $\beta < 1$, set α' = the lesser of 1.0 or $\frac{1}{\delta}(\frac{\beta}{1 - \beta})$, where δ is the ratio of the net area at the bolt line to the gross area at the face of the stem or angle leg, and d' = width of bolt hole parallel to tee stem,

$$\delta = 1 - \frac{d'}{p} \text{ where } p \text{ is the length of connection tributary to each bolt}$$

Finally, the required flange thickness $t_{\text{req}} = \sqrt{\dfrac{4.44\, r_{ut}b'}{pF_y(1 + \delta\alpha')}}$

The LRFD Manual states that if the prying force q_u is to be reduced to insignificant levels we should let $\alpha' = 0$ and revise the expression for the required thickness to

$$t_{reqd} = \sqrt{\frac{4.44\, r_{ut}b'}{pF_y}}$$

If it is desired, even though not required, to determine the factored prying force per bolt, q_u, it can be done with the formulas presented at the bottom of page 9–11 in the Manual.

Example 13-8

A 10-in-long WT8 × 22.5 (t_f = 0.565 in, t_w = 0.345 in, and b_f = 7.04 in) is connected to a W36 × 150 as shown in Fig. 13.19 with four 7/8-in A325 high-strength bolts. If A36 steel is used is the flange sufficiently thick if prying action is considered?

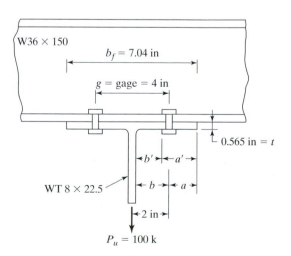

W36 × 150

b_f = 7.04 in

g = gage = 4 in

0.565 in = t

$\leftarrow b' \rightarrow\leftarrow a' \rightarrow$

$\leftarrow b \rightarrow\leftarrow a \rightarrow$

WT 8 × 22.5

$\leftarrow$ 2 in $\rightarrow$

P_u = 100 k

FIGURE 13.19

Solution

$$\phi r_n = \phi F_t A_b = (0.75)(90)(0.6) = 40.5 \text{ k}$$

$$b = \frac{g - t_w}{2} = \frac{4 - 0.345}{2} = 1.827 \text{ in} > 1\tfrac{3}{8} \text{ in}$$

wrench clearance required   (OK)

$$a = \frac{b_f - g}{2} = \frac{7.04 - 4}{2} = 1.52 \text{ in} \le 1.25 b = (1.25)(1.827) = 2.284 \text{ in}\quad \text{(OK)}$$

$$b' = b - \frac{d}{2} = 1.827 - \frac{0.875}{2} = 1.389 \text{ in}$$

$$a' = a + \frac{d}{2} = 1.52 + \frac{0.875}{2} = 1.957 \text{ in}$$

$$\rho = \frac{b'}{a} = \frac{1.389}{1.52} = 0.914$$

$$r_{ut} = \frac{P_u}{\text{number of bolts}} = \frac{100}{4} = 25 \text{ k}$$

$$\beta = \frac{1}{\rho}\left(\frac{\phi r_n}{r_{ut}} - 1\right) = \frac{1}{0.914}\left(\frac{40.5}{25} - 1\right) = 0.678$$

Since $\beta < 1.0$

$\delta = 1 - \frac{d'}{p}$ where with four bolts the maximum effective length is $2g = (2)(4) = 8 \text{ in} < b_f$ of a W36 × 150 = 12.0 in. Therefore, there are 4 in. of tee length tributary to each pair of bolts and $p = 4$ in

$$\delta = 1 - \frac{15/16}{4} = 0.766$$

$$\alpha' = \frac{1}{\delta}\left(\frac{\beta}{1 - \beta}\right) = \frac{1}{0.766}\left(\frac{0.678}{1 - 0.678}\right) = 2.75 > 1.0 \quad \therefore \text{ Use } 1.0$$

Then the required flange thickness is

$$t_{req} = \sqrt{\frac{4.44\, r_{ut} b'}{p F_y (1 + \delta \alpha')}} = \sqrt{\frac{(4.44)(25)(1.389)}{(4)(36)(1 + 0.766 \times 1.0)}}$$
$$= 0.779 \text{ in} > 0.565 \text{ in}\tag{NG}$$

To cause any prying force q_u to be extremely small such that it can be neglected, the required flange thickness is

$$t_{reqd} = \sqrt{\frac{4.44\, r_{ut} b'}{p F_y}} = \sqrt{\frac{(4.44)(25)(1.389)}{(4)(36)}} = 1.035 \text{ in}$$

13.5   HISTORICAL NOTES ON RIVETS

Rivets were the accepted method for connecting the members of steel structures for many years. Today, however, they no longer provide the most economical connections and basically are obsolete. It is doubtful that you could find a steel fabricator who can do riveting. It is, however, desirable for the designer to be familiar with rivets even though he or she will probably never design riveted structures. He or she may have to analyze an existing riveted structure for new loads or for an expansion of the structure. The purpose of these sections is to present only a very brief introduction to the analysis and design of rivets.

The rivets used in construction work were usually made of a soft grade of steel that would not become brittle when heated and hammered with a riveting gun to form the head. The usual rivet consisted of a cylindrical shank of steel with a rounded head on one end. It was heated in the field to a cherry-red color (approximately 1800°F), inserted in the hole, and a head formed on the other end probably with a portable rivet gun operated by compressed air. The rivet gun, which had a depression in its head to give the rivet head the proper shape, applied a rapid succession of blows to the rivet.

For riveting done in the shop the rivets were probably heated to a light cherry-red color and driven with a pressure-type riveter. This type of riveter, usually called a "bull" riveter, squeezed the rivet with a pressure of perhaps as high as 50 to 80 tons (445 to 712 kN) and drove the rivet with one stroke. Because of this great pressure the rivet in its soft state was forced to fill the hole very satisfactorily. This type of riveting was much to be preferred over that done with the pneumatic hammer, but no greater nominal strengths were allowed by riveting specifications. The bull riveters were built for much faster operation than were the portable hand riveters, but the latter riveters were needed for places that were not easily accessible (i.e., field erection).

As the rivet cooled, it shrank or contracted and squeezed together the parts being connected. The squeezing effect actually caused considerable transfer of stress between the parts being connected to take place by friction. The amount of friction was

The U.S. Customs Court, Federal Office Building under construction in New York City. (Courtesy of Bethlehem Steel Corporation.)

not dependable, however, and the specifications did not permit its inclusion in the strength of a connection. Rivets shrink diametrically as well as lengthwise and actually become somewhat smaller than the holes that they are assumed to fill. (Permissible strengths for rivets are actually given in terms of the nominal cross-sectional areas of the rivets before driving.)

Some shop rivets were driven cold with tremendous pressures. Obviously, the cold-driving process worked better for the smaller size rivets (probably 3/4 in in diameter or less) although larger ones were successfully used. Cold-driven rivets fill the holes better, eliminate the cost of heating, and are stronger because the steel is cold worked. There is, however, a reduction of clamping force since the rivets do not shrink after driving.

13.6   TYPES OF RIVETS

The sizes of rivets used in ordinary construction work were 3/4 in and 7/8 in in diameter, but they could be obtained in standard sizes from 1/2 in to 1 1/2 in in 1/8-in increments. (The smaller sizes were used for small roof trusses, signs, small towers, etc., while the larger sizes were used for very large bridges or towers and very tall buildings.) The use of more than one or two sizes of rivets or bolts on a single job is usually undesirable because it is expensive and inconvenient to punch different-size holes in a member in the shop, and the installation of different-size rivets or bolts in the field may be confusing. Some cases arise where it is absolutely necessary to have different sizes, as where smaller rivets or bolts are needed for keeping the proper edge distance in certain sections, but these situations should be avoided if possible.

Rivet heads, usually round in shape, were called *button heads*; but if clearance requirements dictated, the head was flattened or even countersunk and chipped flush. These situations are shown in Fig. 13.20.

The countersunk and chipped-flush rivets did not have sufficient bearing areas to develop full strength and the designer usually discounted their computed strengths by 50 percent. A rivet with a flattened head was preferred over a countersunk rivet, but if a smooth surface was required, the countersunk and chipped-flush rivet was necessary. This latter type of rivet was appreciably more expensive than the button head type, in addition to being weaker; and it was not used unless absolutely necessary.

There are three ASTM classifications for rivets for structural steel applications, as described in the following paragraphs.

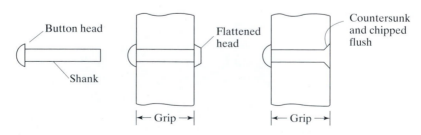

FIGURE 13.20

Types of rivets.

13.6.1 ASTM Specification A502, Grade 1

These rivets were used for most structural work. They had a low carbon content of about 0.80 percent, were weaker than the ordinary structural carbon steel, and had a higher ductility. The fact that these rivets were easier to drive than the higher-strength rivets was the main reason that when rivets were used they probably were A502, grade 1, regardless of the strength of the steel used in the structural members.

13.6.2 ASTM Specification A502, Grade 2

These carbon-manganese rivets have higher strengths than the grade 1 rivets and were developed for the higher-strength steels. Their higher strength permitted the designer to use fewer rivets in a connection and thus smaller gusset plates.

13.6.3 ASTM Specification A502, Grade 3

These rivets have the same nominal strengths as the grade 2 rivets, but they have much higher resistance to atmospheric corrosion equal to approximately four times that of carbon steel without copper.

13.7 STRENGTH OF RIVETED CONNECTIONS—RIVETS IN SHEAR

The factors determining the strength of a rivet are its grade, its diameter, and the thickness and arrangement of the pieces being connected. The actual distribution of stress around a rivet hole is difficult to determine, if it can be determined at all; and to simplify the calculations it is assumed to vary uniformly over a rectangular area equal to the diameter of the rivet times the thickness of the plate.

The strength of a rivet in-single shear is the nominal shearing strength times the cross-sectional area of the shank of the rivet. Should a rivet be in double shear, its shearing strength is considered to be twice its single-shear value.

The nominal tension and shearing strengths for rivets and A307 bolts were given in Table 12.5 (LRFD Table J3.2). These values are for static loads only and are repeated in Table 13.2. Notice that the nominal shear strength of the A307 bolts is not affected if the bolt threads are in the shear plane.

TABLE 13.2   LRFD Nominal Tensile and Shearing Strengths for Rivets and A307 Bolts

Fastener type	Tensile strength (ksi)	Shearing strength in bearing-type connections (ksi)
A502, grade 1, hot-driven rivets	45.0, $\phi = 0.75$	25.0, $\phi = 0.75$
A502, grade 2 or 3, hot-driven rivets	60.0, $\phi = 0.75$	33.0, $\phi = 0.75$
A307 bolts	45.0, $\phi = 0.75$	24.0, $\phi = 0.75$

Examples 13-9 and 13-10 illustrate the calculations necessary to determine the design strengths of existing connections or to design riveted connections. Little comment is made here concerning A307 bolts. The reason is that all the calculations for these fasteners are made exactly as they are for rivets, except that the shearing strengths given by the LRFD Specification are different. Only one brief example with these common bolts (13-11) is included.

Example 13-9

Determine the design strength ϕP_n of the bearing-type connection shown in Fig. 13.21. A36 steel and A502, grade 1 rivets are used in the connection and it is assumed that standard-size holes are used and that edge distances and center-to-center distances are > 1.5 in and 3 in respectively. Neglect block shear.

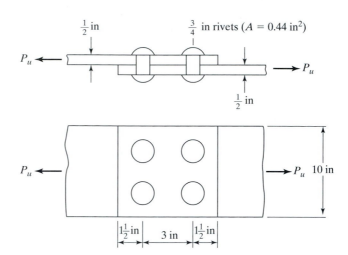

FIGURE 13.21

Solution. Design tensile force applied to plates

$$A_g = \left(\tfrac{1}{2}\right)(10) = 5.00 \text{ in}^2$$
$$A_n = \left[\left(\tfrac{1}{2}\right)(10) - (2)\left(\tfrac{7}{8}\right)\left(\tfrac{1}{2}\right)\right] = 4.125 \text{ in}^2 = A_e$$
$$\phi_t P_n = \phi_t F_y A_g = (0.90)(36)(5.00) = 162 \text{ k}$$
$$\phi_t P_n = \phi_t F_u A_e = (0.75)(58)(4.125) = 179.4 \text{ k}$$

Rivets in single shear and bearing on 1/2 in

Design shearing strength of rivets $= (0.75)(0.44)(25)(4) = 33.0 \text{ k} \leftarrow$

Design bearing strength of rivets with $L_C = 1.5 - \frac{0.875}{2} = 1.06 \text{ in}$

$$\phi R_n = (0.75)\,(1.2)\,(1.06)\,\left(\tfrac{1}{2}\right)(58)\,(4) = 110.7 \text{ k}$$
$$\leq (0.75)\,(2.4)\,\left(\tfrac{3}{4}\right)\left(\tfrac{1}{2}\right)(58)\,(4) = 156.6 \text{ k}$$
$$\phi_t P_n = 33.0 \text{ k}$$

Example 13-10

How many 7/8-in A502, grade 1 rivets are required for the connection shown in Fig. 13.22 if the plates are A36, if standard-size holes are used, and the edge distances and center-to-center distances are > 1.5 in and 3 in, respectively?

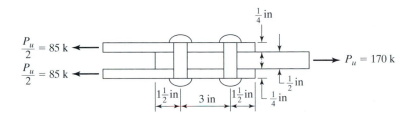

FIGURE 13.22

Solution. Bolts in double shear and bearing on .5 in:

$$\text{Design shear strength of 1 rivet} = (0.75)\,(2 \times 0.6)\,(25) = 22.5 \text{ k} \leftarrow$$

$$\text{Design bearing strength of 1 bolt with } L_C = 1.5 - \frac{1.0}{2} = 1.0$$

$$\phi r_n = (0.75)\,(1.2)\,(1.0)\,\left(\tfrac{1}{2}\right)(58) = 26.1 \text{ k}$$
$$\leq (0.75)\,(2.4)\,\left(\tfrac{7}{8}\right)\left(\tfrac{1}{2}\right)(58) = 45.7 \text{ k}$$

$$\text{No. of rivets required} = \frac{170}{22.5} = 7.56$$

Use eight 7/8-in A502 grade 1 rivets.

Example 13-11

Repeat Example 13-10 using $\tfrac{7}{8}$-in A307 bolts.

Solution. Bolts in double shear and bearing on .5 in:

$$\text{Design shear strength of 1 bolt} = (0.75)\,(2 \times 0.6)\,(24) = 21.6 \text{ k} \leftarrow$$

$$\text{Design bearing strength of 1 bolt with } L_c = 1.5 - \tfrac{1}{2} = 1.0 \text{ in}$$

$$= (0.75)\,(1.2)\,(1.0)\,\left(\tfrac{1}{2}\right)(58) = 26.1 \text{ k}$$
$$\leq (0.75)\,(2.4)\,(0.875)\,\left(\tfrac{1}{2}\right)(58) = 45.7 \text{ k}$$

No. of bolts required $= \dfrac{170}{21.6} = 7.87$

Use eight 7/8 in A307 bolts.

13.8   COMPUTER EXAMPLE

In Example 13-12, the computer program INSTEP 32 is used to determine the ultimate load that an eccentrically loaded connection can support.

Example 13-12

Using the enclosed computer disk, determine the value of P_u for the bearing type connection of Fig. 13.23 if 7/8 in A325 bolts threads excluded from shear plane and A572 Grade 50 steel are used.

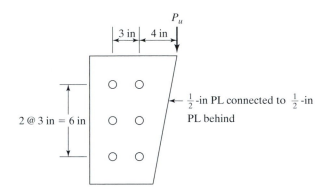

FIGURE 13.23

Solution

Bolted Connection Design

Input:

Factored design loads
Vu = 0 kips
Pu = 0 kips
Vs = 0 kips
Ps = 0 kips
Xo = 5.5 in from CG
Theta = 90 deg

from vertical
Joint parameters
Bolt type A325
Bolt size 0.875 inches
Threads SS bearing, excluded
Joint style Eccentric shear only
t1 = 0.5 in
t2 = 0.5 in

Output:

Bolted Joint Design Summary
CGx: 1.5 in
CGy: 3.0 in
Vu: 65.2152 kips
e: 1.50781 in from CG
Iterations: 14

PROBLEMS

For each of the problems listed, the following information is to be used unless otherwise indicated: (a) A36 steel; (b) standard-size holes; (c) threads of bolts excluded from shear plane.

13-1 to 13-7   Determine the resultant load on the most stressed bolt in the eccentrically loaded connections shown using the elastic method.

13-1.

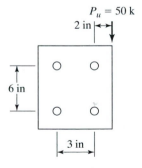

$P_u = 50$ k

2 in

6 in

3 in

FIGURE P13-1 (*Ans.* 21.73 k.)

13-2.

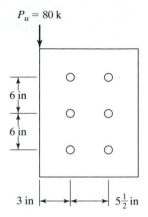

$P_u = 80$ k

6 in

6 in

3 in $5\frac{1}{2}$ in

FIGURE P13-2

13-3.

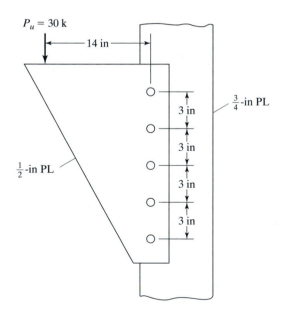

$P_u = 30$ k

14 in

3 in

$\frac{3}{4}$-in PL

3 in

$\frac{1}{2}$-in PL

3 in

3 in

FIGURE P13-3 (*Ans.* 28.64 *k*)

13-4.

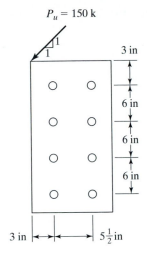

FIGURE P13-4

13-5.

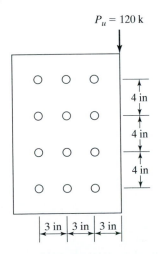

FIGURE P13-5 (*Ans.* 21.87 k)

13-6.

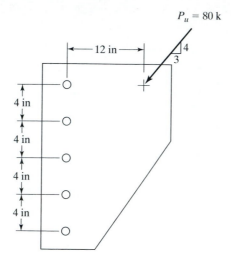

FIGURE P13-6

13-7.

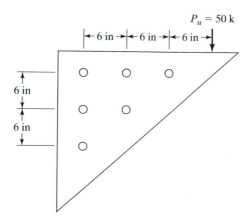

FIGURE P13-7 (*Ans.* 33.74 k)

13-8.   Repeat Prob. 13-2 using the reduced eccentricity method given in Section 13-1 of this chapter.

13-9.   Using the elastic method determine the design strength of the bearing-type connection shown. The bolts are 3/4 in A325 and are in single shear and bearing on 5/8
in (*Ans.* 57.7 *k*)

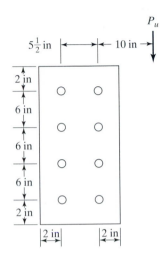

FIGURE P13-9

13-10.   Using the elastic method determine the design strength P_u for the slipcritical connection shown. The 7/8 in A325 bolts are in "double shear." All plates are 1/2 in
thick. Surfaces are Class A.

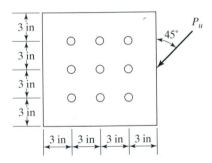

FIGURE P13-10

13-11.   Repeat Prob. 13-9 using the ultimate strength tables entitled "Coefficients C for Eccentrically Loaded Bolt Groups" in Part 7 of the LRFD Manual. (*Ans.* 69.3 k)

13-12.   Repeat Prob. 13-10 using the ultimate strength tables entitled "Coefficients C for
Eccentrically Loaded Bolt Groups" in Part 7 of the LRFD Manual.

13-13. Is the bearing-type connection shown in the accompanying illustration sufficient to resist the 200 k load which passes through the center of gravity of the bolt group? (*Ans.* Yes. $F_t = 67$ ksi > 33.33 ksi)

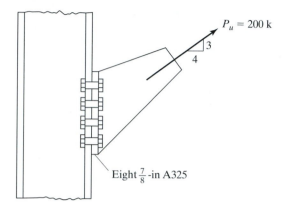

FIGURE P13-13

13-14. Repeat Prob. 13-13 if slip-critical bolts are used and if surfaces are Class A.

13-15. If the load shown in the accompanying bearing-type illustration passes through the center of gravity of the bolt group, how large can it be? (*Ans.* 310 k)

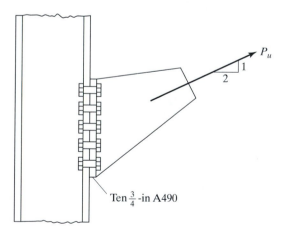

FIGURE P13-15

13-16. Repeat Prob. 13-15 if bolts are A325.

13-17. Determine the number of 3/4-in A325 bolts required in the angles and in the flange of the W shape shown in the accompanying illustration if a bearing-type (snug-tight) connection is used. Use 50 ksi steel $F_u = 65$ ksi, $L_c = 1.0$ in (*Ans.* 6.31 say 7)

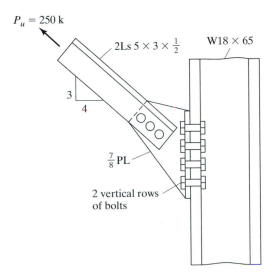

$P_u = 250$ k

2Ls $5 \times 3 \times \frac{1}{2}$

W18 × 65

3

4

$\frac{7}{8}$ PL

2 vertical rows
of bolts

FIGURE P13-17

13-18. Are the bolts shown in this 16 in.-long hanger satisfactory to resist direct tension and prying? There are eight 7/8-in A325 bolts and they are spaced 4 in. on center longitudinally.

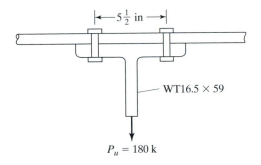

$5\frac{1}{2}$ in

WT16.5 × 59

$P_u = 180$ k

FIGURE P13-18

13-19. Determine the design strength ϕP_n of the connection shown if 3/4-in A502, grade 1 rivets and A36 steel are used. (*Ans.* 99 k)

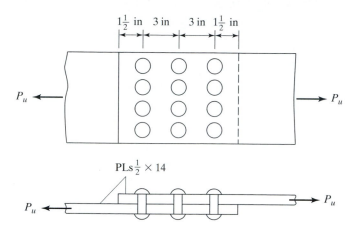

FIGURE P13-19

13-20. The truss tension member shown consists of a single-angle $5 \times 3 \times 5/16$ and is connected to a 1/2-in gusset plate with five 7/8-in A502, grade 1 rivets. Determine P_u if U is assumed to equal 0.9. Neglect block shear. Steel is A36.

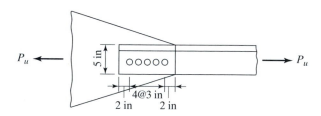

FIGURE P13-20

13-21. How many 3/4-in A502, grade 1 rivets are needed to carry the load shown in the accompanying illustration? (*Ans.* 26.67, use 27 rivets)

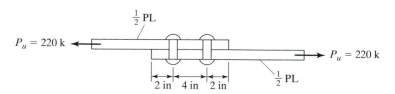

FIGURE P13-21

13-22.   Repeat Prob. 13-21 if A307 bolts are used.

13-23.   How many A502, grade 1 rivets with 1-in diameters need to be used for the butt joint shown? (*Ans.* 6.79 say 7 or 8 rivets.)

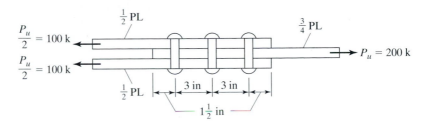

FIGURE P13-23

13-24.   How many 7/8-in A307 bolts are required for the connection shown in the accompanying illustrations? Factored loads are shown.

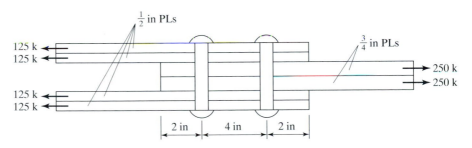

FIGURE P13-24

13-25.   For the connection shown in the accompanying illustration $P_u = 650$ k, determine the number of 7/8-in A502, grade 2 rivets required. (*Ans.* 21.88, use 22 rivets)

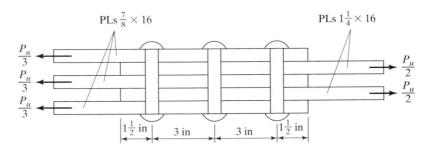

FIGURE P13-25

13-26. For the beam shown in the accompanying illustration, what is the required spacing of 7/8-in A307 bolts if $V_u = 150$ k? Assume $L_c = 1$ in.

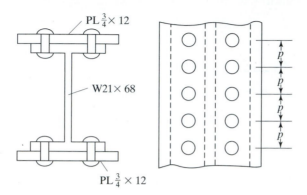

FIGURE P13-26

13-27. Is the connection shown sufficient to resist the 100-k load that passes through the center of gravity of the rivet group? (Ans; No. 13.56 ksi < 22.73 ksi)

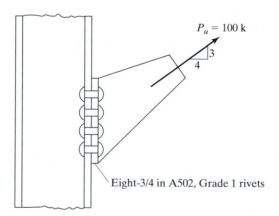

FIGURE P13-27

13-28. Repeat listed using INSTEP32

C H A P T E R   1 4

Welded Connections

14.1 GENERAL

Welding is a process by which metallic parts are connected by heating their surfaces to a plastic or fluid state and allowing the parts to flow together and join (with or without the addition of other molten metal). It is impossible to determine when welding originated, but it was at least several thousand years ago. Metal-working, including welding, was quite an art in ancient Greece three thousand years ago, but welding had undoubtedly been performed for many centuries before that. Ancient welding probably was a forging process in which the metals were heated to a certain temperature (not to the melting stage) and hammered together.

Although modern welding has been available for many years, it has only come into its own in the last few decades for the building and bridge phases of structural engineering. The adoption of structural welding was quite slow for several decades because many engineers thought that welding had two major disadvantages: (1) welds had reduced fatigue strength compared with riveted and bolted connections, and (2) it was impossible to ensure a high quality of welding without unreasonably extensive and costly inspection.

These attitudes persisted for many years, although tests began to indicate that neither reason was valid. Regardless of their validity, these views were widely held and undoubtedly slowed down the use of welding–particularly for highway bridges and, to an even greater extent, railroad bridges. Today, most engineers agree that welded joints have considerable fatigue strength. They will also admit that the rules governing the qualification of welders, the better techniques applied, and the excellent workmanship requirements of the AWS (American Welding Society) specifications make the inspection of welding a much less difficult problem. Furthermore, the chemistry of steels manufactured today is especially formulated to improve their weldability. Consequently, welding is now permitted for almost all structural work other than for some bridges.

On the subject of welding, it is interesting to consider welded ships. Ships are subjected to severe impactive loadings that are difficult to predict, yet naval architects use all-welded ships with great success. A similar discussion can be made for airplanes and

Worker adjusts steel girder. (Courtesy of Bethlehem Steel.)

aeronautical engineers. The slowest adoption of structural welding was for railroad bridges. These bridges are undoubtedly subjected to heavier live loads than highway bridges, larger vibrations, and more stress reversals; but are their stress situations as serious and as difficult to predict as those for ships and planes?

14.2   ADVANTAGES OF WELDING

Today it is possible to make use of the many advantages that welding offers, since the fatigue and inspection fears have been largely eliminated. Following are several of the many advantages that welding offers:

1. To most designers, the first advantage is economic, because the use of welding permits large savings in pounds of steel used. Welded structures allow the elimination of a large percentage of the gusset and splice plates necessary for bolted structures as well as the elimination of bolt heads. In some bridge trusses, it may be possible to save up to 15 percent or more of the steel weight by using welding.

2. Welding has a much wider range of application than bolting. Consider a steel pipe column and the difficulties of connecting it to other steel members by bolting. A bolted connection may be virtually impossible, but a welded connection presents few difficulties. Many similar situations can be imagined in which welding has a decided advantage.

3. Welded structures are more rigid because the members often are welded directly to each other. Frequently, the connections for bolted structures are made through intermediate connection angles or plates that deform due to load transfer, making the entire structure more flexible. On the other hand, greater rigidity can be a disadvantage where simple end connections with little moment resistance are desired. In such cases, designers must be careful as to the type of joints they specify.

4. The process of fusing pieces together creates the most truly continuous structures. Fusing results in one-piece construction, and because welded joints are as strong as or stronger than the base metal, no restrictions have to be placed on the joints. This continuity advantage has permitted the erection of countless slender and graceful statically indeterminate steel frames throughout the world. Some of the more outspoken proponents of welding have referred to bolted structures, with their heavy plates and abundance of bolts, as looking like tanks or armored cars compared with the clean, smooth lines of welded structures. For a graphic illustration of this advantage, compare the moment-resisting connections of Fig. 15.5.

5. It is easier to make changes in design and to correct errors during erection (and at less expense) if welding is used. A closely related advantage has certainly been illustrated in military engagements during the past few wars by the quick welding repairs made to military equipment under battle conditions.

6. Another item that is often important is the relative silence of welding. Imagine the importance of this fact when working near hospitals or schools or when making additions to existing buildings. Anyone with close-to-normal hearing who has attempted to work in an office within several hundred feet of a bolted job can attest to this advantage.

7. Fewer pieces are used, and as a result, time is saved in detailing, fabrication, and field erection.

14.3 AMERICAN WELDING SOCIETY

The American Welding Society's *Structural Welding Code*[1] is the generally recognized standard for welding in the United States. The LRFD Specification clearly states that the provisions of the AWS Code apply under the LRFD Specification with only a few minor exceptions, and these are listed in LRFD Specification J2. Both the AWS and the AASHTO Specifications cover dynamically loaded structures. Normally, however, the AWS Specification is used for designing buildings subject to dynamic loads unless the contract documents state otherwise.

14.4 TYPES OF WELDING

Although both gas and arc welding are available, almost all structural welding is arc welding. Sir Humphry Davy discovered in 1801 how to create an electric arc by bringing close together two terminals of an electric circuit of relatively high voltage. Although he is generally given credit for the development of modern welding, a good many years elapsed after his discovery before welding was actually performed with the electric arc. (His work was of the greatest importance to the modern structural world, but it is interesting to note that many people say his greatest discovery was not the electric arc, but rather a laboratory assistant whose name was Michael Faraday.) Several Europeans formed welds of one type or another in the 1880s with

[1]American Welding Society, *Structural Welding Code-Steel*, AWS D.1.1-00 (Miami: AWS: 2000).

the electric arc, while in the United States the first patent for arc welding was given to Charles Coffin of Detroit in 1889.[2]

The figures shown in this chapter illustrate the necessity of supplying additional metal to the joints being welded to give satisfactory connections. In electric-arc welding, the metallic rod, which is used as the electrode, melts off into the joint as it is being made. When gas welding is used, it is necessary to introduce a metal rod known as a *filler* or *welding rod*.

In gas welding, a mixture of oxygen and some suitable type of gas is burned at the tip of a torch or blowpipe held in the welder's hand or by machine. The gas used in structural welding usually is acetylene, and the process is called *oxyacetylene welding*. The flame produced can be used for flame cutting of metals as well as for welding. Gas welding is fairly easy to learn, and the equipment used is rather inexpensive. It is a slow process, however, compared with other means of welding, and normally it is used for repair and maintenance work and not for the fabrication and erection of large steel structures.

In arc welding, an electric arc is formed between the pieces being welded and an electrode is held in the operator's hand with some type of holder, or by an automatic machine. The arc is a continuous spark that upon contact brings the electrode and the pieces being welded to the melting point. The resistance of the air or gas between the electrode and the pieces being welded changes the electrical energy into heat. A temperature of 6000 to 10,000°F is produced in the arc. As the end of the electrode melts, small droplets or globules of the molten metal are formed and actually are forced by the arc across to the pieces being connected, which penetrate the molten metal to become a part of the weld. The amount of penetration can be controlled by the amount of current consumed. Since the molten droplets of the electrodes actually are propelled into the weld, arc welding can be successfully used for overhead work.

A pool of molten steel can hold a fairly large amount of gases in solution, and if not protected from the surrounding air will chemically combine with oxygen and nitrogen. After cooling, the welds will be relatively porous due to the little pockets formed by the gases. Such welds are relatively brittle and have much less resistance to corrosion. A weld joint can be shielded by using an electrode coated with certain mineral compounds. The electric arc causes the coating to melt and creates an inert gas or vapor around the area being welded. The vapor acts as a shield around the molten metal and keeps it from coming freely in contact with the surrounding air. It also deposits a slag in the molten metal, which has less density than the base metal and comes to the surface to protect the weld from the air while the weld cools. After cooling, the slag can easily be removed by peening and wire brushing (such removal being absolutely necessary before painting or application of another weld layer). A picture showing the elements of the shielded arc welding process is shown in Fig. 14.1. This figure is taken from the *Procedure Handbook of Arc Welding Design and Practice* published by the Lincoln Electric Company. *Shielded metal arc welding* is frequently abbreviated here with the letters SMAW.

The type of welding electrode used is very important because it decidedly affects the weld properties such as strength, ductility, and corrosion resistance. Quite a number

---

[2]Lincoln Electric Company, *Procedure Handbook of Arc Welding Design and Practice*, 11th ed., 1957, Part I.

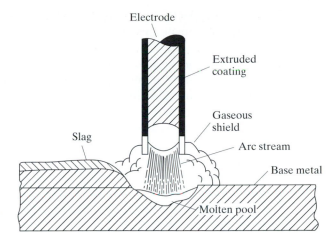

FIGURE 14.1

Elements of the shielded metal arc welding process (SMAW).

of different kinds of electrodes are manufactured, the type to be used for a certain job being dependent upon the metal being welded, the amount of material that needs to be added, the position of the work, etc. The electrodes fall into two general classes—the *lightly coated electrodes* and the *heavily coated electrodes*.

The heavily coated electrodes are normally used in structural welding because the melting of their coatings produces very satisfactory vapor shields around the work as well as slag in the weld. The resulting welds are stronger, more resistant to corrosion, and more ductile than are those produced with lightly coated electrodes. When the

Shielded metal arc welding (SMAW) and electrode just before starting an arc to fillet weld the clip angle to the beam web. (Courtesy of the American Institute of Steel Construction, Inc.)

lightly coated electrodes are used, no attempt is made to prevent oxidation and no slag is formed. The electrodes are lightly coated with some arc-stabilizing chemical such as lime.

Submerged (or hidden) arc welding (SAW) is an automatic process in which the arc is covered with a mound of granular fusible material and thus hidden from view. A bare metal electrode is fed from a reel and melted and deposited as filler material. The electrode, power source, and a hopper of flux are attached to a frame that is placed on rollers and that moves at a certain rate as the weld is formed. SAW welds are quickly and efficiently made and are of high quality, exhibiting high impact strength and corrosion resistance and good ductility. Furthermore, they provide deeper penetration with the result that the area effective in resisting loads is larger. A large percentage of the welding done for bridge structures is SAW. If a single electrode is used, the size of the weld obtained with a single pass is limited. Multiple electrodes may be used, however, permitting much larger welds.

Welds made by the SAW process (automatic or semiautomatic) are consistently of high quality and are very suitable for long welds. One disadvantage is that the work must be positioned for near flat or horizontal welding.

Another type of welding is *flux-cored arc welding* (FCAW). In this process, a flux-filled steel tube electrode is continuously fed from a reel. Gas shielding and slag are formed from the flux. The AWS Specification (4.14) provides limiting sizes for welding electrode diameters and weld sizes, as well as other requirements pertaining to welding procedures.

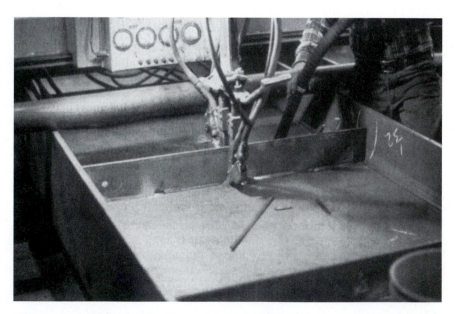

Submerged arc welding (SAW). (Courtesy of the American Institute of Steel Construction, Inc.)

14.5 PREQUALIFIED WELDING

The AWS accepts four welding processes as being prequalified. In this context, the word *prequalified* means that processes are acceptable without the necessity of further proof of their suitability by procedure qualification tests. What we are saying is that, based on many years of experience, sound weld metal with the desired properties can be deposited if the work is performed in accordance with the requirements of the Structural Welding Code of the AWS. The processes that are listed in AWS Specification 1.3.1 are (1) shielded metal arc welding (SMAW), (2) submerged arc welding (SAW), (3) gas metal arc welding (GMAW), and (4) flux cored arc welding (FCAW). The SMAW process is the usual process used for hand welding, while the other three are usually automatic or semiautomatic.

14.6 WELDING INSPECTION

Three steps must be taken to ensure good welding for a particular job: (1) establishment of good welding procedures, (2) use of prequalified welders, and (3) employment of competent inspectors in both the shop and the field.

When the procedures established by the AWS and AISC for good welding are followed, and when welders are used who have previously been required to prove their ability, good results usually are obtained. To make absolutely sure, however, well-qualified inspectors are needed.

Good welding procedure involves the selection of proper electrodes, current, and voltage; the properties of base metal and filler; and the position of welding—to name only a few factors. The usual practice for large jobs is to employ welders who have certificates showing their qualifications. In addition, it is not a bad practice to have each person make an identifying mark on each weld so that those frequently doing poor work can be identified. This practice tends to improve the general quality of the work performed.

14.6.1 Visual Inspection

Another factor that will cause welders to perform better work is simply the presence of an inspector who they feel knows good welding when he or she sees it. A good inspector should have done welding and spent much time observing the work of good welders. From this experience he or she should be able to know if a welder is obtaining satisfactory fusion and penetration. He or she also should be able to recognize good welds in regard to shape, size, and general appearance. For instance, the metal in a good weld should approximate its original color after it has cooled. If it has been overheated, it may have a rusty and reddish-looking color. An inspector can use various scales and gages to check the sizes and shapes of welds.

Visual inspection by a competent person usually gives a good indication of the quality of welds, but is not a perfect source of information, especially regarding the subsurface condition of the weld. It surely is the most economical inspection method and is particularly useful for single-pass welds. This method, however, is only good for picking up surface imperfections. There are several methods for determining the internal soundness of a weld, including the use of penetrating dyes and magnetic particles,

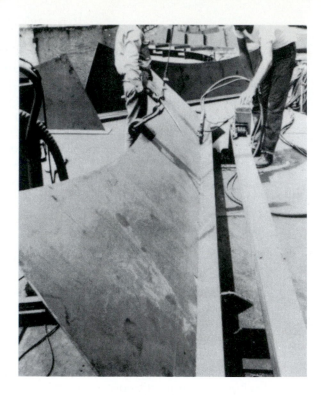

Lincoln ML-3 Squirtwelder mounted on a self-propelled trackless trailer deposits this 1/4-in web-to-flange weld at 28 in/min. (Courtesy of the Lincoln Electric Company.)

ultrasonic testing, and radiographic procedures. These methods can be used to detect internal defects such as porosity, weld penetration, and the presence of slag.

14.6.2 Liquid Penetrants

Various types of dyes can be spread over weld surfaces. These dyes will penetrate into the surface cracks of the weld. After the dye has penetrated into the crack, the excess surface material is wiped off and a powdery developer is used to draw the dye out of the cracks. The outlines of the cracks can then be seen with the eye. Several variations of this method are used to improve the visibility of the defects, including the use of fluorescent dyes. After the dye is drawn from the cracks, they stand out brightly under a black light.[3] Like visual inspection, this method enables us to detect cracks that are open to the surface.

14.6.3 Magnetic Particles

In this method, the weld being inspected is magnetized electrically. Cracks that are at or near the surface of the weld cause north and south poles to form on each side of the cracks. Dry iron powdered filings or a liquid suspension of particles are placed on the weld. These particles form patterns when many of them cling to the cracks, showing the locations of cracks and indicating their size and shape. Only cracks, seams, inclusions,

---

[3]James Hughes, "It's Superinspector," *Steelways* 25, no. 4 (New York: American Iron and Steel Institute, September/October, 1969), pp. 19–21.

etc., within about 1/10 in of the surface can be located by this method. A disadvantage of this method is that if multilayer welds are used, the method has to be applied to each layer.

14.6.4 Ultrasonic Testing

In recent years, the steel industry has applied ultrasonics to the manufacture of steel. Although the equipment is expensive, the method is quite useful in welding inspections as well. Sound waves are sent through the material being tested and are reflected from the opposite side of the material. These reflections are shown on a cathode ray tube. Defects in the weld will affect the time of the sound transmission. The operator can read the picture on the tube and then locate flaws and learn how severe they are. Ultrasonic testing can sucessfully be used to locate discontinuities in carbon and low-alloy steels, but it doesn't work too well for some stainless steels and for extremely coarse-grained steels.

14.6.5 Radiographic Procedures

The more expensive radiographic methods can be used to check occasional welds in important structures. From these tests, it is possible to make good estimates of the percentage of bad welds in a structure. Portable x-ray machines (where access is not a problem) and radium or radioactive cobalt for making pictures are excellent but expensive methods of testing welds. These methods are satisfactory for butt welds (such as for the welding of important stainless steel piping at chemical and nuclear projects), but they are not satisfactory for fillet welds because the pictures are difficult to interpret. A further disadvantage of such methods is the radioactive danger. Careful procedures have to be used to protect the technicians as well as nearby workers. On a construction job, this danger generally requires night inspection of welds, when only a few workers are near the inspection area. (Normally a very large job would be required before the use of the extremely expensive radioactive materials could be justified.)

A properly welded connection can always be made much stronger—perhaps as much as two times stronger—than the plates being connected. As a result, the actual strength is much greater than is required by the specifications. The reasons for this extra strength are as follows: the electrode wire is made from premium steel, the metal is melted electrically (as is done in the manufacture of high-quality steels), and the cooling rate is quite rapid. As a result, it is rare for a welder to make a weld of less strength than required by the design.

14.7 CLASSIFICATION OF WELDS

Three separate classifications of welds are described in this section. These classifications are based on the types of welds made, the positions of the welds, and the types of joints used.

14.7.1 Type of Weld

The two main types of welds are the *fillet welds* and the *groove welds*. In addition, there are plug and slot welds, which are not as common in structural work. These four types of welds are shown in Fig. 14.2.

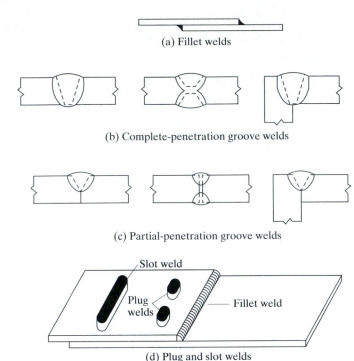

FIGURE 14.2

Four types of structural welds.

The fillet welds will be shown to be weaker than groove welds, although, most structural connections (about 80 percent) are made with fillet welds. Any person who has experience in steel structures will understand why fillet welds are more common than groove welds. Groove welds, (which are welds made in grooves between the members to be joined), are used when the members to be connected are lined up in the same plane. To use them in every situation would mean that the members would have to fit almost perfectly, and unfortunately, the average steel structure does not fit together that way. Have you ever seen steel workers pulling and ramming steel members to get them into position? When members are allowed to lap over each other, larger tolerances are allowable in erection, and fillet welds are the welds used. Nevertheless, groove welds are quite common for many connections, such as column splices, butting of beam flanges to columns, etc., and they make up about 15 percent of structural welding. Groove welds can be either *complete-penetration* welds, which extend for the full thickness of the part being connected, or *partial penetration* welds, which extend for only part of the member thickness.

Groove welds are generally more expensive than fillet welds because of the costs of preparation. In fact, groove welds can cost up to 50–100 percent more than fillet welds.

A plug weld is a circular weld that passes through one member into another, thus joining the two together. A slot weld is a weld formed in a slot or elongated hole that joins one member to the other member through the slot. The slot may be partly or fully filled with weld material. These two expensive types of welds may occasionally be used when members lap over each other and the desired length of fillet welds cannot be obtained. They may also be used to stitch together parts of a member, such as the fastening of cover plates to a built-up member.

A plug or slot weld is not generally considered suitable for transferring tensile forces perpendicular to the faying surface because there is not usually much penetration of the weld into the member behind the plug or slot and the fact is that resistance to tension is provided primarily by penetration.

Structural designers accept plug and slot welds as being satisfactory for stitching the different parts of a member together, but many designers are not happy using these welds for the transmission of shear forces. The penetration of the welds from the slots or plugs into the other members is questionable; in addition, there can be critical voids in the welds that cannot be detected with the usual inspection procedures.

14.7.2   Position

Welds are referred to as *flat, horizontal, vertical,* or *overhead*—listed in order of their economy, with the flat welds being the most economical and the overhead welds being the most expensive. A moderately skilled welder can do a very satisfactory job with a flat weld, but it takes the very best to do a good job with an overhead weld. Although the flat welds often are done with an automatic machine, most structural welding is done by hand. We indicated previously that the assistance of gravity is not necessary for the forming of good welds, but it does speed up the process. The globules of the molten electrodes can be forced into the overhead welds against gravity, and good welds will result; however, they are slow and expensive to make, so it is desirable to avoid them whenever possible. These types of welds are shown in Fig. 14.3.

14.7.3   Type of Joint

Welds can be further classified according to the type of joint used: *butt, lap, tee, edge, corner,* etc. (See Fig. 14.4.)

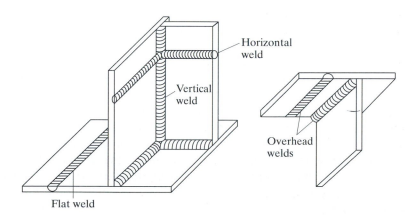

FIGURE 14.3

Weld positions.

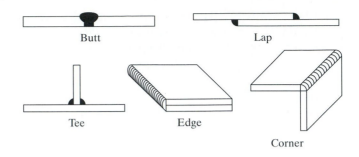

FIGURE 14.4

Types of weld joints.

14.8   WELDING SYMBOLS

Figure 14.5 presents the various welding symbols developed by the American Welding Society. With this excellent shorthand system, a great deal of information can be presented in a small space on engineering plans and drawings. These symbols require only a few lines and numbers and remove the necessity of drawing in the welds and making long descriptive notes. It is certainly desirable for steel designers and draftsmen to use this standardized system. If most of the welds on a drawing are the same size, a note to that effect can be given and the symbols omitted, except for the off-size welds.

The purpose of this section is to give a general idea of the appearance of welding symbols and the information they can convey. (For more detailed information, refer to the LRFD Handbook and other materials published by the AWS.) The information presented in Fig. 14.5 may be quite confusing; for this reason, a few very common symbols for fillet welds are presented in Fig. 14.6, together with an explanation of each.

14.9   GROOVE WELDS

When complete penetration groove welds are subjected to axial tension or axial compression, the weld stress is assumed to equal the load divided by the net area of the weld. Three types of groove welds are shown in Fig. 14.7. The square groove joint, shown in part (a) of the figure, is used to connect relatively thin material up to roughly 5/16 in (8 mm) thickness. As the material becomes thicker, it is necessary to use the single-vee groove welds and the double-vee groove welds illustrated in parts (b) and (c), respectively, of Fig. 14.7. For these two welds, the members are bevelled before welding to permit full penetration of the weld.

The groove welds shown in Fig. 14.7 are said to have *reinforcement*. Reinforcement is added weld metal that causes the throat dimension to be greater than the thickness of the welded material. Because of reinforcement, groove welds may be referred to as 125 percent, 150 percent, etc., according to the amount of extra thickness at the weld. There are two major reasons for having reinforcement: (1) reinforcement gives a little extra strength because the extra metal takes care of pits and other irregularities, and (2) the welder can easily make the weld a little thicker than the welded material. It would be a difficult, if not impossible, task to make a perfectly smooth weld with no places that were thinner or thicker than the material welded.

Reinforcement undoubtedly makes groove welds stronger and better when they are subjected to static loads. When the connection is to be subjected to vibrating loads,

Prequalified Welded Joints

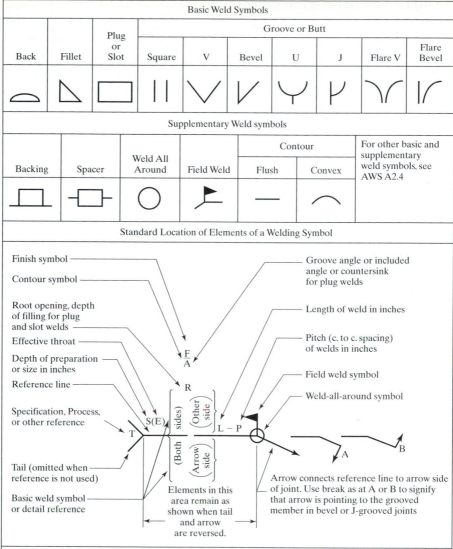

		Plug or Slot	Groove or Butt						
Back	Fillet		Square	V	Bevel	U	J	Flare V	Flare Bevel

Basic Weld Symbols

Supplementary Weld symbols

		Weld All Around	Field Weld	Contour		For other basic and supplementary weld symbols, see AWS A2.4
Backing	Spacer			Flush	Convex	

Standard Location of Elements of a Welding Symbol

Finish symbol

Contour symbol

Root opening, depth of filling for plug and slot welds

Effective throat

Depth of preparation or size in inches

Reference line

Specification, Process, or other reference

Tail (omitted when reference is not used)

Basic weld symbol or detail reference

Groove angle or included angle or countersink for plug welds

Length of weld in inches

Pitch (c. to c. spacing) of welds in inches

Field weld symbol

Weld-all-around symbol

Arrow connects reference line to arrow side of joint. Use break as at A or B to signify that arrow is pointing to the grooved member in bevel or J-grooved joints

Elements in this area remain as shown when tail and arrow are reversed.

Note:
Size, weld symbol, length of weld, and spacing must read in that order, from left to right, along the reference line. Neither orientation of reference nor location of the arrow alters this rule.

The perpendicular leg of ⊾, V, ⊬, ⊮, weld symbols must be at left.

Arrow and other side welds are of the same size unless otherwise shown. Dimensions of fillet welds must be shown on both the arrow side and the other side symbol.

The point of the field weld symbol must point toward the tail.

Symbols apply between abrupt changes in direction of welding unless governed by the "all around" symbol or otherwise dimensioned.

These symbols do not explicitly provide for the case that frequently occurs in structural work, where duplicate material (such as stiffeners) occurs on the far side of a web or gusset plate. The fabricating industry has adopted this convention: that when the billing of the detail material discloses the existence of a member on the far side as well as on the near side, the welding shown for the near side shall be duplicated on the far side.

FIGURE 14.5

Source: American Institute of Steel Construction, *Manual of Steel Construction Load & Resistance Factor Design*, 2nd ed. (Chicago: AISC, 2002), Table 8-3, p. 8-31. Reprint with the permission of the AISC.

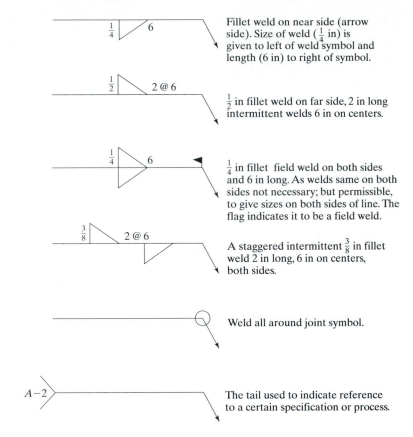

FIGURE 14.6

Sample weld symbols.

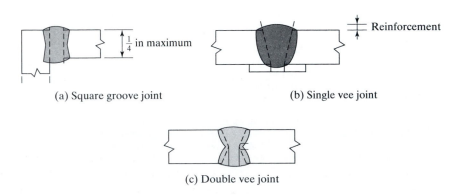

FIGURE 14.7

Groove welds.

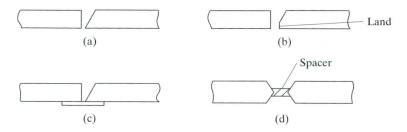

FIGURE 14.8

Edge preparation for groove welds. (a) Bevel with feathered edge. (b) Bevel with a land. (c) Bevel with a backup plate. (d) Double bevel with a spacer.

repeatedly, however, reinforcement is not as satisfactory because stress concentrations develop in the reinforcement and contribute to earlier failure. For these kinds of cases, a common practice is to provide reinforcement and grind it off flush with the material being connected (AASHTO Section 10.34.2.1).

Figure 14.8 shows some of the edge preparations that may be necessary for groove welds. In part (a), a bevel with a feathered edge is shown. When feathered edges are used, there is a problem with burn-through. This may be lessened if a *land* is used, such as the one shown in part (b) of the figure, or a backup strip or backing bar, as shown in part (c). The backup strip is often a 1/4-in copper plate. Weld metal does not stick to copper, and copper has a very high conductivity that is useful in carrying away excess heat and reducing distortion. Sometimes steel backup strips are used, but they will become a part of the weld and are thus left in place. A land should not be used together with a backup strip because there is a high possibility that a gas pocket might be formed, preventing full penetration. When double bevels are used, as shown in part (d) of the figure, spacers are sometimes provided to prevent burn-through. The spacers are removed after one side is welded.

From the standpoints of strength, resistance to impact stress repetition, and amount of filler metal required, groove welds are preferable to fillet welds. From other standpoints, however, they are not so attractive, and the vast majority of structural welding is fillet welding. Groove welds have higher residual stresses, and the preparations (such as scarfing and veeing) of the edges of members for groove welds are expensive, but the major disadvantages more likely lie in the problems involved with getting the pieces to fit together in the field. (The advantages of fillet welds in this respect were described in Section 14-7.) For these reasons, field groove joints are not used often, except on small jobs and where members may be fabricated a little long and cut in the field to the lengths necessary for precise fitting.

14.10 FILLET WELDS

Tests have shown that fillet welds are stronger in tension and compression than they are in shear, so the controlling fillet weld stresses given by the various specifications are shearing stresses. When practical, it is desirable to try to arrange welded connections so they will be subjected to shearing stresses only, and not to a combination of shear and tension or shear and compression.

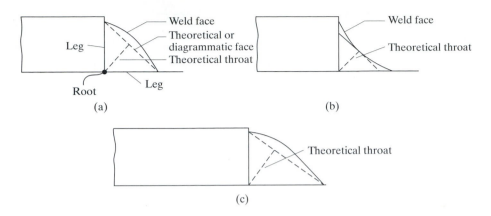

FIGURE 14.9

(a) Convex surface. (b) Concave surface. (c) Unequal leg fillet weld.

When fillet welds are tested to failure with loads parallel to the weld axes, they seem to fail by shear at angles of about 45° through the throat. Their strength is therefore assumed to equal the design shearing stress times the theoretical throat area of the weld. The theoretical throats of several fillet welds are shown in Fig. 14.9. The throat area equals the theoretical throat distance times the length of the weld. In this figure, the root of the weld is the point at which the faces of the original metal pieces intersect, and the theoretical throat of the weld is the shortest distance from the root of the weld to its diagrammatic face.

For the 45° or equal leg fillet, the throat dimension is 0.707 times the leg of the weld, but it has a different value for fillet welds with unequal legs. The desirable fillet weld has a flat or slightly convex surface, although the convexity of the weld does not add to its calculated strength. At first glance, the concave surface would appear to give the ideal fillet weld shape because stresses could apparently flow smoothly and evenly around the corner with little stress concentration. Years of experience, however, have shown that single-pass fillet welds of a concave shape have a greater tendency to crack upon cooling, and this factor has proved to be of greater importance than the smoother stress distribution of convex types.

When a concave weld shrinks, the surface is placed in tension, which tends to cause cracks. When the surface of a convex weld shrinks, it does not place the outer surface in tension; rather, as the face shortens, it is placed in compression.

Another item of importance pertaining to the shape of fillet welds is the angle of the weld with respect to the pieces being welded. The desirable value of this angle is in the vicinity of 45°. For 45° fillet welds, the leg sizes are equal, and such welds are referred to by the leg sizes (e.g., a 1/4-in fillet weld). Should the leg sizes be different (not a 45° weld), both leg sizes are given in describing the weld (e.g., a 3/8-by-1/2 in fillet weld).

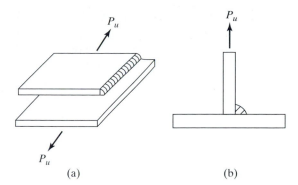

FIGURE 14.10

(a) Longitudinal fillet weld.
(b) Transverse fillet weld.

(a)   (b)

The automatic submerged arc welding (SAW) method provides a deeper penetration than does the usual shielded arc welding process. As a result, the LRFD permits the designer to use a larger throat area for welds made by this process. In Section J2.2a, the LRFD Specification states that the effective throat thickness for SAW fillet welds with legs 3/8 in or less may equal the leg sizes. For larger leg sizes, the effective throat thicknesses are equal to the theoretical throat thicknesses plus 0.11 in.

14.11   STRENGTH OF WELDS

In the discussion that follows, reference is made to Fig. 14.10. The stress in a fillet weld is usually said to equal the load divided by the effective throat area of the weld, with no consideration given to the direction of the load. Tests have shown, however, that transversely loaded fillet welds are appreciably stronger than ones loaded parallel to the weld's axis.

Transverse fillet welds are stronger for two reasons: First, they are more uniformly stressed over their entire lengths, while longitudinal fillet welds are stressed unevenly due to varying deformations along their lengths; second, tests show that failure occurs at angles other than 45°, giving them larger effective throat areas.

The method of determining the strength of fillet welds along their longitudinal axes regardless of the load directions is usually used to simplify computations. It is rather common for designers to determine the strength of all fillet welds by assuming the loads are applied in the longitudinal direction. However, the LRFD Specification in its Appendix Section J2.4 permits a higher strength for fillet welds depending on the angle of the load with respect to the axis of the weld. This specification is used for the problems presented in this book.

14.12   LRFD REQUIREMENTS

When welds are made, the electrode material should have properties of the base metal. If the properties are comparable, the weld metal is referred to as the *matching base metal* (that is, their nominal strengths are similar).

Table 14.1 (which is Table J2.5 of the LRFD Specification) provides nominal strengths for various types of welds, including fillet welds, plug-and-slot welds, and complete-penetration and partial-penetration groove welds.

TABLE 14.1   Design Strength of Welds

Types of Weld and Stress [a]	Material	Resistance Factor ϕ	Nominal Strength F_{BM} or F_W	Filler Metal Requirements [b, c]
Complete-Joint-Penetration Groove Weld				
Tension normal to effective area	Base	0.90	F_y	Matching filler metal shall be used. For CVN requirements see footnote [d].
Compression normal to effective area Tension or compression parallel to axis of weld	Base	0.90	F_y	Filler metal with a strength level equal to or less than matching filler metal is permitted to be used.
Shear on effective area	Base Weld	0.90 0.80	$0.60F_y$ $0.60F_{EXX}$	
Partial-Joint-Penetration Groove Weld				
Compression normal to effective area Tension or compression parallel to axis of weld [e]	Base	0.90	F_y	Filler metal with a strength level equal to or less than matching filler metal is permitted to be used.
Shear parallel to axis of weld	Base Weld	[f] 0.75	[f] $0.60F_{EXX}$	
Tension normal to effective area	Base Weld	0.90 0.80	F_y $0.60F_{EXX}$	
Fillet Welds				
Shear on effective area	Base Weld	[f] 0.75	[f] $0.60F_{EXX}$[g]	Filler metal with a strength level equal to or less than matching filler metal is permitted to be used.
Tension or compression parallel to axis of weld [e]	Base	0.90	F_y	
Plug or Slot Welds				
Shear parallel to faying surfaces (on effective area)	Base Weld	[f] 0.75	[f] $0.60F_{EXX}$	Filler metal with a strength level equal to or less than matching filler metal is permitted to be used.

Source: American Institute of Steel, *Manual of Steel Construction Load & Resistance Factor Design*, 3rd ed. (Chicago, AISC, 2001), Table J 2.5, p. 16–57. Reprinted with permission of AISC.

[a]   For definition of effective area, see Section J2.

[b]   For matching filler metal, see Table 3.1, AWS D1.1.

[c]   Filler metal one strength level stronger than matching filler metal is permitted.

[d]   For T and corner joints with the backing bar left in place during service, filler metal with a classification requiring a minimum Charpy V-notch (CVN) toughness of 20 ft-lb (27 J) @ +40°F(4°C) shall be used. If filler metal without the required toughness is used and the backing bar is left in place, the joint shall be sized using the resistance factor and nominal strength for a partial-joint-penetration weld.

[e]   Fillet welds and partial-joint-penetration groove welds joining component elements of built-up members, such as flange-to-web connections, are not required to be designed with regard to the tensile or compressive stress in these elements parallel to the axis of the welds.

[f]   The design of connected material is governed by Sections J4 and J5.

[g]   For alternative design strength, see Appendix J2.4.

The all-welded 56-story Toronto
Dominion Bank Tower. (Courtesy of the
Lincoln Electric Company).

The design strength of a particular weld is taken as the lower value of ϕF_W (where F_W is the nominal strength of the weld) and ϕF_{BM} (where F_{BM} is the nominal strength of the base material).

For fillet welds, the nominal strength for shear on the effective area of the weld is $0.60\ F_{EXX}$ (where F_{EXX} is the classification strength of the weld metal) and ϕ is 0.75. If we have tension or compression parallel to the axis of the weld, the nominal strength of the base metal F_{BM} is F_y, and ϕ is 0.90. The design shear strength of members being connected is taken as $\phi F_n A_{ns}$, where ϕ is 0.75, F_n is $0.6\ F_u$, and A_{ns} is the net area subject to shear.

The filler metal electrodes for shielded arc welding are listed as E60XX, E70XX, etc. In this classification, the letter E represents an electrode, while the first set of digits (as 60, 70, 80, 90, 100, or 110) indicates the minimum tensile strength of the weld in ksi.

The remaining digits specify the type of coating. Because strength is the most important factor to the structural designer, we usually specify electrodes as E70XX, E80XX, or simply E70, E80, and so on. For the usual situation, E70 electrodes are used for steels with F_y values from 36 to 60 ksi, while E80 is used when F_y is 65 ksi.

In addition to the nominal stresses given in Table 14.1, there are several other provisions applying to welding given in Section J2.2b of the LRFD Specification. Among the more important are the following:

1. The minimum length of a fillet weld may not be less than four times the nominal leg size of the weld. Should its length actually be less than this value, the weld size considered effective must be reduced to one-quarter of the weld length.

2. The maximum size of a fillet weld along edges of material less than 1/4 in thick equals the material thickness. For thicker material, it may not be larger than the material thickness less 1/16 in, unless the weld is specially built out to give a full-throat thickness. For a plate with a thickness of 1/4 in or more, it is desirable to keep the weld back at least 1/16 in from the edge so that the inspector can clearly see the edge of the plate and thus accurately determine the dimensions of the weld throat.

As a general statement, the weldability of a material improves as the thickness to be welded decreases. The problem with thicker material is that thick plates take heat from welds more rapidly than thin plates even if the same weld sizes are used. (The problem can be alleviated somewhat by preheating the metal to be welded to a few hundred degrees Fahrenheit and holding it there during the welding operation.)

3. The minimum permissible size fillet welds of the LRFD Specification are given in Table 14.2 (Table J2.4 of the LRFD Specification). They vary from 1/8 in for 1/4 in or thinner material up to 5/16 in for material over 3/4 in in thickness. The smallest practical weld size is about 1/8 in, and the most economical size is probably about 1/4 or 5/16 in. The 5/16 in. weld is about the largest size that can be made in one pass with the shielded metal arc welded process (SMAW); with the submerged arc process, (SAW), 1/2 in is the largest size.

These minimum sizes were not developed on the basis of strength considerations, but rather because thick materials have a quenching or rapid cooling effect on small welds. If this happens, the result is often a loss in weld ductility. In addition, the thicker

TABLE 14.2   Minimum Size of Fillet Welds

Material Thickness of Thicker Part Joined, in (mm)	Minimum Size of Fillet Weld[a], in (mm)
To $\frac{1}{4}$ (6) inclusive	$\frac{1}{8}$ (3)
Over $\frac{1}{4}$ (6) to $\frac{1}{2}$ (13)	$\frac{3}{16}$ (5)
Over $\frac{1}{2}$ (13) to $\frac{3}{4}$ (19)	$\frac{1}{4}$ (6)
Over $\frac{3}{4}$ (19)	$\frac{5}{16}$ (8)

[a] Leg dimension of fillet welds. Single pass welds must be used.
[b] See Section J2.2b of the LRFD Specification for maximum size of fillet welds.

material tends to restrain the weld material from shrinking as it cools, with the result that weld cracking can result and present problems.

Note that the minimum sizes given in Table 14.2 are dependent on the thicker of the two parts being joined. Regardless of the value given in the table, the minimum size does not have to exceed the thickness of the thinner part. It may be larger, however, if so required by the calculated strength.

4. Sometimes end returns or boxing are used at the end of fillet welds as shown in Figure 14.11. In the past, such practices have been recommended to provide better fatigue resistance and to make sure that weld thicknesses were maintained over the full lengths of welds. Recent research has shown that such returns are not necessary for developing the capacity of such connections. Should, however, cyclic forces be present normal to the outstanding legs of the weld and of a frequency or magnitude that would tend to cause progressive failures beginning from a point of maximum stress at the end of the weld, end returns should be used. For such situations, Section J2.2b of the LRFD Specification requires that the welds be turned around the corners for a distance no less that two times the weld size or the width of the part.

5. When longitudinal fillet welds are used for the connection of plates or bars, their length may not be less than the perpendicular distance between them because of shear lag (discussed in Chapter 3).

6. For lap joints, the minimum amount of lap permitted is equal to five times the thickness of the thinner part joined, but may not be less than 1 in (LRFD J2.2b). The purpose of this minimum lap is to keep the joint from rotating excessively.

7. Should the actual length (L) of an end-loaded fillet weld be greater than 100 times its leg size (w), the LRFD Specification (J2.2b) states that, due to stress variations along the weld, it is necessary to determine a smaller or effective length for strength determination. This is done by multiplying L by the term β as given in the following equation in which w is the weld leg size,

$$\beta = 1.2 - 0.002(L/w) \leq 1.0 \qquad \text{LRFD Equation J2-1}$$

If the actual weld length is greater than 300 w, the value of β is taken to be equal to 0.6.

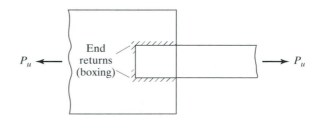

FIGURE 14.11

End returns (or boxing).

14.13   DESIGN OF SIMPLE FILLET WELDS

Examples 14-1 through 14-3 demonstrate the calculations used to determine the strength of various fillet welded connections; Example 14-4 presents the design of such a connection. In these and other problems, weld lengths are selected no closer than the nearest 1/4 in, because closer work cannot be expected in shop or field.

Example 14-1

Determine the design strength of a 1-in length of 5/16-in fillet weld using (a) the shielded metal arc process (SMAW) and (b) the submerged arc process (SAW). Use E70 electrodes with a minimum tensile strength of 70 ksi $= F_{EXX}$ and assume load parallel to weld.

Solution

(a) Shielded metal arc process

$$\text{Effective throat thickness of weld} = (0.707)\left(\tfrac{5}{16}\right) = 0.221 \text{ in}$$

$$\begin{aligned}
\text{Design strength} &= \phi F_w t_e L \\
&= (\phi)(\text{nominal strength of weld } 0.6\, F_{EXX}) \\
&\quad \times (\text{throat } t)(\text{weld length}) \\
&= (0.75)(0.60 \times 70)(0.221)(1.0) = 6.96 \text{ k/in}
\end{aligned}$$

(b) Submerged arc process

From LRFD Section J2.2a the effective throat thickness of weld $= 5/16$ in

$$\begin{aligned}
\text{Design strength} = \phi F_w t_e L &= (0.75)(0.60 \times 70)\left(\tfrac{5}{16}\right)(1.0) \\
&= 9.84 \text{ k/in}
\end{aligned}$$

Fillet welds may not be designed with a stress that is greater than the design stress on the adjacent members being connected. If the external force applied to the member (tensile or compressive) is parallel to the axis of the weld metal, the design strength may not exceed the axial design strength of the member.

Examples 14-2 and 14-3 illustrate the calculations necessary to determine the design strength of plates connected with SMAW and SAW fillet welds. In each of these examples, the shearing strength of the welds per inch controls and is multiplied by the total length of the welds to give the total capacity of the connections.

Example 14-2

What is the design strength of the connection shown in Fig. 14.12 if the plates consist of A572 Grade 50 steel and E70 electrodes are used? The 7/16 in fillet welds shown were made by the SMAW process.

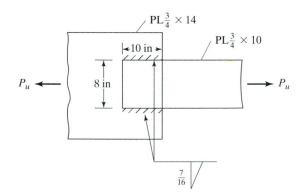

FIGURE 14.12

Solution

$$\text{Effective throat thickness} = (0.707)\left(\tfrac{7}{16}\right) = 0.309 \text{ in}$$

$$\text{Capacity of weld/in} = \phi F_w t_e L = (0.75)(0.60 \times 70)(0.309)(1.0)$$

$$= 9.73 \text{ k/in} \leftarrow$$

$$\text{Total capacity of weld} = \phi R_{nw} = (9.73)(20) = 194.65 \text{ k} \leftarrow$$

$$\text{Design strength of plate} = \phi F_y A_g = (0.90)(50)\left(\tfrac{3}{4} \times 10\right)$$

$$= 337.5 \text{ k}$$

$$\text{Design capacity} = 194.6 \text{ k}$$

Example 14-3

Repeat Example 14-2 if SAW welds are used.

Solution

$$\text{Effective throat thickness} = (0.707)\left(\tfrac{7}{16}\right) + 0.11 = 0.419 \text{ in}$$

$$\text{Capacity of weld/in} = \phi F_w t_e L = (0.75)(0.60 \times 70)(0.419)(1.0)$$

$$= 13.20 \text{ k/in} \leftarrow$$

$$\text{Total capacity of weld} = \phi R_{nw} = (13.20)(20) = 264 \text{ k} \leftarrow$$

$$\text{Design strength of plate} = \phi P_n = \phi F_y A_g = (0.90)(50)\left(\tfrac{3}{4} \times 10\right)$$

$$= 337.5 \text{ k}$$

$$\text{Design capacity} = 264 \text{ k}$$

Example 14-4

Using 50 ksi steel and E70 electrodes, design SMAW fillet welds to resist a full-capacity load on the 3/8 × 6 in member shown in Fig. 14.13. Assume end returns are to be used.

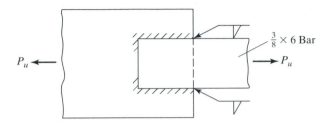

$\frac{3}{8}$ × 6 Bar

P_u

P_u

FIGURE 14.13

Solution

$$\phi_t P_n = \phi_t F_y A_g = (0.90)(50)\left(\tfrac{3}{8} \times 6\right) = 101.2 \text{ k}$$

Maximum weld size $= \frac{3}{8} - \frac{1}{16} = \frac{5}{16}$ in   (LRFD Section J2.2b)

Minimum weld size $= \frac{3}{16}$ in (Table 14.2)

Use $\frac{5}{16}$ in weld

Effective throat thickness of weld $= (0.707)\left(\frac{5}{16}\right) = 0.221$ in

Capacity of weld per in $= \phi F_w t_e L = (0.75)(0.60 \times 70)(0.221)(1.0)$

$$= 6.96 \text{ k/in} \leftarrow$$

Length required $= \frac{101.2}{6.96} = 14.54$ in, or 7.27 in per side use and returns of $2 \times \frac{5}{16} = 1$ in for each side (although they are not particularly effective, according to Section J2.2b of the LRFD Commentary).

Leaving $14.54 - 2.0 = 12.54$ in, or $6\frac{1}{2}$ in for each side

Use $6\frac{1}{2}$-in welds for each side plus end returns

Appendix J2.4 states that the strength of fillet welds loaded transversely in a plane through their centers of gravity may be determined with the following equation in which $\phi = 0.75$ and θ is the angle between the line of action of the load and the longitudinal axis of the weld.

$$\phi F_w = \phi(0.6 \, F_{EXX})(1.0 + 0.50 \sin^{1.5}\theta)(\text{throat } t)(\text{weld length})$$

As the angle θ is increased, the strength of the weld increases. Should the load be perpendicular to the longitudinal axis of the weld, the result will be a 50 percent increase in the computed weld strength. Example 14-5 illustrates the application of this Appendix expression.

Example 14-5

Determine the strength per in. of the 1/4-in SMAW fillets welds formed with E70 elec-trodes shown in part (a) of Fig. 14.14 if the load is applied parallel to the longitudinal axis of the welds. Then determine the strength per in of the welds shown in part (b) of the figure where the load is applied at a 45° angle with the longitudinal axis of the welds.

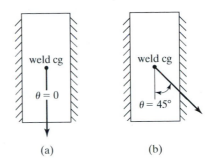

FIGURE 14.14   (a)   (b)

Solution

a. Effective throat thickness $t_e = (0.707)\left(\frac{1}{4}\right) = 0.177$ in

$$\phi F_w t_e L = (0.75)(0.6 \times 70)(0.177)(1.0) = 5.58 \text{ k/in}$$

b. $\phi F_w t_e L = (\phi)(0.6\,F_{EXX})(1.0 + 0.50\sin^{1.5}\theta)(\text{effective throat } t)(\text{length})$

$= (0.75)(0.6)(70)(1.0 + 0.50\sin^{1.5}45°)(0.177)(1.0)$

$= 7.23$ k/in, which is almost a 30 percent increase in strength over the longitudinally loaded case.

On some occasions the lengths available for the usual longitudinal fillet welds are not sufficient for the load to be resisted. For the situation shown in Fig. 14.15 it may be possible to develop sufficient strength by welding along the back of the channel at the edge of the plate if sufficient space is available. The dashed lines in the figure show this weld.

Another possibility is the use of slot welds as illustrated in Example 14-6. There are several LRFD requirements pertaining to slot welds that need to be mentioned here. LRFD Specification J2.3 states that the width of a slot may not be less than the member thickness + 5/16 in (rounded off to the next greater odd 1/16 in, since struc-tural punches are made in these diameter), nor may it be greater than 2.25 times the weld thickness. For members up to 5/8 in thickness, the weld thickness must equal the plate thickness; for members greater than 5/8 in thickness, the weld thickness may not be less than one-half the member thickness (or 5/8 in). The maximum length permitted for slot welds is 10 times the weld thickness. The limitations given in specifications for the maximum sizes of plug or slot welds are caused by the detrimental shrinkage that occurs around these types of welds when they exceed certain sizes. Should holes or

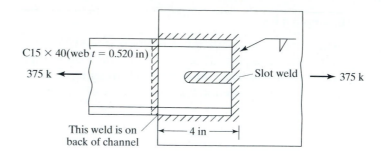

C15 × 40(web $t = 0.520$ in)

375 k ←

Slot weld

→ 375 k

This weld is on
back of channel

|← 4 in →|

FIGURE 14.15

slots larger than those specified be used, it is desirable to use fillet welds around the borders of the holes or slots rather than using a slot or plug weld. Slot and plug welds are normally used in conjunction with fillet welds in lap joints. Sometimes plug welds are used to fill in the holes temporarily used for erection bolts for beam and column connections. They may or may not be included in the calculated strength of these joints.

The strength of a plug or slot weld is equal to its design stress ϕF_w times its nominal area in the shearing plane. This area is equal to the area of contact at the base of the plug or slot. The length of a slot weld can then be determined from the following equation:

$$L = \frac{\text{load}}{(\text{width})(\text{design stress})}$$

Example 14-6 illustrates the design of the welds necessary to connect a channel to a plate. The calculations show that the ordinary side and end fillet welds do not provide sufficient strength in this case because of the limited space available. We then use a slot weld to provide the remaining load resistance needed.

Example 14-6

Design SMAW fillet welds to connect a C15 × 40 to the plate shown in Fig. 14.15. The load to be resisted is 375 k and E70 electrodes are to be used. As shown in the figure, the channel may lap over the plate by only 4 in due to space limitations. Space is not available for welding on back of the channel. Notice the load is applied transversely to the end weld.

Solution. Because of limited space, use

$$\text{Max weld size} = \text{web } t - \tfrac{1}{16} = \tfrac{1}{2} - \tfrac{1}{16} = \tfrac{7}{16} \text{ in}$$

$$\text{Effective throat thickness} = (0.707)\left(\tfrac{7}{16}\right) = 0.309 \text{ in}$$

Capacity of end and side welds

$$= (0.75)(0.60 \times 70)(0.309)(1 + 0.5 \sin^{1.5} 90°)(15)$$

$$+ (0.75)(0.60 \times 70)(0.309)(2 \times 4)$$
$$= 219 + 77.9 = 296.9 \text{ k} < 375 \text{ k}$$

Therefore try a slot weld

$$\text{Min width of slot} = t_w + \tfrac{5}{16} = 0.520 + \tfrac{5}{16} = \tfrac{13}{16} \text{ in}$$

$$\text{Max width} = 2\tfrac{1}{4} \times \text{weld thickness}$$

$$= \left(2\tfrac{1}{4}\right)(\text{web } t \text{ of channel}) = \left(2\tfrac{1}{4}\right)\left(\tfrac{1}{2}\right)$$

$$= 1\tfrac{1}{8} \text{ say } \tfrac{17}{16} \text{ (to odd } \tfrac{1}{16})$$

Use 15/16 in slot width

Since the slot extends to the end of the channel, the length of the fillet weld there will be reduced by the width of the slot.

$$\text{Capacity of } \tfrac{7}{16}\text{-in fillet weld} = \left(\frac{15 - \frac{15}{16}}{15}\right)(219) + 77.9 = 283.2 \text{ k}$$

$$\text{Load to be resisted by slot weld} = 375 - 283.2 = 91.8 \text{ k}$$

$$\text{Length required for slot weld} = \frac{91.8}{\left(\frac{15}{16}\right)(0.75)(0.60 \times 70)} = 3.11 \text{ in}$$

Use $3\tfrac{1}{2}$ in slot length

$$\text{Max length permitted by LRFD} = (10)\left(\tfrac{1}{2}\right)$$
$$= 5.00 \text{ in} > 3\tfrac{1}{2} \text{ in} \qquad \text{(OK)}$$

Use 7/16 in long fillet welds along the end of the channel and 4-in-long welds on each side. Also, use a $\tfrac{15}{16} \times 3\tfrac{1}{2}$ slot weld, E70, SMAW

Alternate Solution Should space have been available on the back of the channel next to the plate, a 7/16-in fillet weld would carry 219 k > 91.8 k.

14.14   DESIGN OF FILLET WELDS FOR TRUSS MEMBERS

Should the members of a welded truss consist of single angles, double angles, or similar shapes and be subjected to static axial loads only, the LRFD Specification (J1.8) permits the connections to be designed by the procedures described in the preceding section. The designers can select the weld size, calculate the total length of the weld required, and place the welds around the member ends as they see fit. (It would not make sense, of course, to place the weld all on one side of a member, such as for the angle of Fig. 14.16, because of the rotation possibility.) Example 14-7 illustrates the simple calculations involved in designing the welds for the ends of a truss member subjected to static axial loads.

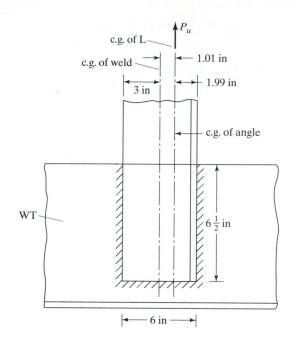

FIGURE 14.16

Example 14-7

Using A36 steel and E70 electrodes, design side and end fillet SMAW welds for the full capacity of a 6 × 4 × 1/2-in angle tension member with the long leg connected. Assume that the load is static.

Solution

Tensile capacity of angle $= \phi_t F_y A_g = (0.90)(36)(4.72) = 152.9 \text{ k}$

A subsequent check of the $\phi_t F_u A_e$ equation shows that it does not control.

Maximum weld size $= \frac{1}{2} - \frac{1}{16} - \frac{7}{16} \text{ in}$

Minimum weld size $= \frac{3}{16} \text{ in. (from Table 14.2).}$

Use $\frac{5}{16}$-in weld as maximum size which can be made in one pass

Effective throat thickness of weld $= (0.707)\left(\frac{5}{16}\right) = 0.221 \text{ in}$

Design strength of end weld $= (0.75)(0.60 \times 70)(0.221)$
$$\times (1 + 0.5 \sin^{1.5} 90°)(6)$$
$$= 62.7 \text{ k}$$

Load to be carried by side welds $= 152.9 - 62.7 = 90.2 \text{ k}$

Design strength/in of side welds $= (0.75)(0.60 \times 70)$
$$\times (0.221)(1) = 6.96 \text{ k/in}$$

$$\text{Length Reqd.} = \frac{90.2}{6.96} = 12.96 \text{ in say } 6\ 1/2 \text{ in each side}$$

Place welds as shown in Fig. 14.16

It should be noted that the centroid of the welds and the centroid of the statically loaded angle do not coincide in the connection selected in Example 14-7 and shown in Fig. 14-16. If a welded connection is subjected to varying stresses (such as those occurring in a bridge member), it generally is desirable to place the welds so that their centroid will coincide with the centroid of the member (or the resulting torsion must be accounted for in design). If the member being connected is symmetrical, the welds will be placed symmetrically; if the member is not symmetrical, the welds will not be symmetrical.

The force in an angle, such as the one shown in Fig. 14.17, is assumed to act along its center of gravity. If the center of gravity of weld resistance is to coincide with the angle force, the weld must be asymmetrically placed, or in this figure L_1 must be longer than L_2. (When angles are connected by bolts, there is usually an appreciable amount of eccentricity, but in a welded joint, eccentricity can be fairly well eliminated.) The information necessary to handle this type of weld design can be easily expressed in equation form, but only the theory behind the equations is presented here.

For the angle shown in Fig. 14.17, the force acting along line L_2 (designated here as P_2) can be determined by taking moments about point A. The member force and the weld resistance should coincide, and the moments of the two about any point must be zero. If moments are taken about point A, the force P_1 (which acts along line L_1) will be eliminated from the equation, and P_2 can be determined. In a similar manner, P_1 can be determined by taking moments about point B or by $\Sigma V = 0$. Example 14-8 illustrates the design of fillet welds of this type. A similar problem is handled in Example 14-9 except an end fillet weld is included, thus permitting a shorter connection. The center of gravity and resistance of the end weld are known and can be easily included in the moment equations.

There are other possible solutions for the design of the welds for the angle considered in these two examples. Although the 7/16-in weld is the largest one permitted at the edges

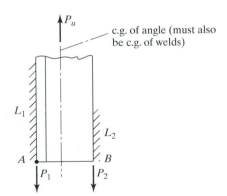

FIGURE 14.17

of the 1/2-in angle and at its end, a larger weld could be used on the other side next to the outstanding leg. From a practical point of view, however, the welds should be the same size because different size welds slow the welder down due to the need to change electrodes to make different sizes. Recall from our discussion of fatigue in Chapter 4 that if the estimated number of loading cycles during the structure's estimated life exceeds 20,000, it will be necessary to study the stress range of the connection at service loads. This study may result in larger connections, as required by Appendix K of the LRFD Specification.

Example 14-8

Use $F_y = 50$ ksi and $F_u = 65$ ksi, E70 electrodes, and the SMAW process to design side fillet welds for the full capacity of the $5 \times 3 \times 1/2$-in angle tension member shown in Fig. 14-18. Assume the member is subjected to repeated stress variations, making any connection eccentricity undesirable. Check block shear strength of member. Assume that the WT chord member has adequate strength to develop the weld strengths and that the thickness of its web is 1/2 in.

Solution. Tensile capacity of angle $\phi_t P_n$

$$= \phi_t F_y A_g = (0.9)(50)(3.75) = 168.7 \text{ k}$$

or

$$\phi_t P_n = \phi_t F_u A_g \text{ assuming } U = 0.87$$
$$= (0.75)(65)(0.87)(3.75) = 159.0 \text{ k} \leftarrow$$

Maximum weld size $= \frac{1}{2} - \frac{1}{16} = \frac{7}{16}$ in

Use 5/16-in weld

Effective throat thickness of weld $= (0.707)\left(\frac{5}{16}\right) = 0.221$ in

Capacity of weld/in $= \phi F_w t_e L = (0.75)(0.60 \times 70)(0.221)(1.0)$
$$= 6.96 \text{ k/in}$$

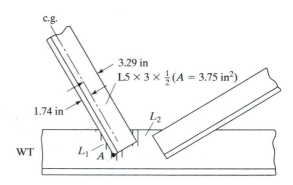

FIGURE 14.18

$$\text{Total weld length required} = \frac{159.0}{6.96} = 22.84 \text{ in}$$

Taking moments about point A (see Fig. 4.18) to determine force P_2

$$(159.0)(1.74) - 5.00\, P_2 = 0$$

$$P_{u2} = 55.33$$

$$P_{u1} = P_u - P_{u2} = 159.0 - 55.33 = 103.7 \text{ k}$$

$$L_1 = \frac{103.7}{6.96} = 14.9 \text{ in} \quad \text{say 15 in}$$

$$L_2 = \frac{55.33}{6.96} = 7.95 \text{ in} \quad \text{say 8 in}$$

Check block shearing strength assuming these dimensions

$$0.6 F_u A_{nv} = (0.6)(65)(15 + 8)\left(\tfrac{1}{2}\right) = 448.5 \text{ k}$$

$$> F_u A_{nt} = (65)\left(5 \times \tfrac{1}{2}\right) = 162.5$$

$$\therefore \phi R_n = 0.75[(0.6)(65)(15 + 8)\left(\tfrac{1}{2}\right) + (50)\left(5 \times \tfrac{1}{2}\right)]$$

$$= 433 \text{ k} > 159.0 \text{ k} \tag{OK}$$

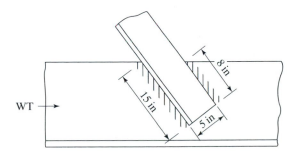

Example 14-9

Rework Example 14-8 using fillet welds along the sides and end of the angle.

Solution. Assuming 5/16-in welds (capacity of side welds $= 6.96$ k/in)

Total strength of end weld

$$= (0.75)(0.60 \times 70)(0.707 \times \tfrac{5}{16})(1 + 0.5 \sin^{1.5} 90°)(5) = 52.2 \text{ k}$$

Taking moments about L_1 to determine force P_2,

$$(159.0)(1.74) - (2.50)(52.2) - 5.00\, P_2 = 0$$

$$P_{u2} = 29.2 \text{ k}$$

$$P_{u1} = 159.0 - 52.2 - 29.2 = 77.6 \text{ k}$$

$$L_1 = \frac{77.6}{6.96} = 11.15 \text{ in} \quad \text{say } 11\tfrac{1}{2} \text{ in}$$

$$L_2 = \frac{29.2}{6.96} = 4.20 \text{ in} \quad \text{say } 4\tfrac{1}{2} \text{ in}$$

It is rather convenient for design purposes to know the strength of a 1/16-in fillet weld 1 in. long. Though this size is below the minimum permissible size given in Table 14.2, the calculated strength of such a weld is useful for determining weld sizes for calculated forces. For a 1-in.-long SMAW weld with the load parallel to weld axis we have

$$\phi F_w t_e L = (0.75)(0.707 \times \tfrac{1}{16} \times 1.0)(0.60 \, F_{EXX}) = 0.02 \, F_{EXX}$$

For an E70 electrode, ϕF_w is $(0.02)(70) = 1.4$ k/in for a 1/16-in weld. If we are designing a fillet weld to carry a factored force of 6.5 k/in, the required weld size will be $6.5/1.4 = 4.64$ sixteenths of an inch, or, say, 5/16-in.

14.15 SHEAR AND TORSION

Fillet welds are frequently loaded with eccentrically applied loads with the result that the welds are subjected to either shear and torsion or to shear and bending. Figure 14.19 is presented to show the difference between the two situations. Shear and torsion, shown in part (a) of the figure, are the subject of this section, while shear and bending, shown in part (b) of the figure, are discussed in Sections 14–16 and 14–17.

As is the case for eccentrically loaded bolt groups (Section 13-1), the LRFD Specification provides the permissible design strength of welds, but it does not specify a method of analysis for eccentrically loaded welds. It's left to the designer to decide which method to use.

14.15.1 Elastic Method

Initially, the very conservative elastic method is presented. In this method, friction or slip resistance between the connected parts is neglected, the connected parts are assumed to be perfectly rigid, and the welds are assumed to be perfectly elastic.

For this discussion, the welded bracket of part (a) of Fig. 14.20 is considered. The pieces being connected are assumed to be completely rigid, as they were in bolted connections. The effect of this assumption is that all deformation occurs in the weld. The weld is subjected to a combination of shear and torsion, as was the eccentrically loaded bolt group considered in Section 13-1. The force caused by torsion can be computed from the following familiar expression:

$$f = \frac{Td}{J}$$

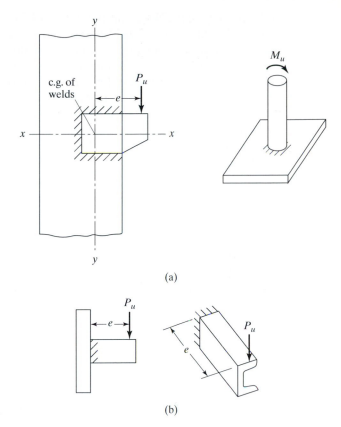

FIGURE 14.19

(a) Welds subjected to shear and torsion.
(b) Welds subjected to shear and bending.

In this expression, T is the torsion, d is the distance from the center of gravity of the weld to the point being considered, and J is the polar moment of inertia of the weld. It is usually more convenient to break the force down into its vertical and horizontal components. In the following expressions, f_h and f_v are the horizontal and vertical components, respectively of the force f:

$$f_h = \frac{Tv}{J} \quad f_v = \frac{Th}{J}$$

Notice that the formulas are almost identical to those used for determining stresses in bolt groups subject to torsion. These components are combined with the usual direct shearing stress which is assumed to equal the reaction divided by the total length of the welds. For design of a weld subject to shear and torsion, it is convenient to assume a 1-in weld and to compute the stresses on a weld of that size. Should the assumed weld be overstressed, a larger weld is required; if it is understressed, a smaller one is desirable.

Although the calculations probably will show the weld to be overstressed or understressed, the calculation does not have to be repeated because a ratio can be set up to give the weld size for which the load would produce a computed stress exactly equal to the design stress. Note that the use of a 1-in weld simplifies the units because 1 in of length of weld is 1 in² of weld, and the computed stresses are said to be either kips per square inch or kips per inch of length. Should the calculations be based on some size other than a 1-in weld, the designer must be very careful to keep the units straight, particularly in obtaining the final weld size. To further simplify the calculations, the welds are assumed to be located at the edges where the fillet welds are placed, rather than at the centers of their effective throats. As the throat dimensions are rather small, this assumption changes the results very little. Example 14-10 illustrates the calculations involved in determining the weld size required for a connection subjected to a combination of shear and torsion:

Example 14-10

For the A36 bracket shown in Fig. 14-20(a), determine the fillet weld size-required if E70 electrodes, the LRFD Specification, and the SMAW process are used.

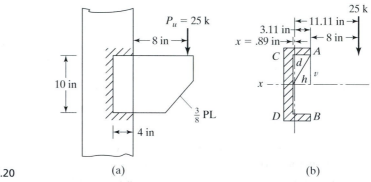

FIGURE 14.20   (a)   (b)

Solution. Assuming a 1-in weld as shown in part (b) of Fig. 14-20.

$$A = 18 \text{ in}^2$$

$$\bar{x} = \frac{(4)(2)(2)}{18} = 0.89 \text{ in}$$

$$I_x = \left(\frac{1}{12}\right)(1)(10)^3 + (2)(4)(5)^2 = 283.3 \text{ in}^4$$

$$I_y = (2)\left(\frac{1}{3}\right)(0.89^3 + 3.11^3) + (10)(0.89)^2 = 28.5 \text{ in}^4$$

$$J = 283.3 + 28.5 = 311.8 \text{ in}^4$$

The 1-in vertical weld has a design strength = $(0.75)(0.60 \times 70) \times (0.707 \times 1.0)(1.0) = 22.27$ k/in, while the horizontal weld has a design strength = $(0.75)(0.60 \times 70)(0.707 \times 1.0)(1 + 0.5 \sin^{1.5} 90°)(1.0) = 33.41$ k/in. Therefore, stress is computed at points A and B in the horizontal weld and at C and D in the vertical weld (see Fig. 14-20) and required weld size determined.

Horizontal Weld	*Vertical Weld*

$$f_h = \frac{(25 \times 11.11)(5)}{311.8} = 4.45 \text{ k/in} \qquad f_k = \frac{(25 \times 11.11)(5)}{311.8} = 4.45 \text{ k/in}$$

$$f_v = \frac{(25 \times 11.11)(3.11)}{311.8} = 2.77 \text{ k/in} \qquad f_v = \frac{(25 \times 11.11)(0.89)}{311.8} = 0.793 \text{ k/in}$$

$$f_s = \frac{25}{18} = 1.39 \text{ k/in} \qquad f_s = \frac{25}{18} = 1.39 \text{ k/in}$$

$$f_r = \sqrt{(2.77 + 1.39)^2 + (4.45)^2} \qquad f_r = \sqrt{(0.793 + 1.39)^2 + (4.45)^2}$$

$$= 6.09 \text{ k/in} \qquad\qquad = 4.96 \text{ k/in}$$

$$\text{Size} = \frac{6.09}{33.41} = 0.182 \text{ in} \qquad \text{Size} = \frac{4.96}{22.27} = 0.223 \text{ in} \leftarrow$$

Use 1/4-in fillet welds, E70, SMAW

14.15.2 Ultimate Strength Method

An ultimate strength analysis of eccentrically loaded welded connections is more realistic than the more conservative elastic procedure just described. For the discussion that follows, the eccentrically loaded fillet weld of Fig. 14.21 is considered. As for eccentrically loaded bolted connections, the load tends to cause a relative rotation and translation between the parts connected by the weld.

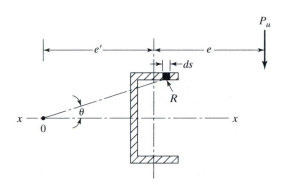

FIGURE 14.21

Even if the eccentric load is of such a magnitude that it causes the most stressed part of the weld to yield, the entire connection will not yield. The load may be increased, the less stressed fibers will begin to resist more of the load, and failure will not occur until all the weld fibers yield. The weld will tend to rotate about its instantaneous center of rotation. The location of this point (which is indicated by the letter O in the figure) is dependent upon the location of the eccentric load, the geometry of the weld, and the deformations of the different elements of the weld.

If the eccentric load P_u is vertical and if the weld is symmetrical about a horizontal axis through its center of gravity, the instantaneous center will fall somewhere on the horizontal x-axis. Each differential element of the weld will provide a resisting force R. As shown in Fig. 14.21, each of these resisting forces is assumed to act perpendicular to a ray drawn from the instantaneous center to the center of gravity of the weld element in question.

Studies have been made to determine the maximum shear forces that eccentrically loaded weld elements can withstand.[4,5] The results, which depend on the load deformation relationship of the weld elements, may be represented either with curves or in formula fashion. The ductility of the entire weld is governed by the maximum deformation of the weld element that first reaches its limit. (The element that is located at the greatest distance from the weld's instantaneous center probably reaches its limit first.)

Just as in eccentrically loaded bolted connections, the location of the instantaneous center of rotation is determined by trial and error. Unlike eccentrically loaded bolt groups, however, the strength and deformation of welds are dependent on the angle θ that the force in each element makes with the axis of that element. The deformation of each element is proportional to its distance from the instantaneous center. At maximum stress, the weld deformation is Δ_m, which is determined with the following expression in which D is the leg size of the fillet weld:

$$\Delta_m = 0.209(\theta + 2)^{-0.32} D$$

The deformation in a particular element is assumed to vary directly in proportion to its distance from the instantaneous center:

$$\Delta_r = \frac{l_r}{l_m} \Delta_m$$

The nominal stress in a particular weld element is computed as

$$R = 0.60 \, F_{EXX}(1.0 + 0.50 \sin^{1.5}\theta)[p(1.9 - 0.9p)]^{0.3}$$

where θ = the angle of loading measured from the weld longitudinal axis, and p = the ratio of the deformation of an element to its deformation at ultimate stress.

We can assume a location of the instantaneous center, determine R values for the different elements of the weld, and compute ΣR_x and ΣR_y. The three equations of

---

[4]L. J. Butler, S. Pal, and G. L. Kulak, "Eccentrically Loaded Weld Connections," *Journal of the Structural Division*, vol. 98, no. ST5, May 1972, pp. 989–1005.

[5]G. L. Kulak and P. A. Timler, "Tests on Eccentrically Loaded Fillet Welds," Dept of Civil Engineering, University of Alberta, Edmonton, December 1984.

TABLE 14.3   Electrode Strength Coefficient C_1

Electrode	F_{EXX} (ksi)	C_1
E60	60	0.857
E70	70	1.00
E80	80	1.03
E90	90	1.16
E100	100	1.21
E110	110	1.34

Source: American Institute of Steel Construction, *Manual of Steel Construction Load & Resistance Factor Design.* 3rd ed. (Chicago: AISC, 2001, p. 8–51. Reprinted with the permission of the AISC.

equilibrium ($\Sigma M = 0$, $\Sigma R_x = 0$, and $\Sigma R_y = 0$) will be satisfied if we have the correct location of the instantaneous center. If they are not satisfied, we will try another location and so on. Finally, when the equations are satisfied, the value of P_u can be computed as $\sqrt{(\Sigma R_x)^2 + (\Sigma R_y)^2}$.

Detailed information on the use of these expressions is given in Part 8 of the LRFD Manual. The LRFD Specification permits weld strengths in eccentrically loaded connections to exceed $0.6\,F_{EXX}$. The solution of these kinds of problems is completely impractical without the use of computers or of computer generated tables such as those provided in Part 8 of the LRFD Manual. Example 14-11 illustrates the use of the LRFD tables to solve a problem like this, while Example 14-18 at the end of the chapter makes use of the computer program INSTEP32 for a solution.

The values given in the tables of Part 8 of the LRFD Manual were developed by this procedure. Using the tables, the ultimate strength P_u of a particular connection can be determined from the following expression in which C is a tabular coefficient that includes a ϕ of 0.75, C_1 is a coefficient depending on the electrode number and given in Table 14.3, D is the weld size in sixteenths of an inch, and l is the length of the vertical weld:

$$\phi R_n = C C_1 Dl$$

The Manual includes tables for both vertical and inclined loads (at angles from the vertical of 0° to 75°). The user is warned not to interpolate for angles in between these values because the results may be non-conservative. Therefore, the user is advised to use the value given for the next lower angle. If the connection arrangement being considered is not covered by the tables, the conservative elastic procedure previously described may be used. Examples 14-11 and 14-12 illustrate the use of these ultimate strength tables.

Example 14-11

Repeat Example 14-10 using the LRFD tables that are based on an ultimate strength analysis. The connection is redrawn in Fig. 14.22.

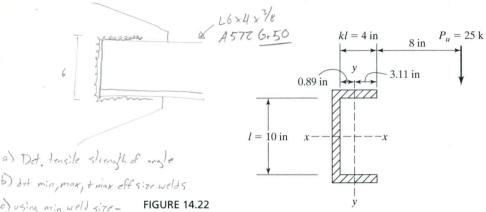

a) Det. tensile strength of angle

b) det min, max, + max eff size welds

c) using min weld size —
 calc design strength of weld/in

FIGURE 14.22

d) Compute total length required using answer from part (a)
 round up to nearest inch

Solution

e) balance welds

a) $F_y = 50$ $F_u = 65$ p20

$e_x = 11.11$ in

$l = 10$ in

$a = \dfrac{11.11}{10} = 1.111$

$k = \dfrac{4}{10} = 0.4$

$\bar{C} = 0.986$ from Table 8-9, in the LRFD Manual for θ

$= 0°$ by straight-line interpolation

$C_1 = 1.0$ from Table 14-3 in this chapter

$D = $ weld size required $= \dfrac{P_u}{CC_1 l}$

$= \dfrac{25}{(0.986)(1.0)(10)} = 2.53$ sixteenths

$= 0.158$ in (as compared with 0.233 in by the elastic method)

Use 3/16 in fillet weld, E70, SMAW

weld → $A_n = A_g$

$A_e = U A_n$

$\boxed{\phi P_n = 148.3 \, kip}$

b) $t = 3/8''$ $F_{E70} = 70 \, ksi$

$a \min = 3/16''$

$a \max = 3/8 - 1/16 = 5/16$

$a \max \, eff = \dfrac{.9(3/8)(.6)(50)}{.75(.707)(.6)(70)} = .455$

$a \max \, eff > a \max$ so

$a \max \, controls = 5/16$

Example 14-12

Determine the weld size required for the situation shown in Fig. 14.23 using the LRFD tables that are based on an ultimate strength analysis (A36 steel, E70 electrodes).

c) $a = a_{min} = 3/16$ in

$\phi R_{n\,weld} = .75(.707)(a)(.6)(F_{EXX})$

$= .75(.707)(3/16)(.6)(70)$

$= 4.176$ kip/in ← controls overall weld strength

check base material

$\phi_{base} = .9$ $t = .375$ $F_y = 50$

$\phi R_n = .9(t)(.6)(F_y)$

$= .9(.375)(.6)(50) = 10.125$ kip/in

d) 148.3 kip ← part (a)

$\dfrac{148.3 \text{ kip}}{\phi R_n \text{ weld}} = \dfrac{148.3 \text{ kip}}{4.176 \text{ kip/in}} = 36$ in

min length $= 4(a) = 4(3/16) = .75$ in → $36 > .75$ so OK

$a_{max\ eff} = \dfrac{.9t(.6F_y)}{.75(.707)(.6)F_{EXX}}$

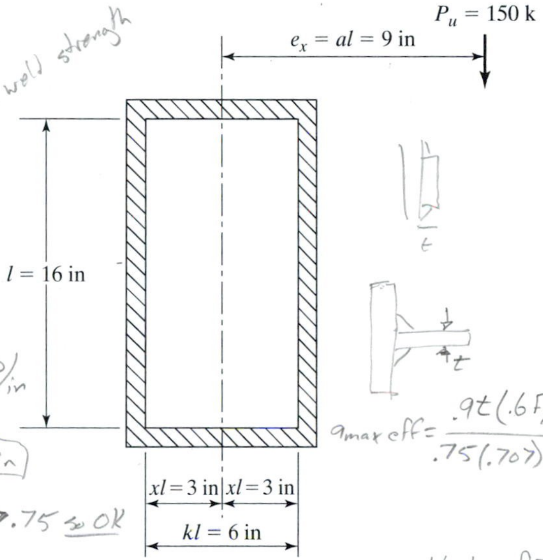

$P_u = 150$ k
$e_x = al = 9$ in
$l = 16$ in
$xl = 3$ in
$xl = 3$ in
$kl = 6$ in

FIGURE 14.23

min Table 14.2 pg 448 ← depends on thicker of 2 parts

max depends on smaller of 2 parts

t	max a
≤ 1/4"	t
> 1/4"	t − 1/16

Solution

$$k = \frac{6}{16} = 0.375$$

$$a = \frac{9}{16} = 0.5625$$

C from table 8-7 for $\theta = 0$ by two-directional, straight-line interpolation $= 2.49$

$C_1 = 1.0$ for E70 electrodes (see Table 14-3)

$$\text{Required weld size} = D = \frac{P_u}{CC_1 l} = \frac{150}{(2.49)(1.0)(16)}$$

$$= 3.77 \text{ sixteenths}$$

Use 1/4-in weld, E70, SMAW

14.16 SHEAR AND BENDING

The welds shown in Fig. 14.19(b) and in Fig. 14.24 are subjected to a combination of shear and bending.

e) y bar = 1.94 in $L_{tr} = 6"$

$\bar{y} = \dfrac{L_{tr}(\frac{6}{2}) + L_{top}(6) + L_{Bt}(0)}{L_{tot}}$

L_{top}, L_{tr}, L_{Bot}, 1.94"

solve for L_{top} ← $L_{tot} = L_{Top} + L_{tr} + L_{Bot}$

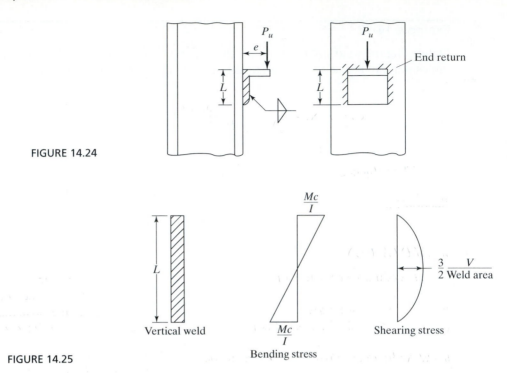

FIGURE 14.24

FIGURE 14.25

For short welds of this type, the usual practice is to consider a uniform variation of shearing stress. If, however, the bending stress is assumed to be given by the flexure formula, the shear does not vary uniformly for vertical welds but as a parabola with a maximum value $1\frac{1}{2}$ times the average value. These stress and shear variations are shown in Fig. 14.25.

The student should carefully note that the maximum shearing stresses and the maximum bending stresses occur at different locations. Therefore, it probably is not necessary to combine the two stresses at any one point. If the weld is capable of withstanding the worst shear and the worst moment individually, it is probably satisfactory. In Example 14-13, however, a welded connection subjected to shear and bending is designed by the usual practice of assuming a uniform shear distribution in the weld and combining the value vectorially with the maximum bending stress.

Example 14-13

Using E70 electrodes, the SMAW process, and the LRFD Specification determine the weld size required for the connection of Fig. 14.24 if $P_u = 45$ k, $e = 2\,1/2$ in, and $L = 8$ in. Assume that the member thicknesses do not control weld size.

Solution. Initially, assume a 1-in weld

$$f_s = \frac{45}{(2)(8)} = 2.81 \text{ k/in}$$

$$f = \frac{(45 \times 2.5)(4)}{(\frac{1}{12})(1)(8)^3(2)} = 5.27 \text{ k/in}$$

$$f_r = \sqrt{(2.81)^2 + (5.27)^2} = 5.97 \text{ k/in} = 5.97 \text{ ksi}$$

$$\text{Weld size required} = \frac{5.97}{(0.707)(1)(0.75)(0.60 \times 70)} = 0.268 \text{ in} \quad (\text{say } \tfrac{5}{16} \text{ in})$$

The subject of shear and bending is a very practical one, as it is the situation commonly faced in moment resisting connections. This topic is continued at length in Section 14.17.

14.17   DESIGN OF MOMENT-RESISTING CONNECTIONS

In this section, we present a brief introduction to moment-resisting connection. It is not our intention to describe in detail all of the possible arrangements of bolted and welded moment-resisting connections at this time, nor to provide a complete design. Instead, we attempt to provide the basic theory of transferring shear and moment from a beam to another member. One numerical example is included. This theory is very easy to understand and should enable the reader to design other moment-resisting connections—regardless of their configurations.

One moment-resisting connection that is popular with many fabricators is shown in Fig. 14.26. There the flanges are groove-welded to the column, while the shear is carried separately by a single plate or shear tab connection (to be described in detail in Section 15-7). The shear also could be carried by a seat angle or by web angles. Several other feasible moment-resisting connections are shown in Fig. 15.5.

The first step in the design is to compute the magnitude of the internal compression and tension forces, T and C. These forces are assumed to be concentrated at the center of each flange, as shown in Fig. 14.27

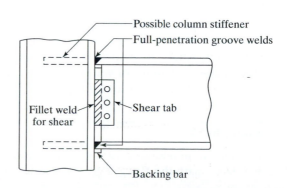

FIGURE 14.26

A moment-resisting connection.

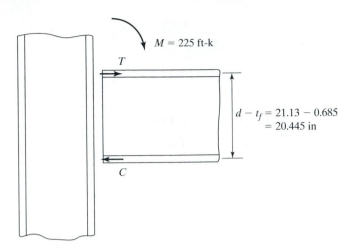

FIGURE 14.27

$$C = T = \frac{M}{d - t_f}$$

Next the areas of the full-penetration welds up against the column are determined. They equal the magnitude of C or T divided by the design stress of a full penetration groove weld as provided in Table 14.1 with $\phi = 0.9$.

$$\text{Area reqd.} = \frac{C \text{ or } T}{\phi F_y}$$

Using this procedure it is theoretically possible to have a weld area larger than the cross-sectional area of the flange. It would then be theoretically necessary to use an auxiliary plate on the flange to resist the extra force. (We may just transfer all of the forces with plates on the flanges. Sometimes the beam flanges are groove-welded flush with the column on one end and connected to the beam on the other end with the auxiliary plates just described. This could help us regarding our problems of fit. The forces are transferred from the beam to the plate with fillet welds and from the plate to the column by groove welds.)

Recent research at the University of California and Lehigh University has shown that the full plastic moment capacity of a beam can be developed with full-penetration welds made only to the flanges.

Example 14-14 illustrates the design of a moment-resisting connection with flange full-penetration groove welds. The reader should understand that this example is not quite complete. As described in the next chapter, it is necessary also to design the shear plate, seat angle, or whatever is used to transfer the shear, and also to check the column for the concentrated T or C force.

Example 14-14

Design a moment-resisting connection for the W21 × 68 beam shown in Fig. 14.27 with the flanges groove-welded to a column. The beam, which consists of

50 ksi steel, has a 60-k factored end reaction and an end moment of 225 ft-k. Use E70 electrodes.

Solution

Using a W21 × 68 (d = 21.1 in, b_f = 8.270 in, t_f = 0.685 in, T = 18.375 in)
Design of moment weld

$$C = T = \frac{(12)(225)}{20.415} = 132.3 \text{ k}$$

$$\text{Area of groove weld reqd} = \frac{132.3}{(0.9)(50)} = 2.94 \text{ in}^2$$

$$\text{Width reqd.} = \frac{2.94}{t_f} = \frac{2.94}{0.685} = 4.29 \text{ in} < b_f$$

Use 5 in wide full penetration groove welds, E70.

Design of shear weld

Try 1/4-in fillet weld on shear tab (or on seat L or web L_g.)

$$\text{Length reqd.} = \frac{60}{(0.707)\left(\frac{1}{4}\right)(0.75)(0.60 \times 70)} = 10.78 \text{ in} > \frac{T}{2} \text{ for W21} \times 68$$

Use 11 in

Note The design is incomplete, as it is necessary to design the shear tab, seat L, or whatever is used, and to check the column for the calculated shear force (C or T) from the beam. We will examine these tasks in the next chapter.

14.18   FULL-PENETRATION AND PARTIAL-PENETRATION GROOVE WELDS

14.18.1 Full-Penetration Groove Welds

When plates with different thicknesses are joined, the strength of a full-penetration groove weld is based on the strength of the thinner plate. Similarly, if plates of different strengths are joined, the strength of a full-penetration weld is based on the strength of the weaker plate. Notice that no allowances are made for the presence of reinforcement—that is, for any extra weld thickness.

Full-penetration groove welds are the best type of weld for resisting fatigue failures. In fact, in some specifications they are the only groove welds permitted if fatigue is possible. Furthermore, a study of some specifications shows that allowable stresses for fatigue situations are increased if the crowns or reinforcement of the groove welds have been ground flush.

14.18.2 Partial Penetration Groove Welds

When we have groove welds that do not extend completely through the full thickness of the parts being joined, they are referred to as *partial-penetration groove welds*. Such

FIGURE 14.28

Partial-penetration groove welds.

welds can be made from one or both sides, with or without preparation of the edges (such as bevels). Partial-penetration welds are shown in Fig. 14.28.

Partial-penetration groove welds often are economical in cases in which the welds are not required to develop large forces in the connected materials, such as for column splices and for the connecting together of the various parts of built-up members.

In Table 14.1, we can see that the design stresses are the same as for full-penetration welds when we have compression or tension parallel to the axis of the welds. When we have tension transverse to the weld axis, there is a substantial strength reduction because of the possiblity of high stress concentrations.

Examples 14-15 and 14-16 illustrate the calculations needed to determine the strength of full-penetration and partial-penetration groove welds. The design strengths of both full-penetration and partial-penetration groove welds are given in Table 14.1 of this chapter (LRFD Table J2.5).

In part (b) of Example 14-16, it is assumed that two W, sections are spliced together with a partial-penetration groove weld, and the design shear strength of the member is determined. To do this, it is necessary to compute the following: (1) the shear rupture strength of the base material as per LRFD Section J4.1, (2) the shear yielding strength of the connecting elements as per LRFD Section J5.3, and (3) the shear strength of the weld as per LRFD Section J2.2 and LRFD Table J2.5. The design shear strength of the member is the least of these three values which are described as follows:

1. Shear fracture of base material $= \phi F_n A_{ns}$ with $\phi = 0.75, F_n = 0.6 F_u,$ and $A_{ns} =$ net area subject to shear.

2. Shear yielding of connecting elements $= \phi R_n = \phi(0.60 A_{vg})F_y$ with $\phi = 0.90$ and $A_{vg} =$ gross area subjected to shear.

3. Shear yielding of the weld $= \phi F_w = \phi(0.60 F_{EXX})A_w$ with $\phi = 0.75$ and $A_w = A_{eff} =$ area of weld.

Example 14-15

    a. Determine the design tensile strength of a full-penetration groove weld for the plates shown in Fig. 14.29. Use 50 ksi steel and E70 electrodes.

    b. Determine the design tensile strength of the plates shown in Fig. 14.29 if a 1/4-in partial-penetration groove weld is used on one side.

FIGURE 14.29

Solution

a. $\phi P_n = \phi F_y A = (0.90)(50)(6 \times \frac{3}{4}) = 202.5 \text{ k}$

b. $\phi P_n = \phi F_y A = (0.90)(50)(6 \times \frac{1}{4}) = 67.5 \text{ k}$

   $\phi P_n = (\phi)(0.60 \, F_{EXX})(A) = (0.80)(0.60 \times 70)(6 \times \frac{1}{4}) = 50.4 \text{ k} \leftarrow$

Example 14-16

a. A full-penetration groove weld made with E70 electrodes is used to splice to-gether the two halves of a 50 ksi ($F_u = 65$ ksi) W21 × 166. Determine the shear design strength of the splice.

b. Repeat part (a) if two vertical partial-penetration welds 50 ksi ($F_u = 65$ ksi) (E70 electrodes) with throat thicknesses of 1/4 in are used.

Solution

Using a W21 × 166 ($d = 22.5$ in, $t_w = 0.750$ in)

a. $V_u = (0.90)(0.60F_y)A_w$

   $= (0.90)(0.60 \times 50)(22.5 \times 0.750) = 455.6$

or

   $V_u = (0.80)(0.60 \, F_{EXX})(A_w)$

   $= (0.80)(0.60 \times 70)(22.5 \times 0.750) = 567 \text{ k}$

b. shear fracture strength
of base material $= \phi F_w A_{nv}$

   $= (0.75)(0.6 \times 65)(22.5 \times 0.750) = 493.6 \text{ k}$

shear yielding of connecting elements $= \phi(0.60 \, A_g)F_y$

   $= (0.90)(0.60)(22.5 \times 0.750)(50)$

   $= 455.6 \text{ k}$

$$\text{shear yielding of weld} = \phi(0.60\,F_{EXX})\,A_w$$
$$= (0.75)\,(0.60 \times 70)\,(22.5 \times 2 \times \tfrac{1}{4})$$
$$= 354.4\ \text{k}$$

14.19 COMPUTER EXAMPLES

Examples 14–17 and 14–18 illustrate the application of the computer program IN-STEP 32 for welded connections.

Example 14-17

Using the computer program INSTEP 32 determine the design strength ϕP_n of the 5/16-in E70 fillet welded connection shown in Fig. 14.30. The steel is A572, Grade 50 and the SMAW process was used.

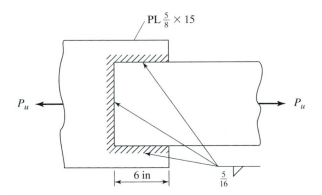

PL $\frac{5}{8} \times 15$

P_u P_u

6 in

$\frac{5}{16}$

FIGURE 14.30

Solution

Input:

Factored design loads	Joint parameters
$P_u = 0$ kips	Electrode E70xx
$e = 0$ in from CG	Process Shielded metal arc
Theta $= 90$ degrees	Weld size 0.3125
from horizontal	Weld type Fillet weld
	Weld style Shear through CG

Output:

Welded Joint Design Summary
Total weld length: 24.00 in
Weld throat: 0.2209 in
Design strength: 208.79 kips

Example 14-18

Using the computer program INSTEP 32 determine the design strength ϕP_n of the 3/8-in eccentrically loaded E70 fillet welded connection of Fig. 14.31. The steel is A572, Grade 50 and the SMAW process was used.

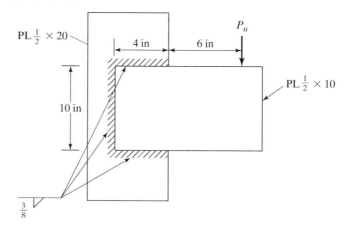

FIGURE 14.31

Input:

Factored design loads	Joint parameters
$P_u = 0$ kips	Electrode E70xx
$e = 9.11$ in from CG	Process Shielded metal arc
Theta = 90 degrees	Weld size 0.375 in
from horizontal	Weld type Fillet weld
	Weld style Eccentric shear

Output:

Welded Joint Design Summary

Total weld length: 18.00 in

Weld throat: 0.2651 in

CGx: 0.8889 in

CGy: 0.0000 in

Pn: 93.67 kips

Pna = 0.75Pn = 70.25 kips

e':1.6076 in from CG
Iterations: 10

PROBLEMS

Unless otherwise noted, A36 steel is to be used for all problems.

14-1. A 1/4 in fillet weld, SMAW process, is used to connect the members shown in the accompanying illustration. Determine the design load that can be applied to this connection according to the LRFD Specification if E70 electrodes are used. (*Ans.* $\phi P_n = 97.2$ k)

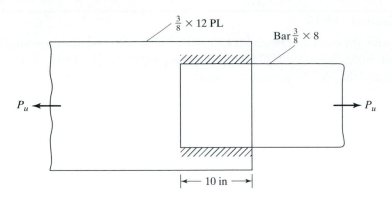

FIGURE P14-1

14-2. Rework Prob. 14-1 if the SAW process is used.

14-3. Rework Prob. 14-1 if A572 grade 65 steel and E80 electrodes are used (*Ans.* $\phi P_n = 127.4$ k)

14-4. Determine the strength of the 5/16-in fillet welds shown if E70 electrodes are used (SMAW Process).

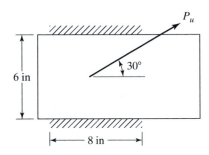

FIGURE P14-4

14-5. a. Repeat Prob. 14-4 if 1/4 in welds are used and if $\theta = 20°$. (*Ans.* $\phi P_n = 98.1$ k)

b. Repeat part (a) if $\theta = 10°$. (*Ans.* $\phi P_n = 92.4$ k)

14-6. Design maximum size fillet welds if $P_u = 150$ k. Use E70 electrodes and the SMAW process.

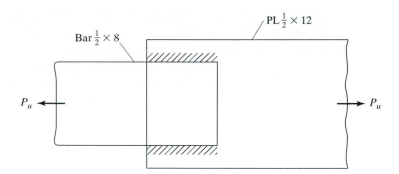

FIGURE P14-6

14-7. Rework Prob. 14-6 if the SAW process is used. (*Ans* Use 7/16-in welds 8 in long each side as required by LRFD Section 12.2b).

14-8. Rework Prob. 14-6 using side welds and a vertical end weld at the end of the $\frac{1}{2} \times 8$ bar. Also use A572 grade 65 steel and E80 electrodes.

14-9. Rework Prob. 14-6 using side welds and welds at the end of the $\frac{1}{2} \times 8$ PL and E70 electrodes (*Ans*: Use 7/16 in end and side welds, side welds are each 2 in long.)

14-10. The 5/8 × 8-in PL shown in the accompanying illustration is to be connected to a gusset plate with 5/16-in SMAW fillet welds. Determine the length L required to develop the full strength of the bar if E70 electrodes are used.

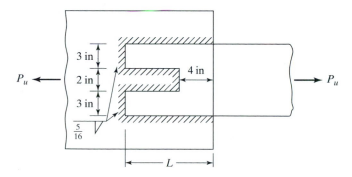

FIGURE P14-10

14-11. Design maximum size side SMAW fillet welds to develop a $P_u = 180$ k for an L6 × 4 × 1/2 using E70 electrodes steel. The member is connected on the sides of the 6-in leg and is subject to alternating loads. (*Ans.* Use 6 in one side and 12 1/2 in the other side.)

14-12. Rework Prob. 14-11 using side welds and a weld at the end of the angle.

14-13. Rework Prob. 14-12 using E80 electrodes. (*Ans.* 1 in one side and 6 1/4 in other side.)

14-14. One leg of an 8 × 8 × 3/4 angle is to be connected with side welds and a weld at the end of the angle to a plate behind to develop a load $P_u = 500$ k. Balance the

SMAW fillet welds around the center of gravity of the angle. Use maximum weld size and assume E70 electrodes.

14-15. It is desired to design 5/16-in SMAW fillet welds necessary to connect a C10 × 30 made from A36 steel to a 3/8-in gusset plate End side and slot welds may be used to develop a P_u = 285.4 k. No welding is permitted on the back of the channel. Use E70 electrodes. It is assumed that due to space limitations the channel can lap over the gusset plate by a maximum of 8 in. (*Ans.* 5/16-in fillet welds and one $1\frac{1}{16} \times 2\frac{1}{2}$ in slot weld)

14-16. Rework Prob. 14-15 using A572 grade 60 steel and E80 electrodes and 5/16-in fillet welds. Use P_u = 446 k.

14-17. Using the elastic method determine the maximum force per inch to be resisted by the fillet weld shown in the accompanying illustration. (*Ans.* 8.83 k/in)

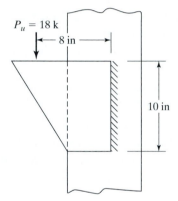

FIGURE P14-17

14-18. Using the elastic method determine the maximum force to be resisted per inch by the fillet weld shown in the accompanying illustration.

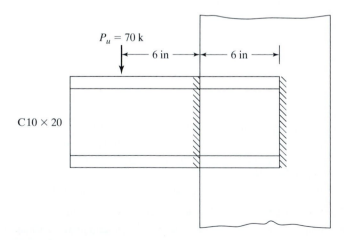

FIGURE P14-18

14-19. Using the elastic method rework Prob. 14-18 if welds are used on the top and bottom of the channel in additional to those shown in the figure. (*Ans.* 6.77 k/in)

14-20. Using the elastic method determine the maximum force per inch to be resisted by the fillet welds shown in the accompanying illustration. Also determine the required weld thickness using E70 electrodes, and SMAW welds.

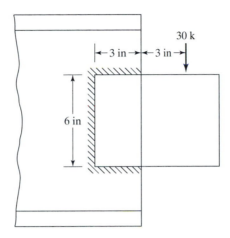

FIGURE P14-20

14-21. Determine the maximum eccentric load ϕP_n that can be applied to the connection shown in the accompanying illustration if 1/4-in SMAW fillet welds are used. Assume plate thickness = 1/2 in and use E70 electrodes. (a) Use elastic method. (b) Use LRFD tables and ultimate strength method. [*Ans.* (a) 25.27 k, (b) 35.84 k]

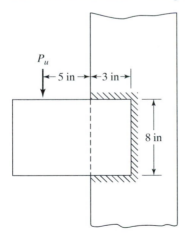

FIGURE P14-21

14-22. Rework Prob. 14-21 if 3/8-in fillet welds are used and the vertical weld is 10 in high.

14-23. Determine the SMAW fillet weld size required for the connection of Prob. 14-17 if the load is equal to 30 k and the height of the weld is 12 in E70. (a) Use elastic method (b) use LRFD tables and ultimate strength method. (*Ans.* (a) 0.463 in say 1/2 in; (b) 0.227 in say 1/4 in)

14-24. Using E70 electrodes and the SMAW process, determine the fillet weld size required for the bracket shown in the accompanying illustration. (a) Use elastic method. (b) Use LRFD Tables and ultimate strength method.

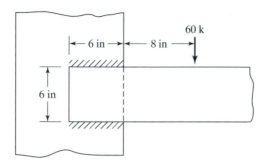

FIGURE P14-24

14-25. Rework Prob. 14-24 if the load is increased from 60 to 80 k and the horizontal weld lengths are increased from 6 to 8 in. [*Ans.* (a) 3/4 in (b) 11/16 in]

14-26. Determine the SMAW fillet weld size required for the connection shown in the accompanying illustration. E70. (a) Use elastic method. (b) Use LRFD tables and ultimate strength method.

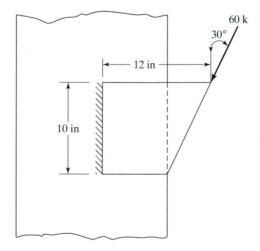

FIGURE P14-26

14-27. Determine the SMAW fillet weld size required for the connection shown in the accompanying illustration. E70. What angle thickness should be used? (a) Use elastic method. (b) Use LRFD tables and ultimate strength method. [*Ans.* $\frac{1}{4}$ in with 3 × 3 × 5/16 Ls (b) 3/16 in with 3 × 3 × 1/4 Ls.]

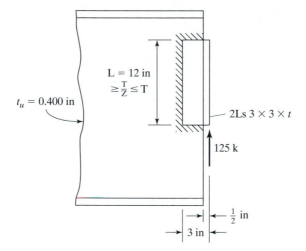

FIGURE P14-27

14-28. Assuming the SMAW process is to be used, determine the fillet weld size required for the connection shown in the accompanying illustration. Use E70 electrodes, and the elastic method.

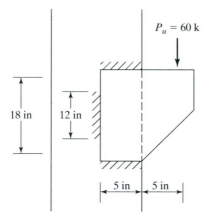

FIGURE P14-28

14-29. Determine the filled weld size required for the connection shown in the accompanying illustration. E70. The SMAW process is to be used. (a) Use elastic method. (b) Use LRFD tables and ultimate strength method. [*Ans.* (a) 7/8 in (b) 11/16 in]

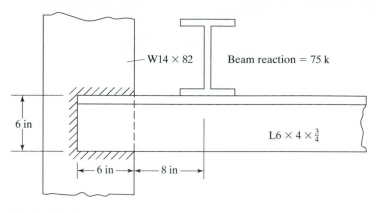

W14 × 82   Beam reaction = 75 k

6 in

L6 × 4 × $\frac{3}{4}$

← 6 in →|← 8 in →|

FIGURE P14-29

14-30. Determine the value of the load P_u that can be applied to the connection shown in the accompanying illustration if 3/8-in fillet welds are used. E70. SAW. Use the elastic method.

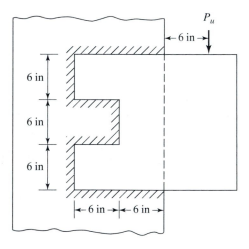

P_u

← 6 in →

6 in

6 in

6 in

|← 6 in →|← 6 in →|

FIGURE P14-30

14-31. Determine the length of 1/4-in SMAW fillet welds 12 in on center required to connect the cover plates for the section shown in the accompanying illustration at a point where the external shear V_u is 225 k. E70. (*Ans.* 8 in)

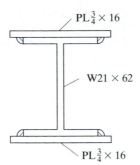

FIGURE P14-31

14-32.   The welded girder shown in the accompanying illustration has an external shear V_u of 900 k at a particular section. Determine the fillet weld size required to fasten the plates to the web if the SMAW process is used. E70.

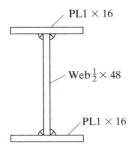

FIGURE P14-32

14-33.   Design SMAW welded moment resisting connections for the ends of a W24 × 76 to resist an R_u of 160 k and an M_u of 250 ft-k. Use A36 steel and E70 electrodes. Assume the column flange is 16 in wide? The moment is to be resisted by full-penetration groove welds in the flanges and the shear as to be resisted by welded clip angles along the web. Assume the beam was selected for bending with $0.9\,F_y$. (*Ans*. 1/4-in fillet welds 14.5 in high on web. Also 6 in wide full-penetration welds in flanges).

14-34.   The beam shown in the accompanying illustration is assumed to be attached at its ends with moment-resisting connections. Select the beam assuming full lateral support, and E70 SMAW electrodes. Use a connection of the type used in Prob. 14-33. $F_y = 50$ ksi.

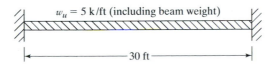

FIGURE P14-34

14-35. The two W16 × 40 beams shown in the accompanying illustration are to be made continuous across the supporting W24 × 84 girder. If the factored end reactions are each 110 k and their factored end moments are 90 ft-k, design the connections using E70 electrodes and SMAW welds. (*Ans.* 8 in shear welds and 10 in moment welds)

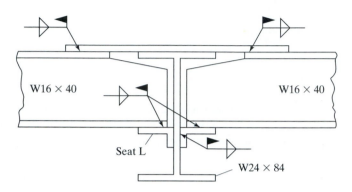

FIGURE P14-35

14-36. a. Assuming the plates in Fig. 14.29 are 8 in wide and 5/8 in thick, determine their design tensile strength if a full-penetration groove weld is used. Use E70 electrodes.

   b. Repeat part (a) if a 5/16-in partial-penetration groove weld is used on one side.

14-37. a. If full-penetration groove welds formed with E70 electrodes are used to splice together the two halves of an A24 × 117, determine the shear design strength of the splice. (*Ans.* 259.8 k)

   b. Repeat part (a) if two vertical partial-penetration groove welds with 1/4-in throat thickness are used. (*Ans.* 259.8 k)

14-38. If the flanges of the spliced beam of Prob. 14-37 are joined with partial-penetration groove welds with 5/16-in throat thickness, determine the design shear strength at the splice.

14-39. to **14-41**. *Solve the problems shown using the enclosed computer diskette, and E70 electrodes.*

14-39. Determine the design strength of the welds shown. (*Ans.* 111.3 k)

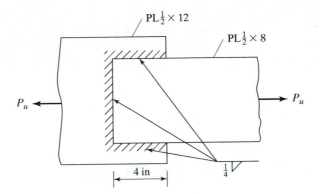

FIGURE P14-39

14-40. Calculate the fillet weld size required for the connection of Prob. 12-39 if the load P_u is equal to 125 k.

14-41. Determine the eccentric load P_u that the connection shown with its 3/8-in fillet weld can support. (*Ans.* 36.56 k)

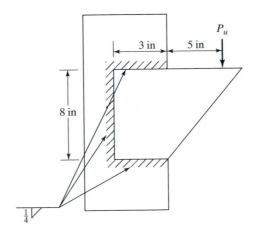

FIGURE P14-41

C H A P T E R   1 5

Building Connections

15.1   SELECTION OF TYPE OF FASTENER

This chapter is concerned with the actual beam-to-beam and beam-to-column connections commonly used in steel buildings. Under present-day steel specifications, four types of fasteners are permitted for these connections: *welds, unfinished bolts, high-strength bolts*, and *rivets*.

Selection of the type of fastener or fasteners to be used for a particular structure usually involves many factors, including requirements of local building codes, relative economy, preference of designer, availability of good welders, loading conditions (as static or fatigue loadings), preference of fabricator, and equipment available. It is impossible to list a definite set of rules from which the best type of fastener can be selected for any given structure. One can only give a few general statements that may be helpful in making a decision. These are listed as follows.

1. Unfinished bolts are often economical for light structures subject to small static loads and for secondary members (such as purlins, girts, bracing, etc.) in larger structures.
2. Field bolting is very rapid and involves less skilled labor than welding. The purchase price of high-strength bolts, however, is rather high.
3. If a structure is later to be disassembled, welding probably is ruled out, leaving the job open to bolts.
4. For fatigue loadings, slip-critical high-strength bolts and welds are very good.
5. Notice that special care has to be taken to properly install high-strength, slip-critical bolts.
6. Welding requires the smallest amounts of steel, probably provides the most attractive-looking joints, and also has the widest range of application to different types of connections.
7. When continuous and rigid fully moment-resisting joints are desired, welding probably will be selected.

8. Welding is almost universally accepted as being satisfactory for shop-work. For fieldwork it is very popular in most areas of the United States, while in a few others it is stymied by the fear that field inspection is rather questionable.

9. To use welds for very thick members requires a great deal of extra care, and bolted connections may very well be used instead. Furthermore, such bolted connections are far less susceptible to brittle fractures.

15.2   TYPES OF BEAM CONNECTIONS

All connections have some restraint—that is, some resistance to changes of the original angles between intersecting members when loads are applied. Depending on the amount of restraint the LRFD Specification (A2.) classifies connections as being fully restrained (Type FR) and partially restrained (Type PR). These two types of connections are described in more detail as follows:

1. Type FR connections are commonly referred to as rigid or continuous frame connections. They are assumed to be sufficiently rigid or restrained to keep the original angles between members virtually unchanged under load.

2. Type PR connections are those that have insufficient rigidity to keep the original angles virtually unchanged under load. Included in this classification are simple and semirigid connections as described in detail in this section.

A *simple connection* is a Type PR connection for which restraint is ignored. It is assumed to be completely flexible and free to rotate, thus having no moment resistance. A *semirigid connection* is also a Type PR connection whose resistance to angle change falls somewhere between the simple and rigid types.

As there are no perfectly rigid connections nor completely flexible ones, all connections really are partly restrained or PR to one degree or another. The usual practice in the past was to classify connections based on a ratio of the moment developed for a particular connection to the moment that would be developed by a completely rigid connection. A rough rule was that simple connections had 0–20 percent rigidity, semirigid connections had 20–90 percent rigidity, and rigid connections had 90–100 percent rigidity. Figure 15.1 shows a set of typical moment-rotation curves for these connections. Notice that the lines are curved because as the moments become larger, the rotations increase at a faster rate.

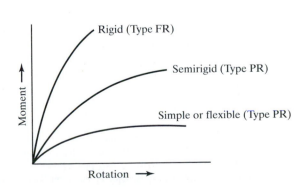

FIGURE 15.1

Typical moment-rotation curves for connections.

During the past few years, quite a few investigators around the world have been trying to develop empirical formulas for describing the rotational characteristics of connections.[1,2,3,4,5,6] Though they have made some progress, the only accurate method for developing such information today involves the actual fabrication of connections followed by load testing. It is very difficult to include in a formula the effects of such things as poor fit, improper tightening of bolts, and so on.

Each of these three general types of connections is briefly discussed in this section, with little mention of the specific types of connectors used. The remainder of the chapter is concerned with detailed designs of these connections using specific types of fasteners. *In this discussion the author probably overemphasizes the semirigid and rigid type connections, because a very large percentage of the building designs with which the average designer works will be assumed to have simple connections.* A few descriptive comments are given in the paragraphs that follow concerning each of these three types of connections.

Simple connections (Type PR) are quite flexible and are assumed to allow the beam ends to be substantially free to rotate downward under load, as true simple beams should. Although simple connections do have some moment resistance (or resistance to end rotation), it is assumed to be negligible and they are assumed to be able to resist shear only. Several types of simple connections are shown in Fig. 15.2. More detailed descriptions of each of these connections and their assumed behavior under load are given in later sections of this chapter. In this figure most of the connections are shown as being made entirely with the same type of fastener, while in actual practice two types of fasteners are often used for the same connection. For example, a very common practice is to shopweld the web angles to the beam web and field-bolt them to the column or girder.

Semirigid connections (Type PR) are those that have appreciable resistance to end rotation, thus developing appreciable end moments. In design practice it is quite common for the designer to assume all connections are either simple or rigid, with no consideration given to those situations in between, thereby simplifying the analysis. Should he or she make such an assumption for a true semirigid connection, he or she may miss an opportunity for appreciable moment reductions. To understand this possibility, the reader is referred to the moment diagrams shown in Fig. 15.3 for a group of uniformly loaded beams supported with connections having different percentages of rigidity. This figure shows that the maximum moments in a beam vary greatly with different types of end connections. For example, the maximum moment in the semirigid connection of part (d) of the figure is only 50 percent of the maximum moment in the

---

[1]R. M. Richard, "A Study of Systems Having Conservative Non-Linearity," Ph.D. Thesis, Purdue University, 1961.

[2]N. Kishi and W.F. Chen, "Data Base of Steel Beam-to-Column Connections," no. CE-STR-86-26, Vols. I and II (West Lafayette, IN: Purdue University, School of Engineering, July, 1986).

[3]L. F. Geschwindner, "A Simplified Look at Partially Restrained Beams," *Engineering Journal*, AISC, 28, no. 2 (2nd quarter, 1991), pp. 73–78.

[4]Wai-Fah Chen et al., "Semi-rigid Connections in Steel Frames," Council on Tall Buildings and Urban Habitat, Committee 43 (McGraw Hill, 1992).

[5]S. E. Kim and W. F. Chen, "Practical Advanced Analysis for semi-rigid Frame Design," *Engineering Journal*, AISC, 33, no. 4 (4th quarter, 1996), pp. 129–141.

[6]J. E. Christopher and R. Bjorhovde, Semi-Rigid Frame Design for Practicing Engineers," *Engineering Journal*, AISC, 36, no. 1 (1st quarter, 1999), pp. 12–28.

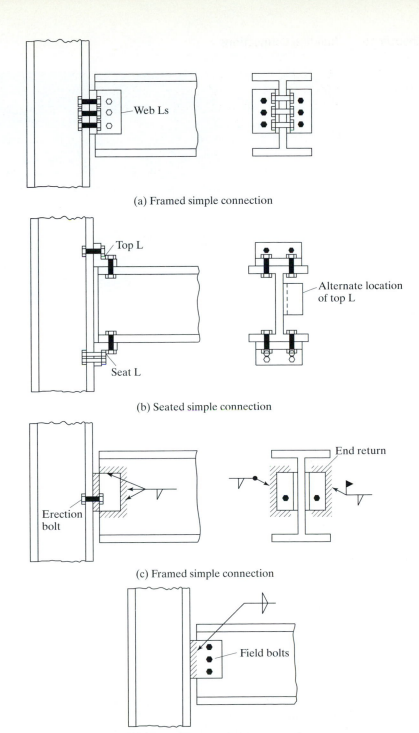

(a) Framed simple connection

(b) Seated simple connection

(c) Framed simple connection

(d) Single-plate or shear tab simple connection

FIGURE 15.2

Some simple connections. Notice how these connections are placed up towards the top flanges so they provide lateral stability at the compression flanges at the beam supports. (a) Framed simple connection. (b) Seated simple connection.
(c) Framed simple connection. (d) Single-plate or shear tab simple connection.

A typical framing angle shop-welded to a beam. It will be field-bolted to another member. (Courtesy American Institute of Steel Construction, Inc.)

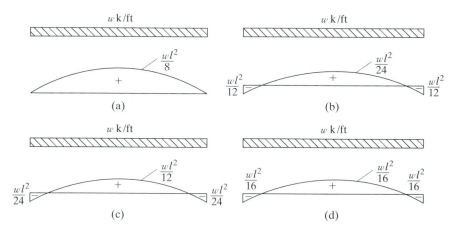

FIGURE 15.3

(a) Simple connections (0 percent rigidity). (b) Rigid connections (100 percent rigidity). (c) Semirigid connections (50 percent rigidity). (d) Semirigid connections (75 percent rigidity).

simply supported beam of part (a) and only 75 percent of the maximum moment in the rigidly supported beam of part (b).

Actual semirigid connections are used fairly often, but usually no advantage is taken of their moment-reducing possibilities in the calculations. Perhaps one factor that keeps the design profession from taking advantage of them more often is the LRFD Specification (Section A2), which states that consideration of a connection as being semirigid is permitted only upon presentation of evidence that it is capable of providing a certain percentage of the end restraint furnished by a completely rigid connection. This evidence must consist of documentation in the technical literature, or must be established by analytical or empirical means.

Three practical semirigid or PR connections capable of providing considerable moment resistance are shown in Fig. 15.4. If the end-plate connection shown in part (a) of the figure is extended above the beam and more bolts are installed, the moment resistance of the connection can be appreciably increased. Part (c) of the figure shows a semirigid connection that is proving to be quite satisfactory for steel-concrete composite floors. Moment resistance in this connection is provided by reinforcing bars placed in the concrete slab above the beam and by the horizontal leg of the seat angle.[7] Another type of semirigid connection is illustrated in an accompanying photograph from the Lincoln Electric Company.

The use of partially restrained connections with roughly 60 to 75 percent rigidities is gradually increasing. When it becomes possible to accurately predict the percentages of rigidity for various connections, and when better design procedures are available, this type of design will probably become even more common.

Rigid connections (Type FR) are those which theoretically allow no rotation at the beam ends and thus transfer close to 100 percent of the moment of a fixed end. Connections of this type may be used for tall buildings in which wind resistance is developed by providing continuity between the members of the building frame. Several Type FR connections that provide almost 100 percent restraint are shown in Fig. 15.5. It will be noticed in the figure that column web stiffeners may be required for some of these connections to provide sufficient resistance to rotation. The design of these stiffeners is discussed in Section 15-12.

The moment connection shown in part (d) is rather popular with steel fabricators, and the end-plate connection of part (e) also has been frequently used in recent years.[8]

You will note the use of *shims* in parts (a) to (c) of Fig. 15.5. Shims are thin strips of steel that are used to adjust the fit at connections. They can be one of two types:

conventional or finger shims. *Conventional shims* are those that are installed with the bolts passing through them, while *finger shims* can be installed after the bolts are in place. You should realize that there is some variation in the depths of beams as they come from the steel mills (see Table 1.54 in Part 1 of the LRFD Manual for permissible tolerances). To provide for such variations, it is common to make the distance between flange plates or angles larger than the nominal beam depths given in the manual.[9]

---

[7]D. J. Ammerman and R. T. Leon, "Unbraced Frames with Semirigid Composite Connections," *Engineering Journal*, AISC, 27, no. 1 (1st quarter 1990), pp. 12–21.

[8]J. D. Griffiths, "End-Plate Moment Connections—Their Use and Misuse," *Engineering Journal*, AISC, vol. 21, no. 1 (first quarter, 1984) pp. 32–34.

[9]W. T. Segui, *Fundamentals of Structural Steel Design*, 2nd. ed. (Boston: PWS-Kent, 1994), pp. 355–356.

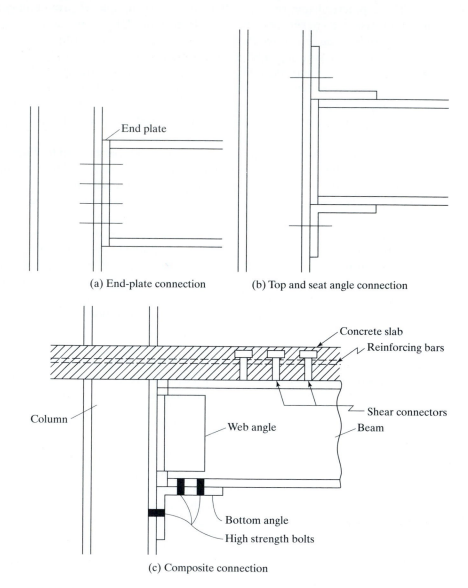

(a) End-plate connection   (b) Top and seat angle connection

(c) Composite connection

FIGURE 15.4

Some semirigid connections.

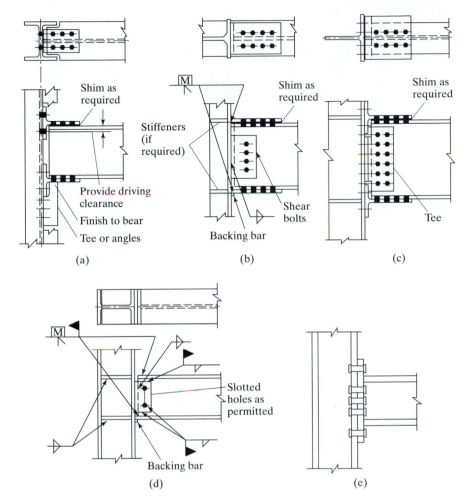

FIGURE 15.5

Moment-resisting connections.

15.3   STANDARD BOLTED BEAM CONNECTIONS

Several types of standard bolted connections are shown in Fig. 15.6. These connections are usually designed to resist shear only, as testing has proved this practice to be quite satisfactory. Part (a) of the figure shows a connection between beams with the so-called *framed connection.* This type of connection consists of a pair of flexible web angles probably shop-connected to the web of the supported beam and field-connected to the supporting beam or column. When two beams are being connected, usually it is necessary to keep their top flanges at the same elevation, with the result that the top flange of one will have to be cut back (called *coping*) as shown in part (b) of the figure. For such connections we must check block shear as discussed in Section 3.7 of this text.

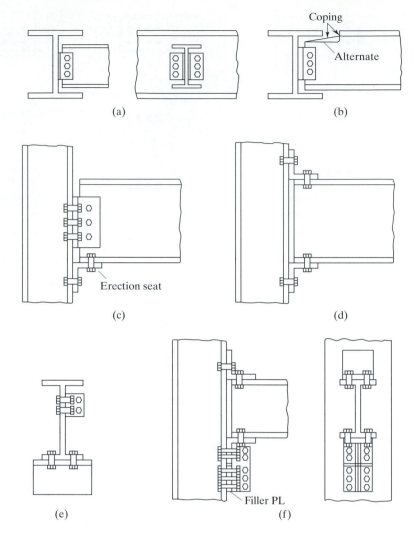

FIGURE 15.6

(a) Framed connection. (b) Framed connection. (c) Framed connection. (d) Seated connection. (e) Seated connection. (f) Seated connection with stiffener angles.

Simple connections of beams to columns can be either framed or seated as shown in Fig. 15.6. In part (c) of the figure, a framed connection is shown in which two web angles are connected to the beam web in the shop, after which bolts are placed through the angles and column in the field. It is often convenient to use an angle called an *erection seat* to support the beam during erection. Such an angle is shown in the figure.

The seated connection has an angle under the beam similar to the erection seat just mentioned, which is shop-connected to the column. In addition, there is another angle—probably on top of the beam—that is field-connected to the beam and column. A seated connection of this type is shown in part (d) of the figure. Should space limitation prove a problem above the beam, the top angle may be placed in the optional location shown

A semirigid beam-to-column connection, Ainsley Building, Miami, FL. (Courtesy of the Lincoln Electric Company.)

in part (e) of the figure. The top angle at either of the locations mentioned is very helpful in keeping the top flange of the beam from being accidentally twisted out of place during construction.

The amount of load that can be supported by the types of connections shown in parts (c), (d), and (e) of Fig. 15.6 is severely limited by the flexibility or bending strength of the horizontal legs of the seat angles. For heavier loads it is necessary to use stiffened seats such as the one shown in part (f) of the figure.

The majority of these connections are selected by referring to standard tables. The LRFD Manual has excellent tables for selecting bolted or welded beam connections of the types shown in Fig. 15.6. After a rolled-beam section has been selected, it is quite convenient to refer to these tables and select one of the standard connections, which will be suitable for the vast majority of cases.

In order to make these standard connections have as little moment resistance as possible, the angles used in making up the connections are usually light and flexible. To qualify as simple end supports the ends of the beams should be as free as possible to rotate downward. Figure 15.7 shows the manner in which framed and seated end connections will theoretically deform as the ends of the beams rotate downward. The designer does not want to do anything that will hamper these deformations if he or she is striving for simple supports.

For the rotations shown in Fig. 15.7 to occur, there must be some deformation of the angles. As a matter of fact, if end slopes of the magnitudes that are computed for

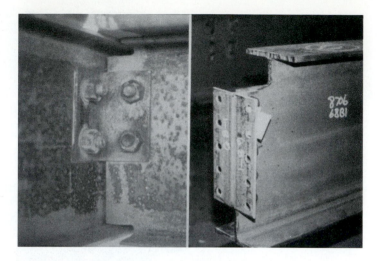

The beam on the right has a coped top flange, while the beam on the left has both flanges coped because its depth is very close to that of the supporting girder. (Courtesy American Institute of Steel Construction, Inc.)

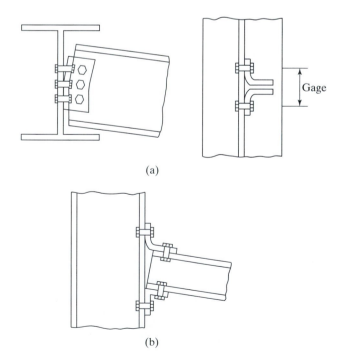

(a)

(b)

FIGURE 15.7

(a) Bending of framed-beam connection.
(b) Bending of seated-beam connection.

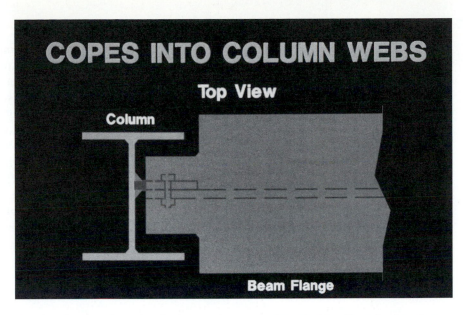

Top view showing a beam framing into a column web where the beam flange is wider than the opening between the column flanges. Thus, both top and bottom flanges are coped by flame-cutting.

simple ends are to occur, the angles will actually bend enough to be stressed beyond their yield points. If this situation occurs, they will be permanently bent and the connections will quite closely approach true simple ends. The student should now see why it is desirable to use rather thin angles and large gages for the bolt spacing if flexible simple end connections are the goal of the designer.

These connections do have some resistance to moment. When the ends of the beam begin to rotate downward, the rotation is certainly resisted to some extent by the tension in the top bolts, even if the angles are quite thin and flexible. Neglecting the moment resistance of these connections will cause conservative beam sizes. If moments of any significance are to be resisted, more rigid-type joints need to be provided than are available with the framed and seated connections.

15.4   LRFD MANUAL STANDARD CONNECTION TABLES

In Part 10 of the LRFD Manual a series of tables are presented that the designer may use to select several different types of standard connections. There are tables for bolted or welded two-angle framed connections, seated beam connections, stiffened seated beam connections, eccentrically loaded connections, single-angle framed connections, and others.

In the next few sections of this chapter (15.5 through 15.8) a few standard connections are selected using the tables in the Manual. The author hopes that these will be sufficient to introduce the reader to the LRFD tables and to enable him or her to make designs using the other tables with little difficulty.

Sections 15.9 to 15.11 present design information for some other types of connections.

15.5   DESIGNS OF STANDARD BOLTED FRAMED CONNECTIONS

For small and low-rise buildings (and that means most buildings), simple framed connections of the types previously shown in parts (a) and (b) of Fig. 15.6 are usually used to connect beams to girders or to columns. The angles used are rather thin (1/2 in is the arbitrary maximum thickness used in the LRFD Manual) so they will have the necessary flexibility shown in Fig. 15.7. The angles will develop some small moments (supposedly not more than 20 percent of full fixed-end conditions), but they are neglected in design.

The framing angles extend out from the beam web by 1/2 in as shown in Fig. 15.8. This protrusion, which is often referred to as the *setback*, is quite useful in fitting members together during steel erection.

In this section, several standard bolted framed connections for simple beams are designed, using the tables provided in Part 10 of the LRFD Manual. In these tables the following abbreviations for different bolt conditions are used:

1. A325-SC and A490-SC (slip-critical connections)
2. A325-N and A490-N (bearing-type connections with threads included in the shear planes)
3. A325-X and A490-X (bearing-type connections with threads excluded from shear planes)

It is thought that the minimum depth of framing angles should be at least equal to one-half of the distance between the web toes of the beam fillets (called the *T* distances and given in the properties tables of Part 1 of the Manual). This minimum depth is used so as to provide sufficient stability during steel erection.

Example 15-1 presents the design of standard framing angles for a simply supported beam using bearing-type bolts in standard-size holes. In this example the design strengths of the bolts and the angles are taken from the appropriate tables. We show how the table values can be checked with the procedures presented in Chapters 12 to 14.

Example 15-1

Select an all-bolted double-angle framed simple end connection for the W30 $\times$ 108 ($t_w = 0.545$ in) shown in Fig. 15.8 if the factored end reaction $R_u = 175$ k and if the connection frames into the flange of a W14 $\times$ 61 column ($t_f = 0.645$ in). Assume that $F_y = 36$ ksi and $F_u = 58$ ksi for the framing angles and 50 ksi and 65 ksi respectively for the beam and column. Use 3/4-in A325-N bolts (bearing type threads included in shear plane) in standard size holes.

Solution. Using double-angle connection with $F_y = 36$ ksi and $F_u = 58$ ksi, we look in the tables for the least number of rows of bolts that can be used with a W30 section. It's 5 rows (page 10-26 Part 10 of the LRFD Manual), but the connection for 3/4-in A325-N bolts will not support a reaction that large. Thus we move to 6 rows on page 10-25 and

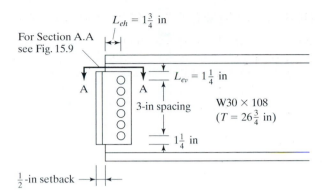

FIGURE 15.8

try a connection with an angle $t = 5/16$ in and with a length $L = 15 + (2)(1\,1/4)$ = 17 1/2 in (This length seems satisfactory when compared to the T of 26 1/2 in for this shape.)

$$\phi R_n = 191 \text{ k} > 175 \text{ k} \qquad \text{(OK)}$$

(Checking the preceding value: the 6 bolts in the beam web are in double shear and bearing on 0.545 in. The shear controls and equals $(0.75)(2 \times 0.4418)$ $(48)(6) = 191$ k.)

Checking supported beam web (bottom of page 10-25 in Part 10 of the LRFD Manual

With $L_{eh} = 1\frac{3}{4}$ in and $L_{ev} = 1\frac{1}{4}$ in (see Fig. 15.8)

$$\phi R_n \text{ for uncoped 0.545-in web} = (0.545)(526) = 287 \text{ k} > 175 \text{ k} \qquad \text{(OK)}$$

Checking supported column flange (bottom of page 10-25 in Part 10 of the LRFD Manual)

Noting there are two angle legs attached to flange

$$\phi R_n = (2)(526)(0.645) = 679 \text{ k} > 175 \text{ k} \qquad \text{(OK)}$$

To select the lengths of the angle legs, it is necessary to study the dimensions given in Fig. 15.9, which is a view along Section A-A in Fig. 15.8. The lower right part of the figure shows the minimum clearances needed for insertion and tightening of the bolts. These are the H_2 and C_1 distances and are obtained for 3/4-in bolts from Table 7.3a in Part 7 of the LRFD Manual.

For the legs bolted to the beam web, a $2\frac{1}{2}$-in gage is used. Using a minimum edge distance of 1 in here we will make these angle legs $3\frac{1}{2}$ in. For the outstanding legs, the minimum gage is $\frac{5}{16} + 1\frac{3}{8} + 1\frac{1}{4} = 2\frac{15}{16}$ in, say 3 in. We will make this a 4-in angle leg.

$$\text{Use 2Ls } 4 \times 3\frac{1}{2} \times \frac{5}{16} \times 1 \text{ ft} - 5\frac{1}{2} \text{ in A36}$$

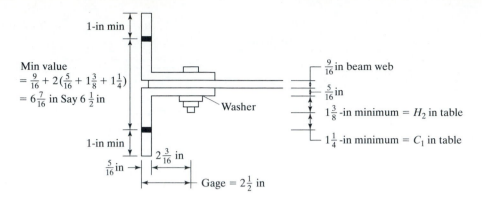

1-in min

Min value
$= \frac{9}{16} + 2(\frac{5}{16} + 1\frac{3}{8} + 1\frac{1}{4})$
$= 6\frac{7}{16}$ in Say $6\frac{1}{2}$ in

Washer

1-in min

$\frac{5}{16}$ in → | $2\frac{3}{16}$ in

Gage $= 2\frac{1}{2}$ in

$\frac{9}{16}$ in beam web

$\frac{5}{16}$ in

$1\frac{3}{8}$ -in minimum $= H_2$ in table

$1\frac{1}{4}$ -in minimum $= C_1$ in table

FIGURE 15.9

Using the tables of Part 10 of the LRFD Manual, framed bolted connections can be easily selected where one or both flanges of a beam are coped. Coping may substantially reduce the design strength of beams and require the use of larger members or the addition of web reinforcing. The design strength of coped members for flexural yielding, local buckling, and lateral torsional buckling is described in Part 10 of the LRFD Manual. Example 10-1 on pages 10-12 to 10-14 illustrates the procedure.

15.6   DESIGNS OF STANDARD WELDED FRAMED CONNECTIONS

Table 10.2 in Part 10 of the LRFD Manual includes the information necessary to use welds instead of bolts as used in Example 15-1. The values in the table are based on E70 electrodes. The table is normally used where the angles are attached to the beams in the shop and then field-bolted to the other member. Should the framing angles be welded to both members the weld values provided in Table 10.3 in Part 10 of the LRFD Manual are used.

A weld used to connect the angles to the beam web is called Weld A as shown in Figure 15.10. If a weld is used to connect the beam to another member, that weld is called Weld B.

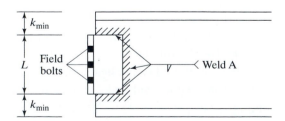

k_{min}

L Field bolts

k_{min}

Weld A

FIGURE 15.10

For the usual situations, $4 \times 3\frac{1}{2}$ in angles are used with the $3\frac{1}{2}$-in legs connected to the beam webs. The 4-in outstanding legs will usually accommodate the standard gages for the bolts going into the other members. The angle thickness selected equals the weld size plus 1/16 in or the minimum value given in Table 10.1 for the bolts. The angle lengths are the same as those used for the nonstaggered bolt cases (that is, $5\frac{1}{2}$ through $35\frac{1}{2}$ in).

The design strengths of the welds to the beam webs (Weld A) given in Table 10.2 in Part 10 of the LRFD Manual were computed using the instantaneous center of rotation method, which we briefly described in Chapter 14. To select a connection of this type, the designer picks a weld size from Table 10.2 and then goes to Table 10.1 to determine the number of bolts required for connection to the other member. This procedure is illustrated in Example 15-2.

Example 15-2

Design a framed beam connection to be welded (SMAW) to a W30 $\times$ 90 beam ($t_w = 0.470$ in and $T = 26\frac{1}{2}$ in) and then bolted to another member. The factored beam reaction R_u is 210 k, the steel for the angles is A36, the weld is E70, and the bolts are 3/4-in A325-N.

Solution. Table 10.2 Weld A Design Strength.

From this table one of several possibilities is a 3/16-in weld $20\frac{1}{2}$ in long with a design strength of 228 k. The depth or length is compatible with the T value of $26\frac{1}{2}$ in. For a 3/16-in weld, the minimum angle thickness is 1/4 in, as per LRFD Specification J2.2b. The minimum web thickness for shear given in Table 10.2 is 0.286 in for $F_y = 36$ ksi and this is less than the 0.470 in furnished.

Table 10.1 Selection of Bolts.

The angle length selected is $20\frac{1}{2}$ in and this corresponds to a 7-row bolted connection in Table 10.1. From this latter table we find that 7 rows of 3/4-in A325-N bolts will support 188 k if the angle thickness is 1/4 in, but R_u is 210 k. If we use a 5/16-in angle thickness the bolts will support 223 k, which is satisfactory.

Use 2Ls $4 \times 3\frac{1}{2} \times \frac{5}{16} \times$ 1 ft $8\frac{1}{2}$ in A36

Table 10.2 in Part 10 of the LRFD Manual provides the information necessary for designing all-welded standard framed connections; that is, with Welds A and B as shown in Fig. 15.11. The design strengths for Weld A were determined by the ultimate strength or instantaneous center or rotation method, while the values for Weld B were determined using the elastic method. A sample design is presented in Example 15-3.

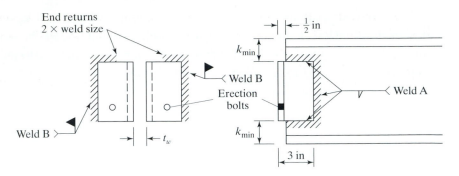

FIGURE 15.11

Example 15-3

Select an all-welded double-angle framed simple end connection for a W30 × 108 beam (t_w = 0.545 in) to a W14 × 61 column (t_f = 0.645 in). Assume F_y = 50 ksi and F_u = 65 ksi for the members and R_u = 200 k.

Solution. Select Weld A for beam

From Table 10.3 one possibility is a 3/16-in weld 20-in long. For a beam with F_y = 50 ksi the minimum web thickness is 0.286 in, which is < t_w of 0.545 in of the beam.

$$\phi R_n = 223 \text{ k} > 200 \text{ k} \qquad \text{(OK)}$$

Select Weld B for column

The 3/16 in welds will not provide sufficient capacity. ∴ Use 1/4 in. For a column with F_y = 50 ksi the minimum flange thickness is 0.26 in, which is < t_f of 0.710 in of the column (ϕR_n = 226 k).

Check minimum angle thickness

$$\min t = \tfrac{5}{16} + \tfrac{1}{16} = \tfrac{3}{8} \text{ in}$$

Use 2Ls 4 × 3 × $\tfrac{3}{8}$ × 1 ft 8 in. A36

15.7   SINGLE-PLATE OR SHEAR TAB FRAMING CONNECTIONS

An economical type of flexible connection for light loads that is being used more and more frequently is the single-plate framing connection, which was illustrated in Fig. 15.2(d). The bolt holes are prepunched in the plate and the web of the beam. The plate is then shop-welded to the supporting beam or column; lastly, the beam is bolted to the plate in the field. Steel erectors like this kind of connection because of its simplicity. They are particularly pleased with it when there is a beam connected to each side of a girder, as shown in Fig. 15.12(a). All they have to do is bolt the beam webs to the single plate on each side of the girder. Should web clip angles be used for such a connection, the bolts will have to pass through the angles on each side of the girder as well as the girder web, as shown in part (b) of the figure. This is a little more difficult field operation. Frequently, one side will have one extra bolt row so that each beam can be erected seperately.

  With the single-plate connection, the reaction or shear load is assumed to be distributed equally among the bolts passing through the beam web. It is also assumed that

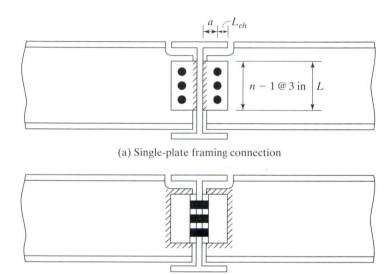

(a) Single-plate framing connection

FIGURE 15.12

(a) Single-plate framing connection.
(b) Simple connection with web angles.

(b) Simple connection with web angles

A shear tab or single-plate framing connection, (Courtesy of American Institute of Steel Construction, Inc.)

relatively free rotation occurs between the end of the member and the supporting beam or column. Because of these assumptions, this type of connection often is referred to as a "shear tab" connection. Various studies and tests have shown that these connections can develop some end moments depending on the number and size of the bolts and their arrangement, the thicknesses of the plate and beam web, the span-to-depth ratio of the beam, the type of loading, and the flexibility of the supporting element.

Tables for shear tab connections are presented in the LRFD Manual. An empirical design procedure using service loads is presented by R. M. Richard et al.[10]

Example 15-4

Design a single-plate shear connection for the W16 × 50 beam to the W14 × 90 column shown in Fig. 15.13 if R_u = 55 k. Use 3/4-in A325-N high-strength bolts and E70 electrodes. The beam and column are to have F_y = 50 ksi and F_u = 65 ksi, while the plate is to have F_y = 36 ksi and F_u = 58 ksi.

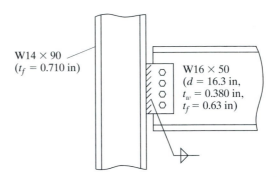

FIGURE 15.13

Solution. Assuming the column provides rigid support: 4 rows of bolts, a 1/4-in plate and 3/16-in fillet welds are selected from Table 10-9 in Part 10 of the LRFD Manual. This problem is Example 10-11 in the LRFD Manual.

$$\phi R_n = 55.5 \text{ k} > 55 \text{ k} \tag{OK}$$

Checking supported beam web

From Table 10.9 for 4 rows of bolts with $L_{ev} = 1\frac{1}{2}$ in and $L_{eh} = 1\frac{1}{2}$ in and an uncoped beam.

ϕR_n for beam web design strength
(Table 10.1) = (0.380) (351) = 133 k > 55 k   (OK)

Use $PL\frac{1}{4} \times 4\frac{1}{2} \times$ 1ft 0 in A36 steel with 4 rows of 3/4-in A325-N bolts and 3/16-in fillet welds, E70, SMAW

---

[10]R. M. Richard et al. "The Analysis and Design of Single-Plate Framing Connections," *Engineering Journal*, AISC, vol. 17, no. 2 (second quarter, 1980), pp. 38–52.

Should the beam be coped it will be necessary to also check flexural yielding and local web buckling at the cope. This process is illustrated in Example 10-12 in the LRFD Manual, pages 10-114 to 10-116.

15.8   END-PLATE SHEAR CONNECTIONS

Another type of connection is the *end-plate* connection. It consists of a plate shop-welded flush against the end of a beam and field-bolted to a column or another beam. To use this type of connection it is necessary to carefully control the length of the beam and the squaring of its ends, so that the end plates are vertical. Camber must also be considered in its effect on the position of the end plate. After a little practice in erecting members with end-plate connections, fabricators seem to like to use them. Nonetheless, there still is trouble in getting the dimensions just right, and they are not as commonly used as the single-plate connectors.

Part (a) of Fig. 15.14 shows an end-plate connection that is satisfactory for PR situations. End-plate connections are illustrated in Part 10 of the LRFD Manual. Should the end plate be extended above and below the beam, as shown in part (b) of Fig. 15.14, appreciable moment resistance will be achieved.

Part 12 of the LRFD Manual provides tables and a procedure for designing extended end-plate connections. These connections may be designed as being FR ones only for statically loaded structures and for buildings in seismic zone 1 areas and for unimportant buildings in seismic zone 2 areas.

15.9   DESIGNS OF WELDED SEATED BEAM CONNECTIONS

Another type of fairly flexible beam connection is obtained by using a beam seat such as the one shown in Fig. 15.15. Beam seats obviously offer an advantage to the worker performing the steel erection. The connections for these angles may be bolts or welds, but only the welded type is considered here. For such a situation, the seat angle would usually be shop-welded to the column and field-welded to the beam. When welds are used, the seat angles, also called *shelf angles*, may be punched for erection bolts as

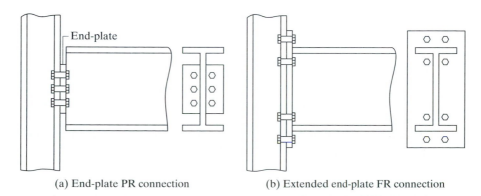

(a) End-plate PR connection   (b) Extended end-plate FR connection

FIGURE 15.14

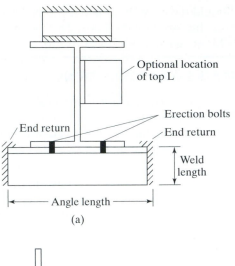

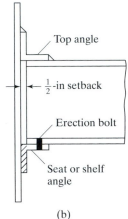

FIGURE 15.15   (b)

shown in the figure. These holes can be slotted if desired to permit easy alignment of the members.

    A seated connection may be used only when a top angle is used, as shown in Fig. 15.15. This angle provides lateral support for the beam and may be placed on top of the beam, or at the optional location shown on the side of the beam in part (a) of the figure. As the top angle is not usually assumed to resist any of the load, its size probably is selected by judgment. Fairly flexible angles are used that will bend away from the column or girder to which it is connected when the beam tends to rotate downward under load. This desired situation was illustrated in Fig. 15.7(b). A common top angle size selected is $4 \times 4 \times \frac{1}{4}$.

    As will be seen in the LRFD tables, unstiffened seated beam connections of practical sizes can support only fairly light factored loads—up to as high as 128 k if 1-in-thick seat angles of A36 steel are used. For loads of this size, two vertical end welds on the seat are sufficient. The top angle is welded only on its toes, so when the beam tends to rotate, this thin flexible angle will be free to pull away from the column.

The seat design strengths given in Tables 10.5 and 10.6 of the Manual were developed for seat angles with either 3 to 4 in outstanding legs. The steel used for the angles has $F_y = 36$ ksi and $F_u = 58$ ksi.

The design strengths in the tables were obtained giving consideration to both shear and flexural yielding of the outstanding leg of the seat angle, and crippling of the beam web as well. The values were calculated on the basis of a $\frac{3}{4}$-in setback rather than the nominal $\frac{1}{2}$ in used for web framing angles. This larger value was used to provide for possible mill underrun in the beam lengths. The weld design strengths provided were calculated using the elastic method. Example 15-5 illustrates the use of the Manual tables for designing a welded unstiffened seated beam connection. Other tables are included in the Manual for bolted seated connections as well as bolted or welded stiffened seated connections.

Example 15-5

Design an unstiffened beam connection to support a factored reaction of 70 k from a W24 × 55 beam ($d = 23.6$ in, $t_w = 0.395$ in or 3/8 in as a fraction, $t_f = 0.505$ in, $k = 1.11$ in). The connection is to be made to the flange of a W14 × 68 column ($t_f = 0.720$ in). Use 3/4-in A325-N bolts in standard size holes to connect the beam to the seat and top angle and E70 electrodes to connect those angles to the column flange. The angles are A36 while the beam and column have an F_y of 50 ksi and an F_u of 65 ksi.

Solution. Design seat angle and welds

For local web yielding (ϕR_1 and ϕR_2 were obtained from Table 9.5)

$$N_{\min} = \frac{R_u - \phi R_1}{\phi R_2} = \frac{70 - 71}{19.8} = \text{neg. number} < k = 1.11 \text{ in}$$

For web crippling (ϕR_3 and ϕR_4 obtained from Table 9.5)

$$\frac{N}{d} = \frac{1.11}{23.57} < 0.2$$

$$N_{\min} = \frac{R_u - \phi_r R_3}{\phi_r R_4} = \frac{70 - 63.7}{5.60} = 1.125 \text{ in} > k = 1.11 \text{ in}$$

$$\frac{N}{d} = \frac{1.125}{23.6} = 0.048 < 0.2$$

$$\therefore N_{\min} = 1.125 \text{ in}$$

From Table 10.5, an 8-in angle length with a 3/4 in thickness and a $3\frac{1}{2}$ in minimum outstanding leg will provide

$$\phi R_n = 117 \text{ k} > 70 \text{ k} \tag{OK}$$

Try L 8 × 4 × 3/4, 8 in long with 3/8-in fillet welds Strength from Table 10.6

$$\phi R_m = 80.1 \, \text{k} > 70 \, \text{k} \qquad \text{(OK)}$$

Use 2-$\frac{3}{4}$-in A325-N bolts to connect beam to seat angle

Select top angle, bolts and welds
Use $L4 \times 4 \times \frac{1}{4}$ with 2-$\frac{3}{4}$ in A325-N bolts through supported leg of angle.

Use minimum size weld = 3/16 in specified in LRFD Specification Table J2.4.

15.10 STIFFENED SEATED BEAM CONNECTIONS

When beams are supported by seated connections and when the factored reactions become fairly large, it is necessary to stiffen the seats. These larger reactions cause moments in the outstanding or horizontal legs of the seat angles that cannot be supported by standard thickness angles unless they are stiffened in some manner. Typical stiffened seated connections are shown in Fig. 15.6(f) and in Fig. 15.16.

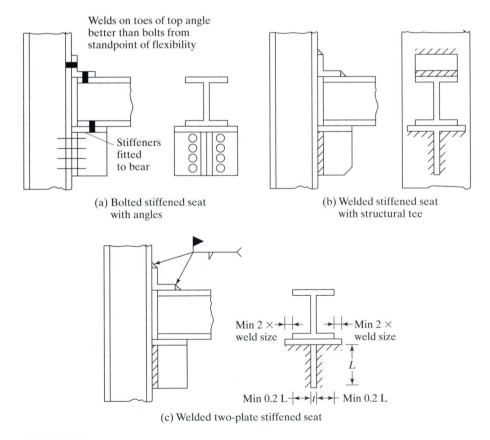

FIGURE 15.16

(a) Bolted stiffened seat with angles. (b) Welded stiffened seat with structural tee. (c) Welded two-plate stiffened seat.

Stiffened seats may be bolted or welded. Bolted seats may be stiffened with a pair of angles as shown in part (a) of Fig. 15.16. Structural tee stiffeners either bolted or welded may be used. A welded one is shown in part (b) of the same figure. Welded two-plate stiffeners, such as the one shown in part (c) of the figure, also are commonly used.

Examples 10-9 and 10-10 in Part 10 of the LRFD Manual illustrate the design of stiffened seated beam connections. These designs make use of Tables 10.7 and 10.8 in the Manual.

15.11   DESIGN OF MOMENT-RESISTING CONNECTIONS

In Chapter 14, a moment-resisting connection was designed (Example 14-14) where the flanges were welded to a column and a shear tab was used to transfer shear between the beam web and the column. Several other types of moment-resisting connections were shown in Fig. 15.5.

As the general procedure used in Example 14-15 is applicable to all moment-resisting connections, no additional numerical examples are presented in this text. The moment to be resisted is divided by the distance between the cgs. of the top and bottom parts of the couple (*C* and *T*), and then welds or bolts are selected that will provide the necessary design strengths so determined. Next, a shear tab, or a pair of framing angles, or a beam seat are selected to resist the shear force. Finally, as described in the next section, it may be necessary to provide stiffeners for the column web, or to select a larger column section.

Another moment-resisting connection is presented in Fig. 15.17. In this particular connection, the *T* and *C* values are transferred by fillet welds into the plates and by groove welds from the plates to the columns. For easier welding, these plates may be tapered as shown in the bottom of the figure. You may have noticed such tapered plates used for facilitating welding in other situations.

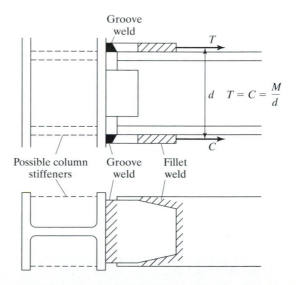

FIGURE 15.17

For a rigid or continuous connection of the type shown in Fig. 15.17, we must be careful to check the strength of the top and bottom plates. Should the plates be bolted, this involves the tensile strength of the top plate, including the effect of the bolt holes and block shear. The design compressive strength of the other plate must also be checked. Examples 12-1 and 12-2 in the Manual illustrate such calculations.

15.12 COLUMN WEB STIFFENERS

If a column to which a beam is being connected bends appreciably at the connection, the moment resistance of the connection will be reduced, regardless of how good the connection may be. Furthermore, if the top connection plate in pulling away from the column tends to bend the column flange as shown in part (a) of Fig. 15.18, the middle part of the weld may be greatly overstressed (like the prying action for bolts we discussed in Chapter 13).

When there is a danger of the column flange bending as described here, we must make sure that the desired moment resistance of the connection is provided. This may be done either by using a heavier column with stiffer flanges or by introducing column web stiffener plates as shown in part (b) of Fig. 15.18. *It is almost always desirable to use a heavier column because column web stiffener plates are quite expensive and a nuisance to use.*

Column web stiffener plates are somewhat objectionable to architects because they find it convenient to run pipes and conduits inside their columns; this objection can easily be overcome, however. First, if the connection is to only one column, flange, the stiffener does not have to run for more than half the column depth, as shown in part (b) of Fig. 15.18. If connections are made to both column flanges, the column

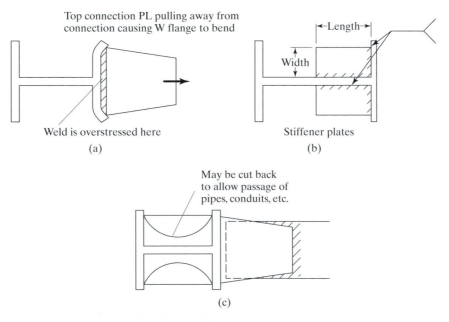

FIGURE 15.18

stiffener plates may be cut back to allow the passage of pipes, conduits, etc., as shown in part (c) of the figure.

For this discussion, the factored force applied from the beam flange to the column is referred to as P_{bf}. Should its value be larger than any one of the following resisting forces, it will be necessary to provide column web stiffeners, says LRFD Specification K1. In the expressions to follow h is the depth of the column web between the web toes of the fillets and t_f is the thickness of the beam flange or connection plate from which the concentrated force is being applied.

$$\phi R_n = P_{fb}$$

$$\phi = 0.90$$

$$\phi R_n = P_{fb} = 6.25\, t_f^2\, F_{yf} \qquad \text{(LRFD Equation K1.1)}$$

$$\phi R_n = P_{wi} t_b + P_{wo} = \text{local web yielding strength}$$
$$\text{considered in Chapter 10}$$

$$\phi = 1.0, P_{wi} = \phi F_{yw} t_w,\ P_{wo} = \phi 5 F_{yw} t_w k \qquad \text{(LRFD Equation K1.2)}$$

$$\phi R_n = P_{wb} = \text{compression buckling strength}$$
$$\text{of an unstiffened member web}$$

$$\phi = 0.9$$

$$P_{wb} = \phi \frac{24\, t_w^3 \sqrt{E\, F_{yw}}}{h} \qquad \text{(LRFD Equation K1.8)}$$

The values of P_{wb}, P_{fb}, etc., for the W shapes normally used as columns have been computed and recorded in the column tables of Table 4.2 of the LRFD Manual for steels with $F_y = 50$ ksi.

The LRFD Manual presents a set of suggested rules for the design of column web stiffeners. These are as follows:

1. The width of the stiffener plus one-half of the column web thickness should not be less than one-third the width of the beam flange or of the moment connection plate that applies the concentrated force.
2. The stiffener thickness should not be less than $t_b/2$ or half the thickness of the moment connection plate delivering the concentrated load and not less than $1.79\sqrt{\frac{F_y}{F}}$.
3. If there is a moment connection applied to only one flange of the column, the length of the stiffener plate does not have to exceed one-half the column depth.
4. The stiffener plate should be welded to the column web with a sufficient strength to carry the force caused by the unbalanced moment on the opposite sides of the column.

For the column given in Example 15-7, it is necessary to use column web stiffeners or to select a larger column. These alternatives are considered in the solution.

Example 15-7

It is assumed that a particular column is a W12 × 87 consisting of 50 ksi steel and subjected to 215-k C and T forces transferred by an FR type connection through the flanges of a W18 × 46 beam. It will be found that this column is not satisfactory to resist these forces. (a) Select a larger W12 column section that will be satisfactory. (b) Using a W12 × 87 column, design column web stiffeners including the stiffener connections using E70 SMAW welds.

Solution

$\quad\quad$ Beam is a W18 × 46(b_f = 6.06 in, t_b = t_f = 0.605 in)
$\quad\quad$ Column is a W12 × 87(d = 12.5 in, t_w = 0.515 in, t_f = 0.810 in)

$\quad$ Checking to see if forces transferred to column are too large. Using LRFD column tables for the W12 × 87,

$$P_{fb} = 185 \text{ k} < 215 \text{ k} \quad\quad\quad\text{(NG)}$$
$$P_{wi}t_b + P_{wo} = (26)(0.605) + 193 = 208.9 \text{ k} < 215 \text{ k} \quad\quad\quad\text{(NG)}$$
$$P_{wb} = 366 \text{ k} > 215 \text{ k} \quad\quad\quad\text{(OK)}$$

$\quad\therefore$ Either a larger column or column web stiffeners are required.

a. Selecting a larger column

$\quad$ Try W12 × 96

$$P_{fb} = 228 \text{ k} > 215 \text{ k} \quad\quad\quad\text{(OK)}$$
$$P_{wi}t_b + P_{wo} = (28)(0.605) + 223 = 239.9 \text{ k} > 215 \text{ k} \quad\quad\quad\text{(OK)}$$
$$P_{wb} = 446 \text{ k} > 215 \text{ k} \quad\quad\quad\text{(OK)}$$

$\quad$ Use W12 × 96 A36

b. Design of web stiffeners using a $W12 \times 87$ column and the suggested LRFD rules presented before this example.

$$\text{Required stiffener area} = \frac{215 - 185}{50} = 0.60 \text{ in}^2$$

$$\text{Min. width of stiffener} = \tfrac{1}{3}b_f - \frac{t_w}{2} = \frac{6.06}{3} - \frac{0.515}{2} = 1.76 \text{ in}$$

$$\text{Min. } t \text{ of stiffeners} = \frac{t_b}{2} = \frac{0.605}{2} = 0.3025 \text{ in or } 1.79\sqrt{\frac{50}{29,000}} = 0.0743 \text{ in}$$

$$\text{Required } t \text{ of stiffeners} = \frac{0.60}{1.76} = 0.341 \text{ in, say } \frac{3}{8} \text{ in}$$

$$\text{Required width} = \frac{0.60}{0.375} = 1.60 \text{ in, say 4 in for practical conditions}$$

$$\text{Min. length} = \frac{d}{2} - t_f = \frac{12.5}{2} - 0.810 = 5.45 \text{ in, say 6 in}$$

Use 2 PLs $\frac{3}{8}$ × 4 × 0 ft 6 in A36

Design of welds for stiffener plates

Min. Weld size as required by LRFD Specification Table J2.4 = 3/16 in based on the column web t_w of 0.515 in.

$$\text{Required length of weld} = \frac{215 - 185}{(0.75 \times 0.60 \times 70)(0.707)\left(\frac{3}{16}\right)}$$

$$= 7.18 \text{ in, say 8 in}$$

15.13 CONNECTION DESIGN AIDS—HANDBOOKS AND COMPUTER PROGRAMS

Several computer programs are available for the design and checking of connections. They are economical to buy, save a great deal of time, and reduce the possibility of numerical mistakes. Anyone who has struggled through connection designs, particularly the selection of dimensions, will appreciate these programs.

One very satisfactory program is CONXPRT. It was developed by Structural Engineers, Inc., of Radford, Virginia, and is marketed by the AISC. It is applicable to double-angle, shear end-plate, and single-plate simple framing connections. Not only is a program such as CONXPRT useful for developing designs, it also is very good for checking existing designs. Imagine that you are working for a consulting engineer who receives the detailed drawings for a large steel building from the steel fabricator, and he or she asks you to check the adequacy of all the connections. If you check them with the handbook and a pocket calculator, you are facing a monumental job. But with a computer program, the work can be handled quickly and accurately.

PROBLEMS

15-1. Determine the maximum end reaction that can be transferred through the A36 web angle connection shown in the accompanying illustration. The beam steel is 50 ksi and the bolts are 3/4-in A325-N and are used in standard-size holes. The beam is connected to the web of a W30 × 90 girder. (*Ans.* $\phi R_m = 136.9$ k)

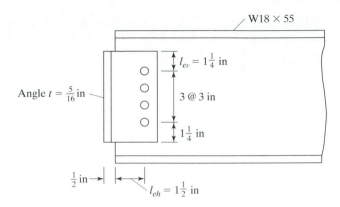

FIGURE P15-1

15-2.   Repeat Prob. 15-1 if the bolts are 7/8-in A325-N.

15-3.   Repeat Prob. 15-1 if the bolts are 1-in A325-X. (*Ans.* $\phi R_m = 114$ k).

15-4.   Using the LRFD Manual select a pair of bolted standard web angles for a W33 × 141 connected to the flange of a W14 × 120 column with a dead load service reaction of 150 k and a live load service reaction of 100 k. The bolts are to be 7/8-in A325-N in standard-size holes and the steel is A36 for the angles and A992 for the W shapes.

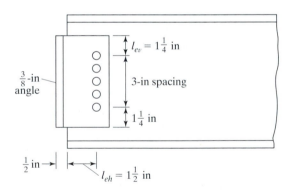

FIGURE P15-4

15-5.   Repeat Prob. 15-4 if 1-in A325-N bolts are to be used. (*Ans.* 8-row connection)

15-6.   Repeat Prob. 15-1 if 3/4-in A325 S.C. bolts are to be used.

15-7.   Design a framed beam connection for a W27 × 84 connected to a W30 × 116 girder web to support a factored beam reaction of 250 k. The bolts are to be 1-in A325 S.C. Class A in standard-size holes. Angles are A36 steel while beams are A992. The edge distances and bolt spacings are the same as those shown in the sketch for Prob. 15-4. (*Ans.* 6-row connection).

15-8. Repeat Prob. 15-1 if the bolts are 1-in A325-S.C. Class A and are used in $1\frac{1}{16} \times 1\frac{5}{16}$ in. short slots with long axes perpendicular to the transmitted force. Angle t is 1/2 in.

15-9. Design a framed beam connection for a W18 × 50 to support a factored beam reaction of 65 k. The beam's top flange is to be coped for a 2-in depth, and 7/8-in A325-X bolts in standard-size holes are to be used. The beam is connected to a W27 × 146 girder. Connection is A36 while W shapes are A992. (*Ans.* 4-row connection).

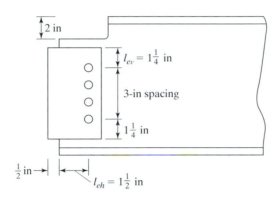

FIGURE P15-9

15-10. Repeat Prob. 15-7 if the factored beam reaction is to be 285 and 1-in A325 N bolts and A572 steel ($F_y = 50$ ksi and $F_u = 65$ ksi) are to be used.

15-11. Select a framed beam connection for a W33 × 130 beam (A992 steel) with a factored beam reaction of 200 k. It is to be connected to the flange of a W36 × 50 column. The A36 web angles are to be welded with E70 electrodes (weld A in the Manual) and are to be field-connected to the girder with $\frac{7}{8}$-in A325-N bolts. (*One ans.* 2Ls 4 × $3\frac{1}{2}$ × $\frac{5}{16}$ × 1 ft $8\frac{1}{2}$ in, $\frac{1}{4}$-in weld A and 7-row bolt connection to girder).

15-12. Repeat Prob. 15-11 using SMAW shop and field welds (welds A and B in the Manual).

15-13. Select an A36 framed beam connection from the LRFD Manual for a W30 × 124 consisting of 50 ksi steel using SMAW E70 shop and field welds. The factored beam reaction is 190 k. The beam is to be connected to the flange of a 50 ksi W14 × 145 column. (*Ans.* 2Ls 4 × 3 × $\frac{3}{8}$ × 1 ft-6 in).

15-14. Repeat Prob. 15-13 if $R_u = 290$ k.

15-15. Select an unstiffened A36 seated beam connection bolted with 7/8-in A325-N bolts in standard-size holes for the following data: Beam is W16 × 67, column is W14 × 82, both consisting of 50 ksi steel, R_u is 80 k, and the column gage is $5\frac{1}{2}$ in. (*Ans.* Use 6 × 4 × $\frac{3}{4}$ seat L and 4 × 4 × $\frac{1}{4}$ top angle).

CHAPTER 16

Composite Beams

16.1 COMPOSITE CONSTRUCTION

When a concrete slab is supported by steel beams, and there is no provision for shear transfer between the two, the result is a noncomposite section. Loads applied to non-composite sections obviously cause the slabs to deflect along with the beams, resulting in some of the load being carried by the slabs. Unless a great deal of bond exists between the two (as would be the case if the steel beam were completely encased in concrete, or where a system of mechanical shear connectors is provided), the load carried by the slab is small and may be neglected.

For many years steel beams and reinforced-concrete slabs were used together with no consideration made for any composite effect. In recent decades, however, it has been shown that a great strengthening effect can be obtained by tying the two together to act as a unit in resisting loads. Steel beams and concrete slabs joined together compositely can often support 33 to 50% or more load than could the steel beams alone in noncomposite action.

Composite construction for highway bridges was given the green light by the adoption of the 1944 AASHTO Specifications, which approved the method. Since about 1950, the use of composite bridge floors has rapidly increased until today they are commonplace all over the United States. In these bridges the longitudinal shears are transferred from the stringers to the reinforced-concrete slab or deck with shear connectors (to be described in Section 16.5), causing the slab or deck to assist in carrying the bending moments. This type of section is shown in part (a) of Fig. 16.1.

The first approval for composite building floors was given by the 1952 AISC Specification; today, they are very common. These floors may either be encased in concrete (very rare due to expense) as shown in part (b) of Fig. 16-1, or be nonencased with shear connectors as shown in part (c) of the figure. Almost all composite building floors being built today are of the nonencased type. If the steel sections are encased in concrete, the shear transfer is made by bond and friction between the beam and the concrete and by the shearing strength of the concrete along the dashed lines shown in part (b) of Fig. 16.1.

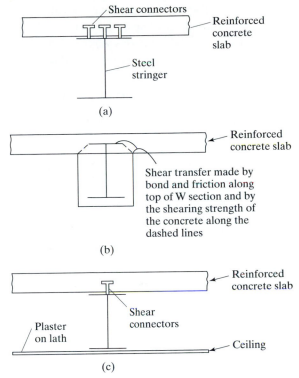

FIGURE 16.1

(a) Composite bridge floor with shear connectors.
(b) Encased section for building floors. (c) Building
floors with shear connectors.

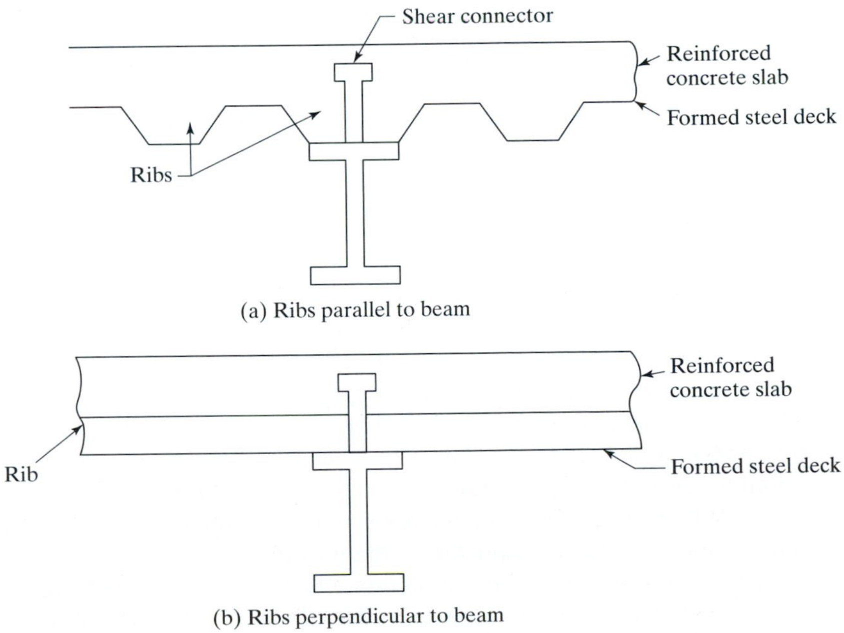

Shear connectors

Reinforced concrete slab

Steel stringer

(a)

Reinforced concrete slab

Shear transfer made by
bond and friction along
top of W section and by
the shearing strength of
the concrete along the
dashed lines

(b)

Reinforced concrete slab

Shear connectors

Plaster on lath

Ceiling

(c)

Shear connector

Reinforced concrete slab

Formed steel deck

Ribs

(a) Ribs parallel to beam

Reinforced concrete slab

Formed steel deck

Rib

(b) Ribs perpendicular to beam

FIGURE 16.2

Composite sections using formed steel deck.

Today formed steel deck (illustrated in Fig. 16.2) is used for almost all composite building floors. The initial examples in this chapter, however, pertain to the calculations for composite sections where formed steel deck is not used. Sections that make use of formed steel deck are described later in the chapter.

16.2   ADVANTAGES OF COMPOSITE CONSTRUCTION

The floor slab in composite construction acts not only as a slab for resisting the live loads, but also as an integral part of the beam. It actually serves as a large cover plate for the upper flange of the steel beam, appreciably increasing the beam's strength.

A particular advantage of composite floors is that they make use of concrete's high compressive strength by putting a large part of the slab in compression. At the same time, a larger percentage of the steel is kept in tension (also advantageous) than is normally the case in steel-frame structures. The result is less steel tonnage required for the same loads and spans (or longer spans for the same sections). Composite sections have greater stiffness than noncomposite sections and they have smaller deflections—perhaps only 20 to 30 percent as large. Furthermore, tests have shown that the ability of a composite structure to take overload is decidedly greater than for a non-composite structure.

An additional advantage of composite construction is the possibility of having smaller overall floor depths—a fact of particular importance for tall buildings. Smaller floor depths permit reduced building heights with the consequent advantages of smaller costs for walls, plumbing, wiring, ducts, elevators, and foundations. Another important advantage available with reduced beam depths is a saving in fireproofing costs, because a coat of fireproofing material is provided on smaller and shallower steel shapes.

It frequently is necessary to increase the load-carrying capacity of an existing floor system. Often, this can be handled quite easily for composite floors by welding cover plates onto the bottom flanges of the beams.

A disadvantage for composite construction is the cost of furnishing and installing the shear connectors. This extra cost usually will exceed the cost reductions mentioned when spans are short and lightly loaded.

16.3   DISCUSSION OF SHORING

After the steel beams are erected, the concrete slab is placed on them. The formwork, wet concrete, and other construction loads must therefore be supported by the beams or by temporary shoring. Should no shoring be used, the steel beams must support all of these loads as well as their own weights. Most specifications say that after the concrete has gained 75 percent of its 28-day strength, the section has become composite and all loads applied thereafter may be considered to be supported by the composite section. When shoring is used, it supports the wet concrete and the other construction loads. It does not really support the weight of the steel beams unless they are given an initial upward deflection (probably impractical). When the shoring is removed (after the concrete gains at least 75 percent of its 28-day strength) the weight of the slab is transferred to the composite section, not just to the steel beams. The student can see that if shoring is used, it will be possible to use lighter and thus cheaper steel beams.

A truss assembly is slipped into place on the Newport Bridge
high above Narragensett Bay in Newport, Rhode Island.
(Courtesy of Bethlehem Steel Corporation.)

The question then arises, "Will the savings in steel cost be greater than the extra cost of shoring?" The answer probably is "no." The usual decision is to use heavier steel beams and to do without shoring for several reasons, including the following:

1. Apart from reasons of economy, the use of shoring is a tricky operation, particularly where settlement of the shoring is possible, as is often the case in bridge construction.

2. Both theory and load tests show that the ultimate strengths of composite sections of the same sizes are the same whether shoring is used or not. If lighter steel beams are selected for a particular span because shoring is used, the result is a smaller ultimate strength.

3. Another disadvantage of shoring is that after the concrete hardens and the shoring is removed, the slab will participate in composite action in supporting the dead loads. The slab will be placed in compression by these long-term loads and will have substantial creep and shrinkage parallel to the beams. The result will be a great decrease in the stress in the slab with a corresponding increase in the steel stresses. The probable consequence is that most of the dead load will be supported by the steel beams anyway and composite action will really apply only to the live loads as though shoring had not been used.

4. Also, in shored construction cracks occur over the steel girders, necessitating the use of reinforcing bars. In fact, we should use reinforcing over the girders in unshored construction, too. Although cracks will be smaller there, they are going to be present nonetheless, and we need to keep them as small as possible.

Nevertheless, shored construction does present some advantages compared with unshored construction. First, deflections are smaller because they are all based on the properties of the composite section. (In other words, the initial wet concrete loads are not applied to the steel beams alone, but rather to the whole composite section.) Second, it is not necessary to make a strength check for the steel beams for this wet load condition. This is sometimes quite important for situations in which we have low ratios of live to dead loads.

The deflections of unshored floors due to the wet concrete sometimes can be quite large. If the beams are not cambered, additional concrete (perhaps as much as 10 percent or more) will be used to even up the floors. If, on the other hand, too much camber is specified, we may end up with slabs that are too thin in those areas where wet concrete deflections aren't as large as the camber.

16.4   EFFECTIVE FLANGE WIDTHS

There is a problem involved in estimating how much of the slab acts as part of the beam. Should the beams be rather closely spaced, the bending stresses in the slab will be fairly uniformly distributed across the compression zone. If, however, the distances between beams are large, bending stresses will vary quite a bit nonlinearly across the flange. The further a particular part of the slab or flange is away from the steel beam, the smaller will be its bending stress. Specifications attempt to handle this problem by replacing the actual slab with a narrower or effective slab that has a constant stress. This equivalent slab is deemed to support the same total compression as is supported by the actual slab. The effective width of the slab b_e is shown in Fig. 16.3.

The portion of the slab or flange that can be considered to participate in the composite beam action is controlled by the specifications. LRFD Specification I3.1 states that the effective width of the concrete slab on each side of the beam center line shall not exceed the least of the values to follow. The following set of rules applies, whether the slab exists on one or both sides of the beam:

1. One-eighth of the span of the beam measured center-to-center of supports for both simple and continuous spans.
2. One-half of the distance from the beam center line to the center line of the adjacent beam.
3. The distance from the beam center line to the edge of the slab.

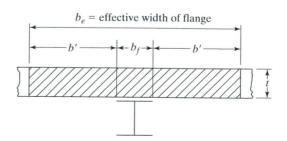

FIGURE 16.3

The AASHTO requirements for determining effective flange widths are somewhat different. The maximum total flange width may not exceed one-fourth of the beam span, twelve times the least thickness of the slab, or the distance center-to-center of the beams. Should the slab exist on only one side of the beam, its effective width may not exceed one-twelfth of the beam span, six times the slab thickness, or one-half of the distance from the center line of the beam to the center line of the adjacent beam.

16.5   SHEAR TRANSFER

The concrete slabs may rest directly on top of the steel beams, or the beams may be completely encased in concrete for fireproofing purposes. This latter case, however, is very expensive and thus is rarely used. The longitudinal shear can be transferred between the two by bond and shear (and possibly some type of shear reinforcing), if needed, when the beams are encased. When not encased, mechanical connectors must transfer the load. Fireproofing is not necessary for bridges, and the slab is placed on top of the steel beams. Bridges are subject to heavy impactive loads and the bond between the beams and the deck, which is easily broken, is considered negligible. For this reason shear connectors are designed to resist all of the shear between bridge slabs and beams.

Various types of shear connectors have been tried, including spiral bars, channels, zees, angles, and studs. Several of these types of connectors are shown in Fig. 16.4. Economic considerations have usually led to the use of round studs welded to the top

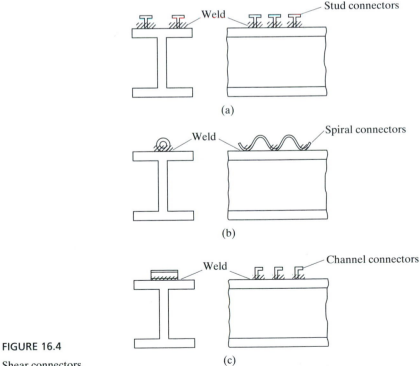

FIGURE 16.4

Shear connectors.

flanges of the beams. These studs are available in diameters from 1/2 to 1 in and in lengths from 2 to 8 in, but the LRFD Specification (I5.1) states that their length may not be less than 4 stud diameters.

They actually consist of rounded steel bars welded on one end to the steel beams. The other end is upset or headed to prevent vertical separation of the slab from the beam. These studs can be quickly attached to the steel beams through the steel decks with stud-welding guns by semiskilled workers. The AISC Commentary (I3.5) describes special procedures needed for 16 gage and thicker decks and for decks with heavy galvanized coatings (>1.25 ounces per sq ft).

Shop installation of shear connectors is initially more economical, but there is a growing tendency to use field installation. There are two major reasons for this trend: the connectors may easily be damaged during transportation and setting of the beams, and they serve as a hindrance to the workers walking along the top flanges during the early phases of construction.

When a composite beam is being tested, failure will probably occur with a crushing of the concrete. At that time it seems reasonable to assume that the concrete and steel will both have reached a plastic condition.

For the following discussion, reference is made to Fig. 16.5. Should the plastic neutral axis (PNA) fall in the slab, the maximum horizontal shear (or horizontal force on the plane between the concrete and the steel) is said to be A_sF_y; and if the plastic neutral axis is in the steel section, the maximum horizontal shear is considered to equal $0.85\,f_c'A_c$ where A_c is the effective area of the concrete slab. (For the student unfamiliar with the strength design theory for reinforced concrete, the average stress at failure on the compression side of a reinforced concrete beam is usually assumed to be $0.85\,f_c'$.)

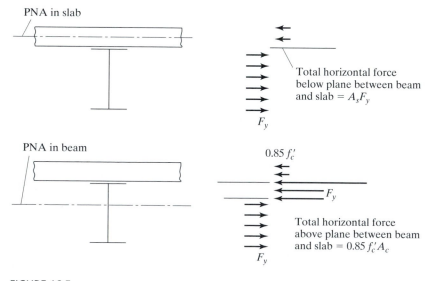

FIGURE 16.5

Channel-section shear connectors, Grand Rapids, Mich. (Courtesy of the Lincoln Electric Company.)

From this information expressions for the shear to be taken by the connectors can be determined: The LRFD (I5.2) says that for composite action the total horizontal shear between the points of maximum positive moment and zero moment is to be taken as the least of the following where ΣQ_n is the total nominal strength of the shear connectors provided as described in Section I6-7.

a. $0.85 f_c' A_c$

b. $A_s F_y$ (for hybrid beams this yield force must be calculated separately for each of the components of the cross section)

c. ΣQ_n

16.6   PARTIALLY COMPOSITE BEAMS

For this discussion it is assumed that we need to select a steel section that will have a design strength of 450 ft-k when made composite with the concrete slab. It is further assumed that when we select a section from the Manual, it has an M_u equal to 510 ft-k (when made composite with the slab). If we now provide shear connectors for full composite action, the section will have a design strength of 510 ft-k. But we need only 450 ft-k.

It seems logical to assume that we may decide to provide only a sufficient number of connectors to develop a design strength of 450 ft-k. In this way we can reduce the number of connectors and reduce costs (perhaps substantially if we repeat this section many times in the structure). The resulting section is a *partially composite section*, one that does not have a sufficient number of connectors to develop the full flexural strength of the composite beam. We will encounter this situation in Examples 16-3 and 16-4.

It is usually felt that the total strength of the shear connectors used in a particular beam should not be less than 25 percent of the shearing strength required for full composite action ($A_s F_y$). Otherwise, our calculations may not accurately depict the stiffness and strength of a composite section.

16.7   STRENGTH OF SHEAR CONNECTORS

For composite sections it is permissible to use normal weight stone concrete (made with aggregates conforming to ASTM C33) or lightweight concrete weighing not less than 90 lb/ft^3 (made with rotary kiln-produced aggregates conforming to ASTM C330).

The LRFD Specification provides strength values for headed steel studs not less than 4 diameters in length after installation and for hot-rolled steel channels. *They do not, however, give resistance factors for the strength of shear connectors.* This is because they feel that the factor used for determining the flexural strength of the concrete is sufficient to account for variations in concrete strength, including those variations that are associated with shear connectors.

16.7.1   Stud Shear Connectors

The nominal shear strength in kips of one stud shear connector embedded in a solid concrete slab is to be determined with the following expression from LRFD Specification I5.3. In this expression A_{sc} is the cross-sectional area of the shank of the connector in square inches and f_c' is the specified compressive stress of the concrete in ksi. E_c is the modulus of elasticity of the concrete in ksi (MPa) and equals $w^{1.5}\sqrt{f_c'}$ in which w is the unit weight of the concrete in lb/ft^3. Finally, F_u is the specified minimum tensile strength of the steel stud in ksi (MPa).

$$Q_n = 0.5 A_{sc} \sqrt{f_c' E_c} \leq A_{sc} F_u \qquad \text{(LRFD Equation I5-1)}$$

TABLE 16.1   Shear Stud Connectors Nominal Shear Strength Q_n, kips[a]

Specified Compressive Strength of Concrete f_c', ksi	Light-Weight Concrete (115 lb/ft^3)				Normal-Weight Concrete (145 lb/ft^3)			
	Nominal Shear Stud Connector Diameter, in				Nominal Shear Stud Connector Diameter, in			
	1/2	5/8	3/4	7/8	1/2	5/8	3/4	7/8
3	7.88	12.3	17.7	24.1	9.35	14.6	21.0	28.6
3.5	8.82	13.8	19.8	27.0	10.5	16.4	23.6	32.1
4	9.75	15.2	21.9	29.9	11.6	18.1	26.1	35.5
4.5	10.7	16.6	24.0	32.6	11.8	18.4	26.5	36.1
5	11.5	18.0	25.9	35.3	11.8	18.4	26.5	36.1
Minimum Stud Length, in	2	$2\frac{1}{2}$	3	$3\frac{1}{2}$	2	$2\frac{1}{2}$	3	$3\frac{1}{2}$

[a] Applicable only to concrete made with ASTM C33 aggregates.

Table16.1, which is Table 5-13 in Part 5 of the Manual, provides Q_n values calculated with the preceding equation for different size studs embedded in concrete slabs. These values are provided for several grades of concrete and for concretes weighing 145 lb/ft^3 and 115 lb/ft^3. (The reader should understand that some reduction of these values may have to be made when connectors are used with formed steel decking as required by LRFD Specification I3.5b. This topic is addressed later in the chapter.)

16.7.2   Channel Shear Connectors

The nominal shear strength in kips for one channel shear connector is to be determined from the following expression from LRFD Specification I5.4 in which t_f and t_w are, respectively, the flange and web thicknesses of the channel and L_c is its length, in in. (mm):

$$Q_n = 0.3 \, (t_f + 0.5t_w) \, L_c \, \sqrt{f_c' E_c} \qquad \text{(LRFD Equation I5-2)}$$

16.7.3   Other Connectors

Should other types of shear connectors be used, the LRFD Specification (I6) states that their nominal strengths are to be determined with suitable tests.

16.8   NUMBER, SPACING, AND COVER REQUIREMENTS FOR SHEAR CONNECTORS

The number of shear connectors to be used between the point of maximum moment and each adjacent point of zero moment equals the horizontal force to be resisted divided by the nominal strength of one connector Q_n.

16.8.1   Spacing of Connectors

Tests of composite beams with shear connectors spaced uniformly, and of composite beams with the same number of connectors spaced in variation with the statical shear, show little difference as to ultimate strengths and deflections at working loads. This situation prevails as long as the total number of connectors is sufficient to develop the shear on both sides of the point of maximum moment. As a result, the LRFD Specification (I5.6) permits uniform spacings of connectors on each side of the maximum moment point. However, the number of connectors placed between a concentrated load and the nearest point of zero moment must be sufficient to develop the maximum moment at the concentrated load.

16.8.2   Maximum and Minimum Spacings

Except for formed steel decks, the minimum center-to-center spacing of shear connectors along the longitudinal axes of composite beams permitted by the LRFD Specification (I5.6) is 6 diameters, while the minimum value transverse to the longitudinal axis is 4 diameters. Within the ribs of formed steel decks, the minimum permissible spacing is 4 diameters in any direction.

When the flanges of steel beams are rather narrow, it may be difficult to achieve the minimum transverse spacings described here. For such situations the studs may be staggered. Figure 16.6 shows possible arrangements.

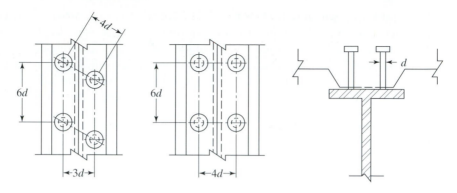

FIGURE 16.6

Connector arrangements.

If the deck ribs are parallel to the axis of the steel beam and more connectors are required than can be placed within the rib, the LRFD Commentary (I5.6) permits the splitting of the deck so that adequate room is made available.

Shear connectors must be capable of resisting both horizontal and vertical movement because there is a tendency for the slab and beam to separate vertically as well as to slip horizontally. The upset heads of stud shear connectors help to prevent vertical separation. The LRFD Specification (I5.6) states that the maximum spacing of shear connectors may not exceed 8 times the total slab thickness.

16.8.3 Cover Requirements

The LRFD Specification (I5.6) requires that there be at least 1 in. of lateral concrete cover provided for shear connectors. This rule does not apply to connectors used within the ribs of formed steel decks because tests have shown that strengths are not reduced, even when studs are placed as close as possible to the ribs.

When studs are not placed directly over beam webs, they have a tendency to tear out of the beam flanges before their full shear capacity is reached. To keep this situation from occurring, the LRFD Specification (I5.6) requires that the diameter of the studs may not be greater than 2.5 times the flange thickness of the beam to which they are welded unless they are located over the web.

When formed steel deck is used, the steel beam must be connected to the concrete slab with stud shear connectors with diameters not larger than 3/4 in. These may be welded through the deck or directly to the steel beam. After their installation they must extend for at least $1\frac{1}{2}$ in above the top of the steel deck and the concrete slab thickness above the steel deck may not be less than 2 in (LRFD Specification I3.5a).

16.9   MOMENT CAPACITY OF COMPOSITE SECTIONS

The nominal flexural strength of a composite beam in the positive moment region may be controlled by the plastic strength of the section or by the strength of the concrete slab or by the strength of the shear connectors. Furthermore, if the web is very slender and if a large portion of the web is in compression, web buckling may possibly limit the nominal strength of the member.

Little research has been done on the subject of web buckling for composite sections, and for this reason the LRFD Specification (I3.2) has conservatively applied the same rules to composite section webs as to plain steel webs. The positive flexural strength ($\phi_b M_n$ with $\phi_b = 0.85$) of a composite section is to be determined assuming a plastic stress distribution if $h/t_w \leq 3.76 \sqrt{E/F_{yf}}$. In this expression h is the distance between the web toes of the fillet, that is, $d - 2k$, t_w is the web thickness, and F_{yf} is the yield stress of the beam flange. All of the rolled W, S, M, HP, and C shapes in the Manual meet this requirement for F_y values up to 65 ksi. (For built-up sections, h is the distance between adjacent lines of fasteners or the clear distance between flanges when welds are used.)

If h/t_w is greater than $3.76 \sqrt{E/F_{yf}}$, the value of $\phi_b M_n$ with $\phi_b = 0.90$ is to be determined by superimposing the elastic stresses. The effects of shoring must be considered for these calculations.

The nominal moment capacity of composite sections as determined by load tests can be estimated very accurately with the plastic theory. With this theory, the steel section at failure is assumed to be fully yielded, and the part of the concrete slab on the compression side of the neutral axis is assumed to be stressed to $0.85 f_c'$. If any part of the slab is on the tensile side of the neutral axis, it is assumed to be cracked and incapable of carrying stress.

The plastic neutral axis (PNA) may fall in the slab or in the flange of the steel section or in its web. Each of these cases is discussed in this section.

16.9.1   Neutral Axis in Concrete Slab

The concrete slab compression stresses vary somewhat from the PNA out to the top of the slab. For convenience in calculations, however, they are assumed to be uniform with a value of $0.85 f_c'$ over an area of depth a and width b_e, determined as described in Section 16-4. (This distribution is selected to provide a stress block having the same total compression C and the same center of gravity for the total force as we have in the actual slab.)

The value of a can be determined from the following expression where the total tension in the steel section is set equal to the total compression in the slab.

$$A_s F_y = 0.85 f_c' a b_e$$

$$a = \frac{A_s F_y}{0.85 f_c' b_e}$$

If a is equal to or less than the slab thickness, the PNA will fall in the slab and the nominal or plastic moment capacity of the composite section may be written as the

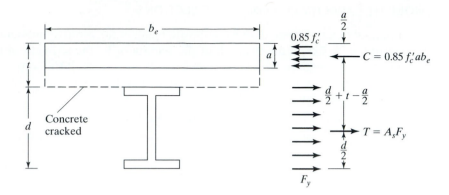

FIGURE 16.7

Plastic neutral axis (PNA) in the slab.

total tension T or the total compression c times the distance between their centers of gravity. Reference is made here to Fig. 16.7.

Example 16-1 illustrates the calculation of $\phi_b M_p = \phi_b M_n$ for a composite section where the PNA falls within the slab.

Example 16-1

Compute $\phi_b M_p = \phi_b M_n$ for the composite section shown in Fig. 16.8 if $f_c' = 4$ ksi and $F_y = 50$ ksi.

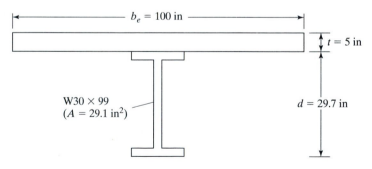

FIGURE 16.8

Solution. Determining $\phi_b M_n$

$$h = d - 2k = 29.7 - (2)(1.32) = 27.06 \text{ in}$$

$$\frac{h}{t_w} = \frac{27.06}{0.520} = 52.04 < 3.76 \sqrt{\frac{E}{F_{yf}}} = 3.76 \sqrt{\frac{29 \times 10^3}{50}} = 90.55$$

$$\therefore \phi_b = 0.85$$

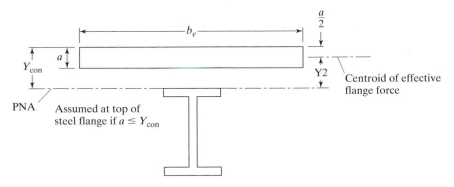

FIGURE 16.9

Locate PNA

$$a = \frac{A_s F_y}{0.85\, f_c' b_e} = \frac{(29.1)\,(50)}{(0.85)\,(4)\,(100)} = 4.28 \text{ in} < 5 \text{ in} \quad \therefore \text{ PNA is in slab}$$

$$M_n = M_p = A_s F_y\left(\frac{d}{2} + t - \frac{a}{2}\right)$$

$$= (29.1)\,(50)\left(\frac{29.7}{2} + 5 - \frac{4.28}{2}\right)$$

$$= 25{,}768 \text{ in-k} = 2147.3 \text{ ft-k}$$

$$\phi_b M_n = (0.85)\,(2147.3) = 1825.2 \text{ ft-k}$$

Note: If the reader refers to Part 5 of the LRFD Manual he or she can determine the ϕM_n value for this composite beam with reference being made to Fig. 16.9. To use the Manual composite tables it is assumed that the PNA is located at the top of the steel flange (TFL) or down in the steel shape. In the LRFD tables Y1 represents the distance from the PNA to the top of the beam flange while Y2 represents the distance from the centroid of the effective concrete flange force to the top flange of the beam $(t - a/2)$.

With the PNA for the preceding example being located at the top of the beam flange, from page 5-139 of the Manual with Y2 = 5 − 4.28/2 = 2.86 in and Y1 = 0 and for a W30 × 99 the value of $\phi M_n = \phi_b M_p$ by interpolation is 1826 ft-k.

16.9.2 Neutral Axis in Top Flange of Steel Beam

If a is calculated as previously described and is greater than the slab thickness t the PNA will fall down in the steel section. If this hsappens it will be necessary to find out whether the PNA is in the flange or below the flange. Suppose we assume that it's at the base of the flange. We can calculate the total compressive force C above = 0.85 $f_c' b_e t + A_f F_y$, where A_f is the area of the flange and the total tensile force below

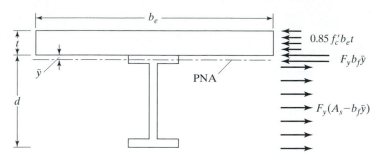

FIGURE 16.10

$T = F_y(A_s - A_f)$. If $C > T$ the PNA will be in the flange. If $C < T$ the PNA is below the flange.

Assuming we find the PNA is in the flange we can find its location letting $\bar{y}$ be the distance to the PNA measured from the top of the top flange by equating C and T as follows:

$$0.85 \, f_c' b_e t + F_y b_f \bar{y} = F_y A_s - F_y b_f \bar{y}$$

From which $\bar{y}$ is

$$\bar{y} = \frac{F_y A_s - 0.85 \, f_c' b_e t}{2 F_y b_f}$$

Then the nominal or plastic moment capacity of the section can be determined from the expression to follow with reference being made to Fig. 16.10. Taking moments about the PNA we get:

$$M_p = M_n = 0.85 \, f_c' b_e t \left(\frac{t}{2} + \bar{y} \right) + 2 F_y b_f \bar{y} \left(\frac{\bar{y}}{2} \right) + F_y A_s \left(\frac{d}{2} - \bar{y} \right)$$

Example 16-2, which follows, illustrates the calculation of $\phi_b M_p = \phi_b M_n$ for a composite section in which the PNA falls in the flange.

Example 16-2

Compute $\phi_b M_p = \phi_b M_n$ for the composite section shown in Fig. 16.11 if 50 ksi steel is used and if f_c' is 4 ksi.

Solution. Determining ϕ_b

$$h = d - 2k = 30.00 - (2)(1.50) = 27.00 \text{ in}$$

$$\frac{h}{t_w} = \frac{27.00}{0.565} = 47.78 < 3.76 \sqrt{\frac{29 \times 10^3}{50}} = 90.55$$

$$\therefore \ \phi_b = 0.85$$

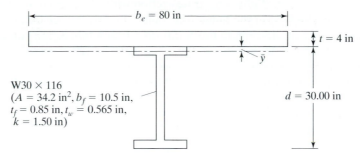

FIGURE 16.11

Is PNA located at top of steel flange?

$$a = \frac{A_s f_y}{0.85\, f'_c b_e} = \frac{(50)\,(34.2)}{(0.85)\,(4)\,(80)} = 6.29 \text{ in} > 4.00 \text{ in}$$

∴ PNA is located down in steel section.

Is PNA in flange or in web? Here we assume it is at base of steel flange.

$$C = 0.85\, f'_c bt + F_y b_f t_f = (0.85)\,(4)\,(80)\,(4) + (50)\,(10.5)\,(0.850) = 1534 \text{ k}$$
$$T = F_y(A_s - b_f t_f) = (50)\,(34.2 - 10.5 \times 0.850) = 1264 \text{ k}$$

Since $C > T$, the PNA falls in the steel flange and can be located as follows:

$$\bar{y} = \frac{F_y A_s - 0.85\, f'_c b_e t}{2\, F_y b_f} = \frac{(50)\,(34.2) - (0.85)\,(4)\,(80)\,(4)}{(2)\,(50)\,(10.5)} = 0.592 \text{ in}$$

Then using the moment expression given just before this example

$$M_n = M_p = (0.85)\,(4)\,(80)\,(4)\left(\frac{4}{2} + 0.592\right)$$

$$+ (2)\,(50)\,(10.5)\,(0.592)\left(\frac{0.592}{2}\right)$$

$$+ (50)\,(34.2)\left(\frac{30.00}{2} - 0.592\right)$$

$$= 27.650 \text{ in-k} = 2304 \text{ ft-k}$$

$$M_u = \phi_b M_p = \phi_b M_n = (0.85)\,(2304) = 1959 \text{ ft-k}$$

By interpolation in the LRFD Composite Design tables with Y1 = 0.593 in and Y2 = 2.0 in we get $1990 - \frac{0.592 - 0.425}{0.638 - 0.425}(1990 - 1950) = 1958.6$

If we have a partially composite section with ΣQ_n less than $A_s F_y$, the PNA will be down in the shape; if in the flange, the value of $\phi_b M_n$ can be determined with the equation used in Example 16-2. In the composite design tables presented in the Manual values of ΣQ_n and $\phi_b M_n$ are shown for seven different PNA locations—top of the flange, quarter points in the flange, bottom of the flange, and two points down in the web. Straight-line interpolation may be used for numbers in between the tabulated values.

16.9.3 Neutral Axis in Web of Steel Section

If for a particular composite section we find that a is larger than the slab thickness, and if we then assume the PNA is located at the bottom of the steel flange and we calculate C and T and find T is larger than C, the PNA will fall in the web. We can go through calculations similar to the ones we used for the case in which the PNA was located in the flange. Space is not taken to show such calculations because the Composite Design tables in Part 5 of the Manual cover most common cases.

16.10 DEFLECTIONS

Deflections for composite beams may be calculated by the same methods used for other types of beams. The student must be careful to compute deflections for the various types of loads separately. For example, there are dead loads applied to the steel section alone (if no shoring is used), dead loads applied to the composite section, and live loads applied to the composite section.

The long-term creep effect in the concrete in compression causes deflections to increase with time. These increases, however, are usually not considered significant for the average composite beam. If the designer feels that it is a matter of importance, he or she should compute long-term deflections using values of about $2n$ for the modular ratio when calculating the composite section properties for the transformed section to be used for deflections.

Should lightweight concrete be used, the actual modulus of elasticity of that concrete E_c (which may be rather small) should be used in calculating the transformed section moment of inertia I_{tr} for deflection computations. For stress calculations we use E_c for normal-weight concrete.

Generally speaking, shear deflections are neglected, although on occasion they can be quite large.[1] The steel beams can be cambered for all or some portion of deflections. It may be feasible in some situations to make a floor slab a little thicker in the middle than on the edges to compensate for deflections.

The designer may want to control vibrations in composite floors subject to pedestrian traffic or other moving loads. This may be the case where we have large open floor areas with no damping furnished by partitions, as in shopping malls. For such cases dynamic analyses should be made.[2]

---

[1] L. S. Beedle et al., *Structural Steel Design* (New York: Ronald Press, 1964), p. 452.
[2] Thomas M. Murray, "Design to Prevent Floor Vibrations," *Engineering Journal*, AISC, 12, no. 3 (3d quarter, 1975), pp. 82–87.

When the LRFD Specification is used to select steel beams for composite sections, the results often will be some rather small steel beams and thus some quite shallow floors. Such floors when unshored frequently will have large deflections when the concrete is placed. As a result designers usually will require cambering of the beams. Other alternatives include the selection of larger beams or the use of shoring.[3,4]

The beams selected must, of course, have sufficient $\phi_b M_n$ values to support themselves and the wet concrete. Nevertheless, their sizes probably are dictated more by wet concrete deflections than by moment considerations.

An alternative solution for these problems involves the use of partly restrained or PR connections (discussed in Chapter 15). When these connections are used the midspan deflections and moments are appreciably reduced, enabling us to use smaller girders. Furthermore, there are reductions in the annoying vibrations that are a problem in shallow composite floors.

When PR connections are used, negative moments will develop at the supports. In Section 16-12 of this chapter it is shown that the LRFD Specification permits the use of negative design moment strength for composite floors, provided certain requirements are met as to shear connectors and development of slab reinforcing in the negative moment region.

16.11   DESIGN OF COMPOSITE SECTIONS

Composite construction is of particular advantage economically when loads are heavy, spans are long, and beams are spaced at fairly large intervals. For steel building frames, composite construction is economical for spans varying roughly from 25 to 50 ft, with particular advantage in the longer spans. For bridges, simple spans have been economically constructed up to approximately 120 ft and continuous spans 50 or 60 ft longer. Composite bridges are generally economical for simple spans greater than about 40 ft and for continuous spans greater than about 60 ft.

Occasionally cover plates are welded to the bottom flanges of steel beams with improved economy. One can see that with the slab acting as part of the beam, there is a very large compressive area available and that by adding cover plates to the tensile flange, a little better balance is obtained.

In tall buildings where headroom is a problem, it is desirable to use the minimum overall floor thicknesses possible. For buildings, minimum depth-span ratios of approximately $\frac{1}{24}$ are recommended if the loads are fairly static and $\frac{1}{20}$ if the loads are of such a nature as to cause appreciable vibration. The thicknesses of the floor slabs are known (from the concrete design), and the depths of the steel beams can be fairly well estimated from these ratios.

Before we attempt some composite designs, several additional points relating to lateral bracing, shoring, estimated steel beam weights, and lower bound moments of inertia are discussed in the paragraphs to follow.

---

[3]R. Leon, "Composite Semi-Rigid Connections," *Modern Steel Construction*, Volume 32, no. 9 (AISC, Chicago: October 1992), pp. 18–23.

[4]"Innovative Design Cuts Costs," *Modern Steel Construction*, Volume 33, no. 4 (AISC, Chicago: April 1993), pp. 18–21.

16.11.1 Lateral Bracing

After the concrete slab hardens, it will provide sufficient lateral bracing for the compression flange of the steel beam. However, during the construction phase before the concrete hardens, lateral bracing may be insufficient and its design strength may have to be reduced depending on the estimated unbraced length. When steel-formed decking or concrete forms are attached to the beam's compression flange they usually will provide sufficient lateral bracing. The designer must be very careful in consideration of lateral bracing for fully encased beams.

16.11.2 Beams with Shoring

If beams are shored during construction, we will assume that all loads are resisted by the composite section after the shoring is removed.

16.11.3 Beams without Shoring

If temporary shoring is not used during construction, the steel beam alone must be able to support all the loads before the concrete is sufficiently hardened to provide composite action.

Without shoring, the wet concrete loads tend to cause large beam deflections, which may lead us to build thicker slabs where the beam deflections are larger. This situation can be counteracted by cambering the beams.

The LRFD Specification does not provide any extra margin against yield stresses occurring in beams during construction of unshored composite floors. Assuming satisfactory lateral bracing is provided, the Specification (F1.2) states that the maximum factored moment may not exceed $0.90 F_y Z$. The 0.90 in effect limits the maximum factored moment to a value about equal to the yield moment $F_y S$.

To calculate the moment to be resisted during construction, it makes sense to count the wet concrete as a live load and to also include some extra live load (perhaps 20 psf) to account for construction activities.

16.11.4 Estimated Steel Beam Weight

As illustrated in Example 16-3, it sometimes may be useful to make an estimate of the weight of the steel beam. The LRFD Manual provides the following empirical formula for this purpose in its Part 5 (page 5-26).

$$\text{Estimated beam weight} = \left[\frac{12 \, M_u}{(d/2 + Y_{\text{con}} - a/2) \, \phi \, F_y} \right] 3.4$$

where M_u = required flexural strength of composite section

d = nominal steel beam depth

Y_{con} = distance from top of steel beam to top of concrete slab

a = effective concrete slab thickness (which can be conservatively estimated as somewhere in the range of about 2 in)

ϕ = 0.85

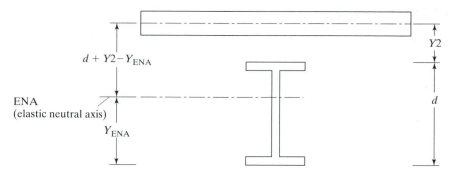

FIGURE 16.12

16.11.5 Lower Bound Moment of Inertia

To calculate the service load deflections for composite sections, a table of lower bound moment of inertia values is presented in Part 5 of the Manual (Table 5.15). These values are computed from the area of the steel beam and an equivalent concrete area of $\Sigma Q_n/F_y$. The remainder of the concrete flange is not used in these calculations. This means that if we have partially composite sections the value of the lower bound moment of inertia will reflect this situation because ΣQ_n will be smaller. The lower bound moment of inertia is computed with the following expression, reference being made to Fig. 16.12 where Y_{ENA} is the distance from the bottom of the beam to the elastic neutral axis (ENA).

$$I = I_x + A_s\left(Y_{ENA} - \frac{d}{2}\right)^2 + \left(\frac{\Sigma Q_n}{F_y}\right)(d + Y_2 - Y_{ENA})^2$$

16.11.6 Extra Reinforcing

For building design calculations, the spans often are considered to be simply supported, but the steel beams generally do not have perfectly simple ends. The result is that some negative moment may occur at the beam ends, with possible cracking of the slab above. To prevent or minimize cracking, some extra steel can be placed in the top of the slab extending 2 or 3 ft out into the slab. The amount of steel added is in addition to the temperature and shrinkage requirements specified by the American Concrete Institute.[5] (See LRFD Commentary I3.2 for detailed comments on reinforcing.)

16.11.7 Example Problems

Examples 16-3 and 16-4 illustrate the designs of two unshored composite sections.

---

[5]*Building Code Requirements for Reinforced Concrete* (ACI std. 318-99). Detroit: American Concrete Institute, 1999, Section 7.12.

Example 16-3

Beams 10 ft on center with 36-ft simple spans are to be selected to support a 4-in-deep lightweight concrete slab on a 3-in-deep formed steel deck with no shoring. The ribs for the steel deck, which are perpendicular to the beam center lines, have average widths of 6 in. If the service dead load (including the beam weight) is to be 0.78 k/ft of length of the beams and the service live load is 2 k/ft, (a) select the beams, (b) determine the number of $\frac{3}{4}$-in-diameter headed studs required, (c) compute the service live load deflection, and (d) check the beam shear. Other data: 50 ksi steel, $f_c' = 4$ ksi, and concrete weight 115 lb/ft³.

Solution. Loads and moments

$$w_u = (1.2)(0.78) + (1.6)(2.0) = 4.136 \text{ k/ft}$$

$$M_u = \frac{(4.136)(36)^2}{8} = 670 \text{ ft-k}$$

Effective flange width b_e

$$b_e = (2)(\tfrac{1}{8} \times 36 \times 12) = 108 \text{ in} \leftarrow$$
$$b_e = (2)(5 \times 12) = 120 \text{ in}$$

a.   Select W section

Y_{con} = distance from top of slab to top of steel flange = 4 + 3 = 7 in

Assume $a = 2$ in < 4 in slab thickness (It's usually quite small, particularly for relatively light sections.)

Y1 is distance from PNA to top flange = 0 in

Y2 is the distance from the center of gravity of the concrete flange force to the top flange of the beam = $7 - a/2 = 7 - 2/2 = 6$ in

Looking through the composite tables of the Manual with $M_u = 670$ ft-k, Y1 = 0, and Y2 = 6 in, we can see that several W 18s (46 lb, 50 lb, and 55 lb) seem reasonable.

We can use the LRFD estimated weight formula for an 18-in-deep section.

$$wt = \left[\frac{(12)(670)}{(18/2 + 7 - 2/2)(0.85)(50)} \right] 3.4 = 42.88 \text{ lb}$$

Try W18 × 46($A = 13.5$ in²),

Assume $\Sigma Q_n = A_s F_y = (13.5)(50) = 675$ k

Thus a required $= \dfrac{\Sigma Q_n}{0.85 \, f_c' b_e} = \dfrac{675}{(0.85)(4)(108)} = 1.84$ in < 4 in

$$Y1 = 0$$

$$Y2 = 7.0 - \frac{1.84}{2} = 6.08 \text{ in}$$

$\phi_b M_n$ from Manual by interpolation $= 719 + \left(\dfrac{0.08}{0.50}\right)(743 - 720)$

$$= 722.7 \text{ k} > M_u = 670 \text{ ft-k} \qquad \text{(OK)}$$

Use W18 × 46($A = 13.5 \text{ in}^2$, $d = 18.00 \text{ in}$, $t_w = 0.360 \text{ in}$)

Now this time we will go to the case where Y1 is the largest possible to provide a $\phi_b M_n$ of about 670 ft-k with Y2 = 6.08 in, because ΣQ_n will be smaller and fewer shear connectors will be required. This will occur with Y1 = 0.151 in, where $\phi_b M_n = 680$ ft-k if Y2 = 6 in

$$\Sigma Q_n = 583 \text{ k}$$

$$a = \frac{583}{(0.85)(4)(108)} = 1.59 \text{ in} < 4 \text{ in}$$

$$Y2 = 7 - \frac{1.59}{2} = 6.20 \text{ in}$$

$$\phi_b M_n = 680 + \left(\frac{0.20}{0.50}\right)(701 - 680)$$

$$= 688 \text{ ft-k} > 670 \text{ ft-k} \qquad \text{(OK)}$$

Use W18 × 46 $F_y = 50$ ksi

b. Design of studs

Since we have steel-formed deck we must compute a stud reduction factor (SRF) as required by Specification I3.5(b).

$h_r = 3$ in

Assume $H_s = 5.0$ in (can't be $> h_r + 3 = 6$ in)

Assume 1 stud in a rib at beam intersection $= N_r$

$w_r = 6$ in

$$\text{SRF} = \frac{0.85}{\sqrt{N_r}}\left(\frac{w_r}{h_r}\right)\left[\frac{H_s}{h_r} - 1.0\right] = \frac{0.85}{\sqrt{1}}\left(\frac{6}{3}\right)\left[\frac{5.0}{3} - 1.0\right] = 1.13 > 1.0$$

$$\text{(LRFD Equation I3-1)}$$

∴ No reduction is necessary.

Q_n from Table 16.1 for 3/4-in headed studs, $f'_c = 4$ ksi, and concrete weight of 115 lb/ft^3 = 21.9 k

ΣQ_n for W18 × 46 with Y1 of 0.15 in = 583 k from Manual Table 5-14

$$\text{Total no. of connectors required} = \frac{2\Sigma Q_n}{Q_n} = \frac{(2)(583)}{21.9} = 53.24$$

Use 54-$\frac{3}{4}$-in studs (27 on each side of point of maximum moment which is £ here).

c. Compute LL deflection

Assume maximum permissible LL deflection

$$\frac{1}{360}\,\text{span} = \left(\frac{1}{360}\right)(12 \times 36) = 1.2 \text{ in}$$

$C_1 = 161$ from Fig. 10-8 of textbook (or Fig. 5-2 of Manual)

$$M_{LL} = \frac{(2.0)(36)^2}{8} = 324 \text{ ft-k}$$

$I_x = $ lower bound moment of inertia from tables in Part 5 of Manual using straight-line interpolation

$$= 2130 + \left(\frac{0.20}{0.50}\right)(2220 - 2130) = 2166 / \text{in}^4$$

$$\Delta_{LL} = \frac{ML^2}{C_1 I_{LB}} = \frac{(324)(36)^2}{(161)(2166)} = 1.20 \text{ in} = 1.2 \text{ in} \qquad\qquad (\text{OK})$$

d. Check beam shear

$$V_u = \frac{(36)(4.136)}{2} = 74.4 \text{ k}$$

$\phi V_n = \phi 0.6\,F_y A_w$ for W18 × 46
$$= (0.9)(0.6)(50)(18.00)(0.360)$$
$$= 175 \text{ k} > 74.4 \text{ k} \qquad\qquad (\text{OK})$$

Use W18 × 46 $F_y = 50$ ksi with fifty four 3/4-in headed studs

---

Example 16-4

Using the same data as for Example 16-3 except that $w_1 = 1.2$ k/ft and $f'_c = 3$ ksi, perform the following tasks:

a. Select steel beam.

b. If the stud reduction factor for metal decks = 1.0, determine the number of 3/4-in headed studs required.

c. Check the beam strength before the concrete hardens.

d. Compute service load deflection before concrete hardens. Assume a construction live load of 20 psf.

e. Determine the service live load deflection after composite action is available.

f. Check shear.

g. Select a steel section to carry all the loads if no shear connectors are used and compute its total service load deflection.

Solution. *From Example 16-3*

$$W_u = (1.2)(0.78) + (1.6)(1.2) = 2.856 \text{ k/ft}$$

$$M_u = \frac{(2.856)(36)^2}{8} = 462.7 \text{ ft-k}$$

a. Select W section

$$Y_{con} = 4 + 3 = 7 \text{ in}$$

Assume $a = 2$ in

$$Y1 = 0$$

$$Y2 = 7 - \tfrac{2}{2} = 6 \text{ in}$$

Try W18 × 35 ($A = 10.3 \text{ in}^2$, $d = 17.70 \text{ in}$, $t_w = 0.300 \text{ in}$)

Assume $\Sigma Q_n = (10.3)(50) = 515 \text{ k}$

$$a = \frac{515}{(0.85)(3)(108)} = 1.87 \text{ in} < 4 \text{ in}$$

$$Y2 = 7 - \frac{1.87}{2} = 6.065 \text{ in}$$

$\phi_b M_n$ from Manual by interpolation

$$= 542 + \left(\frac{0.065}{0.50}\right)(560 - 542) = 544.3 \text{ ft-k} > 462.7 \text{ k} \qquad \text{(OK)}$$

b. Design of studs
Q_n from stud table $= 17.7 \text{ k}$

We can go down in W18 × 35 table to Y1 = 0.213 in and still obtain the M_u required (it's between 487 and 500 ft-k)

$$\Sigma Q_n = 388 \text{ k}$$

$$\text{Number of connectors required} = \frac{2 \times 388}{17.7} = 43.8$$

Use 22 connectors each side of Ł

c. Check strength of W section before concrete hardens
Assume the wet concrete is a LL during construction and also add a 20 psf construction live load.

Concrete wt = wt of slab + wt of ribs

$$= \left(\frac{4}{12}\right)(10)(115) + (3)(6)/144\,(115)(10) = 527$$

Other dead loads $= 780 - 527 = 253$ lb/ft

$$w_u = (1.2)(0.253) + (1.6)(0.020 \times 10 + 0.527) = 1.47 \text{ k/ft}$$

$$M_u = \frac{(1.47)(36)^2}{8} = 238.1 \text{ ft-k}$$

$\phi_b M_p$ from beam tables for W18 $\times$ 35 $= 249$ ft-k > 238.1 ft-k $\qquad$ (OK)

d. Service load deflection before concrete hardens

$$C_1 = 161$$

I_x for W18 $\times$ 35 $= 510$ in^4 (not lower bound I)

$$\text{Use } w_D = 0.78 + (10)(0.02) = 0.98 \text{ k/ft}$$

$$M_{DL} = \frac{(0.98)(36)^2}{8} = 158.8 \text{ ft-k}$$

$$\Delta_{DL} = \frac{(158.8)(36)^2}{(161)(510)} = 2.51 \text{ in}$$

(We might camber beam for this deflection and/or use PR connections.)

e. Service LL deflection after composite action is available

$$M_L = \frac{(1.2)(36)^2}{8} = 194.4 \text{ ft-k}$$

Lower bound I with Y1 = 0.21 in and Y2 = 6.065 in

$$I = 1490 + \left(\frac{0.065}{0.50}\right)(1550 - 1490) = 1498 \text{ in}^4$$

$$\Delta_{LL} = \frac{(194.4)(36)^2}{(161)(1498)} = 1.04 \text{ in} < \frac{L}{360} = 1.2 \text{ in} \qquad (\text{OK})$$

f. Check shear

$$w_u = (1.2)(0.78) + (1.6)(1.2) = 2.856 \text{ k/ft}$$

$$V_u = \frac{(36)(2.856)}{2} = 51.4 \text{ k}$$

$\phi V_m = \phi 0.6 F_y A_w = (0.9)(0.6)(50)(17.70 \times 0.300) = 143.4 \text{ k} > 51.4 \text{ k} \quad (\text{OK})$

g. Select section if we have no composite action

$$M_u = 462.7 \text{ ft-k}$$

$$Z \quad \text{Regd} = \frac{(12)(462.7)}{(0.9)(50)} = 123.4 \text{ in}^3$$

Requires a W21 × 55 ($I_x = 1140 \text{ in}^4$) from Table 5-3 in Manual entitled "w-shapes selection by Z_x."

$$\Delta = \frac{(462.7)(36)^2}{(161)(1140)} = 3.27 \text{ in} > \frac{L}{360} = \frac{(12)(36)}{360} = 1.20 \text{ in}$$

Min I_x to limit deflection to 1.2 in $= \left(\frac{3.27}{1.20}\right)(1140) = 3106 \text{ in}^4$

A W30 × 90 is required by Table 5-2 in Manual to provide such an I_x

16.12   CONTINUOUS COMPOSITE SECTIONS

The LRFD Specification (I3.2) permits the use of continuous composite sections. The flexural strength of a composite section in a negative moment region may be considered to equal $\phi_b M_n$ for the steel section alone, or it may be based upon the plastic strength of a composite section considered to be made up of the steel beam and the longitudinal reinforcement in the slab. For this latter method to be used the following conditions must be met:

1. The steel section must be compact and adequately braced.
2. The slab must be connected to the steel beams in the negative moment region with shear connectors.

Continuous-welded plate girders in Henry Jefferson County, Iowa. (Courtesy of the Lincoln Electric Company.)

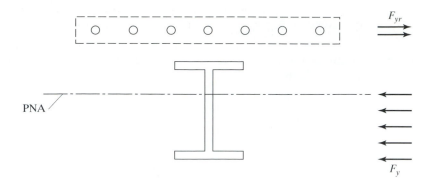

FIGURE 16.13

Stress distribution in negative moment range.

3. The longitudinal reinforcing in the slab parallel to the steel beam and within the effective width of the slab must have adequate development lengths. (*Development length* is a term used in reinforced-concrete design and refers to the length that reinforcing bars have to be extended or embedded in the concrete to properly anchor them or develop their stresses by means of bond between the bars and the concrete.)

For a particular beam, the total horizontal shear force between the point of zero moment and the point of maximum negative moments is to be taken as the smaller of $A_r F_{yr}$ and ΣQ_n, where A_r is the cross-sectional area of the properly developed reinforcing and F_{yr} is the yield stress of the bars. The plastic stress distribution for negative moment in a composite section is illustrated in Fig. 16.13.

16.13   DESIGN OF CONCRETE-ENCASED SECTIONS

For fireproofing purposes, it is possible to completely encase in concrete the steel beams used for building floors. This practice definitely is not economical, because light-weight spray-on fire protection is so much cheaper. Furthermore, encased beams may increase the floor system dead load by as much as 15 percent.

For the rare situation in which encased beams are used, the horizontal shears be-tween the slabs and beams can be considered to be transferred by natural bond and friction if certain conditions are met. The LRFD Specification (I1) states that for the transfer to be permissible the encasing concrete must be placed integrally with the slab concrete and must cover the steel by at least 2 in on the sides and bottom (or soffit). It is further required that the top of the steel section be at least $1\frac{1}{2}$ in below the top of the slab and 2 in above the bottom of the slab. Finally, the encasing concrete must have ad-equate mesh or other reinforcing for its full depth and across the soffit of the beam to prevent spalling of the concrete. The exact amount, which is not specified by the LRFD, can be very nominal in size (from "chicken wire size" on up).

1. With one method, the design strength of the encased section may be based on the plastic moment capacity $\phi_b M_p$ of the steel section alone.
2. By another method, the design strength is based on the first yield of the tension flange, assuming composite action between the concrete which is in compression and the steel section.

If the second method is used and we have unshored construction, the stresses in the steel section caused by the wet concrete and other construction loads are calculated. Then the stresses in the composite section caused by loads applied after the concrete hardens are computed. These stresses are superimposed on the first set of stresses. The total stresses so computed may not exceed $\phi_b F_y$ with $\phi_b = 0.90$. If we have shored construction, all of the loads may be assumed to be supported by the composite section and the stresses computed accordingly. For stress calculations, the properties of a composite section are computed with the transformed area method. In this method the cross-sectional area of one of the two materials is replaced or transformed into an equivalent area of the other. For composite design, it is customary to replace the concrete with an equivalent area of steel, whereas the reverse procedure is used in the working stress design method for reinforced-concrete design.

In the transformed area procedure, the concrete and steel are assumed to be bonded tightly together so that their strains will be the same at equal distances from the neutral axis. The unit stress in either material can then be said to equal its strain times its modulus of elasticity (ϵE_c for the concrete or ϵE_s for the steel). The unit stress in the steel is then $\epsilon E_s / \epsilon E_c = E_s / E_c$ times as great as the corresponding unit stress in the concrete. The E_s / E_c ratio is referred to as the modular ratio n; therefore, n in^2 of concrete are required to resist the same total stress as 1 in^2 of steel; and the cross-sectional area of the slab (A_c) is replaced with a transformed or equivalent area of steel equal to A_c / n.

The American Concrete Institute (ACI) Building Code states that the following expression may be used for calculating the modulus of elasticity of concrete weighing from 90 to 155 lb/ft:^{3}

$$E_c = w_c^{1.5} 33 \sqrt{f_c'}$$

In this expression, w_c is the weight of the concrete in pounds per cubic foot, and f_c' is the 28-day compressive strength in pounds per square inch.

> In SI units with w_c varying from 1500 to 2500 kg/m^3 and with
>
> f_c' in N/mm^2 or MP_a $E_c = w_c^{1.5}(0.043)\sqrt{f_c'}$

There are no slenderness limitations required by the LRFD Specification for either of the two methods because the encasement is effective in preventing both local and lateral buckling.

In Example 16-5, which follows, stresses are computed with the elastic theory assuming composite action as described for the second method. Notice that the author has divided the effective width of the slab by n to transform the concrete slab into an equivalent area of steel.

Example 16-5

Review the encased beam section shown in Fig. 16.14 if no shoring is used and the following data are assumed:

$$\text{Simple span} = 36 \text{ ft}$$
$$\text{Service dead load} = 0.50 \text{ k/ft before concrete hardens plus an}$$
$$\text{additional } 0.25 \text{ k/ft after concrete hardens}$$
$$\text{Construction live loads} = 0.2 \text{ k/ft}$$
$$\text{Service live load} = 1.0 \text{ k/ft after concrete hardens}$$
$$\text{Effective flange width } b_e = 60 \text{ in and } n = 9$$
$$F_y = 50 \text{ ksi}$$

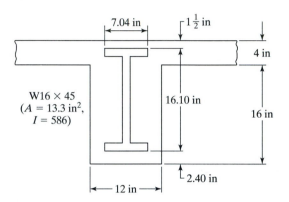

FIGURE 16.14

Solution. Calculated Properties of Composite Section: Neglecting concrete area below the slab.

$$A = 13.3 + \frac{(4)(60)}{9} = 39.96 \text{ in}^2$$

$$y_b = \frac{(13.3)(10.45) + (26.66)(18)}{39.96} = 15.50 \text{ in}$$

$$I = 586 + (13.3)(5.05)^2 + \left(\tfrac{1}{12}\right)\left(\tfrac{60}{9}\right)(4)^3 + (26.66)(2.5)^2$$

$$= 1127 \text{ in}^4$$

Stresses before concrete hardens

Assume wet concrete is a live load

$$w_u = (1.6)(0.5 + 0.2) = 1.12 \text{ k/ft}$$

$$M_u = \frac{(1.12)(36)^2}{8} = 181.4 \text{ ft-k}$$

$$f_t = \frac{(12)(181.4)(8.05)}{586} = 29.90 \text{ ksi}$$

$$< \phi_b F_y = (0.9)(50) = 45 \text{ ksi} \qquad \textbf{(OK)}$$

Stresses after concrete hardens

$$w_u = (1.2)(0.25) + (1.6)(1.0) = 1.9 \text{ k/ft}$$

$$M_u = \frac{(1.9)(36)^2}{8} = 307.8 \text{ ft-k}$$

$$f_t = \frac{(12)(307.8)(15.50 - 2.40)}{1129} = 42.86 \text{ ksi}$$

Total $f_t = 29.90 + 42.86 = 72.76 \text{ ksi} > 45 \text{ ksi} > 0.9F_y = 45 \text{ ksi}$ (NG)

For buildings, continuous composite construction with encased sections is permissible. For continuous construction the positive moments are handled exactly as has been illustrated by the preceding examples. For negative moments, however, the transformed section is taken as shown in Fig. 16-15. The crosshatched area represents the concrete in compression, and all concrete on the tensile side of the neutral axis (that is, above ts he axis) is neglected.

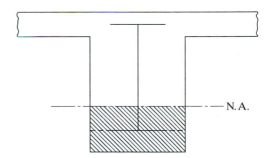

—N.A.

FIGURE 16.15

PROBLEMS

16-1.   Determine $\phi_b M_n$ for the section shown assuming sufficient shear connectors are provided to ensure full composite section. Solve using the procedure presented in Section 16-9 and check answer with tables in Manual. $F_y = 50$ ksi, $f'_c = 3$ ksi. (*Ans.* 425 ft-k)

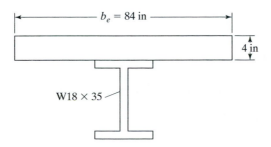

$b_e = 84$ in

4 in

W18 × 35

FIGURE P16-1

16-2.   Repeat Prob. 16-1 if a W18 × 60 is used.

16-3.   Repeat Prob. 16-2 using tables in Manual if it is considered to be partially composite and if ΣQ_n is 486 k. (*Ans.* 658.4 ft-k)

16-4.   Determine $\phi_b M_n$ for the section shown if 50 ksi steel and sufficient shear connectors are used to guarantee full composite action. Use formulas and check with Manual. $f'_c = 4$ ksi

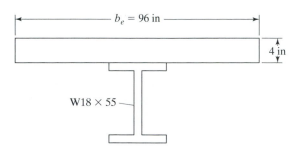

FIGURE P16-4

16-5.   Repeat Prob. 16-4 if a W16 × 36 is used. (*Ans.* 418.2 ft-k)

16-6.   Compute $\phi_b M_n$ for the composite section shown if 50 ksi steel is used and sufficient shear connectors are used to provide full composite action. A $3\frac{1}{4}$-in concrete slab is supported by 3-in-deep composite metal deck ribs perpendicular to beam $f'_c = 4$ ksi. Check answer with Manual.

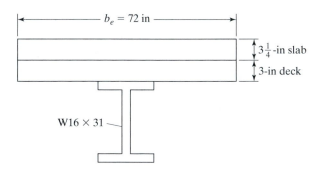

FIGURE P16-6

16-7.   Repeat Prob. 16-6 using Manual tables if ΣQ_n of the connectors is 335 k. (*Ans.* 388.4 ft-k)

16-8. Using the Composite Design Tables of the LRFD Manual, 50 ksi steel, a 145 lb/ft³ concrete slab with $f'_c = 4$ ksi and shored construction, select the steel section, design 3/4-in headed studs, calculate live load service deflection, and check the shear if the service live load is 100 psf.

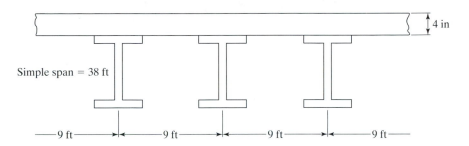

FIGURE P16-8

16-9. Repeat Prob. 16-8 if span is 32 ft and live load is 80 psf (*Ans.* W14 × 22 with $19\frac{3}{4}$-in studs)

16-10. For Prob. 16-9 calculate deflection during construction for wet concrete plus 20 psf live load for construction activities.

16-11. Select a 50 ksi section to support a service dead load of 200 psf and a service live load of 100 psf. The beams are to have 38-ft simple spans and are to be spaced 8 ft 6 in on center. Construction is shored, concrete weighs 115 lb/ft³, f'_c is 3.5 ksi, a metal deck with ribs perpendicular to the steel beams is used together with a 4-in concrete slab. The ribs are 3-in deep and have average widths of 6 in. Design 3/4-in headed studs and calculate live load deflection. (*Ans.* W18 × 46 with $68\frac{3}{4}$-in studs)

16-12. Using the LRFD Manual and 50 ksi steel, design a nonencased unshored composite section for the simple span beams shown in the accompanying figure if a 4-in. concrete slab (145 lb/ft³) with $f'_c = 4$ ksi is used. The total service dead load including the steel beam is 0.6 k/ft of length of the beam and the service live load is 1.25 k/ft.

   a. Select the beams.

   b. Determine the number of 3/4-in-diameter headed studs required.

   c. Compute the service live load deflection.

   d. Check the beam shear.

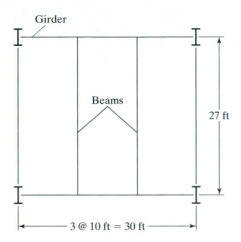

FIGURE P16-12

16-13. Repeat Prob. 16-12 if span is 30 ft and live load is 1 k/ft (*Ans.* W16 × 26 with $19\frac{3}{4}$-in studs)

16-14. If 50 ksi beams 9 ft on center and spanning 40 ft are to be selected to support a 4-in-deep lightweight concrete slab ($f_c' = 4$ ksi, weight 115 lb/ft³) on a 3-in-deep formed steel deck with shoring. The ribs for the steel deck, which are perpendicular to the steel beams, have average widths of 6 in. If the total service dead load including the beam weight is to be 0.80 k/ft of length of the beams and the service live load is 1.25 k/ft, (a) select the beams, (b) determine the number of 3/4-in-diameter headed studs required, (c) compute the service live load deflection, and (d) check the beam shear.

16-15. Repeat Prob. 16-14 if spans are 44 ft. (*Ans.* W21 × 44 with $53\frac{3}{4}$-in studs)

16-16. Repeat Prob. 16-14 if spans are 34 ft.

16-17. Repeat Prob. 16-16 if live load is 2 k/ft. (*Ans.* W18 × 40 with $54\frac{3}{4}$-in studs)

16-18. Using the same data as for Prob. 16-14 except that unshored construction is to be used, perform the following tasks:

    a. Select the steel beam.

    b. If the stud reduction factor for metal decks is 1.0, determine the number of 3/4-in headed studs required.

    c. Check the beam strength before the concrete hardens.

    d. Compute service load deflection before the concrete hardens assuming a construction live load of 25 psf.

    e. Determine the service load deflection after composite action is available.

    f. Check shear.

16-19. Repeat Prob. 16-18 if span is 36 ft and live load is 1.5 k/ft. (*Ans.* W18 × 35 with $48\frac{3}{4}$-in studs)

16-20. Using the transformed area method compute the stresses in the encased section shown in the accompanying illustration if no shoring is used. The section is assumed to be used for a simple span of 30 ft and to have a service dead uniform load of 30 psf applied after composite action develops and a service live uniform load of 120 psf. Assume $n = 9$, $F_y = 50$ ksi, $f'_c = 4$ ksi, and concrete weighing 150 lb/ft³.

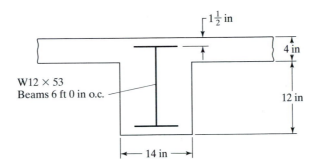

W12 × 53
Beams 6 ft 0 in o.c.

$1\frac{1}{2}$ in

4 in

12 in

14 in

FIGURE P16-20

CHAPTER 17

Composite Columns

17.1   INTRODUCTION

Composite columns are constructed with rolled or built-up steel shapes encased in concrete, or with concrete placed inside steel pipes or tubes. The resulting members are able to support significantly higher loads than reinforced concrete columns of the same sizes.

Several composite columns are shown in Fig. 17.1. In part (a) of the figure, a *W* shape embedded in concrete is shown. The cross sections, which usually are square or rectangular, have one or more longitudinal bars placed in each corner. In addition, lateral ties are wrapped around the longitudinal bars at frequent vertical intervals. Ties are effective in increasing column strengths. They prevent the longitudinal bars from being displaced during construction and they resist the tendency of these same bars to buckle outward under load, which would cause breaking or spalling off of the outer concrete cover. Notice that these ties are all open and U-shaped. Otherwise they could

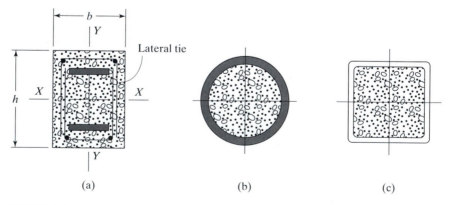

FIGURE 17.1

Composite columns.

not be installed, because the steel column shapes always will have been erected at an earlier time. In parts (b) and (c) of the figure, steel pipe and steel tubing sections filled with concrete are shown.

17.2 ADVANTAGES OF COMPOSITE COLUMNS

For a number of decades, structural steel shapes have been used in combination with plain or reinforced concrete. Originally, the encasing concrete was used to provide only fire and corrosion protection for the steel, with no consideration given to its strengthening effects. More recently, however, the development and increasing popularity of composite frame construction has encouraged designers to include the strength of the concrete in their calculations.[1,2]

Composite columns may be practically used for low-rise and high-rise buildings. For the low-rise types, such as warehouses, parking garages, and so on, the steel columns are often encased in concrete for the sake of appearance, or for protection from fire, corrosion, and (in garages) vehicles. If, we are going to encase the steel in concrete anyway, we may as well take advantage of the concrete and use smaller steel shapes.

For high-rise buildings, the sizes of composite columns often are considerably smaller than is required for reinforced-concrete columns to support the same loads. The results with composite designs are appreciable savings of valuable floor space. Closely spaced composite steel-concrete-columns connected with spandrel beams may be used around the outsides of high-rise buildings to resist lateral loads by the tubular concept (to be described in Chapter 19). Very large composite columns are sometimes placed on the corners of high-rise buildings to increase lateral resisting moments. Also, steel sections embedded within reinforced-concrete shear walls may be used in the central core of high-rise buildings. This also ensures a greater degree of precision in the construction of the core.

With composite construction, the bare steel sections support the initial loads, including the weights of the structure, the gravity, and lateral loads occurring during construction, and the concrete later cast around the W shapes or inside the tube shapes. The concrete and steel are combined in such a way that the advantages of both materials are used in the composite sections. For instance, the reinforced concrete enables the building frame to more easily limit swaying or lateral deflections. At the same time, the light weight and strength of the steel shapes permit the use of smaller and lighter foundations.[3]

---

[1]D. Belford, "Composite Steel Concrete Building Frame," *Civil Engineering* (New York: ASCE, July 1972), pp. 61–65.

[2]Fazlur R. Kahn, "Recent Structural Systems in Steel for High Rise Buildings," BCSA Conference on Steel in Architecture, November 24–26, 1969.

[3]L. G. Griffis, "Design of Encased *W*-shape Composite Columns," *Proceedings 1988 National Steel Construction Conference* (AISC, Chicago, June 8–11, 1988), pp. 20-1–20-28.

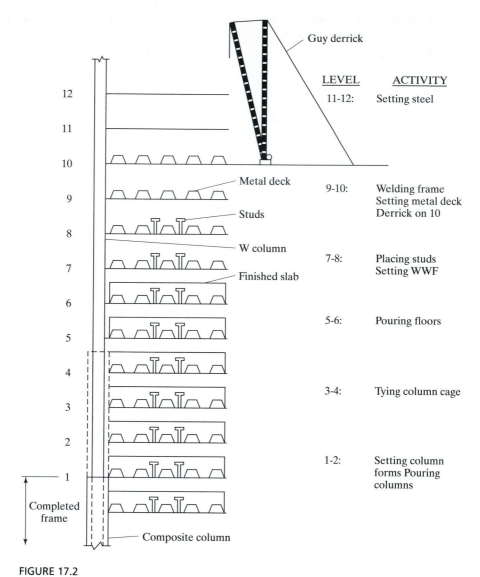

FIGURE 17.2

Sequence of construction operations for a composite building frame. (Courtesy of AISC.)

Composite high-rise structures are erected in a rather efficient manner. There is quite a vertical spread of construction going on at any one time, with numerous trades working simultaneously. This situation, which is pictured in Fig. 17.2, is briefly described here.[4]

---

[4]Griffis, op. cit.

1. One group of workers may be erecting the steel beams and columns for one or two stories on top of the frame.
2. Two or three stories below another group will be setting the metal decking for the floors.
3. A few stories below another group may be placing the concrete for the floor slabs.
4. This continues as we go down the building with one group, tying the column reinforcing bars in cages, while below them others are placing the column forms, placing the column concrete, and so on.

17.3   DISADVANTAGES OF COMPOSITE COLUMNS

As described in the preceding section, composite sections have several important advantages. They also have a few disadvantages. One particular problem with their use in high-rise buildings is the difficulty of controlling their rates and amounts of shortening in relation to shear walls and perhaps adjacent plain steel columns. The accurate estimation of these items is made quite difficult by the different types and stages of construction activities going on simultaneously over a large number of building stories.

If composite columns are used around the outside of a high-rise building, and plain steel sections are used in the building core (or if we have shear walls), the creep in the composite sections can be a problem. The results may be concrete floors that are not very level. Some erectors make very careful elevation measurements at column splices and then try to make appropriate adjustments with steel shims to try to even out the differences between measured elevations and computed elevations.

Another problem with composite columns is the lack of knowledge available concerning the mechanical bond between the concrete and the steel shapes. This is particularly important for the transfer of moments through beam-column joints. It is feared that if large cyclical strain reversals were to occur at such a joint (as in a seismic area) there could be a severe breakdown of the joint.[5]

17.4   LATERAL BRACING

Resistance to lateral loads for the usual structural steel or reinforced-concrete high-rise building is provided as the floors are being constructed. For instance, diagonal bracing or moment-resisting joints may be provided for each floor as a structural steel building frame is being constructed. In a similar manner the needed lateral strength of a reinforced-concrete frame may be provided by the moment resistance obtained with monolithic construction of its members and/or by shear walls.

For composite construction, the desired lateral strength of a building is not obtained until the concrete has been placed around or inside the erected steel members and has sufficiently hardened. This situation is probably being achieved 10 to 18 stories behind the steel erection (see Fig. 17.2).

As we have mentioned, the steel fabricator is used to erecting a steel frame and providing the necessary wind bracing as the floors are erected. The steel frames used

---

[5]Griffis, op. cit.

for high-rise composite buildings, however, do not usually have such bracing and the frames will not have the desired lateral strength. This strength will be obtained only after the concrete is placed and cured for many building stories. Thus, the engineer of record for a composite high-rise building must clearly state the lateral force conditions and what is to be done about them during erection.[6]

17.5   SPECIFICATIONS FOR COMPOSITE COLUMNS

Composite columns theoretically can be constructed with cross sections that are square, rectangular, round, triangular, or any other shape. Practically, however, they usually are square or rectangular, with one reinforcing bar in each column corner. This arrangement enables us to use reasonably simple connections from the exterior spandrel beams and floor beams to the steel shapes in the columns without unduly interfering with the vertical reinforcing.

The LRFD Specification does not provide detailed requirements for reinforcing bar spacings, splices, and so on. It therefore seems logical that the requirements in this regard of the ACI 318 Code[7] should be followed for situations not clearly covered by the LRFD Specification.

Section I2.1 of the LRFD Specification does provide detailed requirements pertaining to cross-sectional areas of steel shapes, concrete strengths, tie areas, and spacings for the vertical reinforcing bars, and so on. This information is listed and briefly discussed in the paragraphs to follow:

1. *The total cross-sectional area of the steel section or sections may not be less than 4 percent of the gross column area.* If the steel percentage is less than 4 percent, the member is classified as a reinforced-concrete column, and its design must be handled by the *Building Code Requirements for Reinforced Concrete* of the American Concrete Institute.

2. *When a steel core is encased in concrete the encasement must be reinforced with longitudinal load-carrying bars (which must be run continuously at framed levels) and with lateral ties spaced no farther apart than 2/3 times the least dimension of the composite member. The area of the ties must not be less than 0.007 sq. in. per inch of bar spacing. There must be at least $1\frac{1}{2}$ in clear cover of concrete outside of any steel (ties or longitudinal bars).* The cover is needed for the protection of the steel from fire or corrosion. The amount of longitudinal and transverse reinforcing required in the encasement is thought to be sufficient to prevent severe spalling of the concrete surface from occurring during a fire.

3. *The specified compression strength of the concrete f_c' must be at least 3 ksi (21 MPa) but not more than 8 ksi if normal weight concrete is used. For lightweight concrete it may not be less than 4 ksi or more than 8 ksi.* The upper limit of 8 ksi is provided because sufficient test data are not available for composite columns with higher-strength concretes at this time. The lower limit of f_c' was specified for

---

[6]Griffis, op. cit.
[7]American Concrete Institute, *Building Code Requirements for Reinforced Concrete*, ACI 318-99 (Detroit: 1999).

the purpose of ensuring the use of good quality but readily available concrete and for the purpose of making sure that adequate quality control is used. This might not be the case if a lower grade of concrete were specified.

4. *The yield stresses of the steel sections and reinforcing bars used may not be greater than 60 ksi (415 MPa). If a steel with a yield stress greater than 60 ksi is actually used in a composite column, only 60 ksi may be used in the calculations.*

The original reason for limiting the value of F_y is given here. One major objective in composite design is the prevention of local buckling of the longitudinal reinforcing bars and the contained steel section. To achieve this objective, the covering concrete must not be allowed to break or spall. It was assumed by the writers of the previous LRFD Specification that such concrete was in danger of breaking or spalling if its strain reached 0.0018. If we take this strain and multiply it by F_s, we get $(0.0018)(29,000) \approx 55$ ksi. Hence, that value was specified as the maximum useable yield stress.

Recent research has shown that due to concrete confinement effects, the 55 ksi value is conservative, and it has been raised to 60 ksi in the current specification. This value corresponds to the yield stress of the reinforcing bars commonly used today: It seems probable that the value will be raised even further in the near future, particularly for tubular composite structures, where it is unduly conservative (see Chapter 19).

5. *The minimum permissible wall thickness of steel tubing filled with concrete is equal to b $\sqrt{F_y/3E}$ for each face of width b of rectangular sections. The minimum thickness for circular sections of outside diameter D is D $\sqrt{F_y/8E}$. These values* are the same as those given in the 1999 ACI Code. It is desired that steel pipe or tubing of sufficient thickness be used so that they will not buckle before yielding.

6. *When composite columns contain more than one steel shape, those shapes have to be connected with lacing, tie plates, or batten plates so that local buckling of the individual shapes is not possible before the concrete hardens.* After the concrete hardens, it is assumed that all the parts of the column act together as a unit in resisting load.

7. *Should the supporting concrete be wider on one or more sides than the loaded area and be otherwise restrained against lateral expansion on the remaining side or sides, the design compressive strength of the composite column resisted by the concrete $\phi_c P_{nc}$ is to be computed as being equal to $\phi_B 1.7 f'_c A_B$ with $\phi_c = 0.65$ for bearing on concrete where A_B is the loaded area.*

17.6   AXIAL DESIGN STRENGTHS OF COMPOSITE COLUMNS

The contribution of each component of a composite column to its overall strength is difficult, if not impossible, to determine. The amount of flexural concrete cracking varies throughout the height of the column. The concrete is not nearly as homogeneous as the steel, and, furthermore, the modulus of elasticity of the concrete varies with time and under the action of long-term or sustained loads. The effective lengths of composite columns in the rigid monolithic structures in which they are frequently used cannot be determined very well. The contribution of the concrete to the total stiffness of a composite column varies, depending on whether it is placed inside a tube or whether it is on the outside of a W section where its stiffness contribution is less.[8]

The preceding paragraph presented some of the reasons it is difficult to develop a useful theoretical formula for the design of composite columns. As a result, a set of empirical formulas is presented in the LRFD Specification for the design of composite columns.

The design strengths of composite columns ($\phi_c P_n$ with $\phi_c = 0.85$ and $P_n = A_g F_{cr}$) are determined much as are the design strengths of plain steel columns. *The formulas to be used for composite columns for F_{cr}, the critical stress, are the same, except that the areas, radii of gyration, yield stresses, and moduli of elasticity are modified in an attempt to account for composite behavior.* The column expressions given in Section E2 of the Specification and previously described in Chapter 5 of this text are listed as follows:

If $\lambda_c \leq 1.5$

$$F_{cr} = (0.658^{\lambda_c^2})F_y \qquad \text{(LRFD Equation E2-2)}$$

If $\lambda_c > 1.5$

$$F_{cr} = \left(\frac{0.877}{\lambda_c^2}\right)F_y \qquad \text{(LRFD Equation E2-3)}$$

where

$$\lambda_c = \frac{KL}{r\pi}\sqrt{\frac{F_y}{E}} \qquad \text{(LRFD Equation E2-4)}$$

The modifications made in these formulas are as follows:

1. Replace A_g with A_s where A_s is the area of the steel shape, tube, or pipe, but not including any reinforcing bars.
2. Replace r with r_m where r_m is the radius of gyration of the steel shapes, pipes, or tubes. For steel shapes encased in concrete it may not be less than 0.3 times the overall thickness of the composite member in the plane of buckling.
3. *Replace F_y with the modified yield stress F_{my} and E with the modified modulus of elasticity E_m.* These values follow:

$$F_{my} = F_y + c_1 F_{yr}\left(\frac{A_r}{A_s}\right) + c_2 f'_c\left(\frac{A_c}{A_s}\right) \qquad \text{(LRFD Equation 12-1)}$$

$$E_m = E + c_3 E_c\left(\frac{A_c}{A_s}\right) \qquad \text{(LRFD Equation 12-2)}$$

In these expressions for F_{my} and E_m the following abbreviations are used:

1. A_c, A_s, and A_r are, respectively, the areas of the concrete, steel section, and reinforcing bars.

---

[8]Task Group 20, Structural Stability Council, "A Specification for the Design of Steel-Concrete Composite Columns," *Engineering Journal*, AISC 16, no. 4 (fourth quarter, 1979), pp. 101–115.

2. E and E_c are, respectively, the moduli of elasticity of the steel and the concrete. LRFD Specification 12.2 states that E_c, the modulus of elasticity of the concrete in ksi, may be computed from $w_c^{1.5}\sqrt{f_c'}$ where w_c is the unit weight of the concrete in lb/ft^3 and f_c' is the specified compression strength of the concrete in ksi.

3. F_y and F_{yr} are the specified minimum yield stresses of the steel section and the reinforcing bars.

4. $c_1, c_2,$ and c_3 are numerical coefficients. For concrete-filled pipes and tubing $c_1 = 1.0$, $c_2 = 0.85$, and $c_3 = 0.4$. For concrete-encased shapes $c_1 = 0.7$, $c_2 = 0.6$, and $c_3 = 0.2$.

The value of $\phi_c P_n$ for a composite column with a W section embedded in concrete is calculated in Example 17-1 in accordance with these LRFD requirements.

Example 17-1

Compute the value of $\phi_c P_n$ for the composite column shown in Fig. 17.3 if 50 ksi steel, 3.5 ksi concrete with a weight of 145 lb/ft^3, and KL of 12 ft are used.

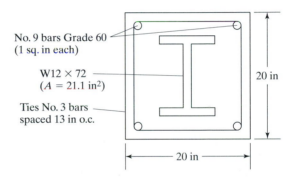

No. 9 bars Grade 60
(1 sq. in each)

W12 × 72
($A = 21.1$ in^2)

Ties No. 3 bars
spaced 13 in o.c.

20 in

20 in

FIGURE 17.3

Solution

$$F_{my} = F_y + c_1 F_{yr}\left(\frac{A_r}{A_s}\right) + c_2 f_c'\left(\frac{A_c}{A_s}\right)$$

$$= 50 + (0.7)(60)\left(\frac{4}{21.1}\right) + (0.6)(3.5)\left(\frac{400 - 21.1}{21.1}\right)$$

$$= 95.67 \text{ ksi}$$

$$E_c = w^{1.5}\sqrt{f_c'} = 145^{1.5}\sqrt{3.5} = 3266.5 \text{ ksi}$$

$$E_m = E + c_3 E_c\left(\frac{A_c}{A_s}\right)$$

$$= 29{,}000 + (0.2)(3266.5)\left(\frac{400 - 21.1}{21.1}\right)$$

$$= 29{,}000 + 11{,}730 = 40{,}730 \text{ ksi}$$

Replace F_y with 95.67 ksi, E with 40,730 ksi, and A_g with $A_s = 21.1$ in^2 r_y of W12 × 72 = 3.04 in, may not be less than $(0.3)(20) = 6.0$ in.

$$\lambda_c = \frac{KL}{r\pi}\sqrt{\frac{F_y}{E}}$$

$$= \frac{(12)(12)}{6.0\pi}\sqrt{\frac{95.67}{40,730}} = 0.370$$

$$F_{cr} = (0.658^{\lambda_c^2})F_y = (0.658^{0.370^2})(95.67) = 90.34 \text{ ksi}$$

$$\phi_c P_n = \phi_c F_{cr} A_s = (0.85)(90.34)(21.1) = 1620 \text{ k}$$

It has been shown that the axial load strength of composite columns using W sections greatly exceeds the axial load strengths of plain W sections. The longer the columns become, the greater the ratio of the strength of composite columns to that of noncomposite ones.

The increasing strength advantage of longer composite columns over that of plain steel columns is clearly shown in Table 17.1. In this table the axial design strength of a 22 in × 22 in column (with $f_c' = 3.5$ ksi and Grade 60 reinforcing bars) composite with a 50 ksi W14 × 90 is compared with the axial design strength of a plain 50 ksi W14 × 90 column. The ratio of strength of the composite section to that of the plain section goes from 1.95 for an effective length of 10 ft to 3.72 for an effective length of 40ft. *Thus we can see that the strength of a composite column falls off at an appreciably lower rate than does that of a plain W column as the effective length increases.*

17.7   LRFD TABLES

In Part 4 of the Manual, a series of tables is presented for HSS sections and steel pipe sections filled with concrete. These tables, numbered 4-12 to 4-17, are set up in exactly the same fashion as the tables for axially loaded plain steel columns, which are also

TABLE 17.1 Axial Design Strengths

Effective Length KL (ft)	Axial design strength of composite section (kips)	Axial design strength of W14 × 90 (kips)	Ratio of strength of composite Section to that of W14 × 90
0	2100	1130	1.86
10	2050	1040	1.97
20	1860	828	2.25
30	1580	564	2.80
40	1260	336	3.75

[A 50 ksi W14 × 90 column compared with strengths of that same section composite with a 22 in. × 22 in. reinforced concrete section ($f_c' = 3.5$ ksi, 4 10 Grade 60 reinforcing bars and 3 ties 14 in o.c.)].

presented in Section 4 of the Manual. The axial strengths are given with respect to the minor axis for a range of $(KL)_y$ values.

We include values for composite HSS square and rectangular sections $(F_y = 46 \text{ ksi})$ for round HSS sections $(F_y = 42 \text{ ksi})$, and for steel pipe sections $(F_y = 35 \text{ ks})$. The tables cover steel sections filled with 4 and 5 ksi concretes. For other grades of concrete, and for steel shapes made composite with encasing concrete, the formulas presented earlier in this chapter may be used to determine $\phi_c P_n$ values.

Examples 17-2 and 17-3 show how the tables can be used to directly determine design strengths for square and rectangular composite HSS sections. Example 17-4 presents the selection of a composite pipe column filled with concrete. In this example, we chose a trial section after a little preliminary scratchwork and then checked its axial design strength using the appropriate LRFD formulas. In addition, the results were checked with the Part 4 tables in the Manual.

Example 17-2

Determine the axial design strength $\phi_c P_n$ of a 46 ksi HSS $12 \times 12 \times \frac{1}{2}$ section filled with 4 ksi concrete if $(KL)_x = (KL)_y = 16$ ft.

Solution. From Table 4.12 for $(KL)_y = 16$ ft

$$\phi_c P_n = 985 \text{ k}$$

Example 17-3

Determine the axial design strength $\phi_c P_n$ of a concrete filled 46 ksi HSS $20 \times 12 \times \frac{1}{2}$ if $f'_c = 5$ ksi and $(KL)_x = 24$ ft and $(KL)_y = 12$ ft.

Solution. From Table 4.13 we find $\dfrac{r_{mx}}{r_{my}} = 1.48$. Then controlling unbraced length is

$$(KL)_y = 12 \text{ ft}$$

or

$$(KL)_{yeq} = \frac{(KL)_x}{\dfrac{r_{mx}}{r_{my}}} = \frac{24}{1.48} = 16.2 \text{ ft} \leftarrow$$

$\phi_c P_n$ by interpolation from Table 5 $= 1574$ k

Example 17-4

Select a round HSS Section ($F_y = 42$ ksi) filled with 4 ksi concrete (145 lb/ft^3) to support a load $P_u = 320$ k if $KL = 14$ ft.

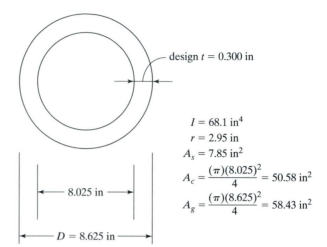

design $t = 0.300$ in

$I = 68.1$ in^4
$r = 2.95$ in
$A_s = 7.85$ in^2
$A_c = \dfrac{(\pi)(8.025)^2}{4} = 50.58$ in^2
$A_g = \dfrac{(\pi)(8.625)^2}{4} = 58.43$ in^2

8.025 in

$D = 8.625$ in

FIGURE 17.4

Properties of round HSS8.625 × 0.322 section

Solution Assume an HSS 8.625 × 0.322 (28.6 lb/ft). Its properties are shown in Fig. 17.4.

Minimum wall t as required by LRFD 12.1 $= D\sqrt{\dfrac{F_y}{8E}}$

$$= 8.625\sqrt{\frac{50}{(8)(29 \times 10^3)}} = 0.127 \text{ in} < 0.322 \text{ in} \qquad \text{(OK)}$$

Checking cross-sectional area of pipe column as a percentage of total composite section area

$$= \frac{7.85}{58.43} = 0.134 > 0.04 \text{ reqd. by LRFD I2.1} \qquad \text{(OK)}$$

Computing modified values of yield stress (F_{my}), modulus of elasticity (F_m), and radius of gyration (r_m). (Note that the second part of LRFD Equation I2-1, which follows, equals zero since there are no longitudinal reinforcing bars and thus $A_r = 0$.)

$$F_{my} = F_y + c_1 F_{yr}\left(\frac{A_r}{A_s}\right) + c_2 f'_c\left(\frac{A_c}{A_s}\right)$$

$$= 42 + 0 + (0.85)(4.0)\left(\frac{50.58}{7.85}\right) = 63.91 \text{ ksi}$$

$$E_c = w_c^{1.5}\sqrt{f_c'} = 145^{1.5}\sqrt{4.0} = 3492 \text{ ksi}$$

$$E_m = E + c_3 E_c\left(\frac{A_c}{A_s}\right) = 29{,}000 + (0.4)(3492)\left(\frac{50.58}{7.85}\right) = 38{,}000 \text{ ksi}$$

$$r_m = r \text{ of pipe } = 2.95 \text{ in, but not less than}$$

$$(0.3D) = (0.3)(8.625) = 2.59 \text{ in per LRFD I2.2.}$$

Slenderness parameter

$$\lambda_c = \frac{KL}{r_m\pi}\sqrt{\frac{F_{my}}{E_m}} = \frac{(12)(14)}{2.95\pi}\sqrt{\frac{63.91}{38{,}000}} = 0.743 < 1.5$$

$$\therefore F_{cr} = (0.658^{\lambda_c^2})F_{my} = 0.658^{(0.743)^2}(63.91) = 50.72 \text{ ksi}$$

Computing $\phi_c P_n$

$$\phi_c P_n = \phi_c A_s F_{cr} = (0.85)(7.85)(50.72) = 338.4 \text{ k} > 320 \text{ k} \qquad \text{(OK)}$$
Note: LRFD Manual in Part 5 (Table 4.14) gives 338 k

The supporting strength provided by the concrete as given in Appendix C of the ACI 318 specification, equals $0.85\phi_B f_c' A_B$ if the supporting concrete area is the same size as the column. In this expression, A_B is the loaded area and $\phi_B = 0.65$. If the supporting concrete is wider than the loaded area on one or more sides, and restrained against lateral expansion on the other sides, the supporting strength is equal to $1.7 \phi_B f_c' A_B$ as given in LRFD Section I2.4.

In Example 17-5 the author computes the bearing area A_B required for load transfer for the composite column of Example 17-1.

Example 17-5

It is assumed that all of the load for the composite column of Example 17-1 is applied to the column at one level. Determine the bearing area A_B of the concrete as required by Section I2.4 of the LRFD Specification. The supporting concrete is wider than the loaded area on all sides.

Solution. From the Prob. 17-1 solution $\phi_c P_n = 1620$ k
From the LRFD column tables a W12 × 72 (50 ksi steel) with $KL = 12$ ft can alone support a value $\phi_c P_{ns} = 761$ k.
The design compressive load resisted by the concrete $\phi_c P_{nc}$ must be developed by direct bearing at the connection

$$\phi_c P_{nc} = \phi_c P_n - \phi_c P_{ns} = 1620 - 761 = 859 \text{ k}$$

Section I2.4 of the LRFD Specification requires that

$$\phi_c P_{nc} \leq 1.7 \phi_B f_c' A_B$$

$$\therefore A_B \geq \frac{\phi_c P_{nc}}{1.7 \phi_B f_c'} = \frac{859}{(1.7)(0.65)(3.5)} = 222.1 \text{ in}^2$$

Our composite column has an area of $20 \times 20 = 400 \text{ in}^2 > 222.1 \text{ in}^2$ (OK)

17.8 FLEXURAL DESIGN STRENGTHS OF COMPOSITE COLUMNS

The nominal flexural strength of composite columns is computed assuming a plastic distribution of stresses. We can locate the plastic neutral axis by equating the tensile forces on one side of the member to the compression force on the other side. On the tensile side, there will be reinforcing bars and part of the embedded steel section stressed to their yield stresses. On the compression side, there will be a compressive force equal to $0.85f_c'$ times the area of an equivalent stress block. The equivalent stress block will have a width equal to the column width and a depth equal to β_1 times the distance to the PNA. (The value of β_1 is provided by the ACI Code.) The nominal flexural strength M_n' then equals the sum of the moments of the axial forces about the PNA.

Values of $\phi_b M_{nx}$ and $\phi_b M_{ny}$ are shown for each of the composite columns in Part 4 of the Manual. These values will be needed for analyzing beam columns as described in the next section.

17.9 AXIAL LOAD AND BENDING EQUATION

The following interaction formulas are used to check plain steel members subject to axial load and bending:
If $P_u/\phi P_n \geq 0.2$,

$$\frac{P_u}{\phi P_n} + \frac{8}{9}\left(\frac{M_{ux}}{\phi_b M_{nx}} + \frac{M_{uy}}{\phi_b M_{ny}}\right) \leq 1.0 \qquad \text{(LRFD Equation H1-1a)}$$

If $P_u/\phi P_n < 0.2$,

$$\frac{P_u}{2\phi P_n} + \left(\frac{M_{ux}}{\phi_b M_{nx}} + \frac{M_{uy}}{\phi_b M_{ny}}\right) \leq 1.0 \qquad \text{(LRFD Equation H1-1b)}$$

These inequalities and their application were discussed in Chapter 11. That presentation included definitions of the various values needed to calculate M_{ux} and $M_{uy}(B_1, M_{nt}, B_2, \text{and } M_{1t})$.

These same interaction inequalities are used to determine the adequacy of composite beam-columns, except that some of the terms are modified, as follows:

1. The Euler elastic buckling loads P_{ex} and P_{ey} that are used in the calculations of the bending factors B_1 and B_2 are to be determined with the expression that follows in which F_{my} is the modified yield stress that was defined in Section 17-6. The values of P_{ex} and P_{ey} times the square of the appropriate effective length in

feet divided by 10^4 are given in the tables for each of the composite columns. Thus, we have

$$P_s = \frac{A_s F_{my}}{\lambda_c^2}$$

2. The resistance factor ϕ_b is to be used as it is in composite beams where it equals 0.85 if $h/t_w \leq 3.76\sqrt{\frac{E}{F_{yf}}}$ and a plastic stress distribution is used to compute M_n; or it is taken as 0.9 if $h/t_w > 3.76\sqrt{\frac{E}{F_{yf}}}$ and M_n determined by superimposing the elastic stresses.

3. The column slenderness parameter λ_c is to be modified as it was for determining the design strengths of axially loaded composite columns in Section 17-6.

17.10   DESIGN OF COMPOSITE COLUMNS SUBJECT TO AXIAL LOAD AND BENDING

This section is devoted to the design of composite columns to resist axial loads and moments. It is a trial-and-error procedure involving the selection of a trial section, the application of the appropriate interaction formula, probably the selection of another trial section, the application of the formula, and so on, until a satisfactory column is obtained.

A perfectly satisfactory design can be achieved by this process, but it can involve quite a few trials if the first estimate is not very good. For this reason, we present a rough method of estimating sizes in the next few paragraphs. This method usually will enable the designer to make a fairly good first size estimate and thus reduce the number of trial designs that need to be made.

For this discussion, it is assumed that a composite column is to be designed to support a certain P_u and a certain M_{ux}, with M_{uy} equal to zero. It is further assumed that LRFD Equation H1-1a applies to the member. If M_{uy} is zero, the formula becomes

$$\frac{P_u}{\phi P_n} + \frac{8}{9}\frac{M_{ux}}{\phi_b M_{nx}} \leq 1.0$$

The designer may estimate the final values of the two parts of this equation. He or she may very well assume the two parts are equal.

$$\frac{P_u}{\phi P_n} = 0.5 \quad \text{and} \quad \frac{8}{9}\frac{M_{ux}}{\phi_b M_{nx}} = 0.5$$

For this discussion, it is assumed that a braced composite column is to be selected consisting of a square HSS section with dimensions roughly equal to 14×14 or 16×16. Other data: $KL = 12$ ft, $F_y = 46$ ksi, $f_c' = 4$ ksi, $\phi_c P_n = 900$ k and $M_{ntx} = 190$ ft-k. If B_{1x} is assumed to equal unity, estimated values of $\phi_c P_n$ and $\phi_b M_{nx}$ for a suitable section can be determined as follows:

$$\frac{P_u}{\phi_c P_n} = 0.5 \qquad \frac{8}{9}\frac{M_{ux}}{\phi_b M_{nx}} = 0.5$$

$$\frac{900}{\phi_c P_n} = 0.5 \qquad\qquad \frac{8}{9}\left(\frac{190}{\phi_b M_{nx}}\right) = 0.5$$

$$\phi_c P_n = 1800 \text{ k} \qquad\qquad \phi_b M_{nx} = 338 \text{ ft-k}$$

Now the designer may go to the composite tables, numbered 4.4 to 4.12 in the Manual, and try a section which has values roughly equal to the ones just estimated (1800 k and 338 ft-k). Looking in the tables he or she will find that a suitable section can probably be found somewhere between the HSS16 × 16 × $\frac{5}{8}$ ($\phi_c P_m$ = 1840 k and $\phi_b M_{nx}$ = 690 ft-k) and the HSS14 × 14 × $\frac{3}{8}$ ($\phi_c P_m$ = 1130 k and $\phi_b M_{nx}$ = 329 ft-k). The authors tried one about halfway in between these two sections. Example 17-6 presents the selection of a section for this situation.

Example 17-6

Select a composite column to resist P_u = 900 k and M_{ux} = 190 ft-k. The column which is braced against sidesway or lateral translation at its ends is to consist of 46 ksi steel and 4 ksi concrete. KL = 12 ft and C_m = 0.85.

Solution

Try HSS14 × 14 × $\frac{5}{8}$ ($\phi_c P_n$ = 1500 k, $\phi_b M_{nx}$ = 521 ft-k, and $\dfrac{P_e(K_x L_x)^2}{10^4}$ = 217 ft²-k)

$$P_{ex} = \frac{(217)(10)^4}{(12)^2} = 15,069 \text{ k}$$

$$B_1 = \frac{C_m}{1 - \dfrac{P_u}{P_{ex}}} = \frac{0.85}{1 - \dfrac{900}{15,069}} < 1.0 \ \therefore \text{ Use } 1.0$$

$$\frac{P_u}{\phi_c P_n} = \frac{900}{1500} > 0.2$$

$\therefore$ Must use LRFD Equation H1-1a

$$\frac{P_u}{\phi_c P_n} + \frac{8}{9}\frac{M_{ux}}{\phi_b M_{nx}} < 1.0$$

$$\frac{900}{1500} + \left(\frac{8}{9}\right)\left(\frac{190}{521}\right) = 0.924 \quad < 1.0$$

(OK)

Use HSS14 × 14 × $\frac{5}{8}$

For Example 17-6, the proportions between the axial load P_u and the bending moment M_{ux} seemed fairly normal. Should the ratio between the two be somewhat different, as when we have a rather large moment compared with the axial load, we might change the estimated values in the interaction equation and have a little better luck in the first trials. For instance, if the moment M_{ux} is rather large compared with P_u, we might assume the following approximate values for the interaction formula and back-figure the values of ϕP_n and $\phi_b M_{nx}$ to look up in the tables:

$$\frac{P_u}{\phi P_n} = 0.3 \qquad \frac{8}{9} \frac{M_{ux}}{\phi_b M_{nx}} = 0.7$$

The rough procedure used here to estimate beam-column sizes does not always work for students as well as it does for the authors. Frequently, several trials will have to be made.

17.11   LOAD TRANSFER AT FOOTINGS AND OTHER CONNECTIONS

A small steel base plate usually is provided at the base of a composite column. Its purpose is to accommodate the anchor bolts needed to anchor the embedded steel shape to the footing for the loads occurring during the erection of the structure before the encasing concrete hardens and composite action is developed. This plate should be sufficiently small to be out of the way of the dowels needed for the reinforced-concrete part of the column.[9]

The LRFD Specification does not provide details for the design of these dowels, but a procedure similar to the one provided by the ACI 318-9 Code seems to be a good one. If the column P_u exceeds $1.7\,\phi_c f'_c A_B$, the excess load should be resisted by dowels. If P_u does not exceed $1.7\phi_c f'_c A_B$, it would appear that no dowels are needed. For such a situation, the ACI Code (Sections 15.8.2.1 and 15.8.2.3) states that a minimum area of dowels equal to 0.005 times the cross-sectional area of the column must be used and those dowels may not be larger than No. 11 bars. This diameter requirement ensures sufficient tying together of the column and footing over the whole contact area. The use of a very few large dowels spaced far apart might not do this very well.

PROBLEMS

For all problems, use 145 lb/ft^3 concrete and 50 ksi steel shapes.

17-1 to 17-3.   Determine $\phi_c P_n$ for each of the encased W sections. The reinforcing bars are grade 60 and $f'_c = 3.5$ ksi for the concrete.

---

[9]Griffis, op. cit.

17-1.

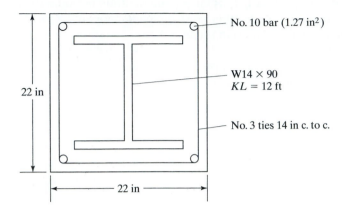

No. 10 bar (1.27 in²)

W14 × 90
$KL = 12$ ft

No. 3 ties 14 in c. to c.

22 in

22 in

FIGURE P17-1 (*Ans.* 2026 k)

17-2.

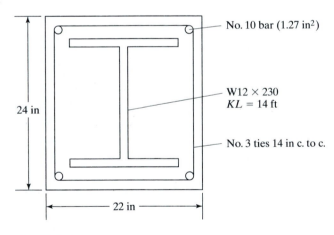

No. 10 bar (1.27 in²)

W12 × 230
$KL = 14$ ft

No. 3 ties 14 in c. to c.

24 in

22 in

FIGURE P17-2

17-3.

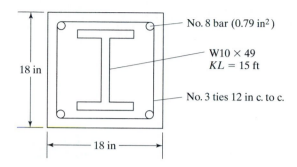

18 in

18 in

No. 8 bar (0.79 in²)

W10 × 49
$KL = 15$ ft

No. 3 ties 12 in c. to c.

FIGURE P17-3 (*Ans.* 1139 k)

17-4 to 17-5. Using the LRFD tables in Section 4 of the manual determine $\phi_c P_n$ values for the concrete filled sections given. They each are braced against sidesway.

17-4.   a. An HSS10 × 8 × $\frac{5}{16}$, $F_y = 46$ ksi, $f'_c = 4$ ksi and $(KL)_x = (KL)_y = 14$ ft.

   b. An HSS14 × 14 × $\frac{1}{2}$, $F_y = 46$ ksi, $f'_c = 5$ ksi and $(KL)_x = 18$ ft and $(KL)_y = 12$ ft.

17-5. A round HSS7.500 × 0.312($F_y = 42$ ksi), $f'_c = 4$ ksi, $(KL)_x = 16$ ft and $(KL)_y = 8$ ft. (*Ans.* 231 k)

17-6. Repeat Problem 17-5 using the LRFD equations.

17-7. Select a square HSS section, $F_y = 46$ ksi, filled with 4 ksi, concrete to resist $P_u = 800$ k and $M_{ux} = 150$ ft-k. The column is to be braced against sidesway at its ends and $(KL)_x = (KL)_y = 14$ ft. $C_m = 0.85$. (*Ans.* HSS 14 × 14 × $\frac{1}{2}$)

17-8. Repeat Problem 17-7 if a round HSS section ($F_y = 42$ ksi) and 5 ksi concrete are used.

C H A P T E R   1 8

Built-Up Beams, Built-Up Wide-Flange Sections, and Plate Girders

18.1 COVER-PLATED BEAMS

Should the largest available *W* section be insufficient to support the loads anticipated for a certain span, several possible alternatives may be taken. Perhaps the most economical solution involves the use of a higher-strength steel *W* section. If this is not feasible, we may make use of one of the following: (1) two or more regular *W* sections side-by-side (an expensive solution), (2) a cover-plated beam, (3) a built-up wide-flange section, (4) a plate girder, or (5) a steel truss. This section discusses the cover-plated beam alternative. The built-up wide flange is presented in Section 18-2, and the remainder of the chapter is devoted to plate girders.

In addition to being practical for cases in which the moments to be resisted are slightly in excess of those that can be supported by the deepest *W* sections, there are other useful applications for cover-plated beams. On some occasions the total depth may be so limited that the resisting moments of *W* sections of the specified depth are too small. For instance, the architect may show a certain maximum depth for beams in his or her drawings for a building. In a bridge, beam depths may be limited by clearance requirements. Cover-plated beams frequently will be a very satisfactory solution for situations like these. Furthermore, there may be economical uses for cover-plated beams where the depth is not limited and where there are standard *W* sections available to support the loads. A smaller *W* section than required by the maximum moment can be selected and have cover plates attached to its flanges. These plates can be cut off where the moments are smaller, with resulting saving of steel. Applications of this type are quite common for continuous beams.

Should the depth be fixed and a cover-plated beam seem to be a feasible solution, the usual procedure will be to select a standard section with a depth that leaves room for top and bottom cover plates. Then the cover plate sizes can be selected.

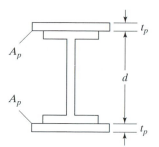

FIGURE 18.1

For this discussion, reference is made to Fig. 18.1. For the derivation to follow, Z is the plastic modulus for the entire built-up section, Z_W is the plastic modulus for the W section, d is the depth of the W section, t_p is the thickness of one cover plate, and A_p is the area of one cover plate. An expression for the required area of one plate can be developed as follows:

$$Z \text{ required} = \frac{M_u}{\phi_b F_y}$$

The total Z of the built-up section must at least equal the Z required. It will be furnished by the W shape and the cover plates as follows:

$$Z \text{ required} = Z_W + Z_{\text{plates}}$$

$$= Z_W + 2 A_p \left(\frac{d}{2} + \frac{t_p}{2} \right)$$

$$A_p = \frac{Z \text{ required} - Z_W}{d + t_p}$$

Example 18-1 illustrates the design of a cover-plated beam. Quite a few satisfactory solutions involving different W sections and varying size cover plates are available other than the one made in this problem.

Example 18-1

Select a beam limited to a maximum depth of 29.50 in. for the loads and span of Fig. 18.2. A 50 ksi steel is used and the beam is assumed to have full lateral bracing for its compression flange.

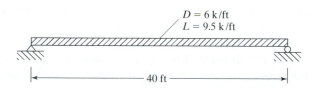

$D = 6 \text{ k/ft}$
$L = 9.5 \text{ k/ft}$

FIGURE 18.2

40 ft

Solution

$$\text{Assume beam wt} = 350 \text{ lb/ft}$$

$$w_u = (1.2)(6.35) + (1.6)(9.5) = 22.82 \text{ k/ft}$$

$$M_u = \frac{(22.82)(40)^2}{8} = 4564 \text{ ft-k}$$

$$Z \text{ required} = \frac{M_u}{\phi_b F_y} = \frac{(12)(4564)}{(0.9)(50)} = 1217 \text{ in}^3$$

The only 50 ksi sections listed in the Manual with depths ≤ 29.50 in and Z values $\geq 1217.1 \text{ in}^3$ are the completely impractical and enormously heavy and expensive W14 × 605, W14 × 665, and heavier W14s. As a result, we use a cover-plated beam here.

Try a W27 × 146 ($d = 27.4$ in, $Z_X = 464 \text{ in}^3$, $b_f = 14.0$ in)
Assume 1-in-thick plates

$$\text{Total depth } d_r = d + 2t_p = 27.40 + (2)(1.00) = 29.40 \text{ in} < 29.50 \text{ in} \qquad \text{(OK)}$$

$$A_p = \frac{Z \text{ required} - Z_W}{d + t_p} = \frac{1217 - 464}{27.4 + 1.0} = 26.51 \text{ in}^2$$

A check of the $\frac{b}{t}$ ratios for the plates, web and flange of this section show them to be satisfactory.

Use W27 × 146 with one PL1 × 28 each flange (Each plate weighs $\left(\frac{28}{144}\right)(490) = 95.3 \text{ lb/ft}$)

Total wt $= 146 + (2)(95.3) = 337 \text{ lb/ft} < 350 \text{ lb/ft}$ \qquad (OK)

18.2   BUILT-UP WIDE-FLANGE SECTIONS

In this section, we initially define the terms *built-up wide-flange sections* and *plate girders* because the reader may quite understandably find it difficult to distinguish between the two.

In Chapter G of the LRFD Specification, a clear distinction is made between beams (whether rolled shapes or built-up wide-flange sections) and plate girders. It is stated that plate girders are to be distinguished from beams on the basis of web slenderness. This slenderness is measured by the ratio h/t_w. The letter h represents (1) the clear distance between flanges less two times the fillet or corner radius for rolled shapes, (2) the distance between adjacent lines of fasteners for built-up sections, or (3) the clear distance between flanges for built-up sections when welds are used. These values are illustrated in Fig. 18.3. Should h/t_w be greater than λ_r (that is, the width-thickness ratio above which a web is classified as being noncompact for flexural

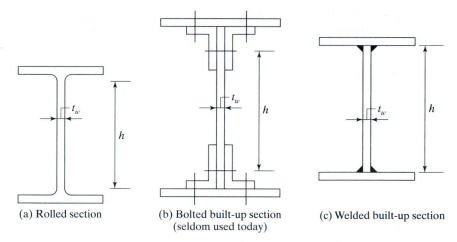

(a) Rolled section   (b) Bolted built-up section   (c) Welded built-up section
                             (seldom used today)

FIGURE 18.3

Values of h.

compression in LRFD Section B5), the member is classified as a plate girder and the provisions of Appendix G of the LRFD Specification apply. (Reference is made here to Fig. 18.4.)

A section is classified as a plate girder if h/t_w for its web is $>5.70\sqrt{E/F_y}$ (Other limiting $\frac{h}{t_w}$ values apply for webs that are in combined flexural and axial compression or for members that have different flange sizes. For such situations it is necessary to see LRFD Section B5 and Appendix B5.1.)

You will note in applying the $5.70\sqrt{E/F_y}$ ratio that we use the yield stress of the flange F_{yf} instead of the yield stress of the web F_{yw} because for hybrid girders inelastic buckling of the web due to bending is dependent upon the strain in the flange.

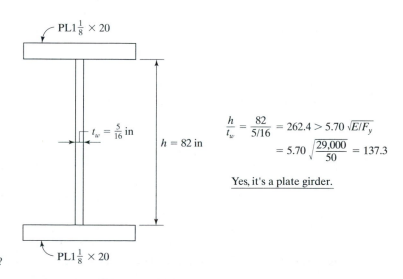

$$\frac{h}{t_w} = \frac{82}{5/16} = 262.4 > 5.70\sqrt{E/F_y}$$

$$= 5.70\sqrt{\frac{29,000}{50}} = 137.3$$

Yes, it's a plate girder.

FIGURE 18.4

Is it a plate girder if $F_{yf} = 50$ ksi?

With the LRFD Specification, overall economy often can be obtained when built-up wide-flange sections are used instead of plate girders. With these sections, the webs selected are sufficiently thick to carry shear without buckling. Even though these sections with their unstiffened webs will be heavier than plate girders for the same spans and loads, their overall costs will often be less because of their smaller fabrication costs. In addition, design calculations will be appreciably reduced.

According to LRFD Specification F1.3, plastic analysis is permitted for compact beams and girders if they meet certain conditions. They must be singly or doubly symmetric and loaded in the plane of symmetry. When bent about their major axes the laterally unbraced lengths of their compression flanges at plastic hinge locations associated with failure mechanisms must not exceed certain values as given in Chapter F of the Specification.

According to Table B5-1 of the LRFD Specification, a section is compact if

$$\frac{b_f}{2t_f} \le 0.38\sqrt{\frac{E}{F_{yf}}} \quad \text{and} \quad \frac{h}{t_w} \le 3.76\sqrt{\frac{E}{F_{yf}}}$$

After a compact web size is selected for a built-up wide-flange member, the next step is to select the flange size. For this discussion reference is made to Fig. 18.5. The total design bending strength of the girder shown equals the bending strength of its web plus the bending strength of its flanges.

The plastic modulus of the entire girder equals the statical moment of the compression and tension areas of the web about the neutral axis plus the statical moment of the areas of both flanges about the neutral axis. After such an expression is written it can be solved for the required area of one of the flanges.

$$\text{Total } Z \text{ required for girder} = \frac{M_u}{\phi_b F_y}$$

$$Z \text{ furnished} = (t_w)\left(\frac{h}{2}\right)\left(\frac{h}{4}\right)(2) + (2A_p)\left(\frac{h}{2} + \frac{t_f}{2}\right) \tag{2}$$

$$\frac{M_u}{\phi_b F_y} = \frac{t_w h^2}{4} + A_p(h + t_f)$$

$$A_p \text{ required} = \frac{M_u}{\phi_b F_y(h + t_f)} - \frac{t_w h^2}{4(h + t_f)}$$

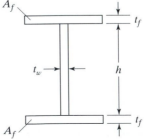

FIGURE 18.5

Example 18-2

Design a 60-in-deep welded built-up wide-flange section with no intermediate stiffeners for a 70-ft simple span to support a service dead load of 1.1 k/ft and a service live load of 3 k/ft. The section is to be framed between columns (thus no end-bearing stiffeners are required as described in Section 18-8) and is to have full lateral bracing for its compression flange. The design is to be made with 50 ksi steel, SMAW fillet welds, and E70 electrodes.

Solution. Maximum shear and moment

$$\text{Assume beam wt} = 200 \text{ lb/ft}$$
$$w_u = (1.2)(1.3) + (1.6)(3) = 6.36 \text{ k/ft}$$
$$M_u = \frac{(6.36)(70)^2}{8} = 3895.5 \text{ ft-k}$$
$$V_u = (35)(6.36) = 222.6 \text{ k}$$

Design for a compact web and flange

$$Z \text{ required} = \frac{M_u}{\phi F_y} = \frac{(12)(3895.5)}{(0.9)(50)} = 1039 \text{ in}^3$$

Web size required

For web to be compact h/t_w must be

$$\leq 5.70\sqrt{\frac{E}{F_{yf}}} = 5.70\sqrt{\frac{29{,}000}{50}} = 137.3$$

Assume $h = 60 - 2 = 58$ in

$$\text{Min. } t_w = \frac{58}{137.3} = 0.422 \text{ in, say } \tfrac{7}{16} \text{ in}$$

Try $\tfrac{7}{16} \times 58$ web ($A_w = 25.38 \text{ in}^2$)

$$\frac{h}{t_w} = \frac{58}{\tfrac{7}{16}} = 132.6$$

Since this value is $>2.45\sqrt{\frac{E}{F_{yf}}} = 59$ (see LRFD Appendix G-4), transverse stiffeners to be discussed in Section 18.8 may be needed. However, if $V_u \leq \phi V_n$ as determined from LRFD Appendix Equation A-F2-3 and A-G3-3, stiffeners will not be necessary. (This topic is discussed in Section 18-5.)

Instead of substituting into the above-mentioned formulas it is possible to determine the value of $\phi_v V_n/A_w$ from Table 9-50 in Part 16 of the Manual with $a/h > 3.0$ and $h/t_w = 132.6$ (a is the clear distance between transverse stiffeners).

$$\frac{\phi_v V_n}{A_w} = 6.77 \text{ ksi}$$

$$\frac{V_u}{A_w} = \frac{222.6}{25.38} = 8.77 \text{ ksi} > 6.77 \text{ ksi}$$

$\therefore$ Intermediate stiffeners are required. However if web t increased from 7/16 in to 1/2 in stiffeners can be shown to be unnecessary. Though this would require several hundred pounds of additional steel in the web it would surely be less expensive than the installation of intermediate stiffeners.

$\therefore$ use $\frac{1}{2} \times 58$ in web

Design of Flange:

$$A_f = \frac{M_u}{\phi_b F_y (h + t_f)} - \frac{t_w h^2}{4(h + t_f)}$$

Assume $\frac{3}{4}$-in plates

$$A_f = \frac{(12)(3895.5)}{(0.9)(50)(58 + 0.75)} - \frac{\left(\frac{1}{2}\right)(58)^2}{(4)(58 + 0.75)} = 10.52 \text{ in}^2$$

Try $3/4 \times 14$ plate each flange (10.5 in^2). Is Flange compact?

$$\frac{b_f}{2t_f} = \frac{14}{(2)(0.75)} = 9.33 > 0.38\sqrt{\frac{E}{F_{yf}}}$$

$$= 0.38\sqrt{\frac{29,000}{50}} = 9.15 \therefore \text{ Its noncompact}$$

Try $13/16 \times 14$ plate each flange (11.375 in^2)

$$\frac{b_f}{2t_f} = \frac{14}{(2)\left(\frac{13}{16}\right)} = 8.62 < 9.15 \tag{OK}$$

Check Z of section

$$Z = (2)\left(\frac{58}{2}\right)\left(\frac{1}{2}\right)\left(\frac{58}{4}\right) + (2)(14)\left(\frac{13}{16}\right)\left(\frac{58}{2} + \frac{13}{32}\right) = 1089.5 \text{ in}^3$$

$$> 1039 \text{ in}^3 \tag{OK}$$

Use $\frac{1}{2} \times 58$ web and one $13/16 \times 14$ plate each flange

18.3   INTRODUCTION TO PLATE GIRDERS

Plate girders are large I-shaped sections built up from plates and perhaps rolled sections. (See Fig. 18.4 of this chapter.) They usually have design strengths somewhere between those of rolled beams and steel trusses. Several possible arrangements are shown in Fig. 18.6. Rather obsolete bolted girders are shown in parts (a) and (b) of the figure, while several welded types are shown in parts (c) through (f). *Since nearly all plate girders constructed today are welded (although they may make use of bolted field splices), this chapter is devoted almost exclusively to welded girders.*

The welded girder of part (d) of Fig. 18.6 is arranged to reduce overhead welding compared with the girder of part (c), but in so doing may be creating a little worse corrosion situation if the girder is exposed to the weather. A box girder, illustrated in part (g), occasionally is used where moments are large and depths are quite limited. Box girders also have great resistance to torsion and lateral buckling. In addition, they make very efficient curved members because of their high torsional strengths.

Plates and shapes can be arranged to form plate girders of almost any reasonable proportions. This fact may seem to give them a great advantage for all situations, but for the smaller sizes the advantage usually is canceled by the higher fabrication costs. For example, it is possible to replace a W36 with a plate girder roughly twice as deep that will require considerably less steel and will have much smaller deflections; however, the higher fabrication costs will almost always rule out such a possibility.

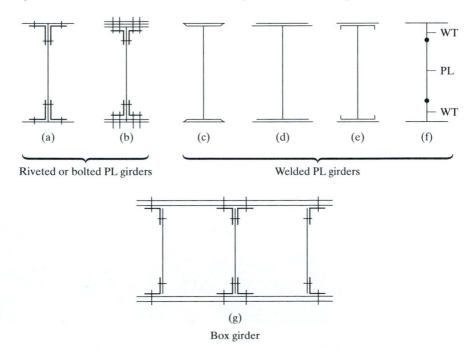

FIGURE 18.6

Most steel highway bridges built today for spans of less than about 80 ft are steel-beam bridges. For longer spans, the plate girder begins to compete very well economically. Where loads are extremely large, such as for railroad bridges, plate girders are competitive for spans as small as 45 or 50 ft.

The upper economical limits of plate girder spans depend on several factors, including whether the bridge is simple or continuous, whether a highway or railroad bridge is involved, and the largest section that can be shipped in one piece.

Generally speaking, plate girders are very economical for railroad bridges in spans of 50 to 130 ft (15 to 40 m), and for highway bridges in spans of 80 to 150 ft (24 to 46 m). However, they are often very competitive for much longer spans, particularly when continuous. In fact, they are actually common for 200-ft (61-m) spans and have been used for many spans in excess of 400 ft (122 m). The main span of the continuous Bonn-Beuel plate girder bridge over the Rhine River in Germany is 643 ft.

Plate girders are not only used for bridges. They also are fairly common in various types of buildings to support heavy concentrated loads. Frequently, a large ballroom or dining room with no interfering columns is desired on a lower floor of a multistory building. Such a situation is shown in Fig. 18.7. The plate girder shown must support some tremendous column loads for many stories above. The usual building plate girder of this type is simple to analyze because it probably does not have moving loads, although some building girders may be called upon to support traveling cranes.

The usual practical alternative to plate girders in the spans for which they are economical is the truss. In general, plate girders have the following advantages, particularly compared with trusses:

Buffalo Bayou Bridge, Houston, TX.—a 270-ft span. (Courtesy of the Lincoln Electric Company.)

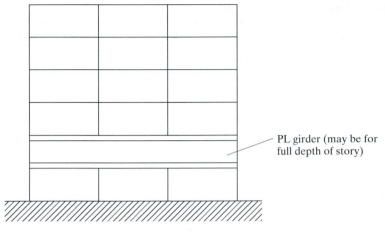

PL girder (may be for full depth of story)

FIGURE 18.7

1. The pound price for fabrication is lower than for trusses, but it is higher than for rolled beam sections.
2. Erection is cheaper and faster than for trusses.
3. Due to their compactness, vibration and impact are not serious problems.
4. Plate girders require smaller vertical clearances than trusses.
5. The plate girder has fewer critical points for stresses than do trusses.
6. A bad connection here or there is not as serious as in a truss, where such a situation could spell disaster.
7. There is less danger of injury to plate girders in an accident, compared with trusses. Should a truck run into a bridge plate girder, it would probably just bend it a little, but a similar accident with a bridge truss member could cause a broken member and perhaps failure.
8. A plate girder is more easily painted than a truss.

On the other hand, plate girders usually are heavier than trusses for the same spans and loads, and they have a further disadvantage in the large number of connections required between webs and flanges.

18.4   PLATE GIRDER PROPORTIONS

18.4.1   Depth

The depths of plate girders vary from about $\frac{1}{6}$ to $\frac{1}{15}$ of their spans, with average values of $\frac{1}{10}$ to $\frac{1}{12}$, depending on the particular conditions of the job. One condition that may limit the proportions of the girder is the largest size that can be fabricated in the shop and shipped to the job. There may be a transportation problem such as clearance requirements that limit maximum depths to 10 or 12 ft along the shipping route.

18.4.2 Web Size

After the total girder depth is estimated, the general proportions of the girder can be established from the maximum shear and the maximum moment. As previously described for I-shaped sections in Section 10-2, the web of a beam carries nearly all of the shearing stress; this shearing stress is assumed by the LRFD Specification to be uniformly distributed throughout the web. The web depth can be closely estimated by taking the total girder depth and subtracting a reasonable value for the depths of the flanges (roughly 1 to 2 in each). The web depths usually are selected to the nearest even inch, because these plates are not stocked in fractional dimensions.

As a plate girder bends, its curvature creates vertical compression in the web, as illustrated in Fig. 18.8. This is due to the downward vertical component of the compression flange bending stress and the upward vertical component of the tension flange bending stress.

The web must have sufficient vertical buckling strength to withstand the squeezing effect shown in Fig. 18.8. This problem is handled in Appendix G1 of the LRFD Specification by providing maximum permissible web slenderness h/t_w values. The values are dependent upon stiffener spacings as represented by a/h ratios, where a is the clear distance between transverse stiffeners. The purpose of these limiting values is to prevent the vertical buckling of a girder flange into the web before the yield stress is reached in the flange.

If transverse stiffeners are used and spaced no more than $1\frac{1}{2}$ times h, the maximum ratio is

For $a/h \leq 1.5$

$$\frac{h}{t_w} \leq 11.7\sqrt{\frac{E}{F_{yf}}} \qquad \text{(LRFD Equation A-G1-1)}$$

If no transverse stiffeners are used, or if they are used but are spaced at greater distances on center than $1\frac{1}{2}$ times h, the maximum ratio is

For $a/h > 1.5$

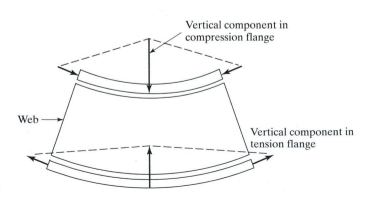

FIGURE 18.8

Squeezing of plate girder web.

$$\frac{h}{t_w} \leq \frac{0.48\,E}{\sqrt{F_{yf}(F_{yf} + 16.5)}} \qquad \text{(LRFD Equation A-G1-2)}$$

In metric units, this equation is

$$\frac{h}{t_w} \leq \frac{0.48\,E}{\sqrt{F_{yf}(F_{yf} + 114)}} \qquad \text{(LRFD Equation A-G1-2M)}$$

For this discussion, it is assumed that stiffener spacing is greater than 1.5 times the distance between the flanges. If we use 50 ksi steel our web may not have a depth-thickness or h/t_w ratio less than $5.70\sqrt{\frac{E}{F_y}} = 5.70\sqrt{\frac{29,000}{50}} = 137.3$ and still be classified as a plate girder, nor may it be larger than

$$\frac{0.48\,E}{\sqrt{F_{yf}(F_{yf} + 16.5)}} = \frac{(0.48)\,(29,000)}{\sqrt{50(50 + 16.5)}} = 241.4.$$

Thus, for a trial girder section using 50 ksi steel, a depth-thickness ratio for the web of somewhere between 137 and 241 is tried. From a corrosion standpoint the usual practice is to use some absolute minimum thickness. For bridge girders, $\frac{3}{8}$ in is a common minimum, while 1/4 or 5/16 in probably are the minimum values for the more sheltered building girders.

The shallower girders probably will be used when the loads are light, and the deeper ones when very large concentrated loads need to be supported, as from the columns in a tall building. If there are no depth restrictions for a particular girder it will probably pay the designer to make rough designs and corresponding cost estimates to arrive at a depth decision. (Computer solutions will be very helpful in preparing these alternate designs.)

18.4.3   Flange Size

After the web dimensions are selected, the next step is to select an area of the flange so that it will not be overloaded in bending. The total bending strength of a plate girder equals the bending strength of the flange plus the bending strength of the web. As almost all of the bending strength is provided by the flange, an approximate expression can be developed to estimate the flange area as follows:

$$A_f F_y h \approx \text{Say } 0.90\,M_n$$

Then

$$M_n \approx \frac{A_f F_y h}{0.90}$$

$$\phi M_n \approx \frac{\phi_b A_f F_y h}{0.90} = A_f F_y h \geq M_u$$

$$A_f \approx \frac{M_u}{F_y h}$$

White hot steel plates pass through giant presses in the plate mill of a Hamilton, Ontario. (Courtesy of Bethlehem Steel Corporation.)

After A_f is computed and trial dimensions are selected, the overall bending strength of the girder is checked as will be described in Sections 18-6 and 18-7.

18.5   DETAILED PROPORTIONS OF WEBS

18.5.1   Buckling Considerations

The depths of plate girders have been previously discussed in Section 18-4 and were said to have average values varying from 1/10 to 1/12 of spans. After the total girder depth is assumed, the web depth can be estimated to be from 2 to 4 in less than the total depth and selected to the nearest inch. A plate-girder web must have sufficient thickness to prevent *vertical buckling of the compression flange*. As the compression flange of a simply supported beam deflects or curves downward, it pushes against the web subjecting it to vertical compression. The amount of this downward force can be estimated to equal the total bending stress in the compression flange times the sine of the angle made by the curved flange with the horizontal.

The total load that can be applied to the web in this manner before the web buckles can be estimated by the Euler formula. Expressing this another way, the Euler formula can be used to determine a limiting height-to-thickness ratio to prevent buckling. Based on such a derivation (including an assumption for residual stress variation) the maximum ratios for webs given in the last section of this chapter were developed.

18.5.2 Web Shear

The LRFD Specification for plate girders permits their design on the basis of postbuckling strength. Designs on this basis provide a more realistic idea of the actual strength of a girder. (Such designs, however, do not necessarily result in better economy because more expensive stiffeners are required.) Should a girder be loaded until initial buckling occurs, it will not collapse because of a phenomenon known as *tension field action*.

After initial buckling a plate girder acts much like a truss. The web acts as a truss with tension diagonals and is able to resist additional shear. A diagonal strip of the web acts similarly to the diagonal of a parallel chorded truss. (See Fig. 18.9.) The stiffeners keep the flanges from coming together and the flanges keep the stiffeners from coming together. The intermediate stiffeners, which before initial buckling were assumed to resist no load, will after buckling resist compressive loads (or will serve as the compression verticals of a truss) due to diagonal tension. The result is that a plate-girder web probably can resist loads equal to two or three times those present at initial buckling before complete collapse will occur.

Until the web buckles initially, deflections are relatively small. However, after initial buckling, the girder's stiffness decreases considerably, and deflections may increase to several times the values estimated by the usual deflection theory.

The estimated total or ultimate shear that a panel (a part of the girder between a pair of stiffeners) can withstand equals the shear initially causing web buckling plus the shear that can be resisted by tension field action. The amount of tension field action is dependent on the proportion of the panels.

The design shear strength of a plate girder is equal to $\phi_v V_n$ with $\phi_v = 0.90$ and V_n equal to the appropriate value determined from the various formulas given in LRFD Appendix G3 as follows:

If $h/t_w \leq 1.10\sqrt{k_v E/F_{yw}}$

$$V_n = 0.6\,F_{yw}A_w \qquad\qquad \text{(LRFD Equation A-G3-1)}$$

If $h/t_w > 1.10\sqrt{k_v E/F_{yw}}$

$$V_n = 0.6F_{yw}A_w\left[C_v + \frac{1 - C_v}{1.15\sqrt{1 + (a/h)^2}}\right] \qquad \text{(LRFD Equation A-G3-2)}$$

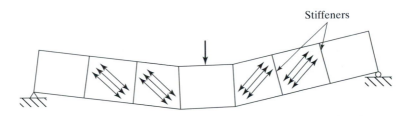

Stiffeners

FIGURE 18.9

Tension field action in plate-girder web (noting that end panels cannot develop tension field action).

For the end panels of nonhybrid girders, for all the panels of hybrid girders, for web-tapered girders, and when $a/h > 3.0$ or $> [260(h/t_w)]^2$, tension field action is not permitted and

$$V_n = 0.6 F_{yw} A_w C_v \qquad \text{(LRFD Equation A-G3-3)}$$

The terms k_v and C_v used in the preceding formulas are defined as follows:

1. k_v is a web plate-buckling coefficient

$$= 5 + \frac{5}{(a/h)^2} \qquad \text{(LRFD Equation A-G3-4)}$$

except that it is to be used $= 5.0$ if $a/h > 3.0$ or $> [260/(h/t_w)]^2$

2. C_v is the ratio of the "critical" web stress (according to the linear buckling theory) to the shear yield stress of the web material as follows:

If $1.10 \sqrt{\dfrac{k_v E}{F_{yw}}} \le \dfrac{h}{t_w} \le 1.37 \sqrt{\dfrac{k_v E}{t_{yw}}}$

$$C_v = \frac{1.10 \sqrt{k_v E/F_{yw}}}{h/t_w} \qquad \text{(LRFD Equation A-G3-5)}$$

If $\dfrac{h}{t_w} > 1.37 \sqrt{\dfrac{k_v E}{F_{yw}}}$

$$C_v = \frac{1.51 k_v E}{\left(\dfrac{h}{t_w}\right)^2 F_{yw}} \qquad \text{(LRFD Equation A-G3-6)}$$

18.6   DESIGN OF PLATE GIRDERS WITH SLENDER WEBS, BUT WITH FULL LATERAL BRACING FOR THEIR COMPACT COMPRESSION FLANGES

In Example 18-3, a plate girder with full lateral bracing for its compression flange is selected. As the girder proportions are not compact, a plastic analysis is not permissible. Transverse stiffeners are considered in detail later in this chapter, and thus their design is not included in this example or the next.

If we have a plate girder with a slender web (that is, with $h/t_w > 5.70 \sqrt{E/F_y}$) its flexural strength is the lesser value computed from the limit states of tension-flange yielding and buckling, as determined by the formulas to follow:

$$M_u = \phi_b M_n \quad \text{with} \quad \phi_b = 0.9$$

For tension-flange yielding

$$M_n = S_{xt} R_e F_{yt} \qquad \text{(LRFD Equation A-G2-1)}$$

For compression flange buckling

$$M_n = S_{xc} R_{PG} R_e F_{cr} \qquad \text{(LRFD Equation A-G2-2)}$$

Swinging a plate girder into place on Goat Island Bridge across the Niagara River, Niagara Falls, NY. (Courtesy of Bethlehem Steel Corporation.)

in which

$$R_{PG} = 1 - \frac{a_r}{1,200 + 300a_r}\left(\frac{h_c}{t_w} - 5.70\sqrt{\frac{E}{F_{cr}}}\right) \qquad \text{(LRFD Equation A-G2.3)}$$

R_e is a hybrid girder factor given in Appendix G2 of the Manual. It is to be taken as 1.0 for nonhybrid girders.

a_r is the ratio of the web area to the compression flange area ≤ 10.

F_{cr} is the critical compression flange stress as determined in Appendix G-2 of the Specification. It equals F_{yf} if the compression flange is adequately braced laterally and if the compression flange is compact, that is, if $\lambda \leq \lambda_p$ for lateral-torsional buckling and for local buckling respectively.

S_{xc} and S_{xt} are the section moduli values referred to the compression and tension flanges, respectively.

h_c = two times the distance from the centroid of the girder to the nearest line of fasteners at the compression flange, or the inside face of the compression flange if welds are used.

Example 18-3

Design a welded plate girder of 50 ksi steel for a 70-ft span to support a uniform service dead load of 2.5 k/ft and a uniform service live load of 5.7 k/ft. The compression flange will be laterally braced for its entire length. The girder is to be simply supported at its ends and is not framed into columns. Do not consider stiffener design.

Solution. Maximum shear and moment

Assume plate girder weight = 250 lb/ft

$$w_u = (1.2)(2.75) + (1.6)(5.7) = 12.42 \text{ k/ft}$$

$$M_u = \frac{(12.42)(70)^2}{8} = 7607 \text{ ft-k}$$

$$V_u = (35)(12.42) = 434.7$$

Preliminary web design

Assume girder depth = $\left(\frac{1}{10}\right)(12 \times 70) = 84$ in

Assume web depth = $84 - 2 = 82$ in

Minimum web t_w if we assume $a/h > 1.5$.

$$= \frac{82}{241.4} = 0.340 \text{ in}$$

The 241.4 was previously determined for 50 ksi steel in the web size part of section 18-4. Maximum t_w for noncompact web (to be classified as plate girder) $82/137.3 = 0.597$ in Select minimum $t_w = 3/8$ in minimum)

$$\frac{h}{t_w} = \frac{82}{3/8} = 218.7$$

Preliminary flange design

Use the following formula to approximate flange area

$$A_f \approx \frac{M_u}{F_y h} \approx \frac{(12)(7607)}{(50)(82)} = 22.26 \text{ in}^2$$

Try $1\frac{1}{8} \times 20$ plate ($A_f = 22.50 \text{ in}^2$). Is it compact?

$$\frac{b_f}{2t_f} = \frac{20.00}{(2)(1\frac{1}{8})} = 8.89 < \frac{65}{\sqrt{50}} = 9.19 \qquad \text{(YES)}$$

Checking girder weight

$$\text{Cross sectional area} = (2)(22.50) + \left(\frac{3}{8}\right)(82) = 75.75 \text{ in}^2$$

$$\text{Weight per ft} = \left(\frac{75.75}{144}\right)(490) = 258 \text{ lb/ft} = \text{assumed value} \qquad \text{(OK)}$$

Check trial girder section

$$I = \left(\tfrac{1}{12}\right)\left(\tfrac{3}{8}\right)(82)^3 + (2)\left(1\tfrac{1}{8} \times 20\right)(41.56)^2 = 94{,}956 \text{ in}^4$$

$$S_{xc} = S_{xt} = \frac{94{,}956}{42.125} = 2254 \text{ in}^3$$

$$\text{As } \frac{h}{t_w} = 218.7 > 5.70\sqrt{\frac{E}{F_y}} = 5.70\sqrt{\frac{29{,}000}{50}} = 137.2$$

$$a_r = \frac{\left(\tfrac{3}{8}\right)(82)}{\left(1\tfrac{1}{8}\right)(20)} = 1.367 < 10$$

$$R_{PG} = 1 - \frac{1.367}{1{,}200 + (300)(1.367)}\left(\frac{82}{3/8} - 5.70\sqrt{\frac{29{,}000}{50}}\right)$$

$$= 0.931 < 1.0$$

Design flexural strength for compression-flange buckling, $F_{cr} = F_{yf}$

$$M_n = S_{xc}R_{PG}R_eF_{cr}$$
$$= (2254)(0.931)(1.0)(50) = 104{,}924 \text{ in-k} = 8744 \text{ ft-k}$$

$$M_u = \phi_b M_n = (0.9)(8744) = 7870 \text{ ft-k} > 7607 \text{ ft-k} \qquad \text{(OK)}$$

Use $\tfrac{3}{8} \times 82$ web with one $1\tfrac{1}{8} \times 20$ plate for each flange $F_y = 50$ ksi

Note: A check of shear values for this girder will show that transverse stiffeners are required.

18.7   DESIGN OF PLATE GIRDERS WITH NONCOMPACT FLANGES AND WITHOUT FULL LATERAL BRACING FOR COMPRESSION FLANGES

As indicated in Section 18-6, the critical flange stress F_{cr} for plate girders with slender webs ($h/t_w > 970/\sqrt{F_{yf}}$), compact flanges, and with full lateral bracing supplied for the compression flange equals F_{yf}. Should these latter two conditions not be provided, it is necessary to consider the limit states of lateral-torsional buckling and local flange buckling as described below and to use *the lower value obtained*.

If $\lambda \leq \lambda_p$

$$F_{cr} = F_{yf} \qquad \text{(LRFD Equation A-G2-4)}$$

If $\lambda_p < \lambda \leq \lambda_r$

$$F_{cr} = C_bF_{yf}\left[1 - \frac{1}{2}\left(\frac{\lambda - \lambda_p}{\lambda_r - \lambda_p}\right)\right] \leq F_{yf} \qquad \text{(LRFD Equation A-G2-5)}$$

If $\lambda > \lambda_r$

$$F_{cr} = \frac{C_{PG}}{\lambda^2} \qquad \text{(LRFD Equation A-G2-6)}$$

For the preceding equations, it is necessary to determine the slenderness parameter for the limit state of lateral-torsional buckling as well as the limit state of local buckling. The slenderness parameter that results in the lowest value of F_{cr} governs.

For the limit state of lateral torsional buckling the following values are to be used in Formulas A-G2-5 and A-G2-6:

$$\lambda = \frac{L_b}{r_T} \qquad\qquad \text{(LRFD Equation A-G2-7)}$$

$$\lambda_p = 1.76\sqrt{\frac{E}{F_{yf}}} \qquad\qquad \text{(LRFD Equation A-G2-8)}$$

$$\lambda_r = 4.44\sqrt{\frac{E}{F_{yf}}} \qquad\qquad \text{(LRFD Equation A-G2-9)}$$

$$C_{PG} = \text{plate girder coefficient}$$

$$= 286,000C_b \qquad\qquad \text{(LRFD Equation A-G2-10)}$$

$$\text{Metric } C_{PG} = 1\,970\,000\,C_b \qquad\qquad \text{(LRFD Equation A-62-10M)}$$

$$C_b = \frac{12.5M_{\max}}{2.5M_{\max} + 3M_A + 4M_B + 3M_C} \qquad\qquad \text{(LRFD Equation F1-3)}$$

r_T = radius of gyration of compression flange plus one-third of the compression part of the web.

For the limit state of flange local buckling the following values are to be used:

$$\lambda = \frac{b_f}{2t_f} \qquad\qquad \text{(LRFD Equation A-G2-11)}$$

$$\lambda_p = 0.38\sqrt{\frac{E}{F_{yf}}} \qquad\qquad \text{(LRFD Equation A-G2-12)}$$

$$\lambda_r = 1.35\sqrt{\frac{E}{F_{yf}/k_c}} \qquad\qquad \text{(LRFD Equation A-G2-13)}$$

$$C_{PG} = 26,200\,k_c \qquad\qquad \text{(LRFD Equation A-G2-14)}$$

$$\text{Metric } C_{PG} = 180\,650\,K_C \qquad\qquad \text{(LRFD Equation A-62-14M)}$$

$$C_b = 1.0$$

In these expressions $k_c = 4/\sqrt{h/t_w}$ and $0.35 \le k_c \le 0.763$. (The LRFD Specification Appendix G2 states that the limit state of flange local buckling is not applicable in this case.)

Example 18-4 illustrates the selection of the cross section for a plate girder for which full lateral bracing is not provided for the compression flange.

Example 18-4

The cross section for a welded plate girder is to be selected for a 54-ft span with simple end supports to support a uniform service dead load of 2 k/ft and concentrated service live loads of 200 k each at its one-third points. The design is to be made with 50 ksi steel assuming that lateral bracing is provided for the compression flange only at the ends and one-third points. Do not design stiffeners.

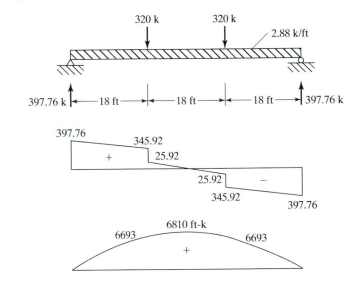

FIGURE 18.10

Solution. Maximum shear and moment

Assume beam weight = 400 lb/ft

$$w_u = (1.2)(2.4) = 2.88 \text{ k/ft}$$
$$P_u = (1.6)(200) = 320 \text{ k}$$

Drawing shear and moment diagrams, (Fig. 18-10)

Preliminary web design

Assume girder depth = $\left(\frac{1}{10}\right)(12 \times 54) = 64.8$ in,   say 64 in

Assume web depth = $64 - 2 = 62$ in

Minimum web t_w if we assume $\dfrac{a}{h} > 1.5 = \dfrac{62}{241.4} = 0.257$ in

Maximum t_w for noncompact web (to be classed as plate girder) = $\dfrac{62}{137.3} = 0.452$ in

Select $t_w = 3/8$ in noting a corrosion minimum of about 5/16 in

$$\frac{h}{t_w} = \frac{62}{3/8} = 165.3$$

Preliminary flange design
Use the following formula to approximate flange area

$$A_f = \frac{M_u}{F_{yf}h} = \frac{(12)(6810)}{(50)(62)} = 26.36 \text{ in}^2$$

Try $1\frac{3}{8} \times 20$ plate ($A_f = 27.5 \text{ in}^2$). Is it compact?

$$\frac{b_f}{2t_f} = \frac{20}{(2)(1.375)} = 7.27 < 0.38\sqrt{\frac{29{,}000}{50}} = 9.15 \qquad \text{(YES)}$$

Check trial girder section
Computing section modulus

$$I = \left(\tfrac{1}{12}\right)\left(\tfrac{3}{8}\right)(62)^3 + (2)\left(1\tfrac{3}{8}\right)(20)(31.689)^2 = 62{,}678 \text{ in}^4$$

$$S_{xc} = S_{xt} = \frac{62{,}678}{32.375} = 1936 \text{ in}^3$$

$$\frac{h}{t_w} = \frac{62}{3/8} = 165.3 > 5.70\sqrt{\frac{29{,}000}{50}} = 137.3$$

$\therefore$ Appendix G applies

Arc-welded plate girders, Louisiana State Fair Grounds, Shreveport, LA. (Courtesy of the Lincoln Electric Company.)

Properties of compression flange plus one-third of compression part of web

$$A_f + \tfrac{1}{6}A_w = (1\tfrac{3}{8})(20) + (\tfrac{1}{6})(62)(\tfrac{3}{8}) = 31.37 \text{ in}^2$$
$$I_y \approx (\tfrac{1}{12})(1.375)(20)^3 = 917 \text{ in}^4$$
$$r_T = \sqrt{\frac{917}{31.37}} = 5.41 \text{ n.}$$

Checking limitations of Appendix G. If $a/h \le 1.5$,

$$\frac{h}{t_w} \text{ maximum} = 11.7\sqrt{\frac{29{,}000}{50}} = 281.8 \ge 137.3 \qquad \text{(LRFD Equation A-G1-1)}$$

Checking bending design strength of center 18-ft section.

For the limit state of lateral-torsional buckling

$$\lambda = \frac{L_b}{r_T} = \frac{(12)(18)}{5.41} = 39.93 \qquad \text{(LRFD Equation A-G2-7)}$$

λ_p = limiting slenderness parameter for compact elements

$$= 1.76\sqrt{\frac{29{,}000}{50}} = 42.39 > 39.93 \qquad \text{(LRFD Equation A-G2-8)}$$

$$\therefore F_{er} = F_{yf} = 50 \text{ ksi}$$

For the limit state of flange local buckling

$$\lambda = \frac{b_f}{2t_f} = \frac{20}{(2)(1.375)} = 7.27 \qquad \text{(LRFD Equation A-G2-11)}$$

$$\lambda_p = 0.38\sqrt{\frac{29{,}000}{50}} = 9.15 > 7.27 \qquad \text{(LRFD Equation A-G2-12)}$$

$$\therefore F_{er} = 50 \text{ ksi}$$

Design flexural strength

$$a_r = \frac{\text{web area}}{\text{compression flange area}} = \frac{(3/8)(62)}{(1.375)(20)} = 0.845 < 10$$

$F_{cr} = 50$ as per LRFD Equation A-G2-4

$$R_{PG} = 1 - \frac{0.845}{1200 + (300)(0.845)}\left(\frac{2 \times 31}{0.375} - 5.70\sqrt{\frac{29{,}000}{50}}\right) = 0.984 < 1.0$$
$$\text{(LRFD Equation A-G2-3)}$$

$$M_n = (1936)(0.984)(1.0)(50) = 95{,}251 \text{ in-k} = 7938 \text{ ft-k}$$
$$\text{(LRFD Equation A-G2-2)}$$

$$M_u = \phi_b M_n = (0.9)(7938) = 7144 \text{ ft-k} > 6810 \text{ ft-k} \qquad \text{(OK)}$$

Check bending strength of end 18-ft section

$$M_{\max} = M @ 18 \text{ ft} = 6693 \text{ ft-k}$$
$$M_A = M @ 4.5 \text{ ft} = (397.76)(4.5) - (4.5)(2.88)(2.25) = 1761 \text{ ft-k}$$
$$M_B = M @ 9.0 \text{ ft} = (397.76)(9.0) - (9.0)(2.88)(4.5) = 3463 \text{ ft-k}$$
$$M_C = M @ 13.5 \text{ ft} = (397.76)(13.5) - (13.5)(2.88)(6.75) = 5107 \text{ ft-k}$$

$$C_b = \frac{(12.5)(6693)}{(2.5)(6693) + (3)(1761) + (4)(3463) + (3)(5107)}$$
$$= 1.63 \qquad \text{(LRFD Equation F1-3)}$$

Obviously, bending strength in this segment is satisfactory

For the limit state of lateral-torsional buckling

$$\lambda = \frac{L_b}{r_T} = \frac{(12)(18)}{5.41} = 39.93 \qquad \text{(LRFD Equation A-G2-7)}$$

$$\lambda_p = 1.76\sqrt{\frac{29{,}000}{50}} = 42.39 > 39.93 \qquad \text{(LRFD Equation A-G2-8)}$$

$$\therefore F_{cr} = 50 \text{ ksi}$$

For the limit state of flange local buckling. As it was for the center 18-ft section $F_{cr} = F_{yf} = 50$ ksi and $R_{PG} = 0.9$

$$\phi_b M_n = 7144 \text{ ft-k} > 6810 \text{ ft-k} \qquad \text{(OK)}$$

Use $\frac{3}{8} \times 62$ web and one $1\frac{3}{8} \times 20$ plate each flange, $F_y = 50$ ksi.

18.8   DESIGN OF STIFFENERS

As previously indicated, it usually is necessary to stiffen the high thin webs of plate girders to keep them from buckling. If the girders are bolted, the stiffeners probably will consist of a pair of angles bolted to the girder webs. If the girders are welded (the normal situation), the stiffeners probably will consist of a pair of plates welded to the girder webs. Figure 18-11 illustrates these types of stiffeners.

Stiffeners are divided into two groups: *bearing stiffeners*, which transfer heavy reactions or concentrated loads to the full depth of the web, and *intermediate or non-bearing stiffeners*, which are placed at various intervals along the web to prevent buckling due to diagonal compression. An additional purpose of bearing stiffeners is the transfer of heavy loads to plate girder webs without putting all the loads on the flange connections.

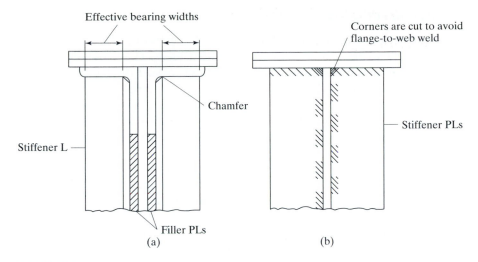

FIGURE 18.11

(a) Angle bearing stiffeners. (b) Plate-bearing stiffeners.

As described in Appendix F2.3 of the LRFD Specification, transverse stiffeners may be either single or double. They do not have to be connected to the flanges except in the following situations:

1. Where bearing strength is needed to transmit concentrated loads or reactions.
2. Where single stiffeners are used and the flange of the girder consists of a rectangular plate. For such a situation the stiffener must be attached to the flange to resist any possible uplift tendency that may be caused by torsion in the flange.
3. Where lateral bracing is attached to a stiffener or stiffeners. Such a stiffener must be connected to the compression flange with a strength sufficient to transmit no less than 1 percent of the total flange stress unless the flange is composed only of angles.

Welds used to attach stiffeners to girder webs must be terminated at distances not less than 4 times the web thickness nor more than 6 times the web thickness from the near toe of the flange to the web weld.

If bolts are used to connect stiffeners to the web, they may not be spaced further apart than 12 inches. If intermittent fillet welds are used, the clear distance between them may not exceed the lesser of 16 times the web thickness or 10 inches.

Stiffeners that are located at concentrated loads or reactions have some special requirements because of the possibility of web crippling or compression buckling of the web. In these situations the stiffeners need to be designed as columns. Should the load or reaction be tensile, it will be necessary to weld the stiffeners to the loaded flange. If the force is compressive the stiffener may either bear against the loaded flange or be welded to it.

18.8.1 Bearing Stiffeners

Bearing stiffeners are placed in pairs on the webs of plate girders at unframed girder ends and where required for concentrated loads. They should fit tightly against the flanges being loaded and should extend out toward the edges of the flange plates or angles as far as possible. If the load normal to the girder flange is tensile, the stiffeners must be welded to the loaded flange. If the load is compressive it is necessary to obtain a snug fit, that is, a good bearing between the flange and the stiffeners. To accomplish this goal the stiffeners may be welded to the flange or the outstanding legs of the stiffeners may be milled.

A bearing stiffener is a special type of column that is difficult to accurately analyze because it must support the load in conjunction with the web. The amount of support provided by these two elements is difficult to estimate. The LRFD Specification (K1.9) states that the factored load or reaction may not exceed the design strength of a column consisting of the stiffener effective area plus a portion of the web equal to $12\,t_w$ at girder ends and $25\,t_w$ at interior concentrated loads. Only the part of the stiffeners outside of the fillets of the flange angles or the parts outside of the flange to web welds (see Fig. 18.11) are to be considered effective to support the bearing loads. The effective length of these "bearing stiffener columns" is assumed by the LRFD Specification (K1.9) as being equal to $0.75\,h$.

At an unframed girder end an end-bearing stiffener is required if the factored reaction R_u is larger than ϕR_n. If an interior load or reaction is larger than the same ϕR_n an interior bearing stiffener is needed.

If the concentrated force is applied at a distance from the member end greater than the member depth d

$$R_n = (5k + N)F_{yw}t_w \qquad \text{(LRFD Equation K1-2)}$$

If the concentrated force is applied at a distance from the member end less than or equal to the member depth d

$$R_n = (2.5k + N)F_{yw}t_w \qquad \text{(LRFD Equation K1-3)}$$

In these expressions

k = distance from outer face of flange to web toe of fillet, in.
N = length of bearing (not less than k for end reaction), in.
ϕ = 1.0

18.8.2 Intermediate Stiffeners

Intermediate or nonbearing stiffeners that also are called stability or transverse intermediate stiffeners are not required by LRFD Specification Appendix G4 if $h/t_w \leq 2.45\sqrt{\frac{E}{F_{yw}}}$ or if the factored shear V_u is $\leq 0.6\phi_v A_w F_{yw}C_v$ where C_v is computed with the appropriate expression to follow, using a value of $k_v = 5.0$ and $\phi_v = 0.90$.

If $1.10\sqrt{\dfrac{k_v E}{F_{yw}}} \leq \dfrac{h}{t_w} \leq 1.37\sqrt{\dfrac{k_v E}{F_{yw}}}$

$$C_v = \frac{1.10\sqrt{k_v E/F_{yw}}}{h/t_w} \qquad \text{(LRFD Equation A-G3-5)}$$

If $\dfrac{h}{t_w} > 1.37\sqrt{\dfrac{k_v E}{F_{yw}}}$

$$C_v = \frac{1.51\, k_i\, E}{(h/t_w)^2\, F_{yw}} \qquad \text{(LRFD Equation A-G3-6)}$$

The LRFD Specification also imposes some additional arbitrary limits on panel aspect ratios (a/h) for plate girders, even where shear stresses are small. The purpose of these limitations is to facilitate the handling of girders during fabrication and erection. The Specification (Appendix G3) states that tension field action is not permitted for end panels in nonhybrid plate girders, for all panels in hybrid and web tapered plate girders, and when $a/h > 3.0$ or $> [260/(h/t_w)]^2$ and that

$$V_n = 0.6\, F_{yw} A_w C_v \qquad \text{(LRFD Equation A-G3-3)}$$

You should note here that as the intermediate stiffener spacing is decreased, C_v will become larger, as will the shear capacity of the girder.

The LRFD Specification (Appendix F2.3) states that the moment of inertia of a transverse intermediate stiffener about an axis at the center of the girder web if a pair of stiffeners is used, or about the face in contact with the web when single stiffeners are used, may not be less than

$$I_{st}\ \text{minimum} = a t_w^3 j$$

$$\text{where } j = \frac{2.5}{(a/h)^2} - 2 \geq 0.5 \qquad \text{(LRFD Equation A-F2-4)}$$

The Appendix to the LRFD Specification (G4) states that, when designing for tension field action, the stiffener area A_{st} may not be less than the value given by the following formula:

$$A_{st}\ \text{minimum} = \frac{F_{yw}}{F_{yst}}\left[0.15\, D h t_w (1 - C_v)\frac{V_u}{\phi_v V_n} - 18\, t_w^2 \right] \geq 0$$

$$\text{(LRFD Equation A-G4-1)}$$

In the preceding formula, D is to be taken equal to 1.0 for stiffeners in pairs, 1.8 for single-angle stiffeners, and 2.4 for single-plate stiffeners. The terms C_v and V_n are defined in Appendix G3, while V_u is the factored shear at the section in question.

The values given for I_{st} minimum, j, and A_{st} minimum are the same as those required by the AASHTO in its specification for load factor design for steel bridges.[1]

---

[1]AASHTO LRFD Bridge Design Specifications, 1994 (Washington, DC: American Association of State Highway and Transportation Officials); p.6–98 to 6–99.

Highway bypass bridge, Stroudsburg, PA. (Courtesy of Bethlehem Steel Corporation.)

18.8.3 Longitudinal Stiffeners

Longitudinal stiffeners, though not as effective as transverse ones, frequently are used for bridge plate girders because many designers feel they are more attractive. As such stiffeners are seldom used for plate girders in buildings, they are not covered in this textbook.

18.9 FLEXURE–SHEAR INTERACTION

When a plate girder whose web depends on tension field action is subjected to relatively large shear and bending moment at the same location, the girder cannot provide its full capacity either in shear or in moment. As a result, an empirical interaction equation is used to check the adequacy of the girder. This equation, which follows, agrees quite well with test results.

If $0.6\,\phi V_n \leq V_u \leq \phi V_n$, and if $0.75\,\phi M_n \leq M_u \leq \phi M_n$ with $\phi = 0.90$, plate girders designed for tension field action also must satisfy the following interaction equation.

$$\frac{M_u}{\phi M_n} + 0.625\,\frac{V_u}{\phi V_n} \leq 1.375 \quad \text{(LRFD Equation A-G5-1)}$$

In the preceding equation, M_n is the nominal flexural strength of the girder as determined by Appendix G2, ϕ is 0.90, and V_n is the nominal shear strength of the girder as determined by Appendix G3. The value of M_u may not exceed ϕM_n, nor may V_u exceed ϕV_n, with $\phi = 0.90$ in both cases.

Bearing and intermediate stiffeners are designed in Example 18-5 for the plate girder of Example 18-4. In addition, the girder is checked for flexure-shear interaction at several locations.

Example 18-5

Design bearing and intermediate stiffeners for the plate girder of Example 18-4.

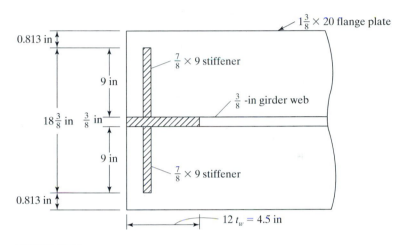

FIGURE 18.12

End bearing stiffeners.

Solution. End Bearing Stiffeners

    a.   Are end bearing stiffeners required?

Assume point bearing of the end reaction (that is, $N = 0$) and assume a $\frac{5}{16}$-in web-to-flange fillet weld.

    Check local web yielding

$$k = \text{distance from outer edge of flange to web end of fillet weld}$$
$$= 1\tfrac{3}{8} + \tfrac{5}{16} = 1.69 \text{ in}$$
$$R_n = (5k + N)F_{yw}t_w \qquad\qquad\qquad \text{(LRFD Equation K1-2)}$$
$$\phi R_n = (1.0)(5 \times 1.69 + 0)(50)\left(\tfrac{3}{8}\right) = 158.4 \text{ k}$$
$$< 397.76 \text{ k} \therefore \text{ End bearing stiffeners are required}$$

If local web yielding check had been satisfactory (that is, $\phi R_n \geq 397.76$ k), it would have then been necessary also to check the web crippling criteria set forth in Sections K1.4 and K1.5 of the LRFD Specification before we could definitely say that end bearing stiffeners were not required.

b. Design of end bearing stiffeners

Try two plate stiffeners 7/8 × 9 as shown in Fig. 18.12.

Check width-thickness ratio (Table B5.1 Specification)

$$\frac{9.00}{0.875} = 10.29 < 0.56\sqrt{\frac{29{,}000}{50}} = 13.49 \qquad \text{(OK)}$$

Check column strength of stiffener (area is shown crosshatched in following sketch)

$$I \approx \left(\tfrac{1}{12}\right)\left(\tfrac{7}{8}\right)(18.375)^3 = 452 \text{ in}^4$$

(neglecting the almost infinitesimal contribution of the web)

$$A \text{ of column} = (2)\left(\tfrac{7}{8}\right)(9) + (4.5)\left(\tfrac{3}{8}\right) = 17.44 \text{ in}^2$$

$$r = \sqrt{\frac{452}{17.44}} = 5.09 \text{ in}$$

$$KL = (0.75)(62) = 46.5 \text{ in}$$

$$\frac{KL}{r} = \frac{46.5}{5.09} = 9.14$$

$\phi F_{cr} = 42.29$ ksi from Table 3-50 of Part 16 of the Manual or can be determined using LRFD Formulas E2-2 or E2-3 as applicable.

$$\phi P_n = (42.29)(17.44) = 738 \text{ k} > 397.76 \text{ k} \qquad \text{(OK)}$$

Check bearing criterion

$$\phi R_n = (0.75)(2.0 \, F_y A_{pd})$$

$A_{pd} =$ projected bearing area of stiffeners subtracting 1/2 in from the 11-in width of each stiffener for the cutouts for the web-to-flange welds as shown in Fig. 18-11 (b).

$$A_{pd} = (2)(9 - 0.5)\left(\tfrac{7}{8}\right) = 14.875 \text{ in}^2$$
$$\phi R_n = (0.75)(2.0)(50)(14.875) = 1116 \text{ k} > 397.76 \text{ k} \qquad \text{(OK)}$$

Use two plates $\tfrac{7}{8} \times 9 \times 5$ ft $1\tfrac{3}{4}$ in for bearing stiffeners   (a   depth   of $61\tfrac{3}{4}$ in is used instead of 62 in for fitting purposes)

c. Design of interior bearing stiffeners

The design procedure for interior bearing stiffeners is the same as for end bearing stiffeners, except a $25\,t_w$ length of the web may be counted as part of the "column," as compared with the $12\,t_w$ value used for the end stiffeners. Here the same size stiffeners are used at the concentrated loads as at the end reactions.

Intermediate Stiffeners

a.   Are intermediate stiffeners required?

Appendix G4 of the LRFD Specification states that intermediate or transverse stiffeners are not required if $\frac{b}{t_w} \leq 2.45\sqrt{\frac{E}{F_{yw}}}$ or if the required shear V_u, as determined by structural analysis for the factored loads, is $\leq 0.6\phi_v F_{yw} A_w C_v$, with C_v determined with $k_v = 5$ and $\phi = 0.9$. We have

$$\frac{h}{t_w} = 165.33 > 1.37\sqrt{\frac{(5)(29{,}000)}{50}} = 73.78$$

$$C_v = \frac{(1.51)(5)(29{,}000)}{(165.33)^2(50)} = 0.160 \qquad \text{(LRFD Equation A-G3-6)}$$

$$V_u @ \text{ end} = 397.76 > (0.6)(0.90)(50)\left(\tfrac{3}{8} \times 62\right)(0.160) = 100.44 \text{ k}$$

∴ Intermediate stiffeners are required

b.   Is tension field action permitted?

Tension field action is not permitted in the end panels of nonhybrid girders according to LRFD Appendix G3.

c.   Spacing of intermediate stiffeners by trial and error

The author assumes a certain distance from the end of the girder to the first stiffener, computes ϕV_n to see if it's sufficiently large to resist the computed shear V_u, tries another spacing, and so on until $\phi V_n \geq V_u$.

After trials of 72 in, 54 in, 48 in, and 45 in (all being unsatisfactory), a 36-in distance is tried below.

Try 36 in to first stiffener

$$\frac{a}{h} = \frac{36}{62} = 0.581$$

$$k_v = 5 + \frac{5}{(0.581)^2} = 19.81 \qquad \text{(LRFD Equation A-G3-4)}$$

$$\text{As } \frac{h}{t_w} = 165.33 > 1.37\sqrt{\frac{(19.81)(29{,}000)}{50}} = 146.84$$

$$C_v = \frac{(1.51)(19.81)(29{,}000)}{(165.33)^2(50)} = 0.635 \qquad \text{(LRFD Equation A-G3-6)}$$

$$V_n = (0.6) \left(\tfrac{3}{8} \times 62\right) (50) (0.635) = 442.91 \; k$$

$$\phi V_n = (0.90) (442.9) = 398.61 > 397.76 \; k \qquad \text{(OK)}$$

Place first stiffener 36 in from girder end

The shear V_u has fallen off slightly in this 36 in from the girder end, as can be seen in Fig. 18.10. [It's $397.76 - \left(\tfrac{36}{12}\right)(2.88) = 389.12$ k.] Thus, we can try a slightly larger spacing and go through these steps again to determine ϕV_n. It is much simpler, however, to use Table 9.50 in Part 16 of the LRFD Manual to determine stiffener spacings, and this is done below.

d. Spacings of intermediate stiffeners using LRFD Manual

$$\frac{\phi_v V_n}{A_w} \geq \frac{V_u}{A_w} = \frac{397.76}{(3/8)(62)} = 17.11 \; \text{ksi}$$

$$\frac{h}{t_w} = \frac{62}{3/8} = 165.33$$

From Table 9.50 in Part 16 of the Manual, $\tfrac{a}{h}$ can be determined to equal 0.586 by double interpolation.

$$a_1 = (0.586)(62) = 36.33 \; \text{in}$$

Use 36 in to first stiffener

Distance to second stiffener

$$V_u \text{ at first interior stiffener} = 397.26 - \left(\tfrac{36}{12}\right)(2.88) = 389.12 \; k$$

$$\frac{V_u}{A_w} = \frac{389.12}{3/8 \times 62} = 16.74 \; \text{ksi}$$

$\dfrac{a_2}{h}$ by double interpolation $= 0.594 \qquad a_2 = (0.594)(62) = 36.8 \; \text{in.}$

$\therefore$ Spacing continues to increase

Place first stiffener at 36 in and remainder of intermediate or transverse stiffeners in the end 18-ft. panel at $\frac{(12)(18) - 36}{5} = 36$ in O.C. (See Fig. 18.13.)

e. Check center 18-ft. panel (as per LRFD Specification Appendix G4).

$$\frac{h}{t_w} = 165.33 > 1.37\sqrt{\frac{(5)(29{,}000)}{50}} = 73.77$$

$$C_v = \frac{(1.51)(5)(29{,}000)}{(165.33)^2(50)} = 0.1602$$

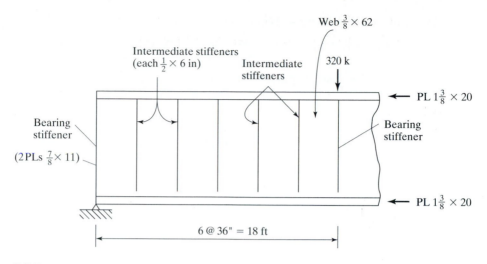

FIGURE 18.13

Stiffener spacing diagram.

$$\phi V_n = 0.6\,\phi_v F_{yw} A_w C_v = (0.6)(0.9)(50)\left(\frac{3}{8} \times 62\right)(.1602)$$

$$= 100.57\ \text{k} > 25.92\ \text{k}\quad\text{from Figure 18.10}$$

$\therefore$ Stiffeners are not required in center panel.

f. Design of intermediate stiffeners

Stiffener design with no tension field action means that we have to comply with the minimum moment of inertia expression, $at^3_w j$ given in LRFD Appendix F2.3.

$$\frac{a}{h} = \frac{36}{62} = 0.581$$

$$j = \frac{2.5}{(a/h)^2} - 2 - 0.5 \qquad\qquad\text{(LRFD Equation A-F2-4)}$$

$$j = \frac{2.5}{(0.581)^2} - 2 = 5.41 > 0.5 \qquad\qquad\text{(OK)}$$

Min $I_{st} = at^3_w j = (36)\left(\frac{3}{8}\right)^3(3.45) = 6.55\ \text{in}^4$

There are many possible satisfactory stiffener sizes. We will try two of them by trial and error.

Try a $\frac{1}{2} \times 6$ single plate stiffener

$$I_{st} = \left(\tfrac{1}{3}\right)\left(\tfrac{1}{2}\right)(6)^3 = 36\ \text{in}^4 > 6.55\ \text{in}^4 \qquad\qquad\text{(OK)}$$

Try a pair of $\frac{1}{2} \times 4\frac{1}{2}$ plate stiffeners

$$I_{st} = \left(\tfrac{1}{12}\right)\left(\tfrac{1}{2}\right)(9.375)^3 = 34.33 \text{ in}^4 > 6.55 \text{ in}^4 \qquad\qquad \text{(OK)}$$

Use $\frac{1}{2} \times 6$ single plate stiffeners

PROBLEMS

18-1.   Select a cover-plated W section limited to a maximum depth of 19.00 in to support the service loads shown in the accompanying figure. Use A36 steel and assume the beam has full lateral bracing for its compression flange. (*Ans.* W16 × 30 with 1 PL $1\frac{1}{2} \times 14$ in each flange. Many other satisfactory designs are available.)

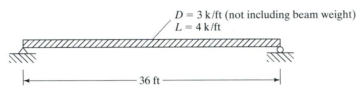

FIGURE P18-1

18-2.   The architects specify that a cover-plated beam no greater than 16.00 in. in depth be designed to support a dead service uniform load of 4 klf, not including the beam weight, and a live service uniform load of 6 klf for a 24-ft simple span. A572 steel with $F_y = 50$ ksi is to be used, and full lateral bracing for the compression flange is to be provided.

18-3.   Design a 48-in-deep welded built-up wide-flange section with no intermediate stiffeners for a 50-ft simple span to support a service dead load of 1 klf (not including the beam weight) and a service live load of 1.8 klf. The section is to be framed between columns and is to have full lateral bracing for its compression flange. The design is to be made with A36 steel. (*Ans.* One solution 5/8 × 10 in PL each flange)

18-4.   Repeat Prob. 18-3 if the span is to be 60 ft and $F_y = 50$ ksi.

18-5.   Design a welded plate girder with A36 steel for a 60-ft span to support a uniform service dead load of 2.0 klf (not including the girder weight) and a service live load of 4.5 klf. The compression flange will be laterally braced for its entire length. The girder is to be simply supported at its ends and is not framed into the columns. Do not design stiffeners. (*Ans.* One solution 5/16 × 70 in web with 1 PL × 22 each flange)

C H A P T E R   1 9

Design of Steel Buildings

19.1 INTRODUCTION TO LOW-RISE BUILDINGS

The material in this section and the next pertains to the design of low-rise steel buildings up to several stories in height, while Sections 19.3 to 19.10 provide information concerning common types of building floors, and Section 19.11 presents common types of roof construction. Sections 19.12 and 19.13 are concerned with walls, partitions, and fireproofing. The sections thereafter present general information relating to high-rise or multistory buildings.

The low-rise buildings considered include apartment houses, office buildings, warehouses, schools, and institutional buildings that are not very tall with respect to their least lateral dimensions. The separating factor between the buildings discussed here and those mentioned in the multistory sections is the matter of wind forces and not the actual height of the building in question. The usual rule of thumb is that if the height of the building is not greater than twice its least lateral dimension, provision for wind forces is unnecessary. For buildings of these dimensions, the walls and partitions probably provide sufficient resistance to wind forces, except in unusual circumstances. Following such a rule of thumb, however, is a dangerous engineering practice. Each structure should be considered on the basis of its dimensions, location, surrounding structures, and so on.

19.2 TYPES OF STEEL FRAMES USED FOR BUILDINGS

Steel buildings usually are classified as being in one of four groups according to their type of construction: *bearing-wall* construction, *skeleton* construction, *long-span* construction, and *combination steel and concrete framing*. More than one of these construction types can be used in the same building. We discuss each of these types briefly in the paragraphs that follow.

Coliseum in Spokane, WA. (Courtesy
of Bethlehem Steel Corporation.)

19.2.1 Bearing-Wall Construction

Bearing-wall construction is the most common type of single-story light commercial
construction. The ends of beams or joists or light trusses are supported by the walls that
transfer the loads to the foundation. The old practice was to rapidly thicken load-bearing
walls as buildings became taller. For instance, the wall on the top floor of a building
might be one or two bricks thick while the lower walls might be increased by one brick
thickness for each story as we come down the building. As a result, this type of construction was usually thought to have an upper economical limit of about two or three
stories although some load bearing buildings were much higher. The tallest load bearing building built in the U.S. in the 19th century was the 17 story Monadnock building
in Chicago. This building completed in 1891 had 72 in. thick walls on its first floor. A
great deal of research has been conducted for load-bearing construction in recent
decades, and it has been discovered that thin load-bearing walls may be quite economical
for many buildings up to 10 or 20 or even more stories.

The average engineer is not very well versed in the subject of bearing-wall
construction, with the result that he or she may often specify complete steel or reinforced-concrete frames where bearing-wall construction might have been just as
satisfactory and more economical. Bearing-wall construction is not very resistant to seismic loadings, and has an erection disadvantage for buildings of more than one story. For
such cases, it is necessary to place the steel floor beams and trusses floor by floor as the
masons complete their work below, thus requiring alternation of the masons and
ironworkers.

Bearing plates usually are necessary under the ends of the beams, or light
trusses that are supported by the masonry walls, because of the relatively low bearing
strength of the masonry. Although theoretically the beam flanges may on many

occasions provide sufficient bearing without bearing plates, plates are almost always used—particularly where the members are so large and heavy that they must be set by a steel erector. The plates usually are shipped loose and set in the walls by the masons. Setting them in their correct positions and at the correct elevation is a critical part of the construction. Should they not be properly set, there will be some delay in correcting their positions. If a steel erector is used, he or she probably will have to make an extra trip to the job.

When the ends of a beam are enclosed in a masonry wall, some type of wall anchor is desirable to prevent the beam from moving longitudinally with respect to the wall. The usual anchors consist of bent steel bars passing through beam webs. These so-called *government anchors* are shown in Fig. 19.1(a). Occasionally, clip angles attached to the web are used instead of government anchors. These are shown in Fig. 19.1(b). Should longitudinal loads of considerable magnitude be anticipated, regular vertical anchor bolts may be used at the beam ends.

For small commercial and industrial buildings, bearing-wall construction is quite economical when the clear spans are not greater than roughly 35 or 40. If the clear spans are much greater, it becomes necessary to thicken the wall and use pilasters to ensure stability. For these cases it may often be more economical to use intermediate columns if permissible.

19.2.2 Skeleton Construction

In skeleton construction the loads are transmitted to the foundations by a frame work of steel beams and columns. The floor slabs, partitions, exterior walls, and so on, all are supported by the frame. This type of framing, which can be erected to tremendous heights, often is referred to as beam-and-column construction.

In beam-and-column construction the frame usually consists of columns spaced 20, 25, or 30 ft apart, with beams and girders framed into them from both directions at each floor level. One very common method of arranging the members is shown in

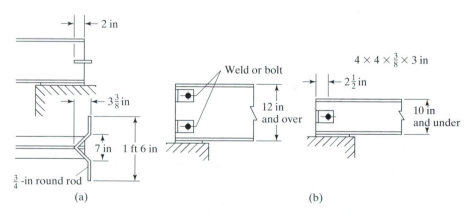

FIGURE 19.1

(a) Government anchor (b) Angle wall anchors. From American Institute Steel Construction, *Manual of Steel Construction Load & Resistance Factor Design*, 2 ed. (Chicago: AISC, 1994), p. 12-24. Reprinted with the permission of AISC.

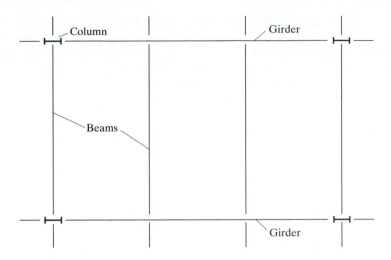

FIGURE 19.2

Beam-and-column construction.

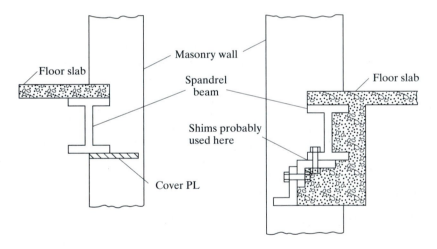

FIGURE 19.3

Spandrel beams.

Fig. 19.2. The girders are placed in the longer direction between the columns, while the beams are framed between the girders in the short direction. With various types of floor construction, other arrangements of beams and girders may be used.

For skeleton framing the walls are supported by the steel frame and generally are referred to as *nonbearing* or *curtain walls*. The beams supporting the exterior walls are called *spandrel beams*. These beams, illustrated in Fig. 19.3, usually can be placed so that they will serve as the lintels for the windows.

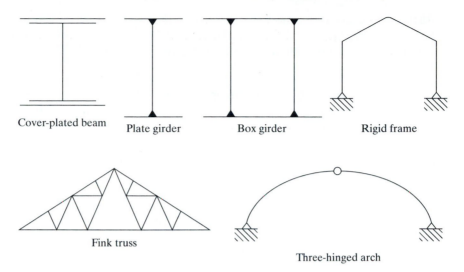

FIGURE 19.4

Long-span structures.

19.2.3 Long-Span Steel Structures

When it becomes necessary to use very large spans between columns—such as for field houses, auditoriums, theaters, hangars, or hotel ballrooms—the usual skeleton construction may not be sufficient. Should the ordinary rolled *W* sections be insufficient, it may be necessary to use cover-plated beams, plate girders, box girders, large trusses, arches, rigid frames, and the like. When depth is limited, cover-plated beams, plate girders, or box girders may be called upon to do the job. Should depth not be so critical, trusses may be satisfactory. For very large spans, arches and rigid frames often are used. These various types of structures are referred to as *long-span structures*. Figure 19.4 shows a few of these types of structures.

19.2.4 Combination Steel and Concrete Framing

A large percentage of the buildings erected today make use of a combination of reinforced concrete and structural steel. When reinforced-concrete columns are used in very tall buildings, they are rather large on the lower floors and take up considerable space. Steel column shapes surrounded by and bonded to reinforced concrete may be used and are referred to as *combination* or *composite* columns. These were discussed in Chapter 17.

19.3   COMMON TYPES OF FLOOR CONSTRUCTION

Concrete floor slabs of one type or another are used almost universally for steel-frame buildings. They are strong and have excellent fire ratings and good acoustic ratings. On the other hand, appreciable time and expense are required to provide the formwork necessary for most slabs. Concrete floors are heavy, they must include some type of reinforcing bars or mesh, and there may be a problem involved in making them watertight. The following are some of the types of concrete floors used today for steel-frame buildings:

1. Concrete slabs supported with open-web steel joists (Section 19.4)
2. One-way and two-way reinforced-concrete slabs supported on steel beams (Section 19.5)
3. Concrete slab and steel beam composite floors (Section 19.6)
4. Concrete-pan floors (Section 19.7)
5. Steel-decking floors (Section 19.8)
6. Flat slab floors (Section 19.9)
7. Precast concrete slab floors (Section 19.10)

Among the several factors to be considered in selecting the type of floor system to be used for a particular building are loads to be supported; fire rating desired; sound and heat transmission; dead weight of floor; ceiling situation below (to be flat or have beams exposed); facility of floor for locating conduits, pipes, wiring, and so on; appearance; maintenance required; time required to construct; and depth available for floor.

You can get a lot of information about these and other construction practices by referring to various engineering magazines and catalogs, particularly *Sweet's Catalog File*, published by McGraw-Hill Information Systems Company. We cannot make too strong a recommendation for these books to help the student see the tremendous amount of data available. The sections to follow present brief descriptions of the floor systems mentioned in this section, along with some discussions of their advantages and chief uses.

19.4   CONCRETE SLABS ON OPEN-WEB STEEL JOISTS

Perhaps the most common type of floor slab in use for small steel-frame buildings is the slab supported by open-web steel joists. The joists are small parallel chord trusses whose members usually are made from bars (hence, the common name *bar joist*) or small angles or other rolled shapes. Steel forms or decks are usually attached to the joists by welding or self-drilling or self-tapping screws, then concrete slabs are poured on top. This is one of the lightest types of concrete floors and also one of the most economical. A sketch of an open-web joist floor is shown in Fig. 19.5.

Open-web joists are particularly well suited to building floors with relatively light loads and for structures where there is not too much vibration. They have been used a great deal for fairly tall buildings, but generally speaking, they are better suited for shorter buildings. Open-web joists are very satisfactory for supporting floor and roof slabs for schools, apartment houses, hotels, office buildings, restaurant buildings, and similar low-level buildings. These joists generally are not suitable for supporting concentrated loads, however, unless they are especially detailed to carry such loads.

Open-web joists must be braced laterally to prevent them from twisting or buckling and to keep the floors from becoming too springy. Lateral support is provided by

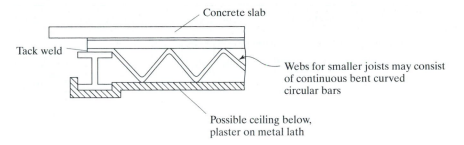

FIGURE 19.5

Open-web joists.

bridging, which consists of continuous horizontal rods fastened to the top and bottom chords of the joists, or of diagonal cross bracing. Bridging is desirably used at spaces not exceeding 7 ft on centers.

Open-web joists are easy to handle, and they are quickly erected. If desired, a ceiling can be attached to the bottom or suspended from the joists. The open spaces in the webs are well suited for placing conduits, ducts, wiring, piping, and so on. The joists should be either welded to supporting steel beams, or well anchored in the masonry walls. When concrete slabs are placed on top of the joists, they are usually from 2 to $2\frac{1}{2}$ in thick. Nearly all of the many concrete slabs cast in place or precast on the market today can be used successfully on top of open-web joists.

Detailed information concerning open-web joists is not included in the LRFD Manual. Such data can be obtained from a publication by the Steel Joist Institute entitled *Standard Specifications, Load Tables, and Weight Tables for Steel Joists and Joist Girders*.[1] At the present time these specifications are based on an allowable stress method rather than LRFD.

Three categories of joists are presented in *Standard Specifications: Open-Web Joists* (called the K-series), *the Longspan Steel Joists* (LH-series) and the *Deep Longspan Steel Joists* (DLH-series). These three types of joists are designed as simply supported trusses which themselves support uniform loads on their top chords. Should concentrated loads need to be supported, a special analysis should be requested from the manufacturer.

One page of the Steel Joist Institute's tables for the K-series joists is presented as Table 19.1 of this chapter. As an illustration of the use of this table, look across the top headings of the columns until you reach the 12K3 joists. This member has a nominal depth of 12 in and an approximate weight of 5.7 lb/ft. If we now move vertically down the column under the 12K3 heading until we reach a 20-ft span, as shown on the left side of the table, we can see that this joist can support a total service load equal to 302 lb/ft. The number given below the 302 is 177 lb/ft. It represents the live uniform

---

[1]Myrtle Beach, SC: Steel Joist Institute, 1996.

Table 19.1 Standard Load Table Open Web Steel Joists, K-Series Based on a Maximum Allowable Tensile Stress of 30,000 psi

Joist Designation	8K1	10K1	12K1	12K3	12K5	14K1	14K3	14K4	14K6	16K2	16K3	16K4	16K5	16K6	16K7	16K9
Depth (in)	8	10	12	12	12	14	14	14	14	16	16	16	16	16	16	16
Approx. Wt. (lb/ft.)	5.1	5.0	5.0	5.7	7.1	5.2	6.0	6.7	7.7	5.5	6.3	7.0	7.5	8.1	8.6	10.0
Span (ft.) ↓																
8	550															
	550															
9	550															
	550															
10	550	550														
	480	550														
11	532	550														
	377	542														
12	444	550	550	550	550											
	288	455	550	550	550											
13	377	479	550	550	550											
	225	363	510	510	510											
14	324	412	500	550	550	550	550	550	550							
	179	289	425	463	463	550	550	550	550							
15	281	358	434	543	550	511	550	550	550							
	145	234	344	428	434	475	507	507	507							
16	246	313	380	476	550	448	550	550	550	550	550	550	550	550	550	550
	119	192	282	351	396	390	467	467	467	550	550	550	550	550	550	550
17		277	336	420	550	395	495	550	550	512	550	550	550	550	550	550
		159	234	291	366	324	404	443	443	488	526	526	526	526	526	526
18		246	299	374	507	352	441	530	550	456	508	550	550	550	550	550
		134	197	245	317	272	339	397	408	409	456	490	490	490	490	490
19		221	268	335	454	315	395	475	550	408	455	547	550	550	550	550
		113	167	207	269	230	287	336	383	347	386	452	455	455	455	455
20		199	241	302	409	284	356	428	525	368	410	493	550	550	550	550
		97	142	177	230	197	246	287	347	297	330	386	426	426	426	426
21			218	273	370	257	322	388	475	333	371	447	503	548	550	550
			123	153	198	170	212	248	299	255	285	333	373	405	406	406
22			199	249	337	234	293	353	432	303	337	406	458	498	550	550
			106	132	172	147	184	215	259	222	247	289	323	351	385	385
23			181	227	308	214	268	322	395	277	308	371	418	455	507	550
			93	116	150	128	160	188	226	194	216	252	282	307	339	363
24			166	208	282	196	245	295	362	254	283	340	384	418	465	550
			81	101	132	113	141	165	199	170	189	221	248	269	298	346
25						180	226	272	334	234	260	313	353	384	428	514
						100	124	145	175	150	167	195	219	238	263	311
26						166	209	251	308	216	240	289	326	355	395	474
						88	110	129	156	133	148	173	194	211	233	276
27						154	193	233	285	200	223	268	302	329	366	439
						79	98	115	139	119	132	155	173	188	208	246
28						143	180	216	265	186	207	249	281	306	340	408
						70	88	103	124	106	118	138	155	168	186	220
29										173	193	232	261	285	317	380
										95	106	124	139	151	167	198
30										161	180	216	244	266	296	355
										86	96	112	126	137	151	178
31										151	168	203	228	249	277	332
										78	87	101	114	124	137	161
32										142	158	190	214	233	259	311
										71	79	92	103	112	124	147

load that will cause this joist to have an approximate deflection equal to 1/360th of the span. The joists given in the Institute tables run all the way from the 8K1 to the 72DLH19. This later joist will support a total load of 497lb/ft for a 120 ft span.

The Steel Joists Institute actually presents a fourth category of joists called Joist Girders. These are quite large joists which may be used to support open web joists.

19.5   ONE-WAY AND TWO-WAY REINFORCED-CONCRETE SLABS

19.5.1   One-Way Slabs

A very large number of concrete floor slabs in old office and industrial buildings consisted of one-way slabs about 4 in. thick, supported by steel beams 6 to 8 ft on centers. These floors were often referred to as *concrete arch* floors because at one time brick or tile floors were constructed in approximately the same shape—that is, in the shape of arches with flat tops.

A one-way slab is shown in Fig. 19.6. The slab spans in the short direction shown by the arrows in the figure. One-way slabs usually are used when the long direction is two or more times the short direction. In such cases the short span is so much stiffer than the long span that almost all of the load is carried by the short span. The short direction is the main direction of bending and will be the direction of the main reinforcing bars in the concrete, but temperature and shrinkage steel is needed in the other direction.

A typical cross section of a one-way slab floor with supporting steel beams is shown in Fig. 19.7. When steel beams or joists are used to support reinforced-concrete floors, it may be necessary to encase them in concrete or other materials to provide the required fire rating. Such a situation is shown in the figure.

It may be necessary to leave steel lath protruding from the bottom flanges or soffits of the beam for the purposes of attaching plastered ceilings. Should such ceilings be required to cover the beam stems, this floor system will lose a great deal of its economy.

One-way slabs have an advantage when it comes to formwork because the forms can be supported entirely by the steel beams with no vertical shoring needed. They have a disadvantage in that they are much heavier than most of the newer lightweight

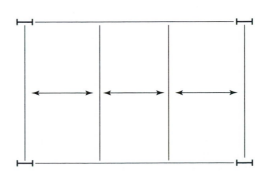

FIGURE 19.6

One-way slab.

Short-span joists in the Univac Center of Sperry-Rand Corporation, Blue Bell, PA. (Courtesy of Bethlehem Steel Corporation.)

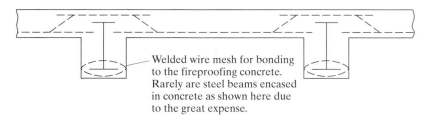

Welded wire mesh for bonding to the fireproofing concrete. Rarely are steel beams encased in concrete as shown here due to the great expense.

FIGURE 19.7

Cross section of one-way slab floor.

floor systems. The result is that they are not used as often as formerly for lightly loaded floors; but when a floor to support heavy loads is needed, or a rigid floor, or a very durable floor, the one-way slab may be the appropriate selection.

19.5.2 Two-Way Concrete Slabs

The two-way concrete slab is used when the slabs are square or nearly so, with supporting beams under all four edges. The main reinforcing runs in both directions. Other characteristics are similar to those of the one-way slab.

19.6   COMPOSITE FLOORS

Composite floors, previously discussed in Chapter 16, have steel beams (rolled sections, cover-plated beams, or built-up members) bonded together with concrete slabs in such a manner that the two act as a unit in resisting the total loads that the beam sections would otherwise have to resist alone. The steel beams can be smaller when composite floors are used, because the slabs act as part of the beams.

A particular advantage of composite floors is that they utilize concrete's high compressive strength by keeping all or nearly all of the concrete in compression, and at the same time stress a larger percentage of the steel in tension than is normally the case with steel-frame structures. The result is less steel tonnage in the structure. A further advantage of composite floors is an appreciable reduction in total floor thickness, which is particularly important in taller buildings.

Two types of composite floor systems are shown in Fig. 19.8. The steel beam can be completely encased in the concrete and the horizontal shear transferred by friction and bond (plus some shear reinforcement if necessary). *This type of composite floor is usually quite uneconomical.* The usual type of composite floor is shown in part (b), where the steel beam is bonded to the concrete slab with some type of shear connectors. Various types of shear connectors have been used during the past few decades, including spiral bars, channels, angles, studs, and so on, but economic considerations usually lead to the use of round studs welded to the top flanges of the beams in place of the other types mentioned. Typical studs are 1/2 to 3/4 in in diameter and 2 to 4 in in length.

Cover plates may occasionally be welded to the bottom flanges of rolled steel sections. The student can see that with the slab acting as a part of the beam there is quite a large area available on the compressive side of the beam. By adding plates to the tensile flange a little better balance is obtained.

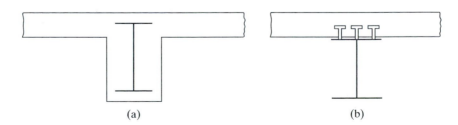

(a)   (b)

FIGURE 19.8

Composite floors. (a) Steel beam encased in concrete (very expensive), (b) steel beam bonded to concrete slab with shear connectors.

The 104-ft joists for the Bethlehem Catholic High School, Bethlehem, PA. (Courtesy of Bethlehem Steel Corporation.)

19.7   CONCRETE-PAN FLOORS

There are several types of pan floors that are constructed by placing concrete in removable pan molds. (Some special light corrugated pans also are available that can be left in place.) Rows of the pans are arranged on wooden floor forms, and the concrete is placed over the top of them, producing a floor cross section such as the one shown in Fig. 19.9. Joists are formed between the pans, giving a tee-beam-type floor.

These floors, which are suitable for fairly heavy loads, are appreciably lighter than the one-way and two-way concrete slab floors. They require a good deal of formwork, including appreciable shoring underneath the stems. Labor in thus higher than for many floors, but saving due to weight reduction and reuse of standard-size pans

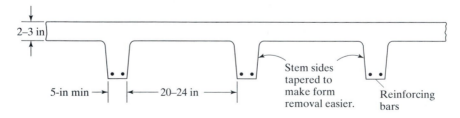

FIGURE 19.9

Concrete-pan floor.

One-piece metal dome pans. (Courtesy of Gateway Erectors, Inc.)

Temple Plaza parking facility, Salt Lake City, UT. (Courtesy of Ceco Steel Products Corporation.)

may make them competitive. If suspended ceilings are required, pan floors will have a decided economic disadvantage.

Two-way construction is available—that is, with ribs or stems running in both directions. Pans with closed ends are used, and the result is a waffle-type floor. This type of floor usually is used when the floor panels are square or nearly so. Two-way construction can be obtained for reasonably economical prices and makes a very attractive ceiling below with fairly good acoustical properties.

19.8 STEEL-DECKING FLOORS

Typical cross sections of steel-decking floors are shown in Fig. 19.10; several other variations are available. *Today, formed steel decking with a concrete topping is by far the most common type of floor system used for office and apartment buildings.* It also is popular for hotels and other buildings where the loads are not very large.

A particular advantage of steel-decking floors is that the decking immediately forms a working platform. The light steel sheets are quite strong and can span up to 20 ft. or more. Due to the considerable strength of the decking, the concrete does not have to be particularly strong, which permits the use of lightweight concrete often as thin as 2 or $2\frac{1}{2}$-in.

The cells in the decking are convenient for conduits, pipes, and wiring. The steel usually is galvanized, and if exposed underneath, it can be left as it comes from the manufacturer or painted as desired. Should fire resistance be necessary, a suspended ceiling with metal lath and plaster may be used. The same is necessary if a flat ceiling is required below for the types of decking shown in Fig. 19.10, but decking is available with a flat soffit.

19.9 FLAT SLABS

Formerly, flat slab floors were limited to reinforced-concrete buildings, but today it is possible to use them in steel-frame buildings. A flat slab is reinforced in two or more directions and transfers its loads to the supporting columns without the use of beams and girders protruding below. The supporting concrete beams and girders are made so wide that they are the same depth as the slab.

Flat slabs are of great value when the panels are approximately square, when more headroom is desired than is provided with the normal beam and girder floors, when heavy loads are anticipated, and when we want to place the windows as near to

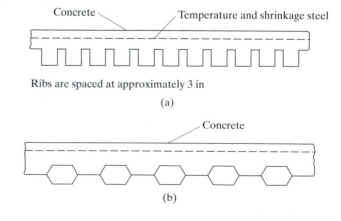

Concrete　Temperature and shrinkage steel

Ribs are spaced at approximately 3 in

(a)

Concrete

(b)

FIGURE 19.10

Steel-decking floors.

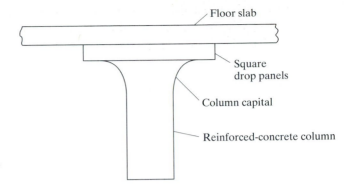

Floor slab

Square drop panels

Column capital

Reinforced-concrete column

FIGURE 19.11

A flat slab floor for a reinforced-concrete building.

the tops of the walls as possible. Another advantage is the flat ceiling produced for the floor below. Although the large amounts of reinforcing steel required cause additional costs, the simple formwork cuts expenses decidedly. The significance of simple formwork will be understood when it is realized that over one-half of the cost of the average poured reinforced-concrete floor slab is in the formwork.

For some reinforced-concrete frame buildings with flat slab floors, it is necessary to flare out the tops of the columns, forming column capitals, and perhaps thicken the slab around the column with the so-called drop panels. These items, which are shown in Fig. 19.11, may be necessary to prevent shear or punching failures in the slab around the column.

It is now possible in steel-frame buildings to use short steel cantilever beams connected to the steel columns and embedded in the slabs. These beams serve the purposes of the flared columns and drop panels in ordinary flat slab construction. This arrangement often is called a *steel grillage* or *column head*.

The flat slab is not a very satisfactory type of floor system for the usual tall building where lateral forces (wind or earthquake) are appreciable, because protruding beams and girders are desirable to serve as part of the lateral bracing system.

19.10   PRECAST CONCRETE FLOORS

Precast concrete sections are more commonly associated with roofs than they are with floor slabs, but their use for floors is increasing. They are quickly erected and reduce the need for formwork. Lightweight aggregates are often used in the concrete, making the sections light and easy to handle. Some of the aggregates used make the slabs nailable and easily cut and fitted on the job. For floor slabs with their fairly heavy loads, the aggregates should be of a quality that will not greatly reduce the strength of the resulting concrete.

The reader again is referred to *Sweet's Catalog File*. In these catalogs a great amount of information is available on the various types of precast floor slabs on the market today. Some of the common types available are listed here, and a cross section of each type mentioned is presented in Fig. 19.12. Due to slight variations in the upper surfaces of precast sections, it usually is necessary to use a mortar topping of 1 to 2 in before asphalt tile or other floor coverings can be installed.

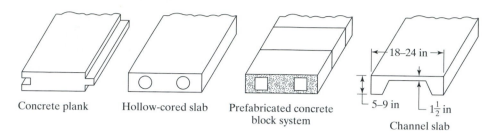

Concrete plank   Hollow-cored slab   Prefabricated concrete   Channel slab
                                       block system
18–24 in
5–9 in $1\frac{1}{2}$ in

FIGURE 19.12

Precast concrete roof and floor slabs.

1. *Precast concrete planks* are roughly 2 to 3 in thick and 12 to 24 in wide and are placed on joists up to 5 or 6 ft on center. They usually have tongue-and-groove edges.
2. *Hollow-cored slabs* are sections roughly 6 to 8 in deep and 12 to 18 in wide and have hollow, perhaps circular, cores formed in a longitudinal direction, thus reducing their weights by approximately 50 percent (as compared with solid slabs of the same dimensions). These sections, which can be used for spans of from roughly 10 to 25 ft, may be prestressed.

101 Hudson St., Jersey City, NJ. (Courtesy of Owen Steel Company, Inc.)

Interns' living quarters, Good Samaritan
Hospital, Dayton, OH. (Courtesy of
Flexicore Company, Inc.)

3. *Prefabricated concrete block systems* are slabs made by tying together precast concrete blocks with steel rods (may also be prestressed). These seldom-used floors vary from 4 to 8 in deep and can be used for spans of roughly 8 to 32 ft.

4. *Channel slabs* are used for spans of approximately 8 to 24 ft. Rough dimensions of these slabs are given in Fig. 19.12.

19.11 TYPES OF ROOF CONSTRUCTION

The types of roof construction commonly used for steel-frame buildings include concrete slabs on open-web joists, steel-decking roofs, and various types of precast concrete slabs. For industrial buildings the roof system in predominant use today is one that involves the use of a cold-rolled steel deck. In addition, the other types mentioned in the preceding sections on floor types are occasionally used, but they usually are unable to compete economically. Among the factors to be considered in selecting the specific types of roof construction are strength, weight, span, insulation, acoustics, appearance below, and type of roof covering to be used.

The major differences between floor-slab selection and roof-slab selection probably occur in the considerations of strength and insulation. The loads applied to roofs generally are much smaller than those applied to floor slabs, thus permitting the use of many types of lightweight aggregate concretes that may be appreciably weaker. Roof slabs should have good insulation properties, or they will have to have insulation materials placed on them and covered by the roofing. Among the many types of lightweight aggregates used are wood fibers, zonolite, foams, sawdust, gypsum, and expanded shale.

Although some of these materials decidedly reduce concrete strengths, they provide very light roof decks with excellent insulating properties.

Precast slabs made with these aggregates are light, are quickly erected, have good insulating properties, and usually can be sawed and nailed. For poured concrete slabs on open-web joists, several lightweight aggregates work very well (zonolite, foams, gypsum, etc.), and the resulting concrete easily can be pumped up to the roofs, thereby facilitating construction. By replacing the aggregates with certain foams, concrete can be made so light it will float in water (for a while). Needless to say, the strength of the resulting concrete is quite low.

Steel decking with similar thin slabs of lightweight and insulating concrete placed on top make very good economical roof decks. A competitive variation consists of steel decking with rigid insulation board placed on top, followed by the regular roofing material. The other concrete-slab types are hard pressed to compete economically with these types for lightly loaded roofs. Other types of poured concrete decks require much more labor.

19.12 EXTERIOR WALLS AND INTERIOR PARTITIONS

19.12.1 Exterior Walls

The purposes of exterior walls are to provide resistance to atmospheric conditions, including insulation against heat and cold, satisfactory sound-absorption and light-refraction characteristics, sufficient strength, and fire ratings. They also should look good and yet be reasonably economical.

For many years, exterior walls were constructed of some type of masonry, glass, or corrugated sheeting. Recently, however, the number of satisfactory materials for exterior walls has increased tremendously. Precast concrete panels, insulated metal sheeting, and many other prefabricated units are commonly used today. Of increasing popularity are the light prefabricated sandwich panels consisting of three layers. The exterior surface is made from aluminum, stainless steel, ceramics, plastics, and other materials. The center of the panel consists of some type of insulating material, such as fiberglass or fiberboard, while the interior surface is made from metal, plaster, masonry, or some other attractive material.

19.12.2 Interior Partitions

The main purpose of interior partitions is to divide the inside space of a building into rooms. Partitions are selected for appearance, fire rating, weight, or acoustical properties. Partitions may be loadbearing or nonloadbearing.

Bearing partitions support gravity loads in addition to their own weights and are thus permanently fixed in position. They can be constructed from wood or steel studs or from masonry units and faced with plywood, plaster, wallboard, or other material.

Nonbearing partitions do not support any loads in addition to their own weights and thus may be fixed or movable. Selection of the type of material to be used for a partition is based on the answers to the following questions: Is the partition to be fixed or movable? Is it to be transparent or opaque? Is it to extend all the way to the ceiling? Is it to be used to conceal piping and electrical conduits? Are there fire-rating and acoustical requirements? The more common types are made of metal, masonry, or concrete. For design of the floor slabs in a building that has movable partitions, some allowance should be given to the fact that the partitions may be moved. Probably the most common practice is to increase the floor-design live load by 15 or 20 psf over the entire floor area.

19.13   FIREPROOFING OF STRUCTURAL STEEL

Although structural steel members are incombustible, their strength is tremendously reduced at temperatures normally reached in fires when the other materials of a building burn. Many disastrous fires have occurred in empty buildings where the only fuel was the buildings themselves. Steel is an excellent heat conductor, and nonfireproofed steel members may transmit enough heat from one burning compartment of a building to ignite materials with which they are in contact in adjoining sections of the building.

The fire resistance of structural steel members can be greatly increased by coating them with fire-protective covers such as concrete, gypsum, mineral fiber sprays, special paints, and other materials. The thickness and kind of fireproofing used depends on the type of structure, the degree of fire hazard, and economics.

In the past, concrete was commonly used for fire protection. Although concrete is not a particularly good insulation material, it is very satisfactory when applied in thicknesses of $1\frac{1}{2}$ to 2 in or more, due to its mass. Furthermore, the water in concrete (16 to 20 percent when fully hydrated) improves its fireproofing qualities appreciably. The boiling off of the water from the concrete requires a great deal of heat. As the water or steam escapes, it greatly reduces the concrete temperature. This is identical to the behavior of steel steam boilers, where the steel temperature is held to maximum values of only a few hundred degrees Fahrenheit due to the escape of the steam. It is true, however, that in very intense fires the boiling off of the water may cause severe cracking and spalling of the concrete.

Although concrete is an everyday construction material and in mass it is a quite satisfactory fireproofing material, its installation cost is extremely high and its weight is large. As a result, for most steel construction, spray-on fireproofing materials have almost completely replaced concrete.

The spray-on materials usually consist of either mineral fibers or cementious fireproofing materials. The mineral fibers formerly used were asbestos. Due to the health hazards associated with this material, its use has been discontinued, and other fibers now are used by manufacturers. The cementious fireproofing materials are composed of gypsum, perlite, vermiculite, and others. Sometimes, when plastered ceilings are required in a building it is possible to hang the ceilings and light-gage furring channels by wires from the floor systems above and to use the plaster as the fireproofing.

The cost of fireproofing structural steel buildings is high and hurts steel in its economic competition with other materials. As a result, a great deal of research is being conducted by the steel industry on imaginative new fireproofing methods. Among these ideas is the coating of steel members with expansive and insulative paints. When heated to certain temperatures, these paints will char, foam, and expand, forming an insulative shield around the members. These swelling or extumescant paints are very expensive.

Other techniques involve the isolation of some steel members outside of the building, where they will not be subject to damaging fire exposure, and the circulation of liquid coolants inside box- or tube-shaped members in the building. It is probable that major advances will be made with these and other fireproofing methods in the near future.[2]

19.14   INTRODUCTION TO HIGH-RISE BUILDINGS

In this introductory text, we do not discuss the design of multistory buildings in detail, but the material of the next few sections gives the student a general idea of the problems involved in the design of such buildings, without presenting elaborate design examples.

Office buildings, hotels, apartment houses, and other buildings of many stories are quite common in the United States, and the trend is toward an even larger number of tall buildings in the future. Available land for building in our heavily populated cities is becoming scarcer and scarcer, and costs are becoming higher and higher. Tall buildings require a smaller amount of this expensive land to provide required floor space. Other factors contributing to the increased number of multistory buildings are new and better materials and construction techniques.

On the other hand, several factors that may limit the heights to which buildings will be erected in the future include the following:

1. Certain city building codes prescribe maximum heights for buildings.
2. Foundation conditions may not be satisfactory for supporting buildings of many stories.
3. Floor space may not be rentable above a certain height. Someone will always be available to rent the top floor or two of a 250-story building, but floor numbers 100 through 248 may not be so easy to rent.
4. There are several cost factors that tend to increase with taller buildings. Among these factors are elevators, plumbing, heating and air-conditioning, glazing, exterior walls, and wiring.

Whether a multistory building is used for an office building, a hospital, a school, an apartment house, or whatever, the problems of design generally are the same. The construction usually is of the skeleton type in which the loads are transmitted to the foundation by a framework of steel beams and columns. The floor slabs, partitions, and exterior walls all are supported by the frame. This type of framing, which can be erected to tremendous heights, may also be referred to as *beam-and-column* construction. Bearing-wall construction is not often used for buildings of more than a few stories, although it has on occasion been used for buildings up to 20 or 25 stories. The columns of a skeleton frame usually are spaced 20, 25, or 30 ft on centers, with beams and girders

---

[2]W. A. Rains, "A New Era in Fire Protective Coatings for Steel," *Civil Engineering* (New York: ASCE, September 1976), pp. 80–83.

framing into them from both directions. (See Fig. 19.2.) On some of the floors, however, it may be necessary to have much larger open areas between columns for dining rooms, ballrooms, and so on. For such cases, very large beams (perhaps plate girders) may be needed to support column loads for many floors above.

It usually is necessary in multistory buildings to fireproof the members of the frame with concrete, gypsum, or some other material. The exterior walls often are constructed with concrete or masonry units, although an increasing number of modern buildings are being erected with large areas of glass in the exterior walls.

For these tall and heavy buildings, the usual spread footings may not be sufficient to support the loads. If the bearing strength of the soil is high, steel grillage footings may be sufficient; for poor soil conditions, pile or pier foundations may be necessary.

For multistory buildings the beam-to-column systems are said to be superimposed on top of each other, story by story or tier by tier. The column sections can be fabricated for one, two, or more floors, with the two-story lengths probably being the most common. Theoretically, column sizes can be changed at each floor level, but the costs of the splices involved usually more than cancel any savings in column weights. Columns of three or more stories in height are difficult to erect. The two-story heights work out very well most of the time.

19.15 DISCUSSION OF LATERAL FORCES

For tall buildings, lateral forces must be considered as well as gravity forces. High wind pressures on the sides of tall buildings produce overturning moments. These moments probably are resisted without difficulty by the axial strengths of the columns, but the

Broadview Apartments, Baltimore, MD. (Courtesy of Lincoln Electric Company.)

horizontal shears produced on each level may be of such magnitude as to require the use of special bracing or moment-resisting connections.

Unless they are fractured, the floors and walls provide a large part of the lateral stiffness of tall buildings. Although the amount of such resistance may be several times that provided by the lateral bracing, it is difficult to estimate and may not be reliable. Today, so many modern buildings have light, movable interior partitions, glass exterior walls, and lightweight floors, that only the steel frame should be assumed to provide the required lateral stiffness.

Not only must a building be sufficiently braced laterally to prevent failure, but it must also be prevented from deflecting so much as to injure its various parts. Another item of importance is the provision of sufficient bracing to give the occupants a feeling of safety. They might not have this feeling in tall buildings that have a great deal of lateral movement in times of high winds. There have actually been tales of occupants of the upper floors of tall buildings complaining of seasickness on very windy days.

The horizontal deflection of a multistory building due to wind or seismic loading is called *drift*. It is represented by Δ in Fig. 19.13. Drift is measured by the *drift index*, Δ/h, where h is the height or distance to the ground.

The usual practice in the design of multistory steel buildings is to provide a structure with sufficient lateral stiffness to keep the drift index between approximately 0.0015 and 0.0030 radians for the worst storms that occur in a period of approximately 10 years. These so-called 10-year storms might have wind in the range of about 90 mph, depending on location and weather records. It is felt that when the index does not exceed these values, the users of the building will be reasonably comfortable.

In addition, multistory buildings should be designed to withstand 50-year storms safely. For such cases, the drift index will be larger than the range mentioned, with the result that the occupants will suffer some discomfort. The recently destroyed 1450-ft World Trade Center in New York City deflected or swayed about 3 ft in 10-year storms (drift index = 0.0021), while in hurricane winds it swayed about 7 ft (drift index = 0.0048).

Most buildings can be designed with little extra expense to withstand the forces caused during an earthquake of fairly severe intensity. On the other hand, earthquakes during recent years have clearly shown that the average building that is not designed for earthquake forces can be destroyed by earthquakes that are not particularly severe. The usual practice is to design buildings for additional lateral loads (representing the

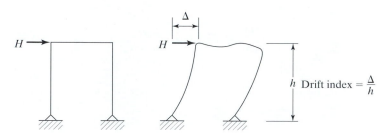

FIGURE 19.13

Lateral drift.

estimate of the earthquake forces) that are equal to some percentage of the weight of the building and its contents.

The lateral forces require the use of more steel, even though load factors are reduced for wind and earthquake forces. Additional steel will be used for bracing or moment-resisting connections, as described in the next section.

19.16   TYPES OF LATERAL BRACING

A steel building frame with no lateral bracing is shown in Fig. 19.14(a). Should the beams and columns shown be connected with the standard ("simple beam") connections, the frame would have little resistance to the lateral forces shown. Assuming the joints to act as frictionless pins, the frame would be laterally deflected as shown in Fig. 19.14(b), eventually collapsing because the structure is unstable.

To resist these lateral deflections, the simplest method from a theoretical standpoint is the insertion of full diagonal bracing, as shown in Fig. 19.14(c). From a practical standpoint, however, the student can easily see that in the average building, full diagonal bracing would often be in the way of doors, windows, and other wall openings. Also, many buildings have movable interior partitions, and the presence of interior cross bracing would greatly reduce this flexibility. Usually, diagonal bracing is convenient in solid walls in and around elevator shafts, stairwells, and other walls in which few or no openings are planned.

Another method of providing resistance to lateral forces involves the use of moment-resisting connections as illustrated in Fig. 19.15. This *bracket-type bracing* may be used economically to provide lateral bracing for lower-rise buildings. As buildings become taller, however, bracket-type bracing is not very economical, nor is it very satisfactory in limiting lateral deflections.

Some ways in which we can transmit lateral forces to the ground for buildings in the 20- to 60-story range are shown in Fig. 19.16. The X bracing system of part (a) works very well, except that it might get in the way. Furthermore, as compared with the systems shown in Fig. 19.16(b) and (c), the floor beams have longer spans and will have to be larger.

The K bracing system (b) provides more freedom for the placing of openings than does the full X bracing. K braces are connected at midspan, with the result that

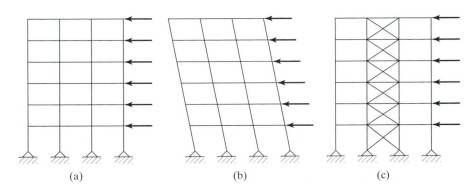

(a)   (b)   (c)

FIGURE 19.14

Chase Manhattan Bank Building, New York City. (Courtesy of Bethlehem Steel Corporation.)

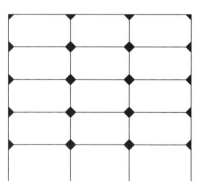

FIGURE 19.15

Bracket-type bracing.

[3]E. H. Gaylord, Jr., and C. N. Gaylord, *Structural Engineering Handbook*, 2d ed. (New York: McGraw-Hill, 1979), pp. 19–77 to 19–111.

FIGURE 19.16

(a) X brace. (b) K brace.
(c) Full-story knee brace.

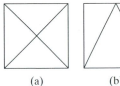
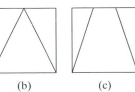

(a)   (b)   (c)

the floor beams will have smaller moments. The K bracing system (like the knee bracing system) uses less material than does X bracing. If we need more room than is available with the K bracing system, we can go to the full-story knee bracing system shown
in Fig. 19.16(c).[3]

In the usual building the floor system (beams and slabs) is assumed to be rigid in
the horizontal plane, and the lateral loads are assumed to be concentrated at the floor
levels. Floor slabs and girders acting together provide considerable resistance to lateral forces. Investigation of steel buildings that have withstood high wind forces have
shown that the floor slabs distribute the lateral forces so that all of the columns on a
particular floor have essentially equal deflections as long as twisting of the structure
does not occur. When rigid floors are present, they spread the lateral shears to the
columns or walls in the building. When lateral forces are particularly large, as in very
tall buildings or where seismic forces are being considered, certain specially designed
walls may be used to resist large parts of the lateral forces. These walls are called *shear
walls*.

It is not necessary to brace every panel in a building. Usually the bracing can be
placed in the outside walls with less interference than in the inside walls, where
movable partitions may be desired. Probably the bracing of the outside panels alone is
not enough, and some interior panels may need to be braced. It is assumed that the
floors and beams are sufficiently rigid to transfer the lateral forces to the braced panels. Three possible arrangements of braced panels are shown in Fig. 19.17 for lateral
forces in one direction. A symmetrical arrangement probably is desirable to prevent
uneven lateral deflection in the building and thus torsion.

Bracing around elevator shafts and stairwells usually is permissible, while for
other locations it often will interfere with windows, doors, movable partitions, glass
exterior walls, open spaces, and so on. Should bracing be used around an elevator shaft
as shown in Fig. 19.18(a), and should calculations show that the drift index is too large,
it may be possible to use a *hat truss* on the top floor, as shown in part (b). Such a truss
will substantially reduce drift. If a hat truss is not feasible because of interference with
other items, one or more *belt trusses* may be possible, as shown in Fig. 19.18(c). A belt
truss will appreciable reduce drift, although not as much as a hat truss.

The bracing systems described so far are not efficient for buildings above approximately 60 stories. These taller buildings have very large lateral wind and perhaps
seismic forces applied hundreds of feet above the ground. The designer needs to develop
a system that will resist these loads without failure and in such a manner that lateral

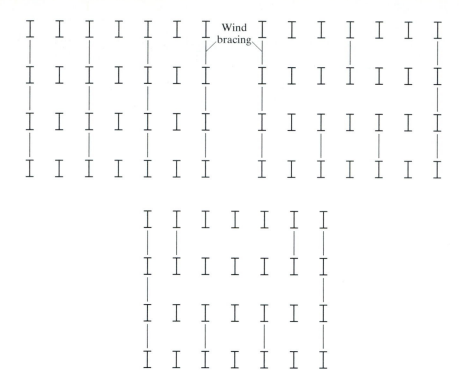

FIGURE 19.17

deflections are not so great as to frighten the occupants. The bracing methods used for these buildings usually are based on a tubular frame concept.

With the tubular system, a vertical tubular cantilever frame is created much like the tube of Fig. 19.19. The tube consists of the building columns and girders in both the longitudinal and transverse directions of the building. The idea is to create a tube that will act like a continuous chimney or stack.

To build the tube, the exterior columns are spaced close together—from 3 or 4 ft up to 10 or 12 ft on center. They are connected with spandrel beams at the floor levels, as shown in Fig. 19.20(a).

Further improvements of the tubular system can be made by cross bracing the frame with X bracing over many stories, as illustrated in Fig. 19.20(b). This latter system is very stiff and efficient, and is very helpful in distributing gravity loads rather uniformly over the exterior columns.

Another variation on this system is the tube-within-a-tube system. Interior columns and girders are used to create additional tubes. In tall buildings it is common to group elevator shafts, utility shafts, and stairs, and these systems can include shear walls and braced bents.[4]

---

[4]R. N. White, P. Gergely, and R. G. Sexsmith, *Structural Engineering*, vol. 3 (New York: Wiley 1974), pp. 537–546.

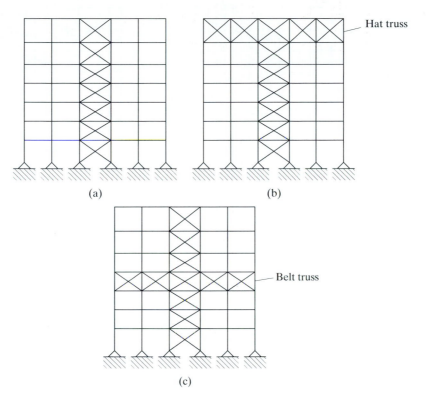

(a)   (b)

Hat truss

Belt truss

(c)

FIGURE 19.18

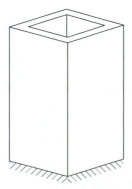

FIGURE 19.19
Solid-wall tube.

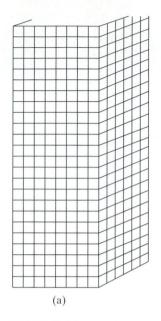
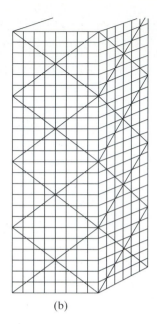

(a)   (b)

FIGURE 19.20

(a) Tubular frame, (b) Cross-braced tubular frame.

19.17 ANALYSIS OF BUILDINGS WITH DIAGONAL WIND BRACING FOR LATERAL FORCES

Full diagonal cross bracing has been described as being an economical type of wind bracing for tall buildings. Where this type of bracing cannot practically be used due to interference with windows, doors, or movable partitions, moment-resisting brackets often are used. However, should very tall narrow buildings (with height–least-width ratios of 5.0 or greater) be constructed, lateral deflections may become a problem with moment-resisting joints. The joints can satisfactorily resist the moments, but the deflections may be excessive.

Should maximum wind deflections be kept under 0.002 times the building height, there is little chance of injury to the building says ASCE Subcommittee 31.[5] The deflection is to be computed neglecting any resistance supplied by floors and walls. To keep lateral deflections within this range, it is necessary to use deep knee bracing, K bracing, or full diagonal cross bracing when the height–least-width ratio is about 5.0; and for greater values of the ratio, full diagonal cross bracing or some other method such as the tubular frame is needed.[6]

The student often may see bracing used in buildings for which he or she would think wind stresses were negligible. Such bracing stiffens up a building appreciably and serves the useful purpose of plumbing the steel frame during erection. Before bracing is installed, the members of a steel building frame may be twisted in many directions. Connecting the diagonal bracing should pull members into their proper positions.

---

[5]"Wind Bracing in Steel Buildings," *Transactions ASCE* 105 (1940), pp. 1713–1739.
[6]L. E. Grinter, *Theory of Modern Steel Structures* (New York: Macmillan, 1962), p. 326.

Transfer truss, 150 Federal Street, Boston, MA. (Courtesy of Owen Steel Company, Inc.).

When diagonal cross bracing is used, it is desirable to introduce initial tension into the diagonals. This prestressing will make the building frame tight and reduce its lateral deflection. Furthermore, these light diagonal members can support compressive stress due to their pretensioning. Since the members can resist compression, the horizontal shear to be resisted will be assumed to split equally between the two diagonals. For buildings with several bay widths, equal shear distribution usually is assumed for each bay.

Should the diagonals not be initially tightened, rather stiff sections should be used so they will be able to resist appreciable compressive forces. The direct axial forces in the girders and columns can be found for each joint from the shear forces assumed in the diagonals. Usually, the girder axial forces so computed are too small to consider, but the values for columns can be quite important. The taller the building becomes, the more critical are the column axial forces caused by lateral forces.

For types of bracing other than cross bracing, similar assumptions can be made for analysis. The student is referred to pages 333–339 of *Theory of Modern Steel Structures* by L. E. Grinter (New York: Macmillan, 1962) for discussion of this subject.

19.18 MOMENT-RESISTING JOINTS

For a large percentage of buildings under eight to ten stories, the beams and girders are connected to each other and to the columns with simple end-framed connections of the types described in Chapter 15. As buildings become taller, it is absolutely necessary to use a definite wind-bracing system or moment-resisting joints. Moment-resisting joints also may be used in lower buildings where it is desired to take advantage of continuity and the consequent smaller beam sizes and depths and shallower floor construction. Moment-resisting brackets also may be necessary in some locations for loads that are applied eccentrically to columns.

Two types of moment-resisting connections that may be used as wind bracing are shown in Fig. 19.21. The design of connections of these types was also presented in

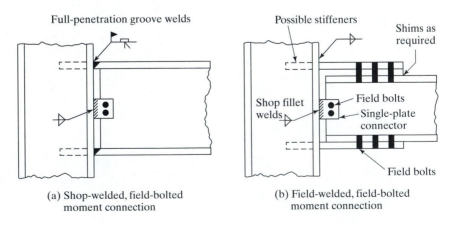

(a) Shop-welded, field-bolted moment connection

(b) Field-welded, field-bolted moment connection

FIGURE 19.21

Popular moment-resisting connections.

Chapter 14. The average design company through the years probably will develop a file of moment-resisting connections from their previous designs. When they have a wind moment of a certain value, they merely would refer to their file and select one of their former designs that would provide the required moment resistance.

These are the two most commonly used moment-resisting connections today. Most fabricators select the type shown in 19.21(a) as being the most economical. Single-plate (or shear tab) connectors are shown, but framing angles may be used instead. Column stiffeners (shown as dashed lines in the figures) sometimes are required by LRFD Specification K1 for local flange bending, local web yielding, sidesway buckling of the web, and so on. Stiffeners are a nuisance and cost an appreciable amount of

The 20-story United Founders Life Tower in Oklahoma City, OK. (Courtesy of Lincoln Electric Company.)

money. As a result, we will, when the specification shows they are needed, try to avoid them by increasing column sizes.

A very satisfactory variation of these last two connections involves an end plate as shown in Figures 12-6 and 12-7 in Part 12 of the LRFD Manual. This type of connection may, however, be used only for static load situations.

19.19   ANALYSIS OF BUILDINGS WITH MOMENT-RESISTING JOINTS FOR LATERAL LOADS

Approximate methods have been very popular for many years for analyzing multistory buildings for lateral forces. Among the several reasons for this popularity have been the following:

1. Large building frames are statically indeterminate to a very high degree, and their analysis by an "exact" method using a pocket calculator is a lengthy and difficult problem.
2. The resistance to lateral forces supplied by the walls and floors in a tall building is difficult to estimate accurately, and the results of analysis by an "exact" method therefore are not very precise.
3. Before the members of a statically indeterminate structure can be designed, their forces have to be determined, but these forces are dependent upon their sizes. Application of an approximate analysis should yield forces from which very good estimates of member sizes can be made for preliminary design.

The analysis of multistory buildings is a very difficult problem, and past practice has been to use one of the several approximate methods of analysis available. Today, however, with the availability of computers, it is feasible to make more precise analyses in appreciably less time than required to make the approximate analyses (without the use of computers). The more accurate values obtained permit the use of smaller members. Computer usage saves money in analysis time and in the use of smaller members.

In the pages that follow, a brief review of the portal and cantilever approximate methods is presented. No consideration is given in either of these methods to the elastic properties of the members. These omissions can be very serious in unsymmetrical frames and in very tall buildings. For example, consider the changes in member sizes in a very tall building. In such a building there probably will not be a great deal of change in beam sizes from the top floor to the bottom floor. For the same loadings and spans, the changed sizes would be due to the large wind moments in the lower floors. The change, however, in column sizes from top to bottom would be tremendous. The result is that the relative sizes of columns and beams on the top floors are entirely different from their relative sizes on the lower floors. When this fact is not considered, it causes large errors in the analysis.

If the height of a building is roughly five or more times its least lateral dimension, generally it is felt that a more precise method of analysis should be used than the methods described here. There are several excellent approximate methods that make use of the elastic properties and that give values closely approaching the results of the "exact" methods. These include the Factor method, the Witmer method of K percent-ages, and the Spurr method. Should an exact method be desired, the slope-deflection and moment-distribution methods are available. If the slope-deflection procedure is used, the designer will have the problem of solving a large number of simultaneous equations. The problem is not so serious, however, if a digital computer is used.

19.19.1 The Portal Method

The most common approximate method of analyzing tall building frames for lateral loads is the *portal method*. The chief advantage of this method, which is said to be sat-isfactory for most buildings up to 25 stories,[7] is its simplicity. A detailed description of this method, including the basis for the assumptions made, can be found in most struc-tural analysis textbooks. Only a brief listing of the steps involved in its application is given here.

1. The horizontal shears on each level are arbitrarily distributed between the columns. One commonly used procedure is to assume the shear divides between the columns in the ratio of one part to exterior columns and two parts to interior columns. Another common distribution (and the one used in the illustrative problem to follow) is to assume the shear V taken by each column is in propor-tion to the floor area it supports.

2. The moment M in each column is assumed to equal the column shear times half the column height (thus assuming a point of contraflexure at middepth).

3. The girder moments M are determined by joints, by noting that the sum of the girder moments at any joint equals the sum of the column moments at that joint. These calculations are easily made by starting at the upper left joint and working joint by joint across to the right; after which the next level is handled left to right, and so on.

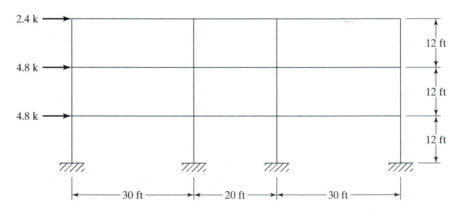

FIGURE 19.22

[7]"Wind Bracing in Steel Buildings," *Transactions ASCE* 105 (1940), p. 1723.

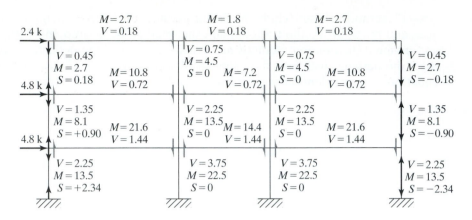

FIGURE 19.23

Frame analysis by the portal method.

4. The shear V in each girder is assumed to equal its moment divided by half the girder length.

5. Finally, the column axial forces S are determined by summing up the beam shears and other column axial forces at each joint. These calculations again are handled conveniently by working from left to right and from the top floor down.

Figure 19.22 shows a building frame that is to be analyzed by the portal method. The frames are assumed to be placed 20 ft 0 in. on centers and the exterior walls are assumed to be subjected to a wind pressure of 20 lb per vertical square foot. From these data the horizontal loads shown at each floor level are calculated.

This frame is analyzed in Fig. 19.23 by the portal method. The arrows shown on the figure give the direction of the girder shears and the column axial forces. The student can visualize the stress condition of the frame if he or she assumes the frame is tending to be pushed over from left to right by the wind, stretching the left exterior columns and compressing the right exterior columns.

19.19.2 The Cantilever Method

Another simple method of analyzing building frames for lateral forces is the *cantilever method*. This method is said to be a little more desirable for high narrow buildings than the portal method and may be used satisfactorily for buildings with heights not in excess of 25 to 35 stories.[8] It is not as popular as the portal method.

Instead of initially assuming the horizontal shears to be divided between the columns on each level in some proportion, the first assumption in the cantilever method pertains to the column axial forces. The axial force in a particular column is assumed to vary in direct proportion to its distance from the center of gravity of the group of columns on that level. With the wind blowing from left to right, the columns to the left of the center of gravity will be in tension while those to the right will be in

---

[8]Ibid.

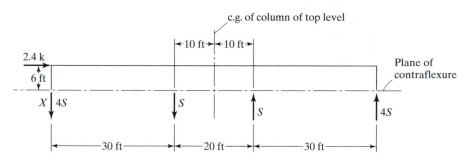

FIGURE 19.24

compression. The steps involved in analyzing a building frame for lateral forces by the cantilever method are as follows:

1. To obtain the column axial forces, moments are taken above an assumed plane of contraflexure through the middepth of the columns on each level. As an illustration of this procedure, moments are taken here to determine the axial forces in the columns on the top level of the building frame of Fig. 19.22. A free body is drawn in Fig. 19.24 of the part of the building above the assumed plane of contraflexure on the top level. The value of S is determined:

$$\Sigma M_x = 0$$

$$(2.4)(6) + (30)(S) - (50)(S) - (80)(4S) = 0$$

$$S = 0.0424 \text{ k}$$

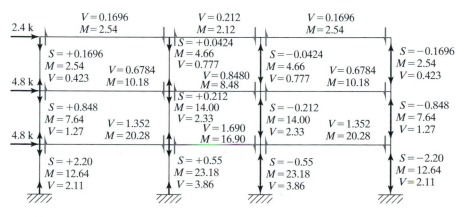

FIGURE 19.25

Frame analysis by the cantilever method.

2. The girder shears are determined joint by joint from the column axial forces.
3. The girder moments are determined by multiplying the girder shears by the half-girder lengths.
4. The column moments are found joint by joint from the girder moments.
5. The column shears are obtained by dividing the column moments by the half-column heights.

The analysis of the frame of Fig. 19.22 by the cantilever method is illustrated in Fig. 19.25. Arrows representing the directions of the column axial forces and the girder shears, are shown again.

19.20 ANALYSIS OF BUILDINGS FOR GRAVITY LOADS

19.20.1 Simple Framing

If *simple framing* is used, the design of the girders is less complex because the shears and moments in each girder can be determined by statics. The gravity loads applied to the columns are relatively easy to estimate, but the column moments may be a little more difficult. If the girder reactions on each side of the interior column of Fig. 19.26 are equal, theoretically no moment will be produced in the column at that level. This situation probably is not realistic, however, because it is highly possible for the live load to be applied on one side of the column and not on the other (or at least be unequal in magnitude). The results will be column moments. If the reactions are unequal, the moment produced in the column will equal the difference between the reactions times the distances to the center of gravity of the column.

Exterior columns often may have moments due to spandrel beams opposing the moments caused by the floor loads on the inside of the column. Nevertheless, gravity loads generally will cause the exterior columns to have larger moments than the interior columns.

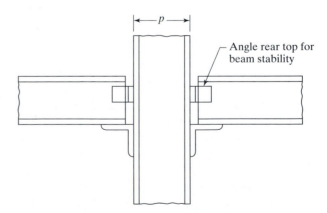

FIGURE 19.26

Simple framing of interior column.

"Topping out" of Blue Cross–Blue Shield Building in Jacksonville, FL. (Courtesy of Owen Steel Company, Inc.) The Christmas tree is an old north European custom used to ward off evil spirits. It also is used today to show that the steel frame was erected with no lost-time accidents to personnel.

To estimate the moment applied to a column above a certain floor, it is reasonable to assume that the unbalanced moment at that level splits evenly between the column above and below. In fact, such an assumption is often on the conservative side, as the column below may be larger than the one above.

19.20.2 Rigid Framing

For buildings with moment-resisting joints, it is a little more difficult to estimate the girder moments and make preliminary designs. If the ends of each girder are assumed to be completely fixed, the moments for uniform loads are as shown in Fig. 19.27(a). Conditions of complete fixity probably are not realized, with the result that the end moments are smaller than shown. There is a corresponding increase in the positive centerline moments approaching the simple moment $(wL^2/8)$ shown in Fig. 19.27(b). Probably a moment diagram somewhere in between the two extremes is more realistic. Such a moment diagram is represented by the dotted line of Fig. 19.27(a). A reasonable procedure is to assume a moment in the range of $wL^2/10$, where L is the clear span.

When the beams and girders of a building frame are rigidly connected to each other, a continuous frame is the result. From a theoretical standpoint, an accurate analysis of such a structure cannot be made unless the entire frame is handled as a unit.

Before this subject is pursued further, it should be realized that in the upper floors of tall buildings the wind moments will be small, and the framed and seated

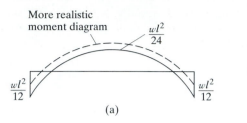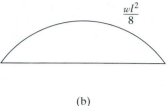

FIGURE 19.27

(a) Fixed-end beam. (b) Simple beam.

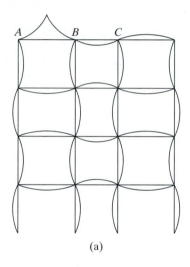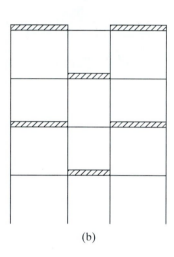

FIGURE 19.28

connections described in Chapter 15 will provide sufficient moment resistance. For this reason, the beams and girders of the upper floors may very well be designed on the basis of simple beam moments, while those of the lower floors may be designed as continuous members with moment-resistant connections due to the larger wind moments.

From a strictly theoretical viewpoint, there are several live-load conditions that need to be considered to obtain maximum shears and moments at various points in a continuous structure. For the building frame shown in Fig. 19.28, it is desired to place live loads to cause maximum positive moment in span AB. A qualitative influence line for positive moment at the centerline of this span is shown in part (a). This influence line shows that, to obtain maximum positive moment at the centerline of span AB, the live loads should be placed as shown in Fig. 19.28(b).

To obtain maximum negative moment at point B, or maximum positive moment in span BC, other loading situations need to be considered. With the availability of computers, more detailed analyses are being done every day. The student can see, however, that unless the designer uses a computer, he or she probably will not have sufficient time to go through all of these theoretical situations. Furthermore, it is doubtful if the accuracy of our analysis methods would justify all of the work anyway. Nonetheless, it often is feasible to take out two stories of the building at a time as a free body and analyze that part by one of the "exact" methods such as the successive correction method of moment distribution.

Before such an "exact" analysis can be made, it is necessary to make an estimate of the member sizes, probably based on the results of analysis by one of the approximate methods.

19.21   DESIGN OF MEMBERS

The girders can be designed for the load combinations mentioned in Section 19.20. For the top several floors the second condition will not control design, but farther down in the building it will control the sizes for the reasons described in the next paragraph.

Each of the members of a building frame must be designed for all of the dead loads that it supports, but it may be possible to design some members for lesser loads than their full theoretical live-load values. For example, it seems unlikely in a building frame of several stories that the absolutely maximum-design live load will occur on every floor at the same time. The lower columns in a building frame are designed for all of the dead loads above, but probably for a percentage of live loads appreciably less than 100 percent. Some specifications require that the beams supporting floor slabs be designed for full dead and live loads, but permit the main girders to be designed under certain conditions for reduced live loads. (This reduction is based on the improbability that a very large area of a floor would be loaded to its full live-load value at any one time.) The ASCE Standard 7 provides some very commonly used reduction expressions.[9]

Should simple framing be used, the girders will be proportioned for simple beam moments plus the moments caused by the lateral loads. For continuous framing the girders will be proportioned for $wL^2/10$ (for uniform loads) plus the moments caused by the lateral loads. An interesting comparison of designs by the two methods is shown in pages 717–719 of Beedle et al. *Structural Steel Design* (New York: Ronald, 1964). It may be necessary to draw the moment diagram for the two cases (gravity loads and lateral loads) and add them together to obtain the maximum positive moment out in the span. The value so computed may very well control the girder size in some of the members.

A major part of the design of a multistory building is involved in setting up a column schedule that shows the loads to be supported by the various columns story by story. The gravity forces can be estimated very well, while the shears, axial forces, and moments caused by lateral forces can be roughly approximated for the preliminary design from some approximate method (portal, cantilever, factor, or other).

---

[9]*American Society of Civil Engineers Minimum Design Loads for Buildings and Other Structures.* ASCE 7–98 (New York: ASCE), Section 4.8.

If both axes of the columns are free to sway, the column effective lengths theoretically must be calculated for both axes as described in Chapter 5. If diagonal bracing is used in one direction, thus preventing sidesway, the K value will be less than 1.0 in that direction. If these frames are braced also in the narrow direction, it is reasonable to use $K = 1.0$ for both axes.

After the sizes of the girders and columns are tentatively selected for a two-story height an "exact" analysis can be performed and the members redesigned. The two stories taken out for analysis often are referred to as a *tier*. This process can be continued tier by tier down through the building. Many tall buildings have been designed and are performing satisfactorily in which this last step (the two-story "exact" analysis) was omitted.

CHAPTER 20

Design of Steel Building Systems

20.1 INTRODUCTION

Up to this point in our discussion about designing with structural steel, we have been predominantly concerned with the design of individual components in the system. Even when we looked at pieces of a frame to determine the end-effects for designing columns, we mostly were designing the components rather than the system. Even in the last chapter, elements of the building frame were the topic of discussion rather than the system. In this chapter, we begin to change our focus. We now begin to investigate the design of a structural system. Through this investigation, we will try to develop an understanding of the relationship between structural analysis and structural design, and the impact of design decisions on the overall behavior of a structural system.

Figure 20.1 shows a flowchart of the process of structural system design. Notice in this illustration that we must develop an acceptable system for both strength requirements and serviceability requirements. The strength requirements are that the structure must support the loads to which it is subjected. Using the procedures discussed throughout this text, design for strength is based on limit states theory. Serviceability requirements, on the other hand, are usually requirements based on elastic response, and especially upon elastic deformation response. These requirements would include excessive live load deflection and cracking of finish surfaces. Observe also from the illustration that the selection of loads and the geometry of the structure are very important considerations when designing a system. Further, unless the structure is statically determinate, changes in design probably will affect the distribution of force throughout the system. We must evaluate these effects after designing the system; structural design consists of several design–analysis cycles.

Another equally important part of structural system design is the consideration of design alternatives. As we progress through the design–analysis process, we should always be looking for ways to improve the design. We also should be looking for ways

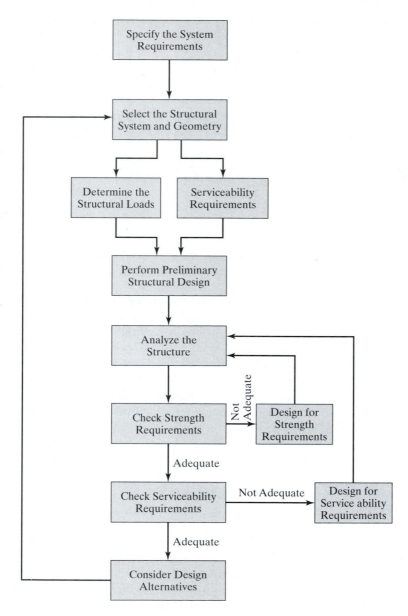

FIGURE 20.1

Flowchart of the Structural System Design Process.

to improve the performance of the system, reduce the overall cost of the system, or simplify construction. Rarely is the design that is actually built the one that resulted from the first pass at the design.

As we proceed through this chapter, we must understand that there is not a single design solution for a given structure or for a particular situation. Rather, when designing a structural system, we should realize that there probably will be a family of equally acceptable solutions. Each solution will have its own merits and weaknesses. A design is acceptable as long as it is safe and the resulting system performs as intended. As we begin, there are several preliminary issues that we need to discuss.

20.1.1 Building Codes and Structural Loads

As discussed in Chapter 2, the minimum design standards of most structures are controlled by building codes. These building codes are based on model building codes, such as the recently introduced *International Building Code*®.[1] The model building codes often specify a particular industry standard for design of particular materials, such as the following paragraph for structural steel from the *International Building Code*®:

> ***2204.1 General***. *The design, fabrication and erection of structural steel for buildings and structures shall be in accordance with either the AISC Load and Resistance Factor Design Specification for Structural Steel Buildings (AISC-LRFD), AISC Specification for Structural Steel Buildings—Allowable Stress Design (AISC-ASD) or AISC Specification for the Design of Steel Hollow Structural Sections (AISC-HSS). Where required, the seismic design of steel structures shall be in accordance with the additional provisions of Section 2212.*[2]

Municipal and state governments adopt the model building codes or develop their own building codes based on the model codes. Once adopted, these codes become legally enforceable laws and the design engineer must comply with them. The intention of the codes is not to limit engineering creativity; rather, they provide a minimum standard for loads and structural performance to safe-guard the health and safety of the public. When necessary, an engineer can always go beyond the minimum specifications in the codes. Also, the codes do not specify structural geometry. They only specify a minimum standard for the design of the system.

20.1.2 Use of Computers in Structural Design

As discussed in other parts of this book, computer software has a very important place in structural design—for both the analysis of the forces in the system and the actual sizing of components. A computer is only a tool in the design process, but it is an indispensable tool in designing systems, because it can perform calculations quickly that would otherwise require a considerable number of hours. Throughout this chapter, we make heavy use of *SAP2000* for analysis as we discuss building system design.

When using computers, an engineer should remember two things. First, all computer programs are based on a set of assumptions that go into formulation of the theoretical basis of the software. Before using any computer program for analysis and

---

[1]2000 International Building Code, *International Code Council, Inc., Falls Church, Virginia 22041-3401, 2000*
[2]*Ibid.* p. 533.

design, a prudent structural engineer will verify the assumptions, and thereby the applicability of the software for the task at hand. Not doing so can quickly lead to disaster if the software was not intended for the purpose for which it is being used.

Secondly, the engineer should guard against being lulled into a false sense of security because of the precision afforded by a computer in calculations. Computers can calculate the result to many significant figures, but a solution with many significant figures is not necessarily representative of system behavior or system safety. The calculated results, regardless of the precision of the calculations, are only as good as the data provided and the assumptions inherent in the solution. Rarely do we know the loads acting on a structure and the material properties to more than two or three significant figures; precision beyond that point in calculations is meaningless and should be ignored.

20.1.3 Relationship between Analysis and Design

Questions about the relationship between structural analysis and structural design are quite old. Hardy Cross wrote an interesting paper on this subject in 1936.[3] The remarks that he made then are still appropriate today, and they are worth quoting at length:

> *Confusion sometimes exists in structural design as to the use to be made of analyses. The designer soon realized that precision is futile in some cases and important in others, and the experienced designer realizes fully that analysis of the conventional type is frequently a poor guide to proper proportions. That analysis shows a certain member to be overstressed commonly indicates that the member should be made larger, but the overstress sometimes has little importance and may be disregarded. In some cases where the over-stress is serious, the best solution is not obvious; sometimes the structural layout should be changed entirely.*[4]

Hardy Cross went on to define four types of structural response:

- *Deformation Stresses*: These stresses are the consequence of strain resulting from internal and external movements in the structure. These are the primary stresses resulting from the applied loads acting on the structure.

- *Participation Stresses*: Today these stresses are generally referred to as *secondary stresses*. We saw this type of stress before when we talked about P-Δ effects. Normally, these stresses are not significant and do not need to be considered in analysis and design. Another example of participation stress is deformation caused by shearing forces in addition to that caused by flexure. In most cases, shear deformation is negligible compared with the deformation caused by flexure and thus can be neglected.

- *Normal Structural Action*: This term is used to describe those parts of the structure whose response is independent of the response of other parts of the structure.

---

[3]From the introduction to Hardy Cross (1936), "The Relation of Analysis to Structural Design," *Transactions of the American Society of Civil Engineers*, Volume 101, pp. 1363–1374.
[4]*Ibid.* p. 1363.

These parts can be designed without consideration (or with very little considera-tion) of the other parts of the structure. Typical of normal structural response is a statically determinate structure.

- *Hybrid Structural Action*: This occurs when action in one part of the structure affects action in another part of the structure. Such action occurs in statically indeterminate structures. It can also happen when sidesway occurs to the extent that P-Δ effects are no longer negligible and must be considered in the design.

Most structural systems that we design will be highly indeterminate structures; thus, they will exhibit hybrid structural action. We must always be cognizant of the effect that design changes in one part of the structure may have on other parts of the struc-ture. We must also be mindful of the conditions under which the participation stress can become significant and affect the load-carrying capacity of the system.

20.1.4   Concept of the Force Envelope

When loads are applied to a structure, the structure responds in reaction to those loads. The forces in a particular component in the system are caused by 1) the loads acting on the structure, 2) where those loads are located, and 3) which loads are acting in combi-nation with each other. From analysis of the response to the various forces acting, we can determine the maximum and minimum forces that can exist in any component. This range of forces is called a *force envelope*.

To illustrate the idea of force envelopes, consider the beam shown in Figure 20.2. This beam is connected to the columns with clip angles and can be assumed to act as a

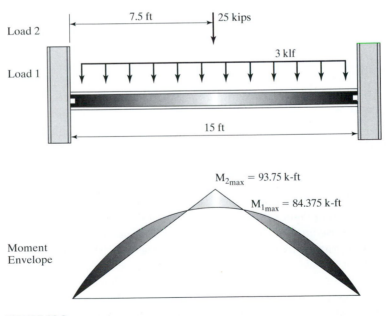

FIGURE 20.2

A moment envelope for a beam.

simply supported beam. It is acted upon by two loads that do not act at the same time.

Using the principles of mechanics, the moment diagrams for these two loads can be determined and plotted as shown in the bottom part of Figure 20.2. The area bounded by these two moment diagrams is the moment envelope for this beam subjected to these loads. This is the shaded area shown in that part of the figure. Such a diagram shows the maximum and minimum forces for which the beam must be designed. Although the forces are all positive in this particular example, very often a force envelope shows both positive and negative forces.

20.1.5 Preparation of Engineering Drawings

An important but often overlooked topic when talking about material design—particularly when talking about design of components—is the preparation of drawings to illustrate the final design. This is an important part of system design because without design drawings the structure cannot be built. Three types of drawings are involved in the system:

- Engineering sketches and drawings to illustrate the design
- Construction drawings from which the building is actually constructed
- Shop drawings from which the individual components in the structure are fabricated.

Of these three types of drawings, we are concerned with those necessary for preparing the construction drawings—the drawings prepared by the design engineer. A draftsperson normally prepares the construction drawings from these drawings. Throughout the discussion in this chapter, we will illustrate some fundamental criteria for preparation of engineering drawings.

20.2 DESIGN OF A STRUCTURAL STEEL BUILDING

In this section, typical components of the structure will be examined so that we can understand the basic principles of design. The design that we prepare here is not a complete one. It is intended to demonstrate the philosophy and principles of system design. We will design a typical frame instead of the entire building.

20.2.1 The Building to be Constructed

The building to be constructed is a three-story office building with the general floor plan shown in Figure 20.3. The architects of this project require a 9-ft ceiling height and intend to use an "open office" concept in all spaces, which divides the workspace using movable partitions rather than fixed walls. The building will be clad with material weighing 50 psf.

The ground floor of the building will be a slab on grade. All heavy storage will be placed on the ground floor; the upper floors will be used only for office and reception space. The geotechnical engineers have determined that a full-moment resisting foundation is possible.

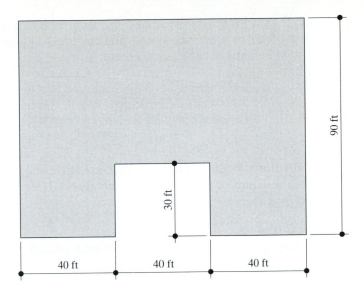

FIGURE 20.3

General building footprint.

20.2.2 Structural Geometry and Primary Framing

As we begin the design process, we must first layout the primary structural frame. The primary structural geometry that we will first use is shown in Figure 20.4. When establishing the geometry, we tried to make the framing as regular as possible. This will facilitate both erection and load transfer throughout the building.

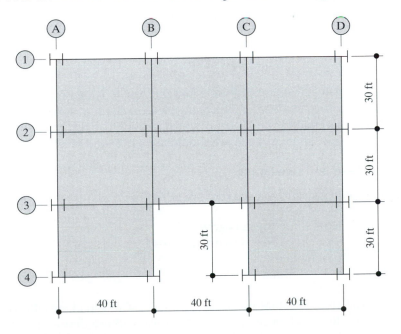

FIGURE 20.4

Primary structural framing at the floor level.

The story height must be sufficient to accommodate the office space, the structural members, and the HVAC and electrical equipment necessary in the building. The story height can be determined as follows:

Clear ceiling height	108 in
Allowance for structure	24 in
Ductwork and cable chases	24 in
Story Height	156 in

The floor-to-floor height of each story, then, will be 156 inches or 13 feet. This should allow adequate space for all of the necessary equipment that must be located in the ceiling spaces.

20.3 LOADS ACTING ON THE STRUCTURAL FRAME

When determining the loads that are acting on the primary structural frame, we must consider the type of occupancy of the building and the framing system used for the floor and roof. We will begin by determining the live loads acting.

20.3.1 Floor Dead and Live Loads

The dead loads acting on the floor are as follows:

Floor finish	2 psf
Suspended electrical and ductwork	10 psf
Ceiling finish	2 psf
Movable partitions	5 psf
Total Dead Load	19 psf

The IBC 2000 building code and the ASCE-7 load specification both suggest a minimum office live load of 50 psf and a load of 80 psf in corridors above the first floor. Because this building will have movable partitions to divide the floor space, the corridors can change location. As such, we should use a live load of 80 psf everywhere.

20.3.2 Roof Dead and Live Loads

The dead loads acting on the roof are as follows:

Roof finish (tar and gravel)	5 psf
Suspended electrical and ductwork	10 psf
Ceiling finish	2 psf
Total Dead Load	17 psf

The IBC 2000 building code and the ASCE-7 load specification both suggest a minimum live load on the roof of 20 psf. We will use that load in this discussion. An engineer should verify that this load would not be exceeded by snow or rain ponding on the roof, or by special equipment or constructions loads.

20.3.3   Roof and Floor Framing

We will support the roof and floor with floor beams spanning between column lines 1–2, 2–3, and 3–4 as shown in Figure 20.5. We choose to let the floor beams span in this direction because that will result in the shortest beams. As bending moment is a function of beam length, this also will result in the lowest bending moment and shearing force in the members. The floor beams will be connected to the girders with a simple connection. The floor beams can thus be treated as simple beams and sized accordingly.

Prior analysis indicates that a 4-inch noncomposite concrete slab will be adequate for the live loads if the floor beams are spaced at 48 inches on-center. A 4-inch concrete slab will impose a 50-psf dead load on the floor beams in addition to the dead load itemized above. The weight of the floor beams will also contribute to the dead load. As such, the floor beams must be designed before the total dead load can be determined. Let us now select the floor beams.

$$M_u = \frac{4[1.2(19 + 50) + 1.6(80)](30^2)}{8} = 94{,}860 \text{ lb-ft} = 94.86 \text{ kip-ft}$$

From the tables in the AISC handbook for $F_y = 50$ ksi, a W12 × 22 will be adequate if the floor slab is constructed to provide continuous bracing for the compression flange. This section will cause a self-weight moment of

$$M_{u\,\text{self-weight}} = \frac{1.2(22)(30^2)}{8} = 2.97 \text{ kip-ft}$$

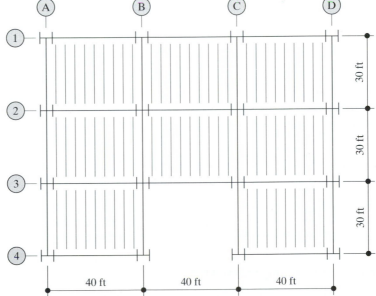

FIGURE 20.5

Floor and roof framing.

The total moment, then is

$$2.97 + 94.86 = 97.8 \text{ kip-ft}$$

This is less than 109.8 kip-ft, which is the capacity of the section. For this section, because of the length of the beam, capacity is controlled by flexure rather than by shear, as indicated in the capacity tables. Further, this section adds an effective dead load of $22/4 = 5.5$ psf to the total dead load of the floor. We will use this same section for the roof beams.

20.3.4 Loads Acting on the Girders

We are now ready to calculate the loads acting on the girders. When doing so, we need to consider girders in both directions, as well as interior and exterior girders. Girders along column lines 1 through 4 carry the floor and roof loads transferred to them from the beams, whereas girders along column lines A through D carry their share of the floor or roof loads.

Interior Girders Along Column Lines 2 and 3

Floor Dead Load:	$W_D = (19 + 50 + 5.5)30 = 2.24$ kips/ft
Floor Live Load:	$W_L = (80)30 = 2.40$ kips/ft
Roof Dead Load:	$W_D = (17 + 5.5)30 = 0.675$ kips/ft
Roof Live Load:	$W_L = (20)30 = 0.60$ kips/ft

Exterior Girders Along Column Lines 1, 3 and 4

Floor Dead Load:	$W_D = (19 + 50 + 5.5)15 = 1.12$ kips/ft
Floor Live Load:	$W_L = (80)15 = 1.20$ kips/ft
Roof Dead Load:	$W_D = (17 + 5.5)15 = 0.338$ kips/ft
Roof Live Load:	$W_L = (20)15 = 0.30$ kips/ft

Interior Girders Along Column Lines B and C

Floor Dead Load:	$W_D = (19 + 50)4 = 0.276$ kips/ft
Floor Live Load:	$W_L = (80)4 = 0.320$ kips/ft
Roof Dead Load:	$W_D = (17)4 = 0.068$ kips/ft
Roof Live Load:	$W_L = (20)4 = 0.080$ kips/ft

Exterior Girders Along Column Lines A, B, C and D

Floor Dead Load:	$W_D = (19 + 50)2 = 0.138$ kips/ft
Floor Live Load:	$W_L = (80)2 = 0.160$ kips/ft
Roof Dead Load:	$W_D = (17)2 = 0.034$ kips/ft
Roof Live Load:	$W_L = (20)2 = 0.040$ kips/ft

20.3.5 Lateral Loads Acting on the Frame

For the purposes of this discussion, we assume that the cladding is attached to the building in such a manner that the loads from the cladding are applied to the structural frame at the intersections of the primary members. We assume also that we have been advised that the lateral loads applied to the cladding are a net 30 psf—a 15 psf pressure load applied to one face and 15 psf suction applied to the opposite face.

For this discussion, we will be designing a typical frame along a column line so we only need to calculate the lateral forces acting on that frame. If we were designing the entire frame, we would have to calculate the lateral loads acting on all of the frames. On the frame under consideration, the lateral loads are computed by the following:

$$\text{Roof Level:} \qquad P_w = \pm 15 A_T = \pm 15 \left[\frac{13}{2}(30) \right] = \pm 2.93 \text{ kips}$$

$$\text{Story Level:} \qquad P_w = \pm 15 A_T = \pm 15 \left[13(30) \right] = \pm 5.85 \text{ kips}$$

These forces act on both faces of the frame. When there is a pressure load acting on one side (indicated by a positive sign), there is a suction load acting on the other side (indicated by a negative sign).

An unfinished skyscraper with stories of bare girders above a (glass, steel) facade.

20.3.6 Cladding Dead Load Acting on the Frame

The cladding in a building must be attached to the structural frame in some manner. The manner in which it is attached will depend upon the type of cladding used and its manufacturer's recommendations. For our purposes, we assume that the cladding is attached to the building in such a manner that the loads from the cladding are applied to the structural frame at the intersections of the primary members. This is consistent with the assumption for transfer of lateral load to the building frame. We also assume that we have been advised that the cladding weighs 15 psf. The cladding dead load acting on the frame we are designing can then be determined from the following:

$$\text{Roof Level:} \qquad P_w = 15A_T = 15\left[\frac{13}{2}(30)\right] = 2.93 \text{ kips}$$

$$\text{Story Level:} \qquad P_w = \pm 15A_T = \pm 15[13(30)] = \pm 5.85 \text{ kips}$$

These forces act downward as they are gravity loads.

20.3.7 Applicable Load Combinations

To obtain the maximum effects in the frame caused by all of the possible loads, staggered or patterned loads need to be considered for the gravity live loads. Pattern loads to obtain the maximum possible forces in a structure probably were discussed when you studied influence lines in your course on structural analysis. For design of the interior frame under consideration, the load patterns shown in Figure 20.6 are used. These loads can act singularly or together, as necessary. These pattern live loads are combined with the other loads on the structure to determine the total design forces.

The AISC requires that the load combinations used for design are those specified in ASCE-7. In accordance with ASCE-7, the following load combinations are used.

1. 1.4Dead
2. 1.2Dead + 1.6LL1 + 1.6LL2 + 0.5LR1 + 0.5LR2
3. 1.2Dead + 1.6LL1 + 0.5LR1
4. 1.2Dead + 1.6LL2 + 0.5LR2
5. 1.2Dead + 1.6LR1 + 1.6LR2 + 0.5LL1 + 0.5LL2
6. 1.2Dead + 1.6LR1 + 0.5LL1
7. 1.2Dead + 1.6LR2 + 0.5LL2
8. 1.2Dead + 1.6LR1 + 1.6LR2 + 0.8Wind
9. 1.2Dead + 1.6LR1 + 0.8Wind
10. 1.2Dead + 1.6LR2 + 0.8Wind
11. 1.2Dead + 1.6Wind + 0.5LL1 + 0.5LL2 + 0.5LR1 + 0.5LR2
12. 1.2Dead + 1.6Wind + 0.5LL1 + 0.5LR1
13. 1.2Dead + 1.6Wind + 0.5LL2 + 0.5LR2
14. 1.2Dead + 0.5LL1 + 0.5LL2
15. 1.2Dead + 0.5LL1

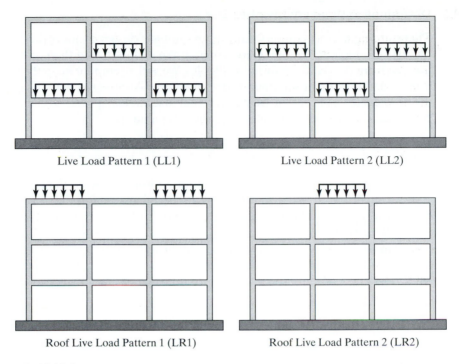

Live Load Pattern 1 (LL1)   Live Load Pattern 2 (LL2)

Roof Live Load Pattern 1 (LR1)   Roof Live Load Pattern 2 (LR2)

FIGURE 20.6

16. 1.2Dead + 0.5LL2
17. 0.9Dead + 1.6Wind

Because the frame is symmetric and the gravity loads acting on it are symmetric, the wind load only needs to be considered in one direction. Frame symmetry will need to be maintained when we select members, so that this limitation in the load combinations remains valid. If symmetry cannot be maintained, load combinations 8 through 13 and 17 will need to be repeated for wind acting from the other direction.

20.4   PRELIMINARY DESIGN AND ANALYSIS

Prior to beginning an analysis of the structure, we must determine the size of the members in the structure. The problem is that we cannot determine the size until we know the forces in the members, and we cannot determine the forces until we know the sizes. To break this cycle, we will estimate the forces in the members and select a size accordingly. We will then analyze the structure, design the structure using these new forces, and then update and reanalyze the structure as necessary.

20.4.1 Determine Preliminary Member Size

There are many ways to perform the preliminary design of the structure. One of the easiest is to look at the total load acting on the members and then select an appropriate size from the AISC uniform load tables. We will attempt to keep the size of the roof girders and the size of the floor girders the same. Although several load combinations are considered in the design, we use the combination that appears to cause the maximum total load on the girders to determine the estimated size:

$$\text{Floor Girder :} \quad W_T = 1.2(2.28)(40) + 1.6(2.4)(40) = 263.0 \text{ kips}$$
$$\text{Roof Girder} \quad W_T = 1.2(0.72)(40) + 1.6(0.60)(40) = 73.0 \text{ kips}$$

Using Table 5-4 in the Manual, we find an appropriate wide-flange section to carry these total loads. A W21×50 appears to be adequate for the roof girders, and a W24×131 appears to be adequate for the floor girders. We will use these as the initial size of the girders.

Similarly, the preliminary column sizes can be determined. For final design, we will try to let the exterior columns be the same size and the interior columns be the same size. To determine the estimated column size, we will estimate the axial load at the bottom level.

$$\text{Interior Columns:} \quad W_T = 1.2(2.28)\left(\frac{40}{2}\right)(2)(2) + 1.6(2.4)\frac{40}{2}(2) = 372.5 \text{ kips}$$

$$\text{Exterior Columns} \quad W_T = 1.2(2.28)\left(\frac{40}{2}\right)(2) + 1.6(2.4)\frac{40}{2} = 186.2 \text{ kips}$$

Using these forces and Table 4.2 in the Manual, we find that a reasonable section for the both interior and exterior columns is a W14 × 68. We obtained this by assuming an effective length of about 20 to 25 feet. This represents an effective length factor approximately in the range of 1.5 to 2.0, which is reasonable for a rigid frame with sidesway.

20.4.2 Model the Structure and Perform Analysis

Next, we need to develop the computer model of this frame, analyze the system, and then perform the design checks. For these tasks, we use the SAP2000 computer software on the CD-ROM included with this textbook.

An excellent discussion about using SAP2000 is included with the software, so we will not discuss it here. This information can be obtained by clicking HELP on the menu bar, then clicking "Search for help on...", and then selecting Contents. The complete data file for this structure is also included on the CD-ROM. You are urged to open the file and to perform the analyses that we have performed here. The fundamental steps, however, are outlined as follows:

1. Define Joint and Member Connectivity: We must define the geometry of the structure. This is accomplished by specifying the location of the joints and then specifying how the members are connected to the joints. A graphical representation of the geometry is very useful to make sure that you have specified the geometry correctly.

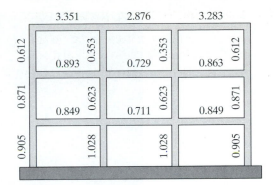

FIGURE 20.7

Design checks resulting from first analysis.

2. Joint Restraints: The structural supports are specified by restraining the joints at the supports. Restraints specify that the displacement in the direction of a degree of freedom is zero—the structure cannot be displaced in that direction.

3. Member Properties: When analyzing a structure we must specify the cross-sectional characteristics and modulus of elasticity for each member.

4. Loads Acting: We must specify all of the loads that are acting on the structure. Loads acting on the joints are specified in the global coordinate system. Loads acting directly on the members are specified in the local coordinate system. We must also specify the cases and combinations in which those loads are acting.

5. Perform Design Checks: At the conclusion of an analysis, we must perform a design check of all the members and make modifications as necessary

The design checks resulting from this first analysis-design cycle are shown in Figure 20.7. These values were calculated by the software when it was asked to Design/Check the structure. They represent the ratio of the force in a member to the capacity of the member. A value less than or equal to unity indicates that the member is acceptable. As we can see from the figure, some of the design unity checks exceed unity, which indicates that the member is underdesigned and thus design changes are necessary.

20.5   REVIEW OF THE RESULTS AND DESIGN CHANGES

We now review the results obtained from the preliminary design and analysis. We can draw several conclusions from the design checks presented in Figure 20.7. The floor girders appear to be sized appropriately, but the roof girders are significantly undersized. The columns at the lower level appear to be marginally underdesigned, those at the middle level appear to be sized reasonably, and those at the upper level appear to be overdesigned. We need to make some design changes.

Let us address the columns first. To maintain as many similarly sized columns as possible, we change our selection philosophy and let the columns of each story level be the same size. This will result in the exterior columns being a little overdesigned. We will next try using a W14 × 74 on the lower story, a W14 × 61 on the middle story, and

A gleaming modern (glass, steel) skyscraper looms over a
lush green park in Regina, Saskatchewan.

a W14 × 48 on the upper story. These new sizes were arrived at by selecting a larger or
smaller section in the same shape family, as appropriate.

We need to significantly increase the size of the roof girders to achieve an acceptable design. For this next trial, let us use a W24 × 84 for the roof girders. The results
of the analysis after this redesign and the design checks are shown in Figure 20.8.

These results are acceptable and show that the structure is designed reasonably well.
Additional study could perhaps be done to see if lighter—and thus more economical—
sections could be used. The final frame is also symmetric, which is consistent with the
load combinations that we used.

20.6 THE DESIGN SKETCHES

The last thing we need to do with this design is prepare the sketches of the system so
that design drawings can be prepared. We will provide sufficient information on those
sketches so that a draftsperson can prepare the design drawings. The two sketches
shown in Figures 20.9 and 20.10 are those needed for the design we have completed.

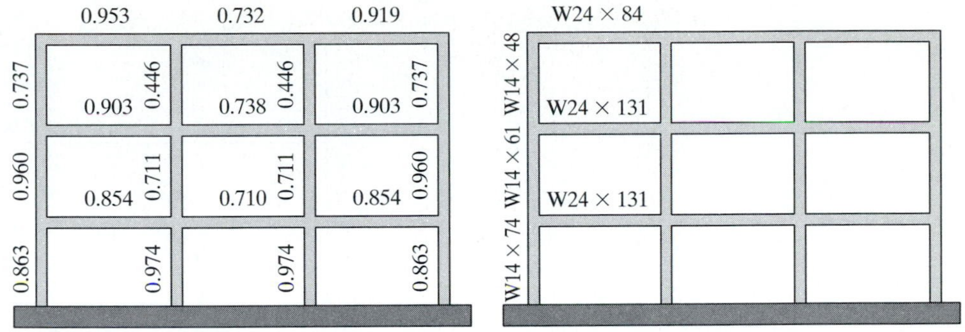

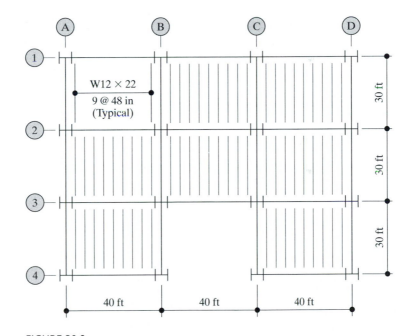

FIGURE 20.8

Final design checks and member sizes.

FIGURE 20.9

Floor and roof deck framing plan.

After the rest of the building is designed, other sketches indicating those designs will need to be prepared. When reviewing the design and construction drawings, we need to ensure that the notes about the connections are satisfied.

20.7 CONCLUSION

Throughout this chapter, we have demonstrated the philosophy and principles of designing an entire structural system. If we were actually designing the structure, there

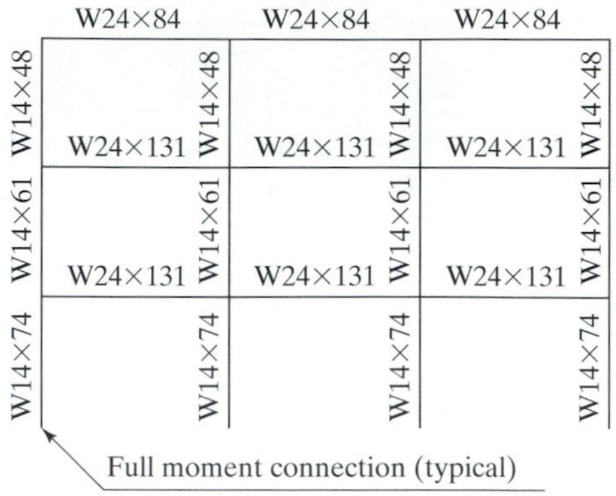

FIGURE 20.10

Note: All connections between the columns and the girders must provide for full transfer of moment.
Column line 2 framing plan.

probably would be more load cases to consider, and we would have to develop a full three-dimensional analysis model of the building instead of looking at a typical frame. We made simplifications here so that we could observe an approach to design without becoming overwhelmed by the complexity. We would also need to design the connections in the building. Typical building connections were discussed in the previous chapters.

PROBLEMS

The homework problems presented here are open-ended. That is to say, there is no single correct solution. Rather there is a family of perfectly acceptable solutions, each of which has its own strengths and weaknesses. To become more aware of the different possibilities that exist, the authors recommend that you work as part of a team of two or three students when performing these designs. Of course, be sure you have the permission of your instructor to do so.

For problems 20.1 to 20.3, use the general layout and framing arrangement for the building from the example in this chapter, as well as the loads acting on the building.

20-1. Design the frame along column line 1 for the building discussed in this chapter.

20-2. Design the frame along column line A for the building discussed in this chapter.

20-3. Design the frame along column line C for the building discussed in this chapter.

For problems 20.4 to 20.6, use the general layout for a two-story office building shown in the figure below. The architects on this project require a 10-ft ceiling height in the open work areas and an 8-ft ceiling height in the fixed offices. The work area will be used for drafting and desk space, and will have movable partitions rather than fixed walls.

The building will be clad with material weighing 45 psf. Assume that the cladding is attached to the building in such a manner that the loads from the cladding are applied to the structural frame at the intersections of the primary members.

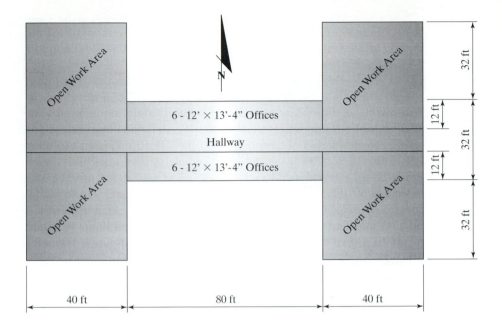

To determine the loads, assume that this building is located in central Illinios. Unless directed by your instructor, you do not need to consider seismic loads. Further, assume that the lateral loads applied to the cladding by wind are a net 30 psf—15 psf pressure load applied to one face and 15 psf pressure load applied to one face and 15 psf suction applied to the opposite face. The ground floor of the building will be a slab on grade. The geotechnical engineers have determined that the foundations are such that a full-moment resisting foundation is possible. As such, the building can be designed as an unbraced frame.

20-4.   Design a typical N-S frame through the open work areas.

20-5.   Design a typical N-S frame through the offices.

20-6.   Design a typical E-W frame along the length of the building along the hallway.

For problems 20.7 to 20.10, use the general layout for a two-story manufacturing facility shown in the figure below. The architects on this project require a 10-ft ceiling height in the manufacturing and storage areas, and an 8-ft ceiling height in the fixed offices.

The building will be clad with material weighing 45 psf. Assume that the cladding is attached to the building in such a manner that the loads from cladding are applied to the structural frame at the intersections of the primary members.

To determine the loads, assume that this building is located in the city where you attend school. The floors can assumed to be 6-in reinforced concrete. Unless directed by your instructor, you do not need to consider seismic loads. The ground. floor of the building will be a slab on grade. The building should be designed as a braced frame.

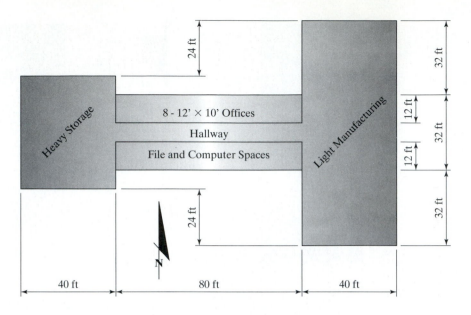

20-7. Design a typical N-S frame through the heavy storage area.

20-8. Design a typical N-S frame through the light manufacturing area.

20-9. Design a typical N-S frame through the office, file, and computer spaces.

20-10. Design a typical E-W frame along the length of the building.

APPENDIX A

Allowable Stress Design

A.1 INTRODUCTION

In the pages that follow, a very brief introduction to the allowable stress design (ASD) method is presented as it applies to tension members, columns, beams, and connections. The authors assume that the reader has previously studied these topics using the LRFD procedure. Therefore, we do not repeat items that were involved in that prior study, such as net areas, effective net areas, effective column lengths, and so on.

With ASD the working or service loads are combined for the various members of a particular structure to obtain the maximum loads. Then, members are selected such that their computed elastic stresses (when the maximum loads are applied) do not exceed certain allowable values. In structural steel design, these stresses usually are given as being equal to some percentage of the yield stress (e.g., $0.60F_y$, $0.66F_y$, etc.).

The ASD design procedure described in these pages is based upon the "Specification for Structural Steel Buildings—Allowable Stress Design and Plastic Design" and the *Manual of Steel Construction Allowable Stress Design* 9th ed., both published in 1989 by the AISC. The Specification is included in the ASD Manual.

A.2 TENSION MEMBERS

The ASD Specification (D-1) provides an allowable tensile stress of $0.60F_y$ on the gross cross-sectional areas of members at sections where there are no holes. For sections where there are holes for bolts, the allowable tensile *stress* is $0.50F_u$, applicable to the effective net areas. Thus, a safety factor of 1.67 is provided against yielding of the gross section ($F_y/0.60F_y = 1.67$), and a value of 2.0 is provided against the fracture of the member through its smallest effective net area ($F_u/0.50F_u = 2.0$). Thus, the allowable capacity of a tension member with bolt holes permitted by the ASD Specification is equal to the smaller of the following two values:

$$T = 0.60F_y A_g \quad \text{or} \quad T = 0.50F_u A_e$$

The preceding value of A_e is the effective net area of a steel section and equals UA_n. *To compute the net area A_n of a bolted member, the diameter of standard holes are assumed to equal the bolt diameters plus 1/8 in.*

The preceding values are not applicable to threaded steel rods or to members with pin holes (as in eyebars). For members with pin holes, an allowable tensile stress of $0.60F_y$ is permitted on the gross area of the member, while a value of $0.45F_y$ is permitted on the net area through the pin hole.

Example A-1 illustrates the calculations necessary for determining the effective net area of a bolted W section connected only to its flanges. In addition the allowable tensile strength of the member is computed.

Example A-1

Determine the allowable tensile load for a W10 × 45 with two lines of $\frac{3}{4}$-in.-diameter bolts in each flange, using A36 steel and the ASD Specification. There are at least three bolts in each line, and the bolts are not staggered with respect to each other. Select U from Table 3-2 on page 80.

Solution. Using a W10 × 45 (A_g = 13.3 in², d = 10.10 in, b_f = 8.020 in, t_f = 0.620 in.)

    a. $T = 0.60F_yA_g = (0.60)(36)(13.3) = 287.3 \text{ k} \leftarrow$

    b. $A_n = 13.3 - (4)\left(\frac{7}{8}\right)(0.620) = 11.13 \text{ in}^2$

           $U = 0.90$ since $b_f > \frac{2}{3}d$, bolts through the flanges and at least 3 bolts in a row

           $A_e = UA_n = (0.90)(11.13) = 10.02 \text{ in}^2$

           $T = 0.50F_uA_e = (0.50)(58)(10.02) = 290.6 \text{ k}$

           Allowable $T = 287.3 \text{ k}$

Example A-2 illustrates the design of a bolted tension W section. As previously stated, the tensile design strength T of a member is the least of (a) $0.60F_gA_g$ or (b) $0.50F_uA_e$.

To satisfy the first of these expressions, the minimum gross area must be at least

$$\min A_g = \frac{T}{0.60F_y}$$

To satisfy the second expression, the minimum value of A_e must be at least

$$\min A_e = \frac{T}{0.50F_u}$$

And since $A_e = UA_n$, the minimum value of A_n is

$$\min A_n = \frac{\min A_e}{U} = \frac{T}{0.50F_uU}$$

Then, the minimum A_g for the second expression must at least equal the minimum value of A_n plus the estimated hole areas.

$$\min A_g = \frac{T}{0.50F_uU} + \text{estimated hole areas}$$

The designer can substitute into Equations (1) and (2), taking the larger value of A_g so obtained for an initial size estimate. However, notice that the maximum preferable slenderness ratio L/r is 300. From this value it is easy to compute the preferable minimum value of r for a particular design—that is, the value of r for which the slenderness ratio will be exactly 300. It is undesirable to consider a section whose least r is less than this value, because its slenderness ratio would exceed the preferable maximum value of 300.

$$\text{Preferable min } r = \frac{L}{300}$$

In Example A-2 a W section is selected for a given tensile load.

Example A-2

Select a 30-ft-long W12 section of A36 steel to support a total tensile load of 220 k. As shown in Fig. A.1, the member is to have two lines of 7/8-in bolts in each flange (at least three in a line). Select U from Table 3-2 on page 80.

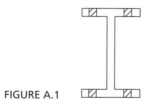

FIGURE A.1

Solution. Computing the minimum A_g required:

$$\text{Min } A_g = \frac{T}{0.60F_y} = \frac{220}{(0.60)(36)} = 10.19 \text{ in}^2$$

$$\text{Min } A_g = \frac{T}{0.50F_uU} + \text{estimated hole areas}$$

Assume $U = 0.90$ and assume flange thickness is about 0.515 in after looking at W12 sections in the ASD Manual, which have areas of about 10.19 in² or a little more.

$$\text{Min } A_g = \frac{220}{(0.50)(58)(0.90)} + (4)(\tfrac{7}{8} + \tfrac{1}{8})(0.515) = 10.49 \text{ in}^2$$

$$\text{Preferable min } r = \frac{L}{300} = \frac{(12)(30)}{300} = 1.2 \text{ in}$$

Try W12 × 40 ($A_g = 11.7 \text{ in}^2$, $d = 11.9$ in,

$b_f = 8.01$ in, $t_f = 0.515$ in, $r_y = 1.94$ in)

Checking:

$$T = 0.60F_yA_g = (0.60)(36)(11.7) = 252.7 \text{ k} > T = 220 \text{ k} \qquad \text{(OK)}$$

$$T = 0.50F_uUA_n, \text{ with } U = 0.90 \text{ since } \frac{b_f}{d} > \frac{2}{3},$$

$$\text{and } A_n = 11.7 - (4)(1.00)(0.515) = 9.64 \text{ in}^2$$

$$T = (0.50)(58)(0.90)(9.64) = 251.6 \text{ k} > 220 \text{ k} \qquad \text{(OK)}$$

$$\frac{L}{r} = \frac{(12)(30)}{1.94} = 186 < 300 \qquad \text{(OK)}$$

A.3   COLUMNS

In the ASD Specification for columns an assumption is made, because of residual stresses, that the upper limit of elastic buckling is defined by an average stress equal to one-half of the yield point $(\frac{1}{2}F_y)$. If this stress is equated to the Euler expression, the value of the slenderness ratio at this upper limit can be determined for a particular steel. This value is referred to as C_c, the slenderness ratio dividing elastic from inelastic buckling, and is determined as follows:

$$\frac{1}{2}F_y = \frac{\pi E}{(L/r)^2} = \frac{\pi E}{C_c^2}$$

$$C_c = \sqrt{\frac{2\pi^2 E}{F_y}}$$

The values of C_c can be computed with little difficulty, but the ASD Manual gives its values for each steel (126.1 for A36, 116.7 for the 42,000 psi yield point steels, etc.). For columns with slenderness ratios less than C_c, a parabolic formula, ASD Equation E2-1, is used.

$$F_a = \frac{\left[1 - \dfrac{(KL/r)^2}{2C_c^2}\right]F_y}{\dfrac{5}{3} + \dfrac{3(KL/r)}{8C_c} - \dfrac{(KL/r)^3}{8C_c^3}} \qquad \text{(ASD Equation E2-1)}$$

For values of KL/r greater than C_c, the Euler formula is used. With a factor of safety of 1.92 (or 23/12), the expression becomes

$$F_a = \frac{12\pi^2 E}{23(KL/r)^2} \qquad \text{(ASD Equation E2-2)}$$

Figure A.2 shows the ranges in which the two ASD column expressions are used. The denominator of Equation E2-1 is the factor of safety to be used and usually gives a value not much greater than the one used for axially loaded tension members.

In Example A-3, the allowable load that a particular column can support is calculated. The value of K, the effective length factor for the column, is determined in the usual manner. (Here, Table C-C2.1 in Part 5 of the ASD Manual is used.) Next, the

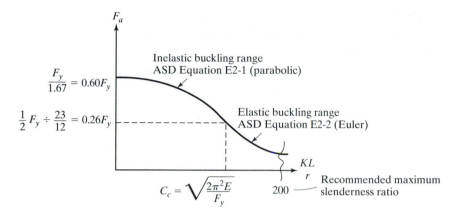

FIGURE A.2

ASD column allowable-stress curve.

slenderness ratio is computed and the allowable design stress F_a is determined from the appropriate column formula, as printed in Table 3-36 of Part 6 of the Manual. Finally, this value is multiplied by the cross-sectional area of the W section to determine the allowable column load.

Example A-3

Using the column design stress values shown in Table 3-36 in the ASD Manual for A-36 steel, determine the allowable compression load P that the axially loaded column shown in Fig. A.3 can support.

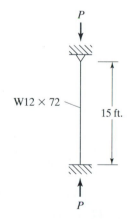

FIGURE A.3

Solution. Using a W12 × 72 $(A = 21.1\text{ in}^2, r_y = 3.04\text{ in})$

$$K = 0.80 \text{ from Table 5-1 in this text (page 136)}$$

$$\frac{KL}{r} = \frac{(0.80)(12 \times 15)}{3.04} = 47.37$$

$$F_a = 18.58 \text{ ksi from Table 3-36 of ASD Manual}$$

$$P = F_a A_g = (18.58)(21.1) = 392 \text{ k}$$

In Example A-4, a column with $KL = 14$ ft is selected using the ASD equations. An effective slenderness ratio of 50 is assumed, the allowable stress for that value is determined from Table C-50 of the ASD Manual, and the resulting stress is divided into the column load to obtain an estimated column area. After a trial section is selected with approximately that area, its actual slenderness ratio and allowable strength are computed. The first estimated size is too small, and the next larger section in that series of shapes is tried and found to be satisfactory.

Example A-4

Select a W14 section for the column and load shown in Fig. A.4 using a steel with $F_y = 50$ ksi.

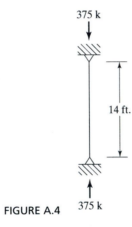

375 k

14 ft.

FIGURE A.4   375 k

Solution

Assume $\dfrac{KL}{r} = 50$

$F_a = 24.35$ ksi from ASD Table C-50

A reqd. $= \dfrac{375}{24.35} = 15.40 \text{ in}^2$

Try W14 × 53 ($A = 15.6 \text{ in}^2$, $r_y = 1.92$ in)

$$\left(\frac{KL}{r}\right)_y = \frac{(1.0)(12 \times 14)}{1.92} = 87.5 > \frac{K_x L_x}{r_x}$$

$F_a = 17.47$ ksi

$P_a = (17.47)(15.6) = 272.5 \text{ k} < P = 375 \text{ k}$ \hfill (NG)

Try W14 × 61 ($A = 17.9 \text{ in}^2$, $r_y = 2.45$ in)

$$\left(\frac{KL}{r}\right)_y = \frac{(1.0)(12 \times 14)}{2.45} = 68.57 > \frac{K_x L_x}{r_x}$$

$F_a = 21.20 \text{ ksi}$

$P_a = (21.20)(17.9) = 379.5 \text{ k} > P = 375 \text{ k}$ (OK)

Use W14 × 61

Note: The same answer can be obtained (and much more quickly) by using the ASD Column Tables in Part 3 of the Manual.

A.4 BEAMS

Included in the items that need to be considered in beam design are moments, shears, crippling, buckling, lateral support, deflection, and perhaps fatigue. Beams are selected that satisfactorily resist the bending moments, and then they are checked to see if any of the other items are critical. To select a beam for a given situation, the maximum moment is calculated for the assumed loading and a section having that much resisting moment is selected from the ASD Manual. It is usually more convenient, however, to work with the section modulus as described herein.

In the flexure formula, f_b is the fiber stress in the outermost fiber at a distance c from the neutral axis, and I is the moment of inertia of the cross section. Remember that this formula is limited to stress situations below the elastic limit because it is based on the usual elastic assumptions: a plane section before bending remains a plane section after bending; stress is proportional to strain; and so on.

If a beam is to be designed for a particular bending moment M and for a certain allowable stress F_b, the section modulus required to provide a beam of sufficient bending strength can be obtained from the flexure formula:

$$S = \frac{I}{C} \geq \frac{M}{F_b}$$

From the table in Part 2 of the ASD Manual, entitled "Allowable Stress Design Selection Table," steel shapes having sufficient section moduli can be quickly selected.

A limiting moment permitted by the allowable-stress method is the moment at which the stress in the outermost fibers first reaches the yield point. The true bending strength of a beam, however, is larger than this commonly used value because the beam will not fail at this condition. The outermost fibers will yield, and the stress in the inner fibers will increase until they reach the yield stress, and so on, until essentially the whole section is plastified.

This plastification process is correct only if the beam remains stable in other ways—that is, it must have sufficient lateral support to prevent lateral buckling of the compression flange, and it must have a sufficiently stocky profile to prevent local buckling.

The ASD Specification (F1) gives different allowable bending stresses for different conditions. For most cases the allowable bending stress is

$$F_b = 0.66F_y \qquad \text{(ASD Equation F1-1)}$$

This expression can be used to determine the allowable bending stress in the extreme fibers of compact hot-rolled shapes and built-up members (not including members consisting of steels with $F_y > 65$ ksi ksi or hybrid girders built up with different yield stress steels) that are symmetrical about and loaded in the plane of their minor axis and that meet the other requirements of Section B5.1 of the ASD Specification for compact sections. One of these requirements is that the flange must be continuously attached to the web. A built-up section with its flanges intermittently welded to the web does not meet this requirement.

Example A-5 illustrates the design of a steel beam whose compression flange has full lateral support, thus permitting the use of the same allowable stress in both its tension and compression flanges.

Example A-5

A 5-in reinforced-concrete slab is to be supported with steel beams 8 ft 0 in on centers. The beams, which will span 20 ft, are assumed to be simply supported. If the concrete slab is designed to support a live load of 100 psf, determine the lightest steel section required to support the slab. The compression flange of the beam will be incorporated in the concrete slab and is thus laterally supported. The concrete weighs 150 lb/ft^3, and the steel has an $F_y = 36$ ksi. Thus, $F_b = (0.66)(36) = 24$ ksi.

Solution

$$\text{Dead load: slab} = (8)\left(\frac{5}{12}\right)(150) = 500 \text{ lb/ft}$$

$$\text{Estimated beam weight} \qquad = \quad 26$$

$$\text{Live load: } 8 \times 100 \qquad = \quad \underline{800}$$

$$\text{Total uniform load} \qquad\qquad = 1326 \text{ lb/ft}$$

$$M = \frac{(1.326)(20)^2}{8} = 66.3 \text{ ft-k}$$

$$S_{\text{reqd.}} \geq \frac{M}{F_b} = \frac{(12)(66.3)}{24} = 33.2 \text{ in}^3$$

Use W12 × 26 ($S_x = 33.4$ in)

The ASD Specification does not require a reduction in the allowable strength of a beam with bolts in either flange, as long as the tensile fracture strength of the net flange area divided by a factor of safety of 2.0 is at least as large as the tensile yield strength of the gross flange area divided by a factor of safety of 1.67. Expressed as an equation, no deduction for bolt or rivet holes in either flange is necessary if

$$0.5F_u A_{fn} \geq 0.6F_y A_{fg} \qquad \text{(ASD Equation B10-1)}$$

In this equation, A_{fn} is the net flange area and A_{fg} is the gross flange area. Substituting into this expression, we find that no deduction is necessary if the net flange area is equal to or greater than 75 percent of the gross flange area for A36 steel or 92 percent for A992 steel. These values are shown in Table A.1.

TABLE A.1 When to Ignore Bolt and Rivet Holes
in Beam and Girder Flanges

Steel	No reduction if $A_{fn}/A_{fg} \geq$
A36	0.75
A992	0.92
A588	0.86

Should $0.5F_uA_{fn}$ be less than $0.6F_yA_{fg}$, the ASD Specification requires that the flexural properties of the section be based on an effective tension flange area A_{fe}, determined as follows:

$$A_{fe} = \frac{5}{6}\frac{F_u}{F_y}A_{fn} \qquad \text{(ASD Equation B10-3)}$$

In other words, to calculate the moment of inertia for the flanges, the transfer expression Ad^2 becomes $A_{fe}d^2$. If the bolt or rivet holes are in the tension flange, we assume they are in the compression flange as well. If they are only in the compression flange, they are ignored because it is felt that the fasteners can adequately transmit compression through the holes.

A beam with holes in its tension flange is examined in Example A-6 in accordance with the provisions of ASD Specification B10.

Example A-6

A W16 × 31 beam section ($d = 15.9$ in, $b_f = 5.53$ in, $t_f = 0.440$ in) consisting of A992 steel has two holes for 7/8-in bolts in its tension flange. Does the flange area need to be reduced for calculating the beam properties? If so, how much should I_x be reduced?

Solution

$A_{fg} = (5.53)(0.440) = 2.433 \text{ in}^2$

$A_{fn} = 2.433 - (2)(1.0)(0.440) = 1.553 \text{ in}^2$

$\dfrac{A_{fn}}{A_{fg}} = \dfrac{1.553}{2.433} = 0.638 < 0.92$ for A572 steel

$\therefore$ A reduction is necessary

$A_{fe} = \dfrac{5}{6}\dfrac{F_u}{F_y}A_{fn} = \left(\dfrac{5}{6}\right)\left(\dfrac{65}{50}\right)(1.553) = 1.682 \text{ in}^2$

Reduction in flange area $= 2.433 - 1.682 = 0.751 \text{ in}^2$

Reduction in $I_x = 2Ad^2$

$\qquad = (2)(0.751)\left(\dfrac{15.9}{2} - \dfrac{0.440}{2}\right)^2 = 89.7 \text{ in}^4$

The ASD Specification (F4) states that the allowable shear stress F_v is $0.40F_y$ if $h/t_w \leq 380/\sqrt{F_y}$. In this expression, h is the clear distance between the beam or girder flanges. Almost all rolled sections fall into this class, and the cross-sectional area effective in resisting shear is considered to be the overall depth of the member times its web thickness. For such cases, the shear stress is calculated as follows:

$$f_v = \frac{V}{dt_w} \leq 0.40F_y$$

If h/t_w is greater than $380/\sqrt{F_y}$, the allowable shear stress F_v becomes $F_y/2.89(C_v) \leq 0.40F_y$, where C_v is given in Section F4 of the Specification and is the ratio of the "critical" web stress to the shear yield stress of the web. This allowable shear stress is applicable to the clear distance between the flanges times the web thickness, as in

$$f_v = \frac{V}{ht_w}$$

The maximum permissible external shear that a particular beam can resist can be calculated by substituting the allowable shear stress F_v into the preceding expression and solving for V:

$$V = F_v dt_w \qquad \text{or} \qquad V = F_v ht_w$$

In the beam tables of the ASD Manual, this value V is tabulated for each of the sections normally used as beams with yield stresses of 36 and 50 ksi. With a glance at these tables, the designer can check the shear in a particular beam. If it is too high, he or she can quickly select another section that has a satisfactory allowable V.

Example A-7

Select a W section for the load and span shown in Fig. A.5. Use 50 ksi steel and assume full lateral bracing for the compression flange. Also check shear.

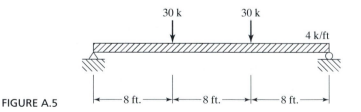

FIGURE A.5

Solution

Assume beam weight = 84 lb/ft

$$M = \frac{(4.084)(24)^2}{8} + (30)(8) = 534 \text{ ft-k}$$

$$S_{reqd.} \geq \frac{M}{F_b} = \frac{(12)(534)}{(0.66)(50)} = 194.2 \text{ in}^3$$

$$W24 \times 84 \ (d = 24.10 \text{ in}, t_w = 0.470 \text{ in}, k = 1.27 \text{ in})$$

Check shear:

$$\frac{h}{t_w} = \frac{24.10 - (2)(1.27)}{0.470} = 45.87 < \frac{380}{\sqrt{50}} = 53.74$$

$$V = (12)(4.084) + 30 = 79.0 \text{ k}$$

$$f_v = \frac{V}{dt_w} = \frac{79.0}{(24.10)(0.470)} = 6.97 \text{ ksi}$$

$$< F_v = 0.4f_y = (0.4)(50) = 20 \text{ ksi} \qquad \text{(OK)}$$

$$\text{Use W24} \times 84$$

A.5 BOLTED CONNECTIONS

A.5.1 Bearing-Type Bolts

The allowable shearing stresses for high-strength bolts are presented in Table J3.2 of the ASD Specification. Upon observing these values, you will note that the allowable shearing stress is reduced when the bolt threads are included in the shear planes because of the reduced cross-sectional area through the threads. For example, the allowable shear stress for A325 bolts for a bearing-type connection and standard-size holes is 21 ksi if threads are not excluded from the shear plane, and 30 ksi if threads are excluded. Should a bolt be in double shear, its shearing strength is considered to be twice its single shear value.

The allowable design strength of a bolt in bearing equals the allowable bearing stress of the connected part in kips per square inch, times the diameter of the bolt, times the thickness of the member that bears against the bolt. (For countersunk bolts and rivets, one-half of the depth of the countersink should be deducted, says ASD Specification J3.3.) When the distance L_e in the direction of the force from the center of a regular or oversized hole (or from the center of the end of a slotted hole) to the edge of a connected part is not less than $1\frac{1}{2}$ times the bolt diameter d, and the distances center-to-center of the holes is not less than $3d$, and two or more bolts are used in the direction of the line of force, the bearing strength is

$$F_p = 1.2F_u \text{ for standard or short-slotted holes} \qquad \text{(ASD Equation J3-1)}$$

$$F_y = 1.0F_u \text{ for long-slotted holes perpendicular to the load} \quad \text{(ASD Equation J3-2)}$$

Should deformations around the hole not be a design consideration, the two preceding expressions may be replaced with

$$F_p = 1.5F_u \qquad \text{(ASD Equation J3-4)}$$

For the examples and home problems in this appendix, F_p is assumed to equal $1.2F_u$. This also is the value used for the examples and tables of Part 4 of the ASD Manual.

Although the allowable bearing stress of a bolt with a small end distance is reduced, *the allowable values of the other bolts in the connection are not reduced.* The value of F_p for a single bolt or for two or more bolts in the line *each with an end distance less than* $1\frac{1}{2}d$ can be determined with the following equation:

$$F_p = \frac{L_e F_u}{2d} \le 1.2F_u \qquad \text{(ASD Equation J3-3)}$$

The allowable load on the edge bolt is the lesser of the allowable shear strength of the bolt on its cross-sectional area and the allowable bearing strength against the side of the hole. Should the allowable bearing strength control, we can increase the allowable strength of the connection by increasing the thickness of the parts being connected and by increasing the edge distance (if edge distance causes a reduction in F_p).

Example A-8 illustrates the calculations involved in determining the strength of a bearing-type connection. Using a similar procedure, the number of bolts required for a certain loading condition is calculated in Example A-9.

The values given for the strength of bolts in this chapter, whether bearing-type or slip-critical, can be obtained from the tables entitled "Bolts, Threaded Parts, and Rivets Shear Allowable Load in Kips" of Part 4 of the ASD Manual.

Example A-8

Determine the allowable design strength P of the bearing-type connection shown in Fig. A.6. The steel F_y is 50 ksi and $F_u = 65$ ksi, the bolts are 7/8-in A325, the holes are standard sizes, the threads are excluded from the shear plane, edge distances are $>1\frac{1}{2}d$, and the distance center-to-center of the holes is $>3d$. $F_p = 1.2F_u$.

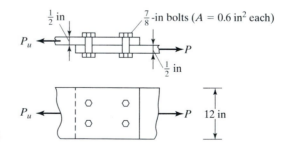

FIGURE A.6

Solution. Allowable tensile force:

$$A_g = \left(\tfrac{1}{2}\right)(12) = 6.0 \text{ in}^2$$

$$A_n = 6.00 - (2)(1.0)\left(\tfrac{1}{2}\right) = 5.0 \text{ in}^2 < 0.85A_g = (0.85)(6.0) = 5.10 \text{ in}^2$$

$$\therefore A_e = 5.0 \text{ in}^2$$

$$P \le 0.60F_y A_g = (0.60)(50)(6.0) = 180 \text{ k}$$

$$P \le 0.50F_u A_e = (0.50)(65)(5.0) = 162.5 \text{ k}$$

Bolts in single shear and bearing on $\frac{1}{2}$ in:

$\quad P$ for single shear $\leq A_b F_v N_b = (0.6)(30)(4) = 72\,k \leftarrow$

$\quad\quad P$ for bearing $= \leq td1.2F_u N_b = (0.5)\left(\frac{7}{8}\right)(1.2)(58)(4) = 121.8\,k$

$\quad$ Design $P = 72\,k$

Example A-9

How many 3/4-in A325 bolts in standard-size holes with threads excluded from the shear plane are required for the bearing-type connection shown in Fig. A-7? Use a steel with $F_y = 50$ ksi and $F_u = 65$ ksi. Assume edge distances and spacing requirements are met, including those for $F_p = 1.2F_u$.

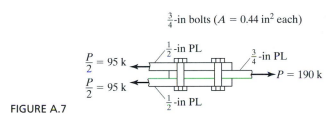

$\frac{3}{4}$-in bolts ($A = 0.44\ \text{in}^2$ each)

$\frac{1}{2}$-in PL

$\frac{P}{2} = 95\,k$

$\frac{3}{4}$-in PL

$\frac{P}{2} = 95\,k$

$P = 190\,k$

$\frac{1}{2}$-in PL

FIGURE A.7

Solution. Bolts in double shear and bearing on 3/4-in:

$\quad$ Design shear strength per bolt $= R = N_g A_b F_v = (2)(0.44)(30) = 26.4\,k \leftarrow$

$\quad$ Design bearing strength per bolt $= R = td\,1.2F_u = \left(\frac{3}{4}\right)\left(\frac{3}{4}\right)(1.2)(58) = 39.1\,k$

$\quad$ Number of bolts required $= N_b = \dfrac{P}{R} = \dfrac{190}{26.4} = 7^+$

$\quad$ Use 8 or 9 bolts (depending on arrangement)

A.5.2   Slip-Critical Connections

If bolts are tightened to their required tensions for slip-critical connections, there is very little chance of their bearing against the plates they are connecting. In fact, tests show that there is very little chance of slip occurring unless there is a calculated shear of at least 50 percent of the total bolt tension. This means that slip-critical bolts are not stressed in shear; however, the ASD Specification provides allowable shear strengths (they really are permissible friction values on the faying surfaces) so that the designer can handle the connections in just about the same manner he or she uses for bearing-type connections. The ASD Specification assumes the bolts are in "shear" and no bearing, and the allowable shear stresses for high-strength bolts are given in their Table J3.2. However, in the event of slippage between the connected parts, bolt strength and plate bearing still need to be confirmed as these criteria are checked for bearing-type connections.

Example A-10 presents the design of a slip-critical connection for a lap joint.

Example A-10

It is desired to design a slip-critical connection for the plates shown in Fig. A.8 to resist an 80-k load using 1-in A325 high-strength bolts with threads excluded from the shear plane and meeting all edge and center-to-center distance requirements.

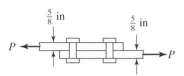

FIGURE A.8

Solution. Slip-critical design, bolts in "single shear," and no bearing:

$$\text{"Single shear"} = R = A_bF_v = (0.785)(17.0) = 13.35/\text{bolt}$$

$$\text{Number of bolts required} = N_b = \frac{P}{R} = \frac{80}{13.35} = 5.99$$

Use six bolts.

A.6   WELDED CONNECTIONS

Table J2.5 of the ASD Specification provides allowable stresses for fillet welds, complete-penetration groove welds, and partial-penetration groove welds. Example A-11 illustrates the calculations necessary to determine the allowable strength of a fillet-welded connection. You will notice that the allowable stress in the weld equals 0.30 times the nominal tensile strength of the weld.

Example A-11

Determine the allowable shear force that may be applied to a 1-in length of a 5/16-in fillet weld using (a) the shielded metal arc process (SMAW) and (b) the submerged arc process (SAW). Use the ASD Specification and E70 electrodes with a minimum tensile strength of 70 ksi

Solution

a.   SMAW process:

Effective throat thickness of weld $= t_e = (0.707)\left(\frac{5}{16}\right) = 0.221$

Allowable capacity $= R = t_e(0.30)F_{EXX} = (0.221)(0.30 \times 70)$

$$= 4.64 \text{ k/in}$$

b.   SAW process:

From ASD (J2.2a) the effective throat thickness of weld $= t_e = \frac{5}{16}$ in

Allowable capacity $= R = t_e(0.30F_{EXX}) = \left(\frac{5}{16}\right)(0.30 \times 70) = 6.56 \text{ k/in}$

Fillet welds may not be designed using an allowable stress that is greater than permitted on the adjacent members being connected. For instance, the allowable shear stress on the effective area of fillet welds is 0.30 times the tensile strength of the electrode, but it may not exceed the allowable stress in the base material ($30F_u$ in shear or $0.60F_y$ in tension). In the ASD Specification, fillet welds are assumed to transmit loads by shear on the effective area regardless of the direction of the loads or of the position of the welds at the connections.

Examples A-11 and A-12 illustrate the calculations necessary to determine the allowable loads that can be applied to plates connected with SMAW and SAW fillet welds. In each of these examples the allowable strength per inch of the welds controls and is multiplied by the total length of the welds to give the total allowable capacity of the connections.

Example A-12

What is the allowable capacity of the connection shown in Fig. A.9 if A36 steel, the ASD Specification, and E70 electrodes are used? The 7/16-in fillet welds shown were made by the SMAW process.

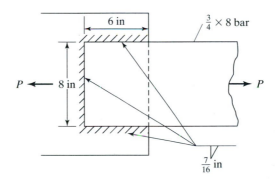

FIGURE A.9

Solution

Effective throat thickness $= t_e = (0.707)\left(\frac{7}{16}\right) = 0.309$ in

Allowable capacity of weld/in $= R_w = t_e(0.30E_{EXX})(0.309)(0.30 \times 70)$

$$= 6.489 \text{ k/in} \leftarrow$$

Base metal strength adjacent to horizontal welds

$$R_{bm} = 0.30F_u t = (0.30)(58)\left(\frac{3}{4}\right) = 13.05 \text{ k/in} > R_w \qquad \text{(OK)}$$

Allowable tensile capacity of weld $= (20)(6.489) = 129.8$ k

Allowable tensile capacity of PL $= (\frac{3}{4} \times 8)(0.60 \times 36) = 129.6 \text{ k} \leftarrow$

For not section fracture at start of weld,

$P_a = 0.50F_u A_m = (0.50)(58)\left(\frac{3}{4}\right)(8) = 174$ k

$P = 129.6$ k

Example A-13

Repeat Example A-11 using SAW welds.

Solution

Effective throat thickness $= t_e = (0.707)\left(\frac{7}{16}\right) + 0.11 = 0.419$ in

Allowable shear capacity of weld/in $= (0.419)(0.30 \times 70)$
$$= 8.799 \text{ k/in} \leftarrow$$

Allowable shear capacity of weld $= (20)(8.799) = 176$ k

Allowable tensile capacity of PL $= \left(\frac{3}{4} \times 8\right)(0.60 \times 36) = 129.6$ k $\leftarrow$

$R = 0.30F_u t = (0.30)(58)\left(\frac{3}{4}\right) = 13.05$ k/in > 8.799 k/in

$P = 129.6$ k

Example A-14 illustrates the design of fillet welds for two plates.

Example A-14

Using A36 steel, the ASD Specification, and E70 electrodes, design SMAW fillet welds to resist a full capacity load on the 3/8 × 6-in member shown in Fig. A.10.

Solution

$P = A_g F_t = \left(\frac{3}{8}\right)(6.0)(22) = 49.5$ k $\leftarrow$

$< P_a = A_e F_t = \left(\frac{3}{8}\right)(6)(0.5)(58) = 65.3$ k

Max weld size $= \frac{3}{8} - \frac{1}{16} = \frac{5}{16}$ in   (ASD Section J2-2b)

Min weld size $= \frac{3}{16}$ in (Table 14.2 in this text)   (ASD Table J2.3)

Use $\frac{5}{16}$-in weld

Effective thickness $= t_e = 0.707a = (0.707)\left(\frac{5}{16}\right) = 0.221$ in

Allowable capacity of weld/in

$$R_w = t_e(0.30F_{EXX}) = (0.221)(0.30 \times 70) = 4.64 \text{ k/in} \leftarrow$$

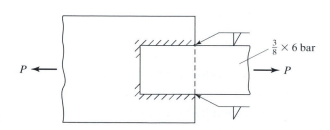

$P \leftarrow$ 　　　　　　　 $\frac{3}{8} \times 6$ bar

　　　　　　　　　　 $\rightarrow P$

FIGURE A.10

Base metal strength

$$R_{bm} = 0.30F_u t = (0.30)(58)(0.375) = 6.53 \text{ k/in}$$

Length required $= l_w = \dfrac{P_a}{R} = \dfrac{49.5}{4.64} = 10.67 \text{ in}$

We may use end return not less than 2a $= (2)\left(\tfrac{3}{16}\right) = \tfrac{3}{8}$ (say 1 in) use $\dfrac{10.67}{2} = 5.33 \text{ in}$ say $5\tfrac{1}{2}$ in each side.

APPENDIX B

Derivation of the Euler Formula

The Euler formula is derived in this section for a straight, concentrically loaded, homogeneous, long, slender, elastic, and weightless column with rounded ends. It is assumed that this perfect column has been laterally deflected by some means as shown in Fig. B.1 and that, if the concentric load P was removed, the column would straighten out completely.

The x- and y-axes are located as shown in the figure. As the bending moment at any point in the column is $-Py$, the equation of the elastic curve can be written as

$$EI\frac{d^2y}{dx^2} = -Py$$

For convenience in integration, both sides of the equation are multiplied by $2dy$:

$$EI2\frac{dy}{dx}d\frac{dy}{dx} = -2Pydy.$$

$$EI\left(\frac{dy}{dx}\right)^2 = -Py^2 + C_1$$

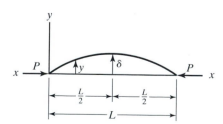

FIGURE B.1

678

When $y = \delta$, $dy/dx = 0$, and the value of C_1 will equal $P\delta^2$ and

$$EI\left(\frac{dy}{dx}\right)^2 = -Py^2 + P\delta^2$$

The preceding expression is arranged more conveniently as follows:

$$\left(\frac{dy}{dx}\right)^2 = \frac{P}{EI}(\delta^2 - y^2)$$

$$\frac{dy}{dx} = \sqrt{\frac{P}{EI}}\sqrt{\delta^2 - y^2}$$

$$\frac{dy}{\sqrt{\delta^2 - y^2}} = \sqrt{\frac{P}{EI}}dx$$

Integrating this expression, the result is

$$\text{arc} \sin\frac{y}{\delta} = \sqrt{\frac{P}{EI}}x + C_2$$

When $x = 0$ and $y = 0$, $C_2 = 0$. The column is bent into the shape of a sine curve expressed by the equation

$$\text{arc} \sin\frac{y}{\delta} = \sqrt{\frac{P}{EI}}x$$

When $x = L/2$, $y = \delta$, resulting in

$$\frac{\pi}{2} = \frac{L}{2}\sqrt{\frac{P}{EI}}$$

In this expression P is the *critical buckling load* or the maximum load that the column can support before it becomes unstable. Solving for P, we have

$$P = \frac{\pi^2 EI}{L^2}$$

This expression is the Euler formula, but usually it is written in a little different form involving the slenderness ratio. Since $r = \sqrt{I/A}$ and $r^2 = I/A$ and $I = r^2A$, the Euler formula may be written as

$$\frac{P}{A} = \frac{\pi^2 E}{(L/r)^2} = F_e$$

APPENDIX C

Slender Compression Elements

Section B5 of the LRFD Specification is concerned with the local buckling of compression elements. In that section, elements are classified as being compact, noncompact, or slender. As compact and noncompact elements have been discussed previously, this appendix is concerned only with the slender element. A brief summary of the LRFD method for determining design stresses for such members is presented in the next few paragraphs.

Should b/t ratios exceed the values given in LRFD Table B5.1 for noncompact elements, those elements will be classified as being slender, and their critical or F_{cr} stresses will have to be reduced. The design strength of an axially loaded compression member with slender elements will be reduced by multiplying it by a reduction factor Q.

The value of Q is equal to the product of two reduction factors Q_s and Q_a. Their values are dependent on whether the member consists of stiffened and/or unstiffened elements. Q_a is a reduction factor for slender stiffened compression elements, while Q_s is a reduction factor for slender unstiffened compression elements. Three cases are considered in Appendix B of the LRFD Specification:

1. For members consisting of unstiffened elements only, Q_s is to be determined with the appropriate formulas presented in this appendix, and $Q_a = 1.0$.
2. For members consisting of stiffened elements only, $Q_s = 1.0$, and Q_a is to be determined with the formulas to be presented in this appendix.
3. For members consisting of some stiffened elements and some unstiffened ones, Q_s and Q_a are determined with the appropriate formulas to be presented in this appendix.

Expressions for computing Q are given in LRFD Appendix B5.3a for the following types of members: (a) single angles; (b) flanges, angles, and plates projecting from rolled beams or columns or other compression members; (c) flanges, angles, and plates projecting from built-up columns or other compression members; and (d) stems of tees. In these equations, the following terms are used:

b = width of unstiffened compression element, in.

t = thickness of unstiffened compression element, in.

F_y = specified minimum yield stress, ksi

Only the Q_s expressions for case (a) are presented here.

When $0.45\sqrt{\dfrac{E}{F_y}} < \dfrac{b}{t} < 0.91\sqrt{\dfrac{E}{F_y}}$

$$Q_s = 1.340 - 0.76\left(\frac{b}{t}\right)\sqrt{\frac{F_y}{E}} \qquad \text{(LRFD Equation A-B5-3)}$$

When $b/t > 0.91\sqrt{\dfrac{E}{F_y}}$

$$Q_s = 0.53E\left[F_y\left(\frac{b}{t}\right)^2\right] \qquad \text{(LRFD Equation A-B5-4)}$$

In Example C-1, the value of Q_s is computed for a pair of angles that are used as a compression member. As the member consists only of unstiffened elements, $Q_a = 1.0$. The computed value for Q_s can be checked in the double-angle tables of Part 1 of the Manual. This same double-angle member is further considered in Appendix D as to lateral-torsional buckling and the Q_s determined here is used there.

Example C-1

The pair of Ls8 × 6 × 1/2, long legs back-to-back separated by 3/8 in, shown in Fig. C.1, is used as a compression member. Compute Q_s for this member, which is assumed to have an $F_y = 50$ ksi.

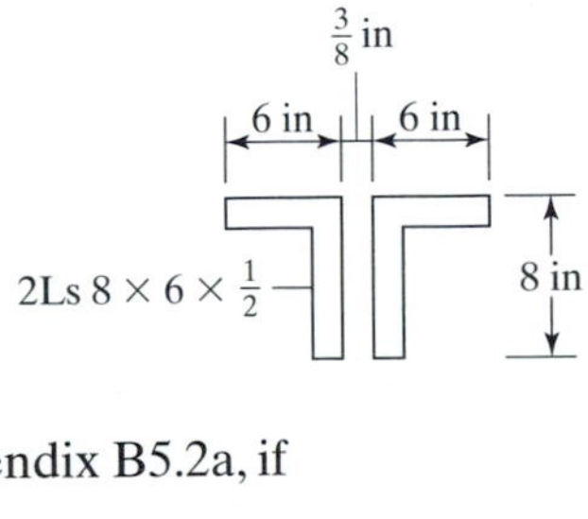

FIGURE C.1

Solution. As shown in LRFD Appendix B5.2a, if

$$0.45\sqrt{\frac{E}{F_y}} < \frac{b}{t} < 0.91\sqrt{\frac{E}{F_y}}$$

then LRFD Equation A-B5-3 is used to compute Q_s for angles projecting from a column:

$$0.45\sqrt{\frac{29{,}000}{50}} = 10.84 < \frac{8}{\frac{1}{2}} = 16 < 0.91\sqrt{\frac{29{,}000}{50}} = 21.92$$

We then have

$$Q_s = 1.340 - 0.76\left(\frac{b}{t}\right)\sqrt{\frac{F_y}{t}} \qquad \text{(LRFD Equation A-B5-3)}$$

$$= 1.340 - (0.76)(16)\sqrt{\frac{50}{29{,}000}}$$

$$= 0.835$$

Checks with value in double angle tables.

If we have slender elements in a compression member, its design compression strength is to be computed as follows:

For $\lambda_c\sqrt{Q} < 1.5$

$$F_{cr} = Q(0.658^{Q\lambda_c^2})F_y \qquad \text{(LRFD Equation A-B5-15)}$$

For $\lambda_c\sqrt{Q} > 1.5$

$$F_{cr} = \left[\frac{0.877}{\lambda_c^2}\right]F_y \qquad \text{(LRFD Equation A-B5-16)}$$

APPENDIX D

Flexural–Torsional Buckling of Compression Members

Usually symmetrical members such as W sections are used as columns. Torsion will not occur in such sections if the lines of action of the lateral loads pass through their shear centers. The *shear center* is that point in the cross section of a member through which the resultant of the transverse loads must pass so that no torsion will occur. The calculations necessary to locate shear centers were presented in Chapter 10. The shear centers of the commonly used doubly symmetrical sections occur at their centroids. This is not necessarily the case for other sections such as channels and angles. Shear center locations for several types of sections are shown in Fig. D.1. Also shown in the figure are the coordinates x_0 and y_0 for the shear center of each section with respect to its centroid. These values are needed to solve the flexural-torsional formulas, presented later in this section.

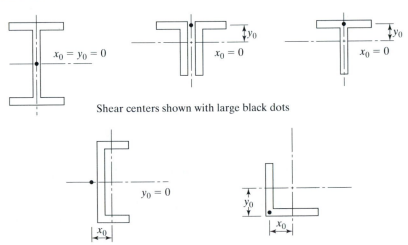

FIGURE D.1

Shear center locations for some common column sections.

Even though loads pass through shear centers, torsional buckling still may occur. If you load any section through its shear center, no torsion will occur, but one still computes torsional buckling strength for these members—that is, buckling load does not depend on the nature of the axial or transverse loading; rather, it depends on the cross-section properties, column length, and support conditions.

The average designer does not consider the torsional buckling of symmetrical shapes or the flexural-torsional buckling of unsymmetrical shapes. They usually think that these conditions don't control the critical column loads, or at least that they don't affect them much. If, however, we have unsymmetrical columns or even symmetrical columns made up of thin plates, we will find that torsional buckling or flexural-torsional buckling may significantly reduce column capacities.

In Appendix E of the LRFD Specification, a long list of formulas is presented for computing the flexural-torsional strength of column sections. The values given for column design strengths ($\phi_c P_n$ values) for double angles, single angles, and tees in Part 4 of the LRFD Manual make use of these formulas.

For flexural-torsion $P_u \leq \phi_c P_n = \phi_c A_g F_{cr}$ with $\phi_c = 0.85$ and F_{cr} to be determined from the formulas to follow from the specification. A list of definitions that are needed for using these formulas also is provided.

If $\lambda_e \sqrt{Q} \leq 1.5$

$$F_{cr} = Q(0.658^{Q\lambda_e^2})F_y \qquad \text{(LRFD Equation A-E3-2)}$$

If $\lambda_e \sqrt{Q} > 1.5$

$$F_{cr} = \left(\frac{0.877}{\lambda_e^2}\right)F_y \qquad \text{(LRFD Equation A-E3-3)}$$

where $Q = 1.0$ for elements meeting the width-thickness ratios λ_r of LRFD section B5.1 (and if not calculated as described in LRFD Appendixes E3 and B5.3).

$$\lambda_e = \sqrt{\frac{F_y}{F_e}} \qquad \text{(LRFD Equation A-E3-4)}$$

F_e = critical flexural-torsional elastic buckling stress

For doubly symmetric shapes,

$$F_e = \left[\frac{\pi^2 E C_w}{(K_z L)^2} + GJ\right]\frac{1}{I_x + I_y} \qquad \text{(LRFD Equation A-E3-5)}$$

For singly symmetric shapes where y is the axis of symmetry,

$$F_e = \frac{F_{ey} + F_{ez}}{2H}\left[1 - \sqrt{1 - \frac{4F_{ey}F_{ez}H}{(F_{ey} + F_{ez})^2}}\right] \qquad \text{(LRFD Equation A-E3-6)}$$

For unsymmetrical section F_e is the lowest root of the following cubic equation

$$(F_e - F_{ex})(F_e - F_{ey})(F_e - F_{ez}) - F_e^2(F_e - F_{ey})\left(\frac{x_0}{r_0}\right)^2$$

$$-F_e^2(F_e - F_{ex})\left(\frac{y_0}{r_0}\right)^2 = 0 \qquad \text{(LRFD Equation A-3E-7)}$$

The following is more convenient form of LRFD Equation A-E3-7:

$$HF_e^3 + \left[\frac{1}{r_0^2}(y_0 F_{ex} + x_0^2 F_{ey}) - (F_{ex} + F_{ey} + F_{ez})\right]F_e^2$$

$$+ (F_{ex}F_{ey} + F_{ex}F_{ez} + F_{ey}F_{ez})F_e - F_{ex}F_{ey}F_{ez} = 0$$

K_z = effective length factor for torsional buckling

G = shear modulus (ksi)

C_w = warping constant (in^6)

J = torsional constant (in^4)

$$\bar{r}_0^2 = x_0^2 + y_0^2 + \frac{I_x + I_y}{A} \qquad \text{(LRFD Equation A-E3-8)}$$

$$H = 1 - \left(\frac{x_0^2 + y_0^2}{\bar{r}_0^2}\right) \qquad \text{(LRFD Equation A-E3-9)}$$

$$F_{ex} = \frac{\pi^2 E}{(KL/r)_x^2} \qquad \text{(LRFD Equation A-E3-10)}$$

$$F_{ey} = \frac{\pi^2 E}{(KL/r)_y^2} \qquad \text{(LRFD Equation A-E3-11)}$$

$$F_{ez} = \left[\frac{\pi^2 E C_w}{(K_z L)^2} + GJ\right]\frac{1}{A\bar{r}_0^2} \qquad \text{(LRFD Equation A-E3-12)}$$

The values of C_w, J, $\bar{r}_0$, and H are provided for many sections in the "Flexural-Torsional Properties" tables of Part 1 of the Manual.

In Example D-1, the authors have gone through all of these formulas for a double-angle column. The resulting value for $\phi_c P_n$ is shown to coincide with the value given in the Manual.

Example D-1

Determine (a) the flexural buckling strength and (b) the flexural-torsional buckling strength of an 18-ft pinned-end column consisting of two Ls 8 × 6 × 1/2 long legs back to back with a 3/8-in gusset plate between them. The angles are made from A572Gr.50 steel and are connected to each other at 6-ft intervals with two fully tensioned high-strength bolts as shown in Fig. D.2. The cross section of the member is shown in Fig. D-3; $G = 11,200$ ksi, and $k = 1.0$.

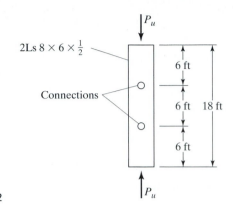

FIGURE D.2

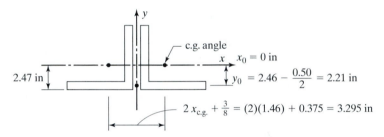

FIGURE D.3

Solution. Using two Ls 8 × 6 × 1/2 long legs back to back, 3/8-in gusset plate in between ($A = 13.6$ in², $r_x = 2.55$ in, r_z for one $L = 1.30$ in, and $r_y = 2.43$ in).

a.   Flexural buckling about the x-axis (perpendicular to the axis of symmetry)

$$\left(\frac{KL}{r}\right)_x = \frac{(1.0)(12 \times 18)}{2.55} = 84.71$$

From Example C-1 $Q_s = 0.834$

$$\lambda_{Lx} = \frac{K_x L_x}{r_x \pi}\sqrt{\frac{F_y}{F}} = \frac{84.71}{\pi}\sqrt{\frac{50}{29,000}} = 1.12$$

$$Q = Q_a Q_s = 1.0 Q_s$$

$$\lambda_{Ly}\sqrt{Q} = 1.018 < 1.12$$

$$F_{cv} = Q(0.658^{Q\lambda_c^2})F_y = 27.01 \text{ ksi}$$

$$Q_c F_{cv} = 0.85 F_{cv} = 22.96 \text{ ksi}$$

$$Q_c P_n = \phi_c F_c A_g = (22.96)(13.6) = 312.3 \text{ k}$$

b.   Flexural-torsional buckling with respect to the y-axis passing through the shear center of the section.

Cross section of member shown in Fig. D.2 together with some properties needed in the calculations.

From flexural-torsional properties tables: $\left.\begin{array}{l} J = 0.584 \text{ in}^4 \\ C_w = 2.280 \text{ in}^6 \end{array}\right\}$ for 1 angle

$\left.\begin{array}{l} r_0 = 4.16 \text{ in} \\ H = 0.709 \end{array}\right\}$ for 2 angles

Checking table values of r_0 and H, we have

$$\bar{r}_0^2 = x_0^2 + y_0^2 + r_x^2 + r_y^2 = 0^2 + 2.21^2 + 2.55^2 + 2.43^2$$

$$\text{(LRFD Equation A-E3-8)}$$

$\bar{r}_0 = 4.16 \text{ in}$

$$H = 1 - \frac{0^2 + 2.21^2}{(4.16)^2} = 0.718 \qquad \text{(LRFD Equation A-E3-9)}$$

Computing values of F_{ey}, F_{ez}, and F_e, we get

$$\frac{K_a}{r_i} \leq \left(\frac{3}{4}\right)\left(\frac{K_y L_y}{r_y}\right)$$

$$\frac{(1.0)(6.0)(12)}{1.30} \leq \left(\frac{3}{4}\right)\frac{(1.0)(1.2)(18.0)}{2.43}$$

$55.38 < 66.67$

$$\left(\frac{KL}{r}\right)_m = \sqrt{\left(\frac{KL}{r}\right)_0^2 + 0.82\frac{\alpha^2}{(1 + \alpha^2)}\left(\frac{a}{r_{ib}}\right)^2}$$

$$\left(\frac{KL}{r}\right)_0 = \left(\frac{KL}{r}\right)_y = \frac{(1.0)(12)(18)}{2.43} = 88.89$$

$$\alpha = \frac{h}{2r_{ib}} = \frac{3.295}{(2)(2.43)} = 0.678$$

$$\left(\frac{KL}{r}\right)_m = \sqrt{(88.89)^2 + 0.82\left[\frac{(0.678)^2}{1 + 0.678^2}\left(\frac{72.0}{2.43}\right)^2\right]} = 90.16$$

$$F_{ey} = \frac{\pi^2 E}{\left(\dfrac{KL}{r}\right)_m^2} = \frac{(\pi)^2(29{,}000)}{(90.16)^2} = 35.21 \text{ ksi} \,(\text{LRFD Equation A-E3-11})$$

$$F_{ez} = \frac{\dfrac{\pi^2 E C_w}{(K_z L_z)^2} + GJ}{A \bar{r}_0^2}$$

$$F_{ez} = \left[\frac{(\pi)^2 (29{,}000)(2 \times 2.28)}{(1.0 \times 12 \times 18)^2} + 11{,}200 \times 2 \times 0.584 \right]$$

$$\times \frac{1}{(13.5)(4.18)^2}$$

$$= 55.58 \text{ ksi} \qquad \text{(LRFD Equation A-E3-12)}$$

$$F_e = \frac{35.21 + 55.58}{(2)(0.709)} \left[1 - \sqrt{1 - \frac{(4)(35.21)(55.58)(0.709)}{(35.21 + 55.58)^2}} \right]$$

$$= 27.43 \text{ ksi} \qquad \text{(LRFD Equation A-E3-6)}$$

$$\lambda_e = \sqrt{\frac{F_y}{F_e}} = \sqrt{\frac{50}{27.43}} = 1.350 \qquad \text{(LRFD Equation A-E3-4)}$$

$$\lambda_e \sqrt{Q} = 1.350\sqrt{0.834} = 1.233 < 1.5$$

$$F_{cv} = Q(0.658^{Q\lambda_e^2})F_y = 0.834(0.658^{(1.233)^2})50 = 22.07 \text{ ksi}$$

$$Q_c F_{cv} = (0.85)(22.07) = 18.76 \text{ ksi}$$

$$\phi_c P_n = (18.76)(13.5) = 253 \, k < \phi_c P_n \text{ of } 310 \, k \text{ for } x\text{-axis}$$

APPENDIX E

Moment-Resisting Column Base Plates

Column bases frequently are designed to resist bending moments as well as axial loads. An axial load causes compression between a base plate and the supporting footing, while a moment increases the compression on one side and decreases it on the other side. For small moments the forces may be transferred to the footing through flexure of the base plate. When they are very large, stiffened or booted connections may be used. For a small moment the entire contact area between the plate and the supporting footing will remain in compression. This will be the case if the resultant load falls within the middle third of the plate length in the direction of bending.

Figures E.1(a) and (b) shows base plates suitable for resisting relatively small moments. For these cases, the moments are sufficiently small to permit their transfer to the footings by bending of the base plates. The anchor bolts may or may not have calculable stresses, but they are nevertheless considered necessary for good construction practice. They definitely are needed to hold the columns firmly in place and upright during the initial steel erection process. Temporary guy cables are also necessary during erection. The anchor bolts should be substantial and capable of resisting unforeseen erection forces. Sometimes these small plates are attached to the columns in the shop, and sometimes they are shipped loose to the job and carefully set to the correct elevations in the field.

Should the eccentricity ($e = M/P$) be sufficiently large that the resultant falls outside the middle third of the plate, there will be an uplift on the other side of the column, putting the anchor bolts on that side in tension.

The moment will be transferred from the column into the footing by means of the anchor bolts, embedded a sufficient distance into the footing to develop the anchor bolt forces. The embedment should be calculated as required by reinforced-concrete design methods.[1] The booted connection shown in Fig. E.1(c) is assumed to be welded

---

[1]*Building Code Requirements for Reinforced Concrete (ACI 318-89) and Commentary (ACI 318R-89)* (Detroit: American Concrete Institute, 1989), pp. 188–189, 250–251.

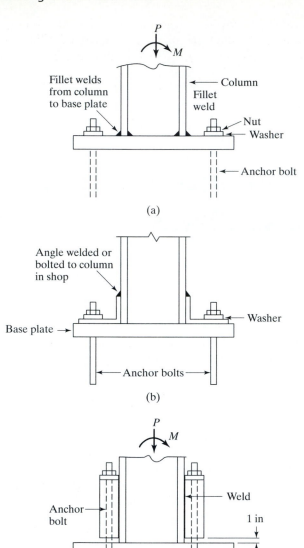

FIGURE E.1

to the column. The boots generally are made of angles or channels and are not, as a rule, connected directly to the base plate. Rather, the tensile force component induced by the moment is transmitted from the column to the foundation by means of the anchor bolts. When booted connections are used, the base plates normally are shipped loose to the job and carefully set to the correct elevations in the field.

The capacity of these connections to resist rotation is dependent on the lengths of the anchor bolts, which are available to deform elastically. This capacity can be increased somewhat by pretensioning the anchor bolts. (This is similar to the prestress discussion for high-strength bolts presented in Section 13-3.) Actually, prestressing is not very dependable and usually is not done because of the long-term creep in the concrete.

When a moment-resisting or rigid connection between a column and its footing is used, it is absolutely necessary for the supporting soil or rock beneath the footing to be appreciably noncompressible, or the column base will rotate as shown in Fig. E.2. If this happens, the rigid connection between the column and the footing is useless. For the purpose of this appendix, the subsoil is assumed to be capable of resisting the moment applied to it without appreciable rotation.

Quite a few methods have been developed through the years for designing moment-resisting base plates. One rather simple procedure used by many designers is presented here. As a first numerical example, a column base plate is designed for an axial load and a relatively small bending moment such that the resultant load falls between the column flanges. Assumptions are made for the width and length of the plate, after which the pressures underneath the plate are calculated and compared with the permissible value. If the pressures are unsatisfactory, the dimensions are changed and the pressures recalculated, and so on, until the values are satisfactory. The moment in the plate is calculated, and the plate thickness is determined. The critical section for bending is assumed to be at the center of the flange on the side where the compression is highest. Various designers will assume the point of maximum moment is located at some other point, such as at the face of the flange or the center of the anchor bolt.

The moment is calculated for a 1-in.-wide strip of the plate and is equated to its resisting moment. The resulting expression is solved for the required thickness of the plate as follows:

$$M_u \le \phi_b M_n = \frac{\phi_b F_y I}{c} = \frac{\phi_b F_y (\frac{1}{12})(1)(t)^3}{t/2}$$

$$t \ge \sqrt{\frac{6M_u}{\phi_b F_y}} \quad \text{with} \quad \phi_b = 0.9$$

From Section J9 of the LRFD Specification,

$$P_p = 0.85 f_c' A_1 \sqrt{\frac{A_2}{A_1}}$$

If we assume $\sqrt{\dfrac{A_2}{A_1}} \ge 2$, then

$$P_p = 1.7 f_c' A_1$$

$$\phi_c P_p = \phi_c 1.7 f_c' A_1 \quad \text{with} \quad \phi_c = 0.60$$

FIGURE E.2

Example E-1

Design a moment-resisting base plate to support a W14 × 120 column with an axial load of 620 k and a bending moment of 225 ft-k. Use A36 steel with $F_y = 36$ ksi and a concrete footing with $f'_c = 3.0$ ksi. $\phi_c F_p = (0.6)(1.7)(3.0) = 3.06$ ksi

Solution.   Using a W14 × 120($d = 14.48$ in., $t_w = 0.590$ in., $b_f = 14.670$ in., $t_f = 0.940$ in.)

$$e = \frac{(12)(225)}{620} = 4.35 \text{ in.}$$

∴ The resultant falls between the column flanges and within the middle third of the plate.

Try a 20 × 28 in. plate (after a few trials)

$$f = -\frac{P_u}{A} \pm \frac{P_u ec}{I} = -\frac{620}{(20)(28)} \pm \frac{(620)(4.35)(14)}{(\frac{1}{12})(20)(28)^3}$$

$$= -1.107 \pm 1.032 \begin{cases} -2.139 < \phi_c P_n = 3.06 \text{ ksi} \\ -0.075 \text{ ksi (still compression)} \end{cases} \qquad \text{(OK)}$$

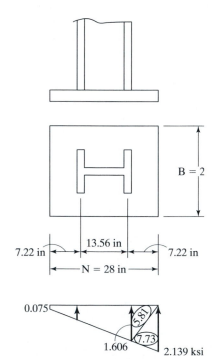

7.22 in   13.56 in   7.22 in

B = 2

N = 28 in

0.075   5.81   7.73   1.606   2.139 ksi

FIGURE E.3

Taking moments to right at center of right flange (see Fig. E.3):

$$M_u = (5.81)\left(\frac{7.22}{3}\right) + (7.73)\left(\frac{2}{3} \times 7.22\right) = 51.19 \text{ in.-k}$$

$$t \geq \sqrt{\frac{6M_u}{\phi_b F_y}} = \sqrt{\frac{(6)(51.19)}{(0.9)(36)}} = 3.08 \text{ in.}$$

Checking bending in transverse direction

$$n = \frac{B - 0.806}{2} = \frac{20 - (0.80)(14.7)}{2} = 4.12 \text{ in.}$$

$$\text{Average } f_p = \frac{0.075 + 2.139}{2} = 1.107 \text{ ksi}$$

$$M_u = (1.107)(4.12)\left(\frac{4.12}{2}\right) = 9.40 \text{ in.-k} < 51.19 \text{ in.-k} \qquad \text{(OK)}$$

Use PL3$\frac{1}{4}$ × 20 × 2 ft 4 in. A36

The moment considered in Example E-2 is of such a magnitude that the resultant load falls outside the column flange. As a result, there will be uplift on one side, and the anchor bolt will have to furnish the needed tensile force to provide equilibrium.

In this design the anchor bolts are assumed to have no significant tension due to tightening. As a result, they are assumed not to affect the force system. As the moment is applied to the column, the pressure shifts toward the flange on the compression side. It is assumed that the resultant of this compression is located at the center of the flange.

Example E-2

Repeat Example E-1 with the same column and design stresses, but with the moment increased from 225 ft-k to 460 ft-k. Refer to Fig. E.4.

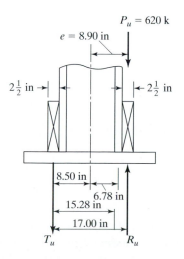

FIGURE E.4

Solution. Using a W14 × 120($d = 14.5$ in., $t_w = 0.590$ in., $b_f = 14.7$ in., $t_f = 0.940$ in.)

$$e = \frac{(12)(460)}{620} = 8.90 \text{ in.} > \frac{d}{2} = 7.25 \text{ in.}$$

∴ The resultant falls outside the column flange. Taking moments about the center of the right flange:

$$(620)(8.90 - 6.78) - 15.26 \, T_u = 0$$
$$T_u = 86.13 \text{ k}$$

$$\text{Anchor bolt } A_{\text{reqd.}} = \frac{T_u}{\phi_t 0.75 F_u} = \frac{86.13}{\phi 0.75 F_u} = \frac{86.13}{(0.75)(0.75)(58)}$$

$$= 2.64 \text{ in.}^2$$

Use two $1\frac{3}{8}$-in. diameter bolts each side

Approximate plate size, assuming a triangular pressure distribution

$$R_u = P_u + T_u = 620 + 86.13 = 706.13 \text{ k}$$

A of plate reqd. $\geq \dfrac{R_u}{\text{Avg } \phi_c F_p} = \dfrac{R_u}{\phi_c F_p / 2}$ with $\phi_c F_p$ given $= 3.06$ ksi in statement of Example E-1.

$$A \text{ reqd.} \geq \frac{706.13}{3.06/2} = 461.52 \text{ in.}^2$$

Try a 24-in.-long plate

The load is located $\frac{24}{2}$ in. minus the distance from the column c.g. to the column flange c.g. $= \frac{24}{2} - 6.78 = 5.22$ in. from edge of the plate. Thus the pressure triangle will be $3 \times 5.22 = 15.66$ in. long and the required plate length B will equal

$$B = \frac{706.13}{\frac{1}{2} \times 3.06 \times 15.66} = 29.47 \text{ in.}$$

Try a 30-in.-long plate

The load R_u is located $\frac{30}{2} - 6.78 = 8.22$ in. from the edge of the plate. The triangular pressure length will be $(3)(8.22) = 24.66$ in. long and the required plate width will be

$$B = \frac{706.13}{(\frac{1}{2})(3.06)(24.66)} = 18.72 \text{ in.}$$

If the plate is made 20 in. wide, the pressure zone will have an area $= 20 \times 24.66 = 493.2$ in.^{2} and the maximum pressure will be twice the average pressure, or

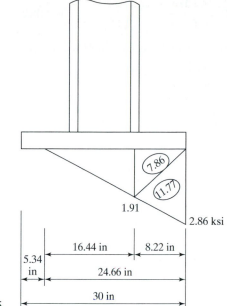

FIGURE E.5

$$\frac{706.13}{493.2} \times 2 = 2.86 \text{ ksi} < 3.06 \text{ ksi} \qquad \text{(OK)}$$

Taking moments to right at center of right column flange (see Fig. E.5)

$$M_u = (7.86)\left(\frac{8.22}{3}\right) + (11.77)\left(\tfrac{2}{3} \times 8.22\right) = 86.04 \text{ in.-k}$$

$$t = \sqrt{\frac{(6)(86.04)}{(0.9)(36)}} = 3.99 \text{ in.}$$

Use PL4 $\times$ 20 $\times$ 2 ft 6 in. A36

Design of weld from column to base PL (see Fig. E.6)

Total length of fillet weld each flange

$$= (2)(14.5) - 0.590 = 28.41 \text{ in.}$$

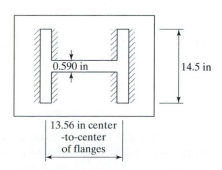

FIGURE E.6

Fillet welds from column to base plate.

$$C = T = \frac{M_u}{d - t_f} = \frac{(12)(460)}{13.56} = 407.08k$$

Strength of 1-in. long 1-in. fillet weld using E70 electrodes

$$\phi R_{nw} = \phi(0.60F_{EXX})(0.707)(a) = (0.75)(0.60 \times 70)(0.707)(1.0)$$
$$= 22.3 \text{ k/in.}$$

$$\text{Weld size required} = \frac{407.08}{(28.41)(22.3)} = 0.643 \text{ in.}$$

Use $\frac{11}{16}$-in. fillet welds, E70 electrode, SMAW

Note: Tension flange design strength $= \phi R_n$

$$= \phi F_y A_f = (0.9)(36)(14.7)(0.940) = 448 \text{ k} > T_u \qquad \text{(OK)}$$

In these examples, the variation of stress in the concrete supporting the columns has been assumed to vary in a triangular or straight-line fashion. It is possible to work with an assumed ultimate concrete theory where the concrete in compression under the plate is assumed to fail at a stress of $0.85f_c'$. Examples of such designs are available in several texts.[2] More detailed information concerning moment-resisting bases is available in several places.[3,4]

---

[2]W. McGuire, *Steel Structures* (Englewood Cliffs, N.J.: Prentice Hall, 1968), pp. 987–1004.

[3]C. G. Salmon, L. Schenker, and B. G. Johnston, "Moment-Rotation Characteristics of Column Anchorages," *Transactions ASCE*, 122 (1957), pp. 132–154.

[4]J. T. DeWolf and E. F. Sarisley, "Column Base Plates with Axial Loads and Moments," *Journal of Structural Division*, ASCE, 106, ST11 (November 1980), pp. 2167–2184.

APPENDIX F

Ponding

It has been claimed that almost 50 percent of the lawsuits faced by building designers are concerned with roofing systems.[1] Ponding, a problem with many flat roofs, is one of the common subjects of such litigation. If water accumulates more rapidly on a roof than it runs off, ponding results because the increased load causes the roof to deflect into a dish shape that can hold more water, which causes greater deflections, and so on. This process continues until equilibrium is reached, or until collapse occurs. Ponding can be caused by increasing deflections, clogged roof drains, settlement of footings, warped roof slabs, and so on.

The best way to prevent ponding is to use appreciable roof slopes ($\frac{1}{4}$ in. per ft or more) together with good drainage facilities. Supposedly more than two-thirds of the flat roofs in the United States have slopes less than $\frac{1}{4}$ in. per ft. The construction of roofs with slopes this large will increase building costs by only a few percent compared with perfectly flat roofs. The supporting girders for flat roofs with long spans should definitely be cambered to reduce the possibility of ponding (as well as the sagging that is so disturbing to the people occupying a building).

Section K2 of the LRFD Specification states that, unless roof surfaces have sufficient slopes to areas of free drainage or sufficient individual drains to prevent water accumulation, the strength and stability of the roof systems during ponding conditions must be investigated. The very detailed work of Marino[2] forms the basis of the ponding provisions of the LRFD Specification. Many other useful references also are available.[3,4,5]

---

[1]Gary Van Ryzin, "Roof Design: Avoid Ponding by Sloping to Drain," *Civil Engineering* (New York: ASCE. January 1980). pp. 77–81.

[2]F. J. Marino, "Ponding of Two-Way Roof System," *Engineering Journal*, AISC, 3rd quarter, no. 3 (1966), pp. 93–100.

[3]L. B. Burgett, "Fast Check for Ponding," *Engineering Journal*, AISC, 10, no. 1 (1st quarter, 1973), pp. 26–28.

[4]J. Chinn, "Failure of Simply-Supported Flat Roofs by Ponding of Rain," *Engineering Journal*, AISC, no. 2, 2nd quarter (1965), pp. 38–41.

[5]J. L. Ruddy, "Ponding of Concrete Deck Floors," *Engineering Journal*, AISC, 23, no. 2 (3rd quarter, 1986), pp. 107–115.

The amount of water that can be retained on a roof depends on the flexibility of the framing. The specifications state that a roof system can be considered stable and not requiring further investigation if we satisfy the expressions

$$C_p + 0.9C_s \leq 0.25 \qquad \text{(LRFD Equation K2-1)}$$

$$I_d \geq 25(S^4)10^{-6} \qquad \text{(LRFD Equation K2-2)}$$

where

$C_p = 32\, L_s L_p^4/10^7 I_p$
$C_s = 32\, SL_s^4/10^7 I_s$
L_p = length of primary members, ft
L_s = length of secondary members, ft
S = spacing of secondary members, ft
I_p = moment of inertia of primary members, in.^{4}
I_s = moment of inertia of secondary members, in.^{4}
I_d = moment of inertia of the steel deck (if one is used) supported on the secondary members, in.^{4} per ft

Should steel roof decks be used, their I_d must at least equal the value given by Equation K2-2. If the roof decking is the secondary system (i.e., no secondary beams, joists, etc.), it should be handled with Equation K2-1.

Some other LRFD requirements in applying these expressions follow:

1. The moment of inertia I_s must be decreased by 15 percent for trusses and steel joists.
2. Steel decking is considered to be a secondary member supported directly by the primary members.
3. Stresses caused by wind or seismic forces do not have to be considered in the ponding calculations.

Should moments of inertia be needed for open-web joists, they can be computed from the member cross sections, or perhaps more easily backfigured from the resisting moments and allowable stresses given in the joist tables. (Since $M_R = FI/c$, we can compute $I = M_R c/F$.)

In effect, these equations reflect *stress indexes* or percentage stress increases. For instance, here we are considering the percentage increase in stress in the steel members caused by ponding. If the stress in a member increases from $0.60F_y$ to $0.80F_y$, we say that the stress index is given by

$$U = \frac{0.80F_y - 0.60F_y}{0.60F_y} = 0.33$$

The terms C_p and C_s are, respectively, the approximate stiffnesses of the primary and secondary support systems. LRFD Equation K2-1 ($C_p + 0.9C_s \leq 0.25$), which gives us an approximate stress index during ponding, is limited to a maximum value of 0.25.

Should we substitute into this equation and obtain a value no greater than 0.25, ponding supposedly will not be a problem. Should the index be larger than 0.25, however, it will be necessary to conduct a further investigation. One method of doing this is presented in the LRFD Commentary and will be described later in this section.

Example F-1 presents the application of LRFD Equation K2-1 to a roofing system.

Example F-1

Check the roof system shown in Fig. F.1 for ponding, using the LRFD Specification and A36 steel.

Solution.

$$C_p = \frac{32\, L_s L_p^4}{10^7 I_p} = \frac{(32)\,(48)\,(36)^4}{(10^7)\,(1830)} = 0.141$$

$$C_s = \frac{32\, S L_s^4}{10^7 I_s} = \frac{(32)\,(6)\,(48)^4}{(10^7)\,(518)} = 0.197$$

$$C_p + 0.9 C_s = 0.141 + (0.9)\,(0.197) = 0.318 > 0.25$$

Indicates insufficient strength and stability, and thus a more precise method of checking should be used.

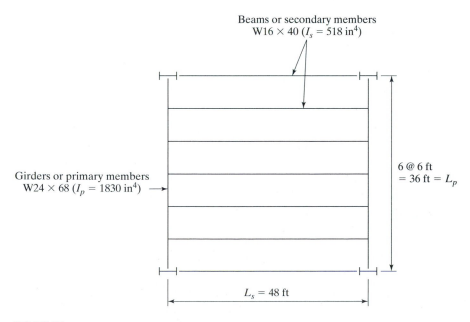

Beams or secondary members
W16 × 40 (I_s = 518 in⁴)

Girders or primary members
W24 × 68 (I_p = 1830 in⁴) →

6 @ 6 ft
= 36 ft = L_p

L_s = 48 ft

FIGURE F.1

The curves in the LRFD Commentary (K2) provide a design aid for use when we need to compute a more accurate flat-roof framing stiffness than is given by the specification provision that $C_p + 0.9C_s \leq 0.25$.

The following stress indexes are computed for the primary and secondary members:

$$U_p = \left(\frac{F_y - f_o}{f_o}\right)_p$$

$$U_s = \left(\frac{F_y - f_o}{f_o}\right)_p$$

In these expressions, f_o represents the stress due to $1.2D + 1.2R$ (where D = nominal dead load and R = nominal load due to rainwater or ice exclusive of ponding contribution). These loads should include any snow that is present, although most ponding failures have occurred during torrential summer rains.

We enter Fig. A-K2.1 in Section K of the LRFD Appendix with our computed U_p and move horizontally to the calculated C_s value of the secondary members. Then we go vertically to the abscissa scale and read the upper limit for the flexibility constant C_p. If this value is more than our calculated C_p value computed for the primary members, the stiffness is sufficient. The same process is followed by Fig. A-K2.2, where we enter with our computed U_s and C_p values and pick from the abscissa the flexibility constant C_s, which should be no less than our C_s value. This procedure is illustrated in Example F-2.

Example F-2

Recheck the roof system considered in Example F-1 using the LRFD curves. Assume that as ponding begins, f_o is 20 ksi. in both girders and beams.

Solution. Checking girders:

$$U_p = \frac{F_y - f_o}{f_o} = \frac{36 - 20}{20} = 0.80$$

For $U_p = 0.80$ and $C_s = 0.197$, we read in Fig. A-K2.1 $C_p = 0.28 >$ our calculated C_p of 0.141.

$\therefore$ Girders (OK)

Checking beams:

$$U_s = \frac{F_y - f_o}{f_o} = \frac{36 - 20}{20} = 0.80$$

For $U_s = 0.80$ and $C_p = 0.141$, we read in Fig. A-K2.2 $C_s = 0.29 >$ our calculated C_s of 0.197.

$\therefore$ Beams (OK)

Glossary

Amplification Factor A multiplier used to increase the computed moment or deflection in a member to account for the eccentricity of the load.

Annealing A process in which steel is heated to an intermediate temperature range, held at that temperature for several hours, and then allowed to slowly cool off to room temperature. The resulting steel has less hardness and brittleness, but more ductility.

Aspect Ratio The ratio of the lengths of the sides of a rectangular panel to each other.

Bar Joist See *open web-joist.*

Bays The areas between columns in a building.

Beam A member that supports loads transverse to its longitudinal axis.

Beam-Column A column that is subjected to axial compression loads as well as bending moments.

Bearing Wall Construction Building construction where all the loads are transferred to the walls and thence down to the foundations.

Block Shear A fracture type shear where fracture may occur on either the tension plane or the shear plane followed by yielding on the other plane (see Fig. 3.16).

Braced Frame A frame that has resistance to lateral loads supplied by some type of auxiliary bracing.

Brittle Fracture Abrupt fracture with little or no prior ductile deformation.

Buckling Load The load at which a straight compression member assumes a deflected position.

Built-Up Member A member made up of two or more steel elements bolted or welded together to form a single member.

Camber The construction of a member bent or arched in one direction so that it won't look so bad when the service loads bend it in the opposite direction.

Cast Iron An iron with a very low carbon content.

Charpy V-Notch Test A test used for measuring the fracture toughness of steel by fracturing it with a pendulum swung from a certain height.

Cladding The exterior covering of the structural parts of a building.

Cold-Formed Light-Gage Steel Shapes Shapes made by cold bending thin sheets of carbon or low-alloy steels into desired cross sections.

Column A structural member whose primary function is to support compressive loads.

Compact Section A section that has a sufficiently stocky profile so that it is capable of developing a fully plastic stress distribution before buckling.

Composite Beam A steel beam made composite with a concrete slab by providing shear transfer between the two (see Fig. 16.).

Composite Column A column constructed with rolled or built-up steel shapes, encased in concrete or with concrete placed inside steel pipes or tubes (see Fig. 17.1).

Coping The cutting back of the flanges of a beam to facilitate its connection to another beam (see Fig. 10.7).

Dead loads   Loads of constant magnitude that remain in one position. Examples: weights of walls, floors, roofs, fixtures, structural frames, and so on.

Drift   Lateral deflection of a building.

Drift Index   The ratio of lateral deflection of a building to its height.

Ductility   The property of a material by which it can withstand extensive deformation without failure under high tensile stress.

Effective Length   The distance between points of zero moment in a column; that is, the distance between its inflection points.

Effective Length Factor _K_   A factor that when multiplied by the length of a column will provide its effective length.

Elastic Design   A method of design that is based on certain allowable stresses.

Elastic Limit   The largest stress that a material can withstand without being permanently deformed.

Elasticity   The ability of a material to return to its original shape after it has been loaded and then unloaded.

Elastic Strain   Strain that occurs in a member under load before its yield stress is reached.

Endurance Limit   The maximum fatigue-type stress in a material for which the material seems to have an infinite life.

Euler Load   The compression load at which a long and slender member will buckle elastically.

Eyebar   A pin-connected tension member whose ends are enlarged with respect to the rest of the member so as to make the strength of the ends approximately equal to the strength of the rest of the member.

Factored Load   A nominal load multiplied by a load factor.

Fasteners   A term representing bolts, welds, rivets, or other connecting devices.

Fatigue   A fracture situation caused by changing stresses.

Faying Surface   The contact or shear area between members being connected.

Fillet Weld   A weld placed in the corner formed by two overlapping parts in contact with each other (see Fig. 14.2).

First-Order Analysis   Analysis of a structure in which equilibrium equations are written based on an assumed nondeformed structure.

Floor Beams   The larger beams in many bridge floors that are perpendicular to the roadway of the bridge and that are used to transfer the floor loads from the stringers to the supporting girders or trusses.

Fracture Toughness   The ability of a material to absorb energy in large amounts. For instance, steel members can be subjected to large deformations during erection and fabrication without failure, thus allowing them to be bent, hammered, sheared, and have holes punched in them.

Gage   Transverse spacing of bolts measured perpendicular to the long direction of the member (see Fig. 12.4).

Girder   A rather loosely used term usually indicating a large beam and perhaps one into which smaller beams are framed.

Girts   Horizontal members running along the sides of industrial buildings used primarily to resist bending due to wind. They often are used to support corrugated siding.

Government Anchors   Bent steel bars used when ends of steel beam are enclosed by concrete or masonry walls. The bars pass through the beam webs parallel to the walls and are enclosed in the walls. They keep beams from moving longitudinally with respect to the walls.

Groove Welds   Welds made in grooves between members that are being joined. They may extend for the full thickness of the parts (complete-penetration groove welds) or they may extend for only a part of the member thickness (partial-penetration groove welds) (see Fig. 14.2).

Hybrid Member   A structural steel member made from parts that have different yield stresses.

Impact Loads   The difference between the magnitude of live loads actually caused and the magnitude of those loads had they been applied as dead loads.

Inelastic Action   The deformation of a member that does not disappear when the loads are removed.

Influence Line   A diagram whose ordinates show the magnitude and character of some function of a structure (shear, moment, etc.) as a unit load moves across the structure.

Instability   A situation occurring in a member where increased deformation of that member causes a reduction in its load-carrying ability.

Intermediate Columns   Columns that fail both by yielding and buckling. Their behavior is said to be inelastic. Most columns fall in this range, where some of the fibers reach the yield stress and some do not.

Ironworker   A person performing steel erection (it's a name carried over from the days when iron structural members were used).

Joists   The closely spaced beams supporting the floors and roofs of buildings.

Jumbo Sections   Very heavy steel W sections (and structural tees cut from those sections) in Groups 4 and 5 of the LRFD Manual. Serious cracking problems sometimes occur in these sections when welding or thermal cutting is involved.

Killed Steel   Steel that has been deoxidized to prevent gas bubbles and to reduce its nitrogen content.

Lamellar Tearing   A separation in the layers of a highly restrained welded joint caused by "through the thickness" strains produced by shrinking of the weld metal.

Limit State   A condition at which a structure or some point of the structure ceases to perform its intended function either as to strength or as to serviceability.

Lintels   Beams over openings in masonry walls such as windows and doors.

Live Loads   Loads that change position and magnitude. They move or are moved. Examples: trucks, people, wind, rain, earthquakes, temperature changes, and so on.

Load Factor   A number almost always larger than 1.0 used to increase the estimated loads a structure has to support to account for the uncertainties involved in estimating loads.

Local Buckling   The buckling of the part of a larger member that precipitates failure of the whole member.

Long Columns   Columns that buckle elastically and whose buckling loads can be predicted accurately with the Euler formula if the axial buckling stress is below the proportional limit.

Malleability   The property of some metals by which they may be hammered, pounded or rolled into various shapes particularly thin sheets.

Mild Steel   A ductile low-carbon steel.

Milled Surfaces   Those surfaces that have been accurately sawed or finished to a smooth or true plane.

Mill Scale   An iron oxide that forms on steel when it is reheated for rolling.

Modulus of Elasticity or Young's Modulus   The ratio of stress to strain in a member under load. It is a measure of the stiffness of the material.

Net Area   Gross cross-sectional area of a member minus any holes, notches, or other indentations.

Nominal Loads   The magnitudes of loads specified by a particular code.

Nominal Strength   The theoretical ultimate strength of a member or connection.

Noncompact Section   A section that cannot be stressed to a fully plastic situation before buckling occurs. The yield stress can be reached in some but not all of the compression elements before buckling occurs.

Open-Web Joist   A small parallel chord truss whose members are often made from bars (hence the common name *bar joist*) or small angles or other rolled shapes. These joists are very commonly used to support floor and roof slabs (see Fig. 19.5).

P-Delta Effect   Changes in column moments and deflections due to lateral deflections.

Partially Composite Section   A section whose flexural strength is governed by the strength of its shear connectors.

Pitch   The longitudinal spacing of bolts measured parallel to the long direction of a member (see Fig. 12.4).

Plane Frame   A frame that for purposes of analysis and design is assumed to lie in a single (or two-dimensional) plane.

Plastic Design   A method of design that is based on a consideration of failure conditions.

Plastic Modulus   The statical moment of the tension and compression areas of a section taken about the plastic neutral axis.

Plastic Moment   The yield stress of a section times its plastic modulus. It's the nominal moment that the section can theoretically resist if it is braced laterally.

Plastic Strain   The strain that occurs in a member after its yield stress is reached with no increase in stress.

Plate Girder   A built-up steel beam (see Fig. 18.4).

Poisson's Ratio   The ratio of lateral strain to axial or longitudinal strain in a member under load.

Ponding   A situation on a flat roof where water accumulates faster than it runs off.

Post-Buckling Strength   The load a member or frame can support after buckling occurs.

Proportional Limit   Largest strain for which Hooke's law applies, or the highest point on the straight-line portion of the stress-strain diagram.

Purlins   Roof beams that span between trusses (see Fig. 4.4).

Quenching   Rapid cooling of steel with water or oil.

Residual Stresses   The stresses that exist in an unloaded member after it's manufactured.

Resistance Factorϕ   A number almost always less than 1.0 multiplied by the ultimate or nominal strength of a member or connection to take into account the uncertainties in material strengths, dimensions, and workmanship. Also called *overcapacity factor*.

Rigid Frame   A structure whose connectors keep substantially the same angles between members before and after loading.

Sag Rods   Steel rods that are used to provide lateral support for roof purlins. They also may be used for the same purpose for girts on the sides of buildings (see Fig. 4.5).

St. Venant Torsion   The part of the torsion in a member that produces only shear stresses in the member.

Scuppers   Large holes or tubes in walls or parapets that enable water above a certain depth to quickly drain from roofs.

Second-Order Analysis   Analysis of a structure for which equilibrium equations are written that include the effect of the deformations of the structure.

Section Modulus   The ratio of the moment of inertia taken about a particular axis of a section divided by the distance to the extreme fiber of the section measured perpendicular to the axis in question.

Seismic   Of or having to do with an earthquake.

Serviceability   The ability of a structure to maintain its appearance, comfort, durability, and function under normal loading conditions.

Service Loads   The loads that are assumed to be applied to a structure when it is in service (also called *working loads*).

Shape Factor   The ratio of the plastic moment of a section to its yield moment.

Shear Center   The point in the cross section of a beam through which the resultant of the transverse loads must pass so that no torsion will occur.

Shear Lag   A nonuniformity of stress in the parts of rolled or built-up sections occurring when a tensile load is not applied uniformly.

Shear Wall   A wall in a structure that is specially designed to resist shears caused by lateral forces such as wind or earthquake in the plane of the wall.

Shims   Thin strips of steel that are used to adjust the fitting of connections. Finger shims are shims that are installed after the bolts already are in place.

Short Columns   Columns whose failure stress will equal the yield stress, and for which no buckling will occur. For a column to fall into this class it would have to be so short as to have no practical application.

Sidesway   The lateral movement of a structure caused by unsymmetrical loads or by an unsymmetrical arrangement of building members.

Skeleton Construction   Building construction in which the loads are transferred for each floor by beams to the columns and thence to the foundations.

Slenderness Ratio   The ratio of the effective length of a column to its radius of gyration, both pertaining to the same axis of bending.

Slender Section   A member that will buckle locally while the stress still is in the elastic range.

Slip-Critical Joint   A bolted joint that is designed to have resistance to slipping.

Space Frame   A three-dimensional structural frame.

Spandrel Beams   Beams that support the exterior walls of buildings and perhaps part of the floor and hallway loads (see Fig. 19.3).

Steel   An alloy consisting almost entirely of iron (usually over 98 percent). It also contains small quantities of carbon, silicon, manganese, sulfur, phosphorus, and other materials.

Stiffened Element   A projecting piece of steel whose two edges parallel to the direction of a compression force are supported (see Fig. 5.6).

Stiffener   A plate or an angle usually connected to the web of a beam or girder to prevent failure of the web (see Figs. 18.11, 18.12, and 18.13).

Story Drift   The difference in horizontal deflection at the top and bottom of a particular story.

Strain-Hardening   Range beyond plastic strain in which additional stress is necessary to produce additional strain.

Stringers   The beams in bridge floors that run parallel to the roadway.

Tangent Modulus   The ratio of stress to strain for a material that has been stressed into the inelastic range.

Tension Field Action   The behavior of a plate girder panel which, after the girder initially buckles, acts much like a truss. Diagonal strips of the web act similarly to the diagonals of a parallel chord truss. The stiffeners keep the flanges from coming together and the flanges keep the stiffeners from coming together (see Fig. 18.9).

Toughness   The ability of a material to absorb energy in large amounts. As an illustration, steel members can be subjected to large deformations during fabrication and erection without fracture, thus allowing them to be bent, hammered, sheared, and have holes punched in them without visible damage.

Unbraced Frame   A frame whose resistance to lateral forces is provided by its members and their connections.

Unbraced Length The distance in a member between points that are braced.

Unstiffened Element A projecting piece of steel having one free edge parallel to the direction of a compression force, with the other edge in that direction unsupported (see Fig. 5.6).

Upset Rods Rods whose ends are made larger than the regular bodies of the rods. Threads are cut into the upset ends, but the area at the root of the thread in each rod is larger than that of the regular part of the bar (see Fig. 4.3).

Warping Torsion The part of the resistance of a member to torsion that is provided by the warping resistance of the member cross section.

Weathering Steel A high-strength low-alloy steel whose surface when exposed to the atmosphere (not a marine one) oxidizes and forms a tightly adherent film that prevents further oxidation and thus eliminates the need for painting.

Web Buckling The buckling of the web of a member (see Fig. 10.10).

Web Crippling The failure of the web of a member near a concentrated force (see Fig. 10.10).

Working Loads See *Service Loads.*

Wrought Iron An iron with a very high carbon content.

Yield Moment The moment that will just produce the yield stress in the outermost fiber of a section.

Yield Stress The stress at which there is a decided increase in the elongation or strain in a member without a corresponding increase in stress.

Index

LICENSE AGREEMENT AND LIMITED WARRANTY

READ THIS LICENSE CAREFULLY BEFORE OPENING THE CD SOFTWARE PACKAGE. BY OPENING THE CD SOFTWARE PACKAGE, YOU ARE AGREEING TO THE TERMS AND CONDITIONS OF THIS LICENSE. IF YOU DO NOT AGREE, DO NOT OPEN THE CD SOFTWARE PACKAGE. *THESE TERMS APPLY TO ALL LICENSED SOFTWARE ON THE DISC EXCEPT THAT THE TERMS FOR USE OF ANY SHARE-WARE OR FREEWARE ON THE CD ARE AS SET FORTH IN THE RELEVANT LICENSE LOCATED IN THIS APPENDIX AND/OR IN THE RELEVANT ELECTRONIC LICENSE LOCATED ON THE CD ("FREEWARE LICENSES"):*

1. **GRANT OF LICENSE and OWNERSHIP:** The enclosed computer programs ("Software") are licensed, not sold, to you by Prentice-Hall, Inc. ("We" or the "Company") in consideration of your purchase of this book, and your agreement to these terms. You own only the disc but we and/or our licensors own the Software itself. This license grants to you a nonexclusive license to use and display the enclosed copy of the Software on a single computer (i.e. with a single CPU), for educational use only, so long as you comply with the terms of this Agreement. You may make one copy for back up only. We reserve any rights not granted to you. The applicable Freeware License may grant you additional rights as between you and the shareware authors.

2. **USE RESTRICTIONS:** You may <u>not</u> sell or license copies of the Software or the Documentation to others, except as provided in any applicable Freeware License. You may <u>not</u> transfer, distribute or make available the Software or the Documentation, except as provided in any applicable Freeware License. You may <u>not</u> reverse engineer, disassemble, decompile, modify, adapt, translate or create derivative works based on the Software or the Documentation, except as provided in any applicable Freeware License. You may be held legally responsible for any copying or copyright infringement which is caused by your failure to abide by the terms of these restrictions.

3. **TERMINATION:** This license is effective until terminated. This license will terminate automatically without notice from the Company if you fail to comply with any provisions or limitations of this license. Upon termination, you shall destroy the Documentation and all copies of the Software. All provisions of this Agreement as to limitation and disclaimer of warranties, limitation of liability, remedies or damages, and our ownership rights shall survive termination.

4. **DISCLAIMER OF WARRANTY: THE COMPANY AND ITS LICENSORS MAKE NO WARRANTIES ABOUT THE SOFTWARE, WHICH IS PROVIDED "<u>AS-IS</u>." IF THE DISC IS DEFECTIVE IN MATERIALS OR WORKMANSHIP, YOUR ONLY REMEDY IS TO RETURN IT TO THE COMPANY WITHIN 30 DAYS FOR REPLACEMENT UNLESS THE COMPANY DETERMINES IN GOOD FAITH THAT THE DISC HAS BEEN MISUSED OR IMPROPERLY INSTALLED, REPAIRED, ALTERED OR DAMAGED. THE COMPANY DISCLAIMS ALL WARRANTIES, EXPRESS OR IMPLIED, INCLUDING WITHOUT LIMITATION, THE IMPLIED WARRANTIES OF MERCHANTABILITY AND FITNESS FOR A PARTICULAR PURPOSE. THE COMPANY DOES NOT WARRANT, GUARANTEE OR MAKE ANY REPRESENTATION REGARDING THE ACCURACY, RELIABILITY, CURRENTNESS, USE, OR RESULTS OF USE, OF THE SOFTWARE.**

5. **LIMITATION OF REMEDIES AND DAMAGES: IN NO EVENT, SHALL THE COMPANY OR ITS EMPLOYEES, AGENTS, LICENSORS OR CONTRACTORS BE LIABLE FOR ANY INCIDENTAL, INDIRECT, SPECIAL OR CONSEQUENTIAL DAMAGES ARISING OUT OF OR IN CONNECTION WITH THIS LICENSE OR THE SOFTWARE, INCLUDING, WITHOUT LIMITATION, LOSS OF USE, LOSS OF DATA, LOSS OF INCOME OR PROFIT, OR OTHER LOSSES SUSTAINED AS A RESULT OF INJURY TO ANY PERSON, OR LOSS OF OR DAMAGE TO PROPERTY, OR CLAIMS OF THIRD PARTIES, EVEN IF THE COMPANY OR AN AUTHORIZED REPRESENTATIVE OF THE COMPANY HAS BEEN ADVISED OF THE POSSIBILITY OF SUCH DAMAGES.** SOME JURISDICTIONS DO NOT ALLOW THE LIMITATION OF DAMAGES IN CERTAIN CIRCUMSTANCES, SO THE ABOVE LIMITATIONS MAY NOT ALWAYS APPLY.

6. **GENERAL:** THIS AGREEMENT SHALL BE CONSTRUED IN ACCORDANCE WITH THE LAWS OF THE UNITED STATES OF AMERICA AND THE STATE OF NEW YORK, APPLICABLE TO CONTRACTS MADE IN NEW YORK, AND SHALL BENEFIT THE COMPANY, ITS AFFILIATES AND ASSIGNEES. This Agreement is the complete and exclusive statement of the agreement between you and the Company and supersedes all proposals, prior agreements, oral or written, and any other communications between you and the company or any of its representatives relating to the subject matter. If you are a U.S. Government user, this Software is licensed with "restricted rights" as set forth in subparagraphs (a)–(d) of the Commercial Computer-Restricted Rights clause at FAR 52.227–19 or in subparagraphs (c)(1)(ii) of the Rights in Technical Data and Computer Software clause at DFARS 252.227–7013, and similar clauses, as applicable.

7. **ACKNOWLEDGMENT:** YOU ACKNOWLEDGE THAT YOU HAVE READ THIS AGREEMENT, UNDERSTAND IT, AND AGREE TO BE BUND BY ITS TERMS AND CONDITIONS. YOU ALSO AGREE THAT THIS AGREEMENT IS THE COMPLETE AND EXCLUSIVE STATEMENT OF THE AGREEMENT BETWEEN YOU AND THE COMPANY.

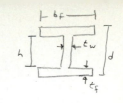

$$\phi_c P_n = \phi_c A_g F_{cr}$$

$$\phi = .85$$

$$P_u \leq \phi P_n$$

③ Slenderness Parameter

$$\lambda_c = \frac{KL}{r\pi} \cdot \sqrt{\frac{F_y}{E}}$$

for $\lambda_c \leq 1.5$ Inelastic Buckling $F_{cr} = \left(.658^{\lambda_c^2}\right) \cdot F_y$

for $\lambda_c > 1.5$ Elastic Buckling $F_{cr} = \left(\frac{.877}{\lambda_c^2}\right) \cdot F_y$

① Flanges $\lambda_r = \frac{b_f}{2t_f} \leq .56\sqrt{\frac{E}{F_y}} = \frac{95}{\sqrt{F_y}}$ E=29000

← check local Buckling

Web $\lambda_r = \frac{h}{t_w} \leq 1.49\sqrt{\frac{E}{F_y}} = \frac{253}{\sqrt{F_y}}$

②

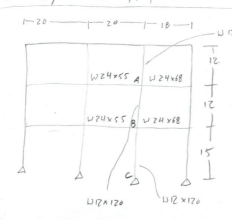

Weak

$12' = 144''$

$18' = 216''$

Strong

$30' = 360''$

$K=1$ $K=1$

$$\frac{KL_y}{r_y} = \frac{216''}{3.04} = \boxed{71.1}$$
controls

$$\frac{KL_x}{r_x} = \frac{360}{5.31} = 67.7$$

$$\frac{KL}{r} < 200 , \quad 71.1 < 200, \text{ so } OK$$

③ Find λ_c, then use approp eqn to find F_{cr}, then find ϕP_n

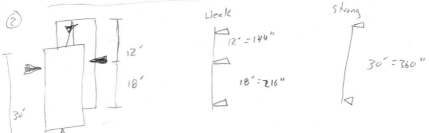

|—20——|—20—|—18—| W 12 x 96

W24x55 A W24x68

W24x55 B W24x68

W12x120   W12x120

Column AB

$$G_A = \frac{833}{12} + \frac{1070}{12}$$ $I_x, W12x96$ $I_x, W12x120$

$$\frac{1360}{20} + \frac{1830}{18}$$ $I_x, W24x55$ $I_x, W24x68$

$G_A = .93$

$$G_B = \frac{\frac{1070}{12} + \frac{1070}{12}}{\frac{1360}{12} + \frac{1830}{18}} = .95$$

$K_x = 1.31$

Joint C is Pin, $G_c = 10$

for unbraced frame,

use fig b

fig 7.2 p 184

$$use \, K \, in \; \lambda_c = \frac{KL}{r}\sqrt{\frac{F_y}{E}} \rightarrow \phi P_n$$

$M_u \equiv \phi M_n \quad \phi = .9$

a) check for compactness p160

b) Def calc $< \dfrac{L}{240}$ ← length

$\phi M_n = .9 \, Z_x \, F_y$

$M_n = M_p = Z F_y \leq 1.5 \, M_y \qquad M_y = F_y \, S \qquad S = \dfrac{27}{d}$

$\quad \phi M_n >$ calc Max Moment

$\phi V_n = .9 \cdot d \cdot t_w (.6 F_y)$

$\quad \phi V_n >$ calc max V

p289   <u>Web yielding</u>

end → $\quad \phi R_n = 1.0 \, (2.5 \, k + N) \, F_{yw} \cdot t_w$

     Bearing Length / F_y for Web

Center $\phi R_n = 1.0 \, (5 k + N) \, F_{yw} \, t_w$

<u>Web crippling</u>

if concentrated load is applied at distance $> d/2$, then d/c

$$\phi = .75$$
$$\phi R_n = (.75)(.4) \, t_w^2 \left[1 + 3 \left(\dfrac{N}{d} \right) \left(\dfrac{t_w}{t_f} \right)^{1.5} \right] \sqrt{\dfrac{E \, F_{yw} \, t_f}{t_w}}$$

if conc load is applied at distance $< d/2$ and if $\dfrac{N}{d} \leq .2$

$$\phi = .75$$
$$\phi R_n = (.75)(.4) \, t_w^2 \left[1 + 3 \left(\dfrac{N}{2} \right) \left(\dfrac{t_w}{t_f} \right)^{1.5} \right] \sqrt{\dfrac{E \, F_{yw} \, t_f}{t_w}}$$

if applied dist $< \dfrac{d}{2}$ and $\dfrac{N}{d} > .2$

$$\phi R_n = (.75)(.4) \, t_w^2 \left[1 + \left(4 \dfrac{N}{d} - .2 \right) \left(\dfrac{t_w}{t_f} \right)^{1.5} \right] \sqrt{\dfrac{E \, F_{yw} \, t_f}{t_w}}$$

$$\phi R_n \leq R_u$$

typically use $N = 3$